Metallomics

Yasumitsu Ogra • Takafumi Hirata
Editors

Metallomics

Recent Analytical Techniques and Applications

Springer

Editors
Yasumitsu Ogra
Graduate School of Pharmaceutical Science
Chiba University
Chiba
Japan

Takafumi Hirata
Graduate School of Science
The University of Tokyo
Tokyo
Japan

ISBN 978-4-431-56461-4 ISBN 978-4-431-56463-8 (eBook)
DOI 10.1007/978-4-431-56463-8

Library of Congress Control Number: 2017931358

Printed on acid-free paper

This Springer imprint is published by Springer Nature
The registered company is Springer Japan KK
The registered company address is: Chiyoda First Bldg. East, 3-8-1 Nishi-Kanda, Chiyoda-ku, Tokyo 101-0065, Japan

Preface

The term "metallomics" was coined by Professor Hiroki Haraguchi (Nagoya University, Japan) in 2002. To develop both the basic principles and the applications of metallomics, the first International Symposium on Metallomics (ISM) was held in Nagoya, Japan, in 2007. Thereafter, ISM was held biennially in different locations around the world, including Cincinnati, USA, in 2009; Münster, Germany, in 2011; Oviedo, Spain, in 2011; and Beijing, China, in 2015. The sixth ISM will be held in Vienna, Austria, in 2017. The dawn of metallomics research was further enhanced by the launch of the journal *Metallomics* in January 2009 by the Royal Society of Chemistry. Metallomics is widely acknowledged as one of the youngest members of omics research.

The history of metallomics is discussed in detail by Professor Haraguchi, the father of metallomics, in Chap. 1 of this book. Given that more than 10 years have passed since its birth, we are of the opinion that metallomics has reached the end of the first stage of growth. In the early days of metallomics, scientists in Japan, the birthplace of metallomics, made exciting contributions to the development of this field. In particular, the development of analytical techniques, pharmacological and medical applications, and applications to environmental sciences was pursued in earnest.

This book contains the latest information on metallomics, including recent analytical techniques and strategies as well as applications in molecular biology, medicine, and pharmaceutical sciences by Japanese pioneers of metallomics. Besides new insights into the study of metallomics, significant effort has been made to include educational contents for the benefit of students, young scientists, and clinical personnel. Thus, this book offers invaluable information for researchers in metallomics.

We believe that the publication of this book is timely for reviewing the accomplishments of metallomics research, which seems to be ready to start the second stage of growth. We are confident that readers will recognize the value of this book in learning, teaching, lectures, and other outreach activities that will help make known to the public the importance of metallomics. Finally, we express our heartfelt thanks to all the authors for their outstanding work and to Springer for giving us the opportunity to publish this book, which is expected to pave the way to achieving our goal.

Chiba, Japan Yasumitsu Ogra
Tokyo, Japan Takafumi Hirata

Contents

Part I Overview

1 **Metallomics: Integrated Biometal Science** 3
Hiroki Haraguchi

Part II Analytical Techniques and Strategies Surrounding Metallomics

2 **Speciation and Identification of Chalcogen-Containing Metabolites** 43
Yasumitsu Ogra

3 **Visualization of Intracellular Elements Using Scanning X-Ray Fluorescence Microscopy** 63
Mari Shimura, Lukasz Szyrwiel, Satoshi Matsuyama, and Kazuto Yamauchi

4 **Quantitative Elemental Bioimaging Protocol Using Femtosecond-Laser Ablation-ICP-Mass Spectrometry Coupled with Glass Standard Reference Material** 93
Yoshiki Makino, Seiya Ohara, Marina Yamada, Sho Mukoyama, Kentaro Hattori, Shuhei Sakata, Yuki Tanaka, Toshihiro Suzuki, Atsuko Shinohara, Takehisa Matsukawa, Kazuhito Yokoyama, and Takafumi Hirata

5 **Single Cell Analysis by Using ICP-MS** 107
Shin-ichi Miyashita, Shin-ichiro Fujii, Kaori Shigeta, and Kazumi Inagaki

6 **Synchrotron Radiation X-Ray Analysis of Metal-Accumulating Plants** 125
Akiko Hokura and Emiko Harada

7 ^{77}Se NMR Spectroscopy for Speciation Analysis of Selenium Compounds 147
Noriyuki Suzuki and Yasumitsu Ogra

8 Protein Quantification and Quantitative Phosphorylation Analysis by the Determination of Hetero Atoms (S and P) by Means of nanoHPLC-ICPMS 157
Yoshinari Suzuki and Naoki Furuta

9 Analysis of Drug Active Pharmaceutical Ingredients and Biomolecules Using Triple Quadrupole ICP-MS 181
Naoki Sugiyama and Yasuyuki Shikamori

10 Highly Sensitive Analysis of Proteins and Metabolites by Metal Tagging Using LC-ICP-MS 197
Daigo Iwahata

Part III Application of Metallomics in Molecular Biology, Medicine and Pharmaceutical Sciences

11 Comprehensive Element Analysis of Prokaryotic and Eukaryotic Cells as well as Organelles by ICP-MS 219
Tomonari Umemura, Yuichiro Matsui, Shinnosuke Sakagawa, Taku Fukai, Eiji Fujimori, Hidetoshi Kumata, and Motohide Aoki

12 Iron Isotope Signature in Red Blood Cell Samples from Japanese Female Donors of Various Ages 239
Yu-ki Tanaka, Kohshiro Takata, Takahiro Kawasaki, Atsuko Shinohara, Kazuko Ishikawa-Takata, and Takafumi Hirata

13 Roles of Zinc Transporters in Cellular Transport of Cadmium and Manganese 265
Seiichiro Himeno and Hitomi Fujishiro

14 Link Between Metal Homeostasis and Neurodegenerative Diseases: Crosstalk of Metals and Amyloidogenic Proteins at the Synapse . . . 285
Dai Mizuno and Masahiro Kawahara

15 New Insight into Metallomics in Cognition 315
Atsushi Takeda and Hanuna Tamano

16 Frontier of Development for Metallodrugs on the Basis of Metallomic Pharmacology and Medicinal Inorganic Chemistry . . . 341
Hiroyuki Yasui and Yutaka Yoshikawa

Part I
Overview

Chapter 1
Metallomics: Integrated Biometal Science

Hiroki Haraguchi

Abstract The historical aspects of metallomics, which were proposed as integrated biometal science in 2004 by the present author, are described. The significant development of analytical atomic spectrometry since the late 1960s allowed the all-element analyses of various biological samples as well as chemical speciation analysis of trace elements. Such a progress of trace metal analysis opened new era of trace metal science in various scientific fields to cooperate with omics sciences such as genomics, proteomics, and metabolomics. Under such situations in life science, it was desired to integrate biometal science as one of omics sciences for further development. So far, the international symposiums of metallomics were held five times around the world since 2007 in Nagoya, and the academic journal of metallomics has been published since 2009 from the Royal Society of Chemistry. Furthermore, essentiality and toxicity of trace metals, a simplified model of the biological system, the research fields in metallomics, and so forth are discussed in this chapter.

Keywords Metallomics • Omics science • Genomics • Proteomics • Life science • Analytical atomic spectrometry • All-Element Present Theory • Chemical speciation • Biological system • Biological essentiality

1.1 Introduction

In this paper, "Metallomics" is proposed as a new scientific field in order to integrate the research field related to biometals. Metallomics should be a scientific field in symbiosis with genomics and proteomics, because syntheses and metabolic functions of genes (DNA and RNA) and proteins cannot be performed without the aid of various metal ions and metalloenzymes. In metallomics, metalloproteins, metalloenzymes and other metal-containing biomolecules are defined as "metallomes", in a similar manner to genome in genomics and proteome in proteomics. Since identification of metallomes and the elucidation of their biological or physiological functions in the biological systems are the main research target of metallomics, chemical speciation for specific identification of bioactive metallomes is one of the most important analytical technologies to establish metallomics as

H. Haraguchi (✉)
Nagoya University, 525-1-3-506, Shinanocho, Totsuka-ku, Yokohama 244-0801, Japan
e-mail: haraguch@gmail.com

Y. Ogra, T. Hirata (eds.), *Metallomics*, DOI 10.1007/978-4-431-56463-8_1

integrated biometal science. In order to rationalize the concept of metallomics, the distributions of the elements in man, human blood serum and seawater, a challenge to all-elements analysis of one biological cell, and some other topics are introduced with emphasis on recent development of chemical speciation of trace metals in some biological samples.

This is the abstract of my paper published under the title of "Metallomics as Integrated Biometal Science" in the *Journal of Analytical Atomic Spectrometry* (*J Anal At Spectromet*) in 2004 [1]. The publication of this paper was an epoch-making event for metallomics, because metallomics was coined as a new academic field from this publication. That is, "metallomics" was proposed for the first time as the academic research field. Since then, metallomics has received great attention as the emerging new scientific field [2], and great progress has been achieved as the one of omics science [3–8]. In the present chapter, thus, the current status and perspective of metallomics will be described together with the research backgrounds before and after the proposal of metallomics.

As will be mentioned later, the advancement of analytical atomic spectrometry such as ICP-AES (inductively coupled plasma atomic emission spectrometry) and ICP-MS (inductively coupled plasma mass spectrometry) was one of the strong driving forces for me to conceive the idea for metallomics. Since the middle of the 1970s, ICP-AES and ICP-MS have been extensively developed as the highly sensitive analytical methods with excellent feasibilities of simultaneous multielement detection [9–11]. As a result, in these days, the distribution analyses of the elements in the biological fluids, cells, organs, and organisms are performed mostly by ICP-AES and ICP-MS [1, 9, 10, 12]. Furthermore, it is now possible to perform all-element analysis of almost all samples by analytical atomic spectrometry. Thus, the present author also proposed the concept of "the Extended All-Present Theory of the Elements" from the consideration on the distributions and cycles of the elements on the earth [1, 9]. In addition, chemical speciation for the identification of chemical species in the biological and environmental samples has been also developed in last two decades and now established as the important analytical technologies [1, 13–16].

It should be further noticed here that genomics [17, 18], proteomics [19], and metabolomics [20] as omics science have been significantly developed in life science and molecular cell biology for last few decades, and now these terms are popularly used, indicating the integrated biological sciences for genes, proteins, and metabolites. In fact, genes and proteins are the key materials to regulate and/or maintain the life systems of animals, plants, and microorganisms. However, the biological functions and physiology of the life system do not work only with genes and proteins, but the aid of various metals is required by the life system in nature. Thus, it was desirable to establish a new scientific field for biometals among omics science. It was "metallomics" [1]. When I conceived the idea of metallomics, I thought that it was really meaningful to integrate (1) metallomics with genomics, proteomics, and other omics sciences, (2) various independent fields in trace metal research, and (3) separated scientific communities. This was the reason why the title of "metallomics as integrated biometal science" was employed in the original paper [1].

1.2 Progress of Analytical Atomic Spectrometry

The present author started the research carrier with analytical atomic spectrometry in 1969 [9] and reached to "metallomics" in 2004 [1]. It was a long way, but it was a great pleasure. Thus, I would like to start this chapter with description of my experience in analytical atomic spectrometry.

In the early 1960s, the instruments of flame atomic absorption spectrometry (FAAS) became commercially available as the sensitive analytical method, and it was extensively applied to trace analysis, especially to environmental analysis [9]. In those days, Japan had serious environmental problems such as Minamata disease caused by mercury and ouch-ouch disease (*itai-itai disease* in Japanese) caused by cadmium. Nowadays, it is well known that Minamata disease as the environmental issue was caused by the poisoning effect of methylmercury, which was bioaccumulated in fishes in the Minamata Bay area. Thus, the environmental criteria of total mercury and methylmercury were enacted to be "0.0005 ppm" and "not detected" by the governmental law in Japan. FAAS was also applied to the analysis of trace elements (metals and metalloids) in various biological samples such as blood, urine, and organs. Thus, FAAS was a powerful analytical tool to elucidate causative heavy metals in environmental pollution [9, 12].

After 1970, various analytical atomic spectrometric methods have been further developed for trace element analysis. Until now, the present author has had a lot of experiences in flame atomic absorption spectrometry (FAAS; 1969–1974), graphite furnace atomic absorption spectrometry (GFAAS;1974–1975), atomic fluorescence spectrometry (AFS; 1975–1977), inductively coupled plasma atomic emission spectrometry (ICP-AES; 1977–2000), and inductively coupled plasma mass spectrometry (ICP-MS; 1990–2010) [1, 3, 9]. During these four decades, the analytical detection limits of those methods have been extensively improved from the ppb (10^{-9} g/mL) level down to the ppt (10^{-12} g/mL) or ppq (10^{-15} g/mL) level. Furthermore, it should be emphasized that ICP-AES and ICP-MS allow simultaneous multielement analysis with wide dynamic ranges of the working calibration curves, while AAS was mostly used for single-element analysis. The instrumental specifications and analytical feasibilities of those methods can be understood from the literatures [9–11]. The analytical characteristics of ICP-AES and ICP-MS will be briefly reviewed in the following paragraphs.

In ICP-AES and ICP-MS, the sample solution is introduced into the high-temperature argon plasma through a nebulizer. The photo of an argon inductively coupled plasma (argon ICP) is shown in Fig. 1.1. The characteristics of the ICP are the formation of the *so-called* doughnut structure, in which the temperature in the central channel of the plasma is lower than that of the surroundings. This doughnut structure is a very important property of the inductively coupled argon plasma for

Fig. 1.1 A photo of an inductively coupled argon plasma used for excitation and ionization sources in ICP-AES and ICP-MS, respectively

the efficient sample introduction into the plasma by pneumatic nebulization. In the case of ICP-AEC, the emission intensities of atomic or ionic lines are observed as the emission signals. Usually, the ionic emission lines provide the better sensitivities than the atomic lines, because most metallic elements are significantly ionized (more than 90%) and efficiently excited in the argon plasma at 6000–8000 K. However, since efficient excitation of ions often causes analytical errors due to serious spectral interferences because of the emission line overlapping [10], the spectral interference correction must be properly performed in the ICP-AES measurements.

In ICP-MS, on the other hand, the ions produced in the argon plasma are let into a mass spectrometer through the sampling cone and the skimmer cone and detected by a secondary electron multiplier [9, 11]. Two types of mass spectrometers such as a low-resolution quadrupole mass spectrometer and a high-resolution (HR) double-focusing mass spectrometer are available as the ICP-MS instruments. The former is called Q-ICP-MS and the latter is HR-ICP-MS. As is well known, many elements are composed of several stable isotopes, and the information for intrinsic isotope ratios corresponding to natural abundances of atoms can be obtained in ICP-MS, since ICP-MS can detect the mass of each isotope (m/z). These isotope ratio measurements, thus, are helpful for the identification of the elements [11]. In addition, isotope dilution mass spectrometry (IDMS) can be performed in ICP-MS by adding some enriched isotopes [16, 21, 22], since IDMS provides most precise and reliable analytical data in quantification of trace metals.

The detection limits obtained by commercially available ICP-AES and ICP-MS instruments are summarized in Fig. 1.2 [23], which were obtained by Q-ICP-MS and HR-ICP-MS. In the figure, the elements are plotted at the positions

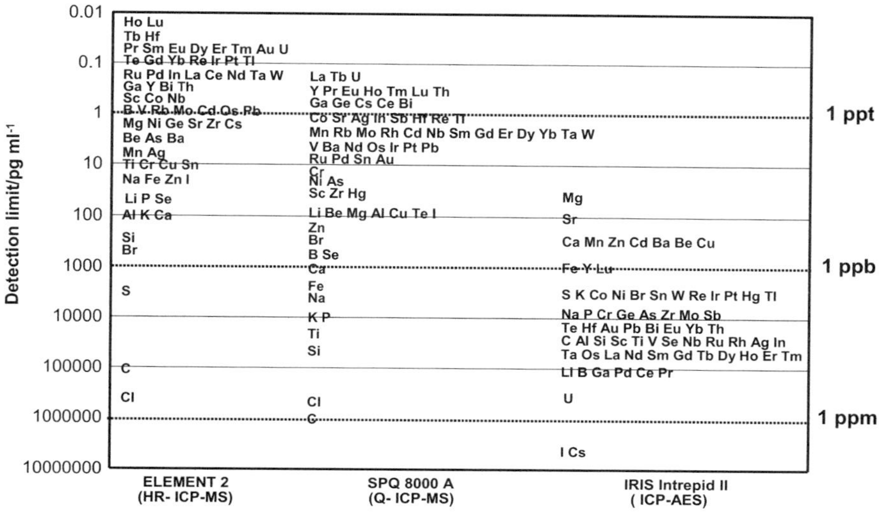

Fig. 1.2 Comparison of the detection limits obtained by HR-ICP-MS, Q-ICP-MS, and ICP-AES [23]. HR-ICP-MS: high-resolution ICP-MS with double-focusing mass spectrometer from Thermo Fisher Scientific Co., Q-ICP-MS (SPQ 8000A): ICP-MS with quadrupole mass spectrometer from Seiko Instruments Co., ICP-AES: ICP-AES with a CID (Charge Injection Device) multielement detector from Thermo Fisher Scientific Co.

corresponding to the detection limits on the vertical axis (concentration unit). It can be seen from Fig. 1.2 that the detection limits at the ppb level are generally obtained by ICP-AES, while the detection limits obtained by ICP-MS are much more improved down to the ppt (pg/mL) or sub-ppt level. Then many elements can be directly detected at the sub-ppt level by HR-ICP-MS.

It is also noted here that ICP-AES and ICP-MS provide the wide linear dynamic ranges of working calibration curves (four to six orders of magnitude), which provide excellent analytical feasibilities for the multielement detection of the elements in the concentration range from ppb up to 1000 ppm in ICP-AES and that from sub-ppt up to 1 ppm in ICP-MS [9]. Thus, if both ICP-AES and ICP-MS instruments are available, analyte elements (metallic and metalloid elements including S, P, Cl, Br, I) in the concentration range from sub-ppt up to 1000 ppm, i.e., over 10^{10} orders of magnitude, can be simultaneously determined for any biological samples. These characteristics of plasma spectrometry make it possible to perform all-element analysis of biological samples as well as geochemical and environmental samples [1, 9], as will be described in the next section.

1.3 All-Element Present Theory

In 1936, I. Noddack [24] published the article entitled "Concerning the ubiquitous nature of the chemical elements" (its original title in Germany was "Über die Allgegenwart der chemischen Elemente") in *Angewandte Chemie* and suggested that all elements in the periodic table might be present in all rock and mineral samples on the earth. Kuroda supported the Noddack's idea (hypothesis) of ubiquitous presence of all elements in geological samples in his book *The Origin of the Chemical Elements and the Oklo Phenomenon*, published in 1982, and he named the Noddack's concept "the All-Present Theory of the Elements" [25].

In the 1930s, the number of elements measurable in rock and minerals was quite limited because the analytical methods available in those days, such as flame emission spectrometry, arc/spark emission spectrometry, electrochemical methods, and so forth, were not sensitive enough to detect the low-abundant elements. Thus, many scientists were thinking that the elements whose concentrations were unknown were not contained in the samples of interest. On the contrary, Noddack believed that even the elements whose analytical values had been unknown might be contained in any geological samples, and she insisted that the existence of all elements in all geochemical samples would be elucidated, when the sensitive analytical techniques for trace elements would be developed in the future.

These days, according to great progress in analytical methodology, as described before, almost all elements are able to be determined or detected not only in geochemical samples but also in biological and environmental samples [9]. Since 1990, thus, the Haraguchi's research group in Nagoya University challenged all-element analyses of the representative samples collected from the atmosphere, lithosphere, hydrosphere, biosphere, and urbanosphere (urban area), for example, airborne particulate matter, rocks, seawater and lake water, biological samples from humans and plants, and bottom/fly ashes from the waste incinerators [1, 9, 12]. Based on such experimental results, Haraguchi proposed a new concept of "the Extended All-Present Theory of the Elements" [9], as follows:

> The elements contained in rocks and minerals are dissolved into water because of the weathering processes on the earth, and plants growing in soil absorb the elements in water for nutrition, and then animals ingest plants for food, and humans drink water and ingest plants and animals as food because humans are at the top of food chain. Consequently, it is obvious to consider that plants and animals, even their organs and blood, contain all elements through the elemental cycles on the earth, in another words, the All-Present Theory of the Element stands up all right not only for rocks and minerals, but also for all biological systems on the earth. Then, the Extended All Present Theory of the Elements is more suitable term to express the situations.

This concept indicates that all elements in the periodic table are ubiquitously contained not only in geochemical samples but also in all biological systems, such as humans, plants, and microorganisms. Humans are of course the object of the all-present theory. Thus, the All-Present Theory of the Elements proposed by Noddack should be extended to the new concept with the wide viewpoint, and then the Extended All-Present Theory of the Elements was newly proposed. Since the newly proposed term is a little long, a slightly simple term of "**the All-Element Present Theory**" is recommended here for the use as the academic term.

Haraguchi is further thinking that the final goal of the All-Element Present Theory is to prove the existence of all elements in single biological cell. This concept "the existence of all elements in single biological cell" may be referred to as "cell microcosm" [1]. If the cell microcosm is scientifically elucidated for the living biological cells, such knowledge of the elemental distributions in the biological cells may provide important information for the study on life science as well as global cycles of the elements in the whole universe including the earth. From the viewpoint of the All-Element Present Theory, the elemental distributions in humans, human blood serum, cherry blossom, and salmon egg cell will be discussed in the following description.

(a) **The elements contained in humans and their components/functions**

In analytical chemistry, the elements in the concentration ranges of 100–1%, 1–0.01%, 0.01–0.0001%, and below 0.0001% are defined as major, minor, trace, and ultratrace elements, respectively. Since 0.0001% is equal to 1 ppm (part per million = μg/g), trace elements are the elements in the concentration range of 1–100 ppm. In the case of metals, they are called "trace metals." It is noted here that, in medical and biological sciences, ultratrace elements are usually considered as the same as trace elements, and thus, all the elements whose concentrations are less than 100 ppm are generally called trace elements hereafter. As will be seen in Table 1.1, the amounts of Fe and Al are about 6 g and 60 mg per 70 kg (the body weight), which correspond to 85.7 μg/g and 0.857 μg/g in the concentration, respectively. According to the definition above, both of Fe and Al are conveniently trace elements, although, exactly to say, Al is ultratrace element.

In Table 1.1, the average amounts of major, minor, trace, and ultratrace elements contained in a human body (70 kg weight) are summarized together with their components or biological functions [1, 26]. As for essential elements, major, minor, and trace (ultratrace) elements established for humans and experimental mammals are marked with *, **, and ***, respectively, in Table 1.1. It is seen from Table 1.1 that a variety of elements in the wide concentration range are contained in our human body and also that many of metallic elements play essential roles as metalloproteins and/or metalloenzymes in the blood and organs of the biological systems (*see* the discussion in "1.8.2 Essentiality and toxicity of the elements"). For example, the antioxidation functions of Se (glutathione peroxidase) and Fe (catalase) are important to protect the biological systems from oxidation caused by peroxides. Superoxide dismutase, which usually contains Mn, Fe, Cu, and/or Zn, acts as an antioxidant against superoxide. In addition, it is known that Na and K act as a Na-K pump across the cell to perform material transportation through cell membranes. Sodium and K are extracellular and intracellular electrolytes, respectively, which means that the Na concentration is higher outside the cell than inside, while the K concentration is higher inside the cell than outside. Such concentration differences of Na and K produce the concentration gradients across the cell membranes, which result in producing osmotic pressure to provide the field for

Table 1.1 The elements contained in humans and their components/functions[a]

	Element[b]	Amounts, (70 kg weight)	Components/functions in body
Major elements	O*	45.5 kg	Water and organics
	C*	12.6	Organics
	H*	7.0	Water and organics
	N*	2.1	Amino acids, proteins, nucleic acids
	Ca*	1.05	Bones, cell membrane, blood
	P*	0.70 (98.5 %)[c]	Bones, nucleic acids, ATP
Minor elements	S*	175 g	Amino acids, Vitamin B_1, heparin
	K*	140	Inner-cell electrolyte, blood
	Na*	105	Outer-cell electrolyte, blood
	Cl*	105	Main anions inside and outside cells
	Mg*	105 (99.4%)[c]	Co-factors of enzymes, chlorophylls
Trace elements	Fe**	6	Transportation and storage of oxygen and iron, catalase
	F***	3	Concomitants of bone and teeth
	Si***	2	*Unknown*
	Zn**	2	Co-factors of enzymes (DNA & RNA polymerases), cell division
	Sr***	320 mg	*Unknown*
	Rb	320	*Unknown*
	Pb***	120	Metabolism of iron, hematopoiesis
	Mn**	100	Co-factors (superoxide dismutase, pyruvate kinase)
	Cu**	80	Redox reactions, electron transfer, oxygen addition
Ultratrace elements	Al	60	Alzheimer disease
	Cd	50	Ouch-ouch disease
	Sn***	20	Redox reactions
	Ba	17	*Unknown*
	Hg	13	Minamata disease (methylmercury)
	Se**	12	Antioxidation (glutathione peroxidase), mercury detoxication
	I**	11	Activation of thyroid gland
	Mo**	10	Co-factors of enzymes (xythanthin oxydase, nitrate reductase),
	Ni***	10	Co-factor of enzyme (urease), stabilization of RNAs
	Cr**	2	Glucose tolerance factor, metabolisms of fats and proteins
	As***	2	Metabolic activation of zinc
	Co**	1.5	Hematopoiesis (vitamin B12)

(continued)

Table 1.1 (continued)

	Element[b]	Amounts, (70 kg weight)	Components/functions in body
	V***	1.5	Metabolism of cholesterols, inhibitors of Na^+- and K^+-ATPase

[a]Cited from Ref. [26] with some revision
[b]*Essential major and minor elements, ** Essential trace elements in humans, *** Essential trace elements in experimental mammals
[c]Cumulative amounts

active or passive transport of various biomaterials across the cell membrane. Although the functions of metallic elements are not explained anymore here, there is no doubt that most of biological systems cannot maintain their activities without the aid of a variety of metallic elements [1, 26, 27]. Thus, metallomics should be systematically established as biometal science in order to appreciate the biological systems from the viewpoint of the biological functions for syntheses and metabolisms.

(b) **The concentrations of the elements in human blood serum**

According to the All-Element Present Theory, human blood serum supposed to contain all elements in the periodic table, too. The concentrations of major-to-ultratrace elements in human blood serum so far reported are summarized in Table 1.2 [1]. It is seen from Table 1.2 that 53 kinds of elements in the concentration range of 12 orders of magnitude are contained in blood serum, where essential elements for humans and animals are indicated in bold letters. In human blood serum, Na, K, Mg, and Ca are contained as major elements. These concentration characteristics of major elements in human blood serum are similar to those in seawater, as will be seen later in (d) of this section. Thus, it is often said that such a similarity of the concentration levels of major elements in blood serum and seawater might be one of evidences for the origin of life in the sea [28].

Here, the concentration levels of essential trace elements in blood serum are compared with those in seawater. It is remarkable that the concentrations of essential trace elements such as Fe, Cu, Zn, Se, Mn, and Co are much higher in blood serum than those in seawater, while the concentrations of I, Mo, and Cr are significantly lower in blood serum than those in seawater. These facts indicate that Fe, Cu, Zn, Se, Mn, and Co are enriched in blood serum by bioaccumulation, because these elements exist as a variety of metalloproteins to express the specific biological roles required by the life systems. Blood serum is also taking important roles to transport such elements to the cells and organs for nutrition and metabolism by circulating our whole body. It is also mentioned here that trace elements such as Pb, Sn, Hg, Cd, and Ag, which are considered to be rather toxic or hazardous elements, are contained at the higher concentration in blood serum rather than in seawater, while these elements does not cause any toxicity at the concentration level in humans and other animals.

Table 1.2 The concentrations of major-to-ultratrace elements in human blood serum[a]

Element[b]	Blood serum concentration (ng/g)	Element[b]	Blood serum concentration (ng/g)
Cl	3,200,000	W	0.34
Na	3,130,000	U	0.31
K	151,000	**Ni**	0.23
P	119,000	*Ce*	0.21
Ca	93,100	Ag	0.20
Mg	17,500	Cd	0.15
Br	4440	**Co**	0.11
Fe	1200	Be	0.09
S	1110	**Cr**	0.069
Cu	750	*La*	0.063
Zn	651	*Nd*	0.034
Rb	170	**V**	0.031
Se	160	*Yb*	0.013
Si	140	*Pr*	0.011
Sr	33	*Dy*	0.0096
Sb	2.3	*Er*	0.0095
B	2.1	*Gd*	0.0072
Al	1.8	**I**	0.007
Pb	1.2	*Sm*	0.0058
Mo	0.95	*Ho*	0.0026
Cs	0.95	*Lu*	0.0025
Y	0.73	*Tm*	0.0017
Mn	0.57	Sc	0.0017
Hg	0.55	Pt	0.0014
Sn	0.51	*Tb*	0.0013
Th	0.50	*Eu*	0.00082
Ba	0.48	Au	0.00003
As	0.45		

[a]Cited from Table 1.4 in Ref. [1], and the references are there
[b]The essential elements for humans and animals are indicated in bold letters (*See* Table 1.1), and rare earth elements are indicated in italic letters

In 2000, Inagaki and Haraguchi reported the existence of all rare earth elements in blood serum [29]. As is seen Table 1.2, the concentrations of rare earth elements are extremely low at the ppt or sub-ppt level, which are almost at the same concentration level as those in seawater. The biological roles or functions of rare earth elements in the life system have not been known so far.

(c) **Multielement analyses of cherry blossom petals and leaves**

The elemental distributions in cherry samples are described as the next topics. It is well known that the cherry blossoms are the symbolic flower of Japan. Usually, the cherry blossoms start flowering late March or early April. Exactly to say, the cherry flowers are cherry blossom petals, which are kept shortly on the tree only for

Table 1.3 Multielement determination of major-to-ultratrace elements in cherry samples by ICP-AES and ICP-MS[a] [30]

Element[b]	Concentration (μg/g) Petals[c] (4/9)	Leaves[c] (8/11)	Element[b]	Concentration (μg/g) Petals[c] (4/9)	Leaves[c] (8/11)
K	19,200	19,800	Pr	0.27	0.23
Ca	2530	11,500	Zr	0.26	0.80
P	2520	1870	Sm	0.17	0.11
Mg	1170	2370	Gd	0.15	0.093
Al	179	37	Sn	0.12	0.032
Fe	115	58.9	Cs	0.11	0.032
Na	69.8	13.5	**Co**	0.102	0.044
Mn	76.7	49.4	Dy	0.093	0.048
B	31.4	35.9	Cd	0.046	0.012
Rb	30.2	29.2	W	0.034	0.011
Zn	22.7	15.2	Eu	0.033	0.024
Cu	7.01	7.01	Th	0.030	0.008
Ba	3.50	25.6	Er	0.027	0.020
Ce	1.85	1.81	Yb	0.019	0.011
Ni	1.60	0.86	Tb	0.019	0.010
La	1.46	1.54	Tl	0.013	0.015
Sr	1.30	18.6	Ho	0.011	0.009
Nd	1.02	0.78	U	0.0073	0.0025
Pb	0.97	0.89	Tm	0.0034	0.0023
Y	0.70	0.34	Lu	0.0028	0.0017
Mo	0.47	0.54			

[a]The elements in bold letters are the essential elements for plants
[b]The samples were digested by using HNO_3 and HF
[c]The sampling dates; petals on April 9, 1996, and leaves on August 11, 1996

about 2 weeks. Soon after the petals fall down, the cherry leaves come out and grow until the end of August. The analytical data for petals collected on April 11 and leaves collected on August 9 are summarized in Table 1.3 [30]. In the present experiment, petals and leaves were collected and analyzed every week from the beginning of April to the end of August in 1996. These samples were analyzed by ICP-AES and ICP-MS after the acid digestion. In addition, the chlorophyll contents in leaves were analyzed every 2 weeks, although the data are not shown. In the year 1996, the cherry trees were almost in full bloom around April 11, while the chlorophyll contents in leaves were almost at the maximum around August 9, indicating that photosynthesis in leaves was most active around the middle of August.

The analytical data were obtained for 41 elements, as is seen in Table 1.3, where the essential elements for plants are indicated in bold letters. The number of the elements measured was limited to about 40, because the quadrupole-type ICP-MS instrument which was in low resolution and less sensitive was only available in this experiment. It is noticeable that the elemental concentrations in short-lived petals were almost the same as those in leaves collected about 4 months later. However, if the data are examined carefully, we can find some characteristic behaviors of the elements by comparing the data between petals and leaves. It is just pointed out here that the concentrations of Mg and Ca are remarkably higher in leaves rather than in petals. In the case of Mg, Mg increased along with the increase of chlorophyll contents in leaves because Mg was a component of chlorophyll. It was also observed that the Mg concentration in leave decreased after the middle of August. This phenomenon may be interpreted by the fact that Mg moved from leaves to other parts of the tree, maybe to trunks and roots for preservation. Calcium also increased in leaves through the experimental period, but its concentration did not decrease even after the middle of August. These results suggest that Ca taken into leaves was accumulated as non-movable components therein, maybe because of some interactions with cell proteins in leaves. As for other elements, their concentrations were almost the same levels in petals and leaves, although some essential elements such as P, Fe, Na, Mn, Zn, and Ni provided the higher concentrations in petals than in leaves.

(d) A challenge to all elements analysis of salmon egg cells

All-element analysis of salmon egg cells was also challenged in order to examine the concept of cell microcosm, i.e., all-element presence in single biological cell. The analytical results are summarized in Table 1.4 [1, 31], together with the concentrations of the elements in seawater. In salmon egg analysis, three or five salmon eggs were digested with nitric acid by using a microwave digestion method. In the present experiment, rare gas elements and natural/artificial radioactive elements were not measured because the specific gas sampling systems as well as the skillful gas treatment techniques were necessary in rare gas analysis and the specially designed experimental facilities for protection from harmful radioactivity were required for treatment of radioactive elements. As a result, 78 elements in the periodic table were the target of all-element analysis in the standard laboratory.

The analytical results for 66 elements among 78 are shown in Table 1.4, where the metallic and metalloid elements were measured by ICP-AES and ICP-MS and nonmetallic elements such as H, C, N, O, and Cl were determined by the conventional elemental analysis method, while other seven elements (Li, Zr, Nb, Hf, Ir, Bi) were just detected because of their low abundances in salmon egg cell. Fluorine, Rh, Te, Ta, and Re in salmon egg cells could not be determined or detected at this experiment, maybe because of their extremely low concentrations. However, these elements except for F were detected by HR-ICP-MS in the recent experiment, although the results have not been reported yet.

Many interesting facts can be found from the results in Table 1.4, but here it is just pointed out that Fe, Zn, Cu, Co, Mn, Se, and P provided the bioaccumulation factors larger than 10,000 [31], which indicated the large enrichment of these elements in salmon egg cell from seawater. In the next section, chemical speciation of the elements in salmon egg cell will be described in relation with the data shown in Table 1.4.

Table 1.4 Concentrations of the elements in salmon egg cells and seawater [31]

Element[a]	Salmon egg cell concentration (ng/g)	Seawater concentration (ng/ml)
O	600,000,000	–
C	250,000,000	–
H	100,000,000	–
N	42,000,000	–
P[†]	3,580,000	62
S	2,990,000	898,000
Cl	2,800,000	19,350,000
K[†]	1,860,000	399,000
Ca[†]	432,000	412,000
Na[†]	247,000	10,780,000
Mg[†]	222,000	1,280,000
Br	20,000	67,000
Zn[†]	13,600	0.35
Fe[†]	10,700	0.03
Cu[†]	8900	0.15
I	3700	58
Sr[†]	3600	7800
Se	1970	0.16
Ti	1420	0.0065
Rb	523	120
Mn[†]	512	0.02
As	192	1.2
Ba	47.3	15
Ni	26.5	0.48
V	18.8	2.0
Co	12.5	0.0012
Hg	12.0	0.00014
Ag	10.7	0.002
Pd	8.0	0.00006
B	7.8	4500
In	7.0	0.0001
Al	6.44	0.31
Mo	6.43	10
Ge	5.83	0.005
Cs	5.58	0.31
Cd	1.1	0.07
Sn	0.96	0.0005
Be	0.80	0.00021
Cr	0.73	0.21
U	0.66	3.2
Pb	0.48	0.0027
Ru	0.19	0.000005

(continued)

Table 1.4 (continued)

Element[a]	Salmon egg cell concentration (ng/g)	Seawater concentration (ng/ml)
Sb	0.12	0.2
Y	0.11	0.017
La	0.075	0.0056
Nd	0.074	0.0033
Au	0.054	0.0066
Tl	0.050	0.013
W	0.033	0.01
Gd	0.016	0.0009
Ce	0.014	0.0007
Pr	0.010	0.0007
Sm	0.010	0.00057
Th	0.009	0.00002
Dy	0.0074	0.0011
Sc	0.0059	0.0007
Pt	0.0043	0.0002
Er	0.0031	0.0012
Yb	0.0027	0.0012
Ga	0.0023	0.0012
Tb	0.0021	0.00017
Os	0.0019	0.000002
Ho	0.0017	0.00036
Eu	0.0011	0.00017
Lu	0.00078	0.00023
Tm	0.00037	0.0002

[a]The elements with † were determined by ICP-AES, and other elements were determined by ICP-MS. In addition, H, C, N, O and Cl were determine by the conventional elemental analysis method

1.4 Chemical Speciation of Trace Metals in Biological Samples

As described earlier, the analytical sensitivities for metallic elements (including metalloids) were substantially improved down to the ppt or sub-ppt level according to the progress of ICP-MS. However, the direct information for the elements obtained by ICP-MS is only their total concentration data, since the molecular information is lost because chemical species are decomposed in the high-temperature argon plasma at around 7000–8000 K. In biological and environmental analysis, however, the information for chemical or molecular species is generally more important than that of total concentrations of the elements, because the impacts and toxicities to the biological systems are dependent on their chemical forms. For example, inorganic arsenic (arsenite and arsenate) are toxic and/or carcinogenic, while methylated arsenicals are nontoxic, as is well known [32]. In the case of chromium, Cr(III) is bio-essential, but Cr(VI) is toxic because of its strong oxidation ability. In addition, the causal substance of Minamata disease was methylmercury, not inorganic mercury. Thus, the identification and quantification of chemical species including oxidation states of the elements are really required to elucidate their biological/physiological functions and activities to the biological systems. The analytical technology for identification and quantification of chemical species is defined as "chemical speciation" or "speciation" in analytical chemistry [13–16].

Trace metals (mostly ionic forms) are usually bound to large molecular proteins to form metalloproteins in biological cells and fluids, but their concentrations in biological samples are generally very low. Then, the highly sensitive and element-selective analytical methods are required for the detection of trace chemical species like metalloproteins. In addition, the chemical compositions of biological samples are very complicated. In another word, trace metallic ions exist in very complex matrices controlled by the complicated chemical equilibria [33], so that the element-selective and highly sensitive detection methods of trace metals are necessary in speciation analysis. At present, in order to fulfill such requirements for chemical speciation analysis, a hyphenated system combined with efficient species separation and highly sensitive element detection (e.g.,HPLC/ICP-MS) is most suitable for speciation analysis. Various types of chromatographic methods are used for efficient separation of chemical species, and ICP-MS is used for the highly sensitive and element-selective detection of metallic and metalloid elements.

In recent years, various combined methods for chemical speciation have been developed and extensively applied to the elucidation of the chemical forms even for biometals. In Fig. 1.3, the present state-of-the-art hyphenated methods for chemical speciation are summarized [34], and there are many excellent review articles concerning with chemical speciation so far published [13–16, 32]. It can be seen in Fig. 1.3 that the analytical systems for chemical speciation generally consist of three main parts such as separation shown in the left-hand side, detection in the right-hand side, and identification of chemical species in the center. Nowadays, a variety of separation methods can be selected, depending on the kinds of chemical species from small molecules to large molecules (proteins and other biopolymers) as well as their chemical properties. As the detection method, ICP-MS is the most frequently used, where a quadrupole-type mass spectrometer (Q-MS), time-of-flight mass spectrometer (TOF-MS), and sector field or high-resolution mass spectrometer (SF-MS or HR-MS) are employed for the detection of analyte elements.

As mentioned before, ICP-MS can provide the element information, but it cannot provide the molecular information. Thus, in chemical speciation using the hyphenated systems, chemical species or molecules are generally identified by referring to the retention times for standard (known) compounds observed in the chromatograms. MALDI-MS (matrix-assisted laser desorption/ionization MS) and ESI-MS (electrospray ionization mass spectrometry) are often used for direct identification of chemical or molecular species, but they are not so sensitive enough to provide the information for trace metals in various biomolecules.

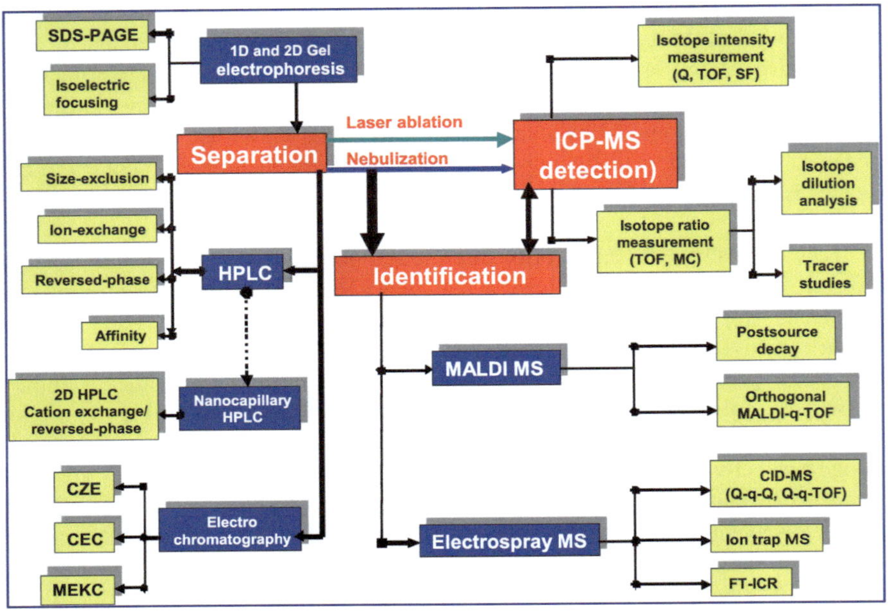

Fig. 1.3 Various hyphenated methods for chemical speciation analysis, cited from Ref. [34] with some revision

As an example, a schematic diagram of SEC (size exclusion chromatography)/ICP-MS system for chemical speciation is shown in Fig. 1.4, where a UV absorption detector is equipped for monitoring protein elution. Such a SEC/ICP-MS system is conveniently used for the identification and molecular weight calibration of metalloproteins.

(a) **Chromatogram of standard proteins**

In Fig. 1.5, a chromatogram of standard proteins obtained by using the SEC column with the UV absorption detection (measured in absorbance) is shown together with the retention times of proteins. The colored zone in Fig. 1.5 indicates the permeation range of the SEC column, and the numbers (1–7) in the chromatogram correspond to proteins examined (proteins and their molecular weights are indicated in the figure caption). In the SEC chromatogram, of course, the larger molecules are eluted earlier than the smaller molecules because of the size effects of the SEC column.

(b) **Chemical speciation of arsenicals in salmon egg cell**

The HPLC chromatograms for arsenicals (arsenic species) in salmon egg cell cytoplasm and membrane are shown in Fig. 1.6a and b, respectively [32], where arsenic was detected at m/z 75 by ICP-MS. Cell cytoplasm (intracellular fluid) and cell membrane were separately collected from whole egg cells by using a Teflon

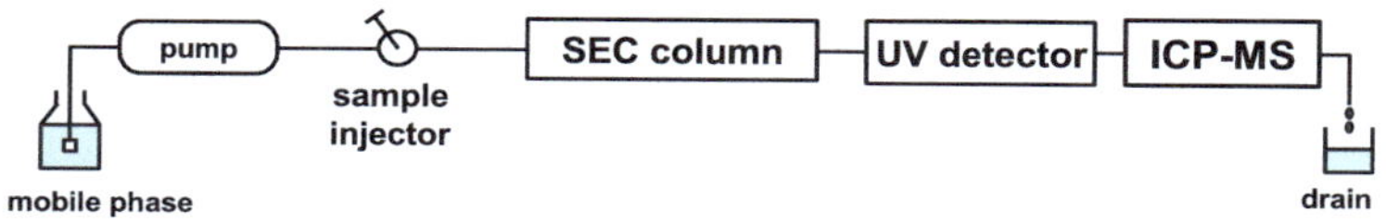

Fig. 1.4 A schematic diagram of SEC/ICP-MS with a UV absorption detector for chemical speciation of metalloproteins

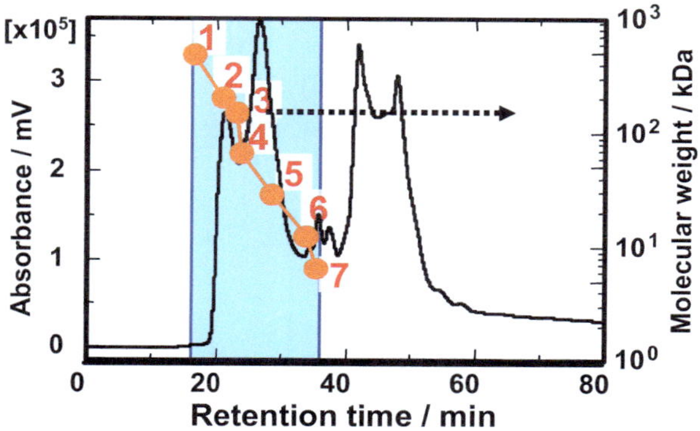

Fig. 1.5 Chromatogram of standard proteins obtained by a hyphenated system of SEC/ICP-MS. Stationary phase: SuperoseTM 12 10/300 GL (molecular permeation range 1–300 kDa); mobile phase, 0.1 M Tris-HCl (pH 8.0); flow rate of mobile phase, 0.5 mL/min; sample injection volume, 100 μL; wavelength of UV absorption detection, 254 nm; standard proteins, (1) ferritin (440 kDa), (2) β-amylase (200 kDa), (3) alcohol dehydrogenase (150 kDa), (4) albumin (66 kDa), (5) carbonic anhydrase (29 kDa), (6) cytochrome c (12 kDa), (7) aprotinin (7 kDa)

tweezers and a Teflon needle. Arsenic species were extracted from cell cytoplasm and cell membrane as follows. Approximately 1 g of cell cytoplasm or cell membrane sample was taken in a centrifugation tube, and 10 mL of 50% methanol was added. The sample was sonicated for 30 min and then centrifuged at 4000 rpm for 5 min. The supernatants as the analysis sample were collected two times, put together in the rotary evaporator, and evaporated almost to dryness at 40°C. The residue was dissolved in 1 mL of HPLC mobile phase solution and then filtered with a membrane filter (pore size; 0.45 μm). The dissolved solution was subjected to the HPLC/ICP-MS analysis. In the case of cell membrane analysis, before the extraction procedure mentioned above, 0.5 mL of TMAH (tetramethylammonium hydroxide) was added as an alkali reagent into the sample to dissolve arsenic species from lipids or proteins in cell membrane.

It is seen in Fig. 1.6a that a large peak of AB (arsenobetaine) and three small peaks of iAs^V (arsenate), DMA (dimethylarsinic acid), and TMAO (trimethylarsine oxide) were observed for cell cytoplasm. The concentrations of these AB, iAs^V, DMA, and TMAO were estimated to be 17.6, 1.7, 0.4, and 3.0 ng/g as As, respectively, on the wet-weight basis. On the other hand, as is seen in Fig. 1.6b, iAs^V, MMA (monomethylarsonic acid), DMA, and AB were found in cell membrane,

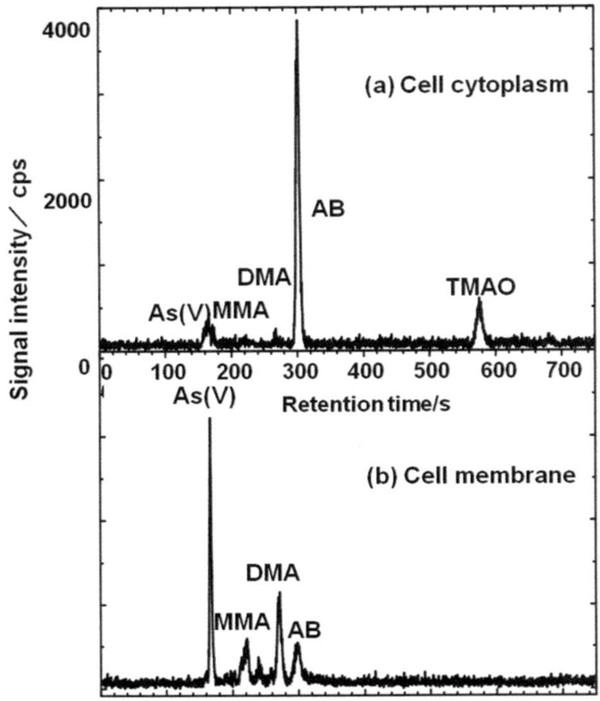

Fig. 1.6 HPLC chromatogram for arsenic species in (**a**) cell cytoplasm and (**b**) cell membrane of salmon egg detected by ICP-MS at m/z 75 [32]. Column: CAPCELL PAK C18 (ODS column); mobile phase, 4 mM malonic acid + 4 mM TMAH + 10 mM 1-butane sulfonic acid sodium salt + 0.05% methanol (pH 3.0); flow rate, 0.75 ml/min; sample injection volume, 10 μL

whose concentrations were 15.4, 8.1, 12.3, and 8.4 ng/g as As, respectively. As a result, the total concentrations of arsenic species in the extracts from cell cytoplasm and cell membrane were 22.7 ng/g and 44.2 ng/g, respectively, which corresponded to ca. 12% and 35% of the total amounts of arsenic in cell cytoplasm and cell membrane, respectively. These results suggest that the rest of arsenic species may exists as large molecular arsenic species such as protein- or sugar-binding arsenic in both cytoplasm and membrane, because only small molecular arsenic species could be identified by HPLC/ICP-MS using the ODS column in ion-pair mode. Inorganic arsenic (iAs^V) was the main arsenic species in cell membrane, while AB (arsenobetaine) was the main species in cell cytoplasm. These facts suggested that methylation of inorganic arsenic to AB occurred on passing through the cell membrane or after penetrating into cell cytoplasm.

(c) **Fractionation analysis of the elements in salmon egg cell cytoplasm**

As one of the examples for chemical speciation, the fractionation analysis of the elements in salmon egg cell cytoplasm was carried out by HPLC/ICP-MS, where the CHAPS (3-[(3-cholamidopropyl)-dimethylammonio]-1-propanesulfonate; zwitterionic surfactant)-coated ODS (octadecyl silica) column (4.6 mm i.d. x 250 mm long) was used in order to examine the protein bindings of the elements [35]. The CHAPS-coated ODS column was prepared by the dynamic coating

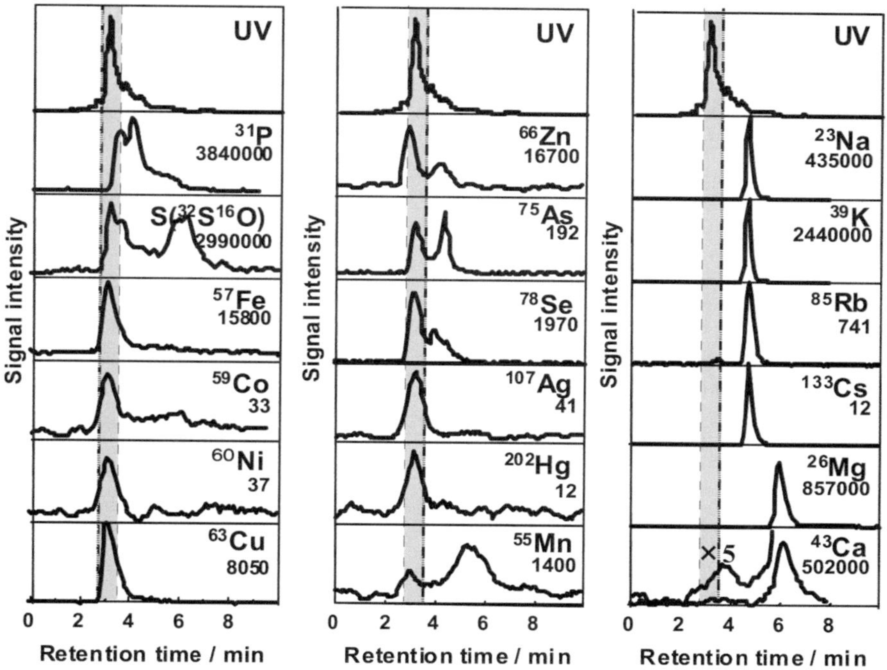

Fig. 1.7 The element-selective chromatograms for salmon egg cytoplasm obtained by HPLC/ICP-MS. [31]. Sample: salmon egg cell cytoplasm diluted five-fold with 0.1 M Tris buffer (pH 7.4); column, CHAPS-coated ODS column; mobile phase, 0.1 M Tris buffer solution (pH 7.4); ICP-MS detection, m/z shown in the figure; UV absorption detection, 254 nm. The numbers below the elements indicate the concentrations (ng/mL) of the elements in salmon egg cell. The retention time range between 3.0 and 3.5 min (*gray* zone) corresponds to the protein elution zone

method, where 20 mM CHAPS aqueous solution was passed through the ODS column at 0.7 ml min^{-1} for 2 h, as described in the previous papers [35, 36]. After coating, the column was rinsed with pure water for 1 h. Since the CHAPS-coated ODS column has unique characteristics for simultaneous separation of ions/small molecules and large molecules (e.g., proteins) [36, 37], the useful information about the elements binding or nonbinding with proteins is easily obtained.

The chromatograms for the elements in salmon egg cell cytoplasm obtained by HPLC/ICP-MS using the CHAPS-coated ODS column are shown in Fig. 1.7 [31]. In Fig. 1.7, the chromatograms with the element-selective detection are illustrated together with those with the UV absorption detection at 254 nm, where the concentrations of the elements determined by ICP-MS are also indicated in each chromatogram. The retention range of 3.0–3.5 min shown as the gray zones in Fig. 1.7, which is almost consistent with the elution zones for proteins detected by UV absorption shown in the upper parts of Fig. 1.7, indicates the elution zone for large molecules with molecular weight (MW) larger than *ca.* 10,000 Da. On the other hand, the retention range after 3.5 min indicates the elution zone for small ions and molecules [37]. However, the information about the molecular weights cannot be estimated from the chromatograms obtained by the CHAPS-coated ODS column, since the CHAPS-coated ODS column has no size exclusion property.

It is seen in Fig. 1.7 that heavy metals such as Fe, Co, Ni, Cu, Ag, and Hg, which are essential or toxic elements, provided single broad peaks within the protein elution zone. These results indicate that these heavy metals mostly exist as protein-binding species. On the other hand, Zn and Mn provided two main peaks within and after the protein elution zone, which suggests that Zn and Mn exist partly as small molecules/ions, maybe free ions or amino acid complexes, in addition to protein-binding species. Furthermore, alkali and alkaline earth elements, such as Na, K, Rb, Cs, and Mg, provided single peaks in the small molecules/ions zone in Fig. 1.7. These experimental results suggest that Na, K, Rb, Cs, and Mg in salmon egg cell exist as the ionic forms. It is interesting that Ca provided small peaks in the large molecular zone (*see* expanded *chromatogram*), which indicated that small amount of Ca was binding with proteins. As for arsenic as well as selenium, two separated peaks in the small and large molecular ranges were observed in a similar manner to the cases of Zn and Mn. These results suggest that arsenic and selenium in salmon egg cell cytoplasm exist not only as small molecules (e.g., arsenate and selenate) but also as protein-binding molecules. Thus, the results in Fig. 1.7 suggest existence of protein-binding molecules of arsenic in cell cytoplasm. In recent years, arsenic-binding proteins have been reported with great attention from the viewpoint of toxicity and drug design [38, 39]. In the case of selenium, glutathione peroxidase, antioxidant, is known as a large molecule (tetramer; MW *ca.* 84 kDa) in animals.

1.5 Human Genome Project and the Rapid Rise of Omics Science

The term of "genomics" emerged as the scientific term in 1988 [40], and in genomics the entirety of genes was defined as "genome" [41]. As is well known, the Human Genome Project had been conducted under the international cooperative research since 1991 in order to determine the sequence of human genome [17, 18]. In 2003, it was declared that the human genome was completely determined.

In the late 1990s, it was considered in the science community that in the early 2000s, the human genome sequence analysis would be completed, and then the post-genome project would be the next big international program, in which protein identification should be performed together with elucidations of their structures and functions [42]. Such a research field was called as "proteomics" in a similar manner to genomics, and the term of "proteomics" appeared as the scientific term in the literature for the first time in 1997 [19]. The entirety of proteins is defined as "proteome," in a similar manner to genome in genomics.

Then, in 2002, "metabolomics" emerged for the first time as the scientific term in the literatures [20]. Metabolomics is defined as the total science of metabolism in biological cell, tissue, organ, or organism (whole body). In metabolomics, the term of "metabolome" is used to express the entirety of all metabolites in analogy to genome in genomics [43]. In the metabolomics study, the LC-MS (liquid chromatography-mass spectrometry) system is extensively used for the profiling analysis of metabolites.

The emergence of genomics, proteomics, and metabolomics in biological science has strongly stimulated to create the new research trends in each field with cross correlation, and they are now called totally as "omics science." The progress of omics science since the late 1980s also gave great impact to the scientific fields other than omics science, of course, including trace metal chemistry, and at last in 2002, the term of "metallomics" as a new scientific field came to my mind, as will be described in the next section.

1.6 Proposal of Metallomics as Integrated Biometal Science—Historical Aspects

The abstract of my paper "Metallomics as Integrated Biometal Science" [1] is shown on the top page of this chapter, because it was the beginning of "metallomics" as the emerging scientific field of biometal science. As is described there, in these last few decades, various scientific fields such as analytical atomic spectrometry, omics science, as well as cell biology have progressed rapidly, and they had given great impacts to promotion of the study on metals in biology. Thinking about such advances of biological science, the idea of "metallomics" came into my mind suddenly one day. It was just like an inspiration, suggesting that metallomics should be a new scientific field among omics science. In such a time, it was very fortunate that good chances to introduce my idea were given to me in the following domestic seminar and international symposiums in 2002. As a result, the year of 2002 was the milestone for the proposal of "metallomics."

The first chance was the Tokushima Seminar on Chemical Engineering, held in Tokushima, Japan, in June 2002. The title of my invited lecture in the seminar was "A Challenge to Pico-World Science and Metallomics: A New Frontier of Trace Element Chemistry." This was the first time to introduce the idea of "metallomics" as the new scientific field, although the lecture was given in Japanese. The abstract of the lecture in the Tokushima Seminar was as follows [44]:

> In recent years, the analytical detection sensitivities have been increasingly improved to pico (10^{-12}) gram in the absolute amount or sub-ppt (10^{-12} g/ml) level in the concentration, according to the development of ICP-MS. As a result, now we have a good chance to challenge the research on the pico-technology or pico-science, which may be called "Pico-World Science".
>
> Such a progress of analytical atomic spectrometry will lead to another interesting and important research on bio-trace elements in the biological systems including our "human

beings", because all-elements might be contained in all the biological systems. This concept is referred to as "Extended All Present Theory of the Elements". Furthermore, various trace elements play important roles in the biological systems, as metalloproteins and/or metalloenzymes. Then, now is the good time to challenge to trace element biochemistry to open our new scientific world "metallomics."

As is seen from the above sentences, my first idea of metallomics was on the basis upon the progress of analytical atomic spectrometry and the concept of the Extended All-Present Theory of the Elements, about which some explanation will be given later. However, my idea was just an inspiration at that time.

The International Symposium on Bio-Trace Elements 2002 (BITREL 2002) was the second chance for proposal of metallomics. This symposium was held as the Joint Symposium of RIKEN (Institute of Physical and Chemical Research) and YIES (Yamanashi Institute of Environmental Sciences) from October 28 to November 2, and it was co-organized by Shuichi Enomoto in RIKEN (he is now a professor in Okayama University) and Yoshiyuki Seko in YIES. The present author delivered the invited lecture with the title of "Trace Element Speciation for Metallomics," and metallomics was formally proposed as a new scientific term on this occasion. The abstract of the invited lecture is shown below, which was cited from the Proceedings of BITREL 2002 [45].

In this paper, "metallomics" is newly proposed as a new scientific field in order to integrate the research fields related to bio-trace metals. Metallomics might be the scientific field of post-genomics and post-proteomics, where metal-containing compounds are defined as metallomes, in a similar manner to genome in genomics and proteome in proteomics. Since the elucidation of the biological or physiological functions of metal-containing species in the biological systems is the main research target of metallomics, elemental speciation is important as one of analytical technologies to promote metallomics.

As is seen in the abstract above, the importance of speciation analysis (species analysis) concerning with biotrace metals in the biological samples as well as in the biological systems was emphasized in the lecture, because most of biotrace metals (actually exist as the ionic forms in the biological systems) are contained in metalloproteins and/or metalloenzymes, and in particular biotrace metals play essential roles as the active centers of metalloenzymes, expressing the biological and physiological functions. On the other hand, it is also known in environmental pollution that some of metals and metalloids are seriously toxic or hazardous to humans and living organisms. Thus, it should be understood that "trace element chemistry" concerns with double natures of the elements; one is essential and the other is toxic.

In BITREL 2002, the distinguished scientists working in the field of biometal science were invited from various countries; they were Ryszard Lobinski (Warsaw University of Technology, Warsaw, Poland; now CNRS, Pau, France), Zhifang Chai (Institute of High Energy Physics, Chinese Academy of Science, Beijing, China), Wolfgang Maret (Harvard Medical School, USA; now Imperial College, London, UK), Joanna Szpunar (Group of Bioinorganic Analytical Chemistry, CNRS, Pau, France), Bibudhendra Sarkar (University Toronto, Ontario, Canada), David Brown (University of Bath, UK), and so forth. In the symposium, some of the participants proposed the term of "metalloproteomics," while I proposed "metallomics." As a result, hot discussion was made about the nomenclature for biometal science.

On the occasion of BITREL 2002, Lobinski asked me to submit a paper about metallomics to *Journal of Analytical Atomic Spectrometry* (J Anal At Spectrom), published from the Royal Society of Chemistry, UK, because the journal was planning to publish the special issue on "Metals in Biology," [46] and he was one of the editorial board members of the special issue. In 2004, then, my paper entitled "Metallomics as Integrated Biometal Science" was published in J Anal At Spectrom [1], which was the summary of the lecture presented in BITREL 2002. The abstract of the paper in J Anal At Spectrom [1] was already introduced on the first page of this chapter, because it provided important suggestion for future direction of biometal science.

Another surprising thing happened in 2009. It was the publication of the journal of *Metallomics* from the Royal Society of Chemistry (frankly, I did not know the plan for publication!). In the first issue of the journal, J. Caruso (chair of the Editorial Board, Cincinnati University) and N. O'Conor (editor of *Metallomics*, RSC) announced their welcome and acknowledge message [2] as follows (*see* also 1.8.1):

> We would like to acknowledge the work of those whose vision has led to the establishment of this field, including Bob Willams (R. J. P. Williams, Coord. Chem. Rev., 2001, 216–217; 583–595), Hiroki Haraguchi (H. Haraguchi, J. Anal. At. Spectrom., 2004, 19, 5–14) and Joanna Szpunar (J. Szpunar, Anal. Bioanal. Chem., 2004, 378, 55–56), and to all those who continue to contribute to the emergence of metallomics, without whom we would not be launching this exciting new journal.

Williams published his paper entitled "Chemical selection of elements by cells" in Coord Chem Rev in 2001 [33]. This paper was the summary of his lecture in the 34th International Conference on Coordination Chemistry (ICCC34) held in the University of Edinburgh, Edinburgh, Scotland, July 09–14, 2000. The abstract of his paper is cited here [33]:

> The selection of the chemical elements by a particular cell from the environment involves a series of steps, the complexity of which depends upon the organism. There are usually two membranes to be crossed (bacteria) but there may be as many as ten (higher animals which distribute elements from intake fluid, through cells of organs to circulating fluids through a further set of cells to fixed locations in particular parts of space). The individual steps can be thermodynamically controlled or kinetically managed. In the second case energy can be used. The elements may remain in relatively fast exchange in their final condition or in non-exchanging chemical combination. The variety of paths which individual elements follow in any organism adds to the specific character of the organism. Clearly the paths have evolved to create an element distribution which we shall call the **metallome**, to parallel the nomenclature of protein distribution, the proteome.

As is seen in the above abstract, William used the term "metallome" for the first time in 2000, but he never mentioned "metallomics" in his papers.

Szpunar published her paper entitled "Metallomics: a new frontier in analytical chemistry" in Anal Bioanal Chem in 2004 [47]. The following sentences are part of her paper [47]:

> Recently, Haraguchi and Matsuura suggested the term"metallomics" to denote metal-assisted function biochemistry and postulated it to be considered at the same level of scientific significance as genomics or proteomics [45]. The metallomic information will comprise the identities of the individual metal species (qualitative metallomics) and their concentrations (quantitative metallomics). As such, metallomics can be considered as a subset (referring to cellular biochemistry) of speciation analysis understood as the identification and/or quantification of elemental species [13]. Species of interest for metallomics will include complexes of trace elements and their compounds (e.g. metalprobes) with endogenous or bio-induced biomolecules such as organic acids, proteins, sugars or DNA fragments. [15]

Szpunar was a participant in the International Symposium of Bio-Trace Elements (BITREL 2002) held in 2002. Then she knew my proposal of metallomics in the symposium as well as the publication of the symposium proceedings before publication of her paper, and so she cited our paper in the proceedings [38]. It is appreciated from these facts that Szpunar just followed my proposal, and she was not a proposer of metallomics. Even so, I would like to say that she recognized the importance of "metallomics" earlier than anyone, because she had been so enthusiastic to develop the chemical speciation methods for the biological systems [14, 15].

1.7 International Symposium on Metallomics

After publication of my paper in 2004, "metallomics" received great attention as the newly emerging scientific field [3, 5–7]. Such situations were the same in Japan, and then in 2004, the present author got the Grant-in-Aid for Specially Promoted Research (2004–2007; 3-year term) with the project title of "Creation of Metallomics as the New Scientific Field," which was supported by the Ministry of Education, Science, Culture, and Sports of Japan. In the proposal of the grant, I promised the organization of the International Symposium of Metallomics, when the project would be terminated. Thus, 3 years later in 2007 (November 28–December 1), I organized as the chairman the International Symposium on Metallomics 2007 (ISM 2007), which was held in Nagoya as the first international symposium on metallomics. Professors Kazuo Suzuki, Hiromu Sakurai, and Naoki Furuta helped me as vice-chairmen of the symposium. In this symposium, about 350 scientists participated and more than 250 lectures and posters were presented [48]. The front cover of the program book for ISM 2007 is shown in Fig. 1.8.

The International Advisory Board members of ISM 2007 are listed in Table 1.5, and they concluded that ISM 2007 was the successful meeting and "metallomics" should be promoted as a newly emerging scientific field thereafter. Furthermore, it was also agreed in the board meeting that the symposium would be held regularly in

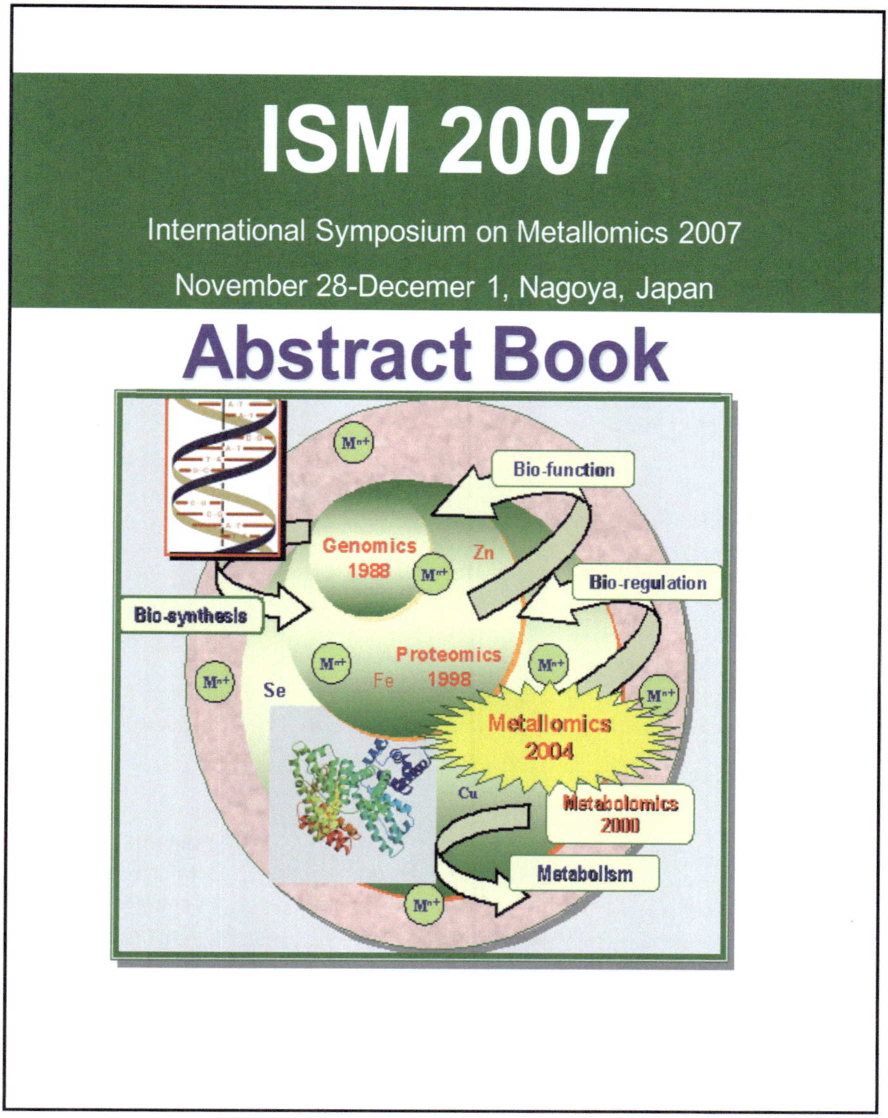

Fig. 1.8 The front cover of the Abstract Book for ISM 2007

every 2 years in the world. In addition, the Proceedings of ISM 2007 was published as the special issue of *Pure and Applied Chemistry* from the International Union of Pure and Applied Chemistry (IUPAC) in 2008 [49].

So far, the ISM has been held five times, that is, at Nagoya, Japan, in 2007; at Cincinnati, USA, in 2009; at Münster, Germany, in 2011; at Oviedo, Spain, in 2011;

Table 1.5 International advisory board members in ISM 2007

Dr. Becker, J. Sabine: Research Center Juelich, Germany
Prof. Caruso, Joseph A: University of Cincinnati, USA
Prof. Chai, Zhifang: Institute of High Energy, Physics, China
Prof. Francescon, Kevin: Karl-Franzens University Graz, Austria
Prof. Hieftje, Gary M: Indiana University, USA
Prof. Qiuquan Wang: Xiamen University, China
Dr. Jakubowski, Norbert: ISAS – Institute for Analytical Sciences, Germany
Prof. Kim, Hasuck: Seoul National University, Korea
Dr. Koppenaal, David W: Pacific Northwest National Laboratory, USA
Prof. Lobinski, Ryszard: CNRS, France
Prof. Lucchini, Roberto: University of Brescia, Italy
Prof. Maret, Wolfgang: The University of Texas Medical Branch, USA
Dr. McArdle, Harry J.: Rowett Research Institute, UK
Prof. McLeod, Cameron W: University of Sheffield, UK
Prof. Natile, Giovanni: Università degli Studi di Bari, Italy
Prof. Prange, Andreas: GKSS Research Centre, Germany
Prof. Sanz-Medel, Alfredo: University of Oviedo, Spain
Prof. Sarkar, Bibudhendra: The Hospital for Sick Children, Canada
Prof. Sun, Hongzhe: The University of Hong Kong, Hong Kong
Prof. Tanner, Scott D:. University of Toronto, Canada
Prof. Haraguchi Hiroki: Nagoya University, Japan---Chairman
Prof. Furuta Naoki: Chuo University, Japan---Vice-chairmen
Prof. Sakurai Hiromu: Kyoto Pharmaceutical University, Japan---Vice-chairmen
Prof. Suzuki Kazuo T.: Chiba University, Japan---Vice-chairmen

Table 1.6 The years, conference place and organizers of the International Symposium on Metallomics

Year	Places	Organizers
2007	Nagoya, Japan	H. Haraguchi
2009	Cincinnati. USA	J. Caruso & G. Hiefje
2011	Munster, Germany	U. Karst & M. Sperling
2013	Oviedo, Spain	A. Sanz-Medel
2015	Beijing, China	Z. Chai & Z. Xinrong
2017	Wien, Austria (scheduled)	G. Köllensperger

and at Beijing, China, 2015. These symposiums are listed in Table 1.6 together with the years and places (It was sad that Prof. Caruso, who organized the second symposium and the first Editorial Board Chair of *Metallomics* journal, passed away in November, 2015.). The proceedings of these symposiums were published as the special issues in the journal *Metallomics* [50–52] from the Royal Society of Chemistry. According to these symposiums, metallomics has been receiving more attention as the emerging scientific field, and in recent years, it has been taken up as the hot topics in other scientific fields, not only in the journals but also in the conferences.

It is additionally mentioned here that in Japan the Metallomics Research Forum has been held every even-numbered year since 2008 as the domestic meetings, and some selected papers presented in the Metallomics Research Forum were published as the themed issues in *Metallomics* [53, 54]. Shuichi Enomoto (RIKEN, Tokyo), Hiroyuki Yasui (Kyoto Pharmaceutical Univ., Kyoto), Yasumitsu Ogura (Showa Pharmaceutical Univ., Kanagawa), and Masahiro Kawahara (Musashino Univ., Tokyo) organized the forums in 2008, 2010, 2012, and 2014, respectively.

1.8 Progress of Metallomics Research

1.8.1 The Scientific Journal of Metallomics

It was a wonderful news and our pleasure that the scientific journal of *Metallomics-Integrated Bimetal Science* (in this chapter it is just called *Metallomics*) was launched in January 2009, from RSC. In the beginning part of the first issue of *Metallomics*, Caruso (the chair of Editorial Board) and O'Connor (editor of RSC) gave the welcome message for celebrating the publication of the new journal as follows [2]:

> Welcome to the first issue of Metallomics: Integrated Biometal Science. The study of metals in biological systems is an increasingly important area of research. Metallomics is a newly emerging scientific field that is receiving great attention as a new frontier in the study of trace elements in the life sciences. It is a global discipline encompassing many areas including biology, chemistry, geology, medicine, physics, and pharmacy.
>
> As this field brings together researchers from such diverse areas, we anticipate that Metallomics will help to bridge the gap between researchers from different backgrounds so that ideas can be shared and the field progresses to the benefit of all. Our journal will serve as a focus for the community of metallomics researchers to come together and gain new perspectives and insights. It is our aim to reflect the interests of the emerging community and to support you as your community grows.

Until 2015, seven volumes (Vol.1–Vol.7) of the journal *Metallomics* have been distributed, in which more than 1000 articles and review papers have been published. It should be highly evaluated that the publication of *Metallomics* has accelerated the progress of the metallomics research as a multidisciplinary academic science. Recently the Editorial Board of *Metallomics* announced the guideline for the publication requirement of the research papers required [8].

1.8.2 Essentiality and Toxicity of the Elements

In trace element chemistry, the terms of essential elements and toxic (or hazardous) elements are often used, when the biological influences/functions of the elements are explained. In such cases, the biological influences of the elements on the

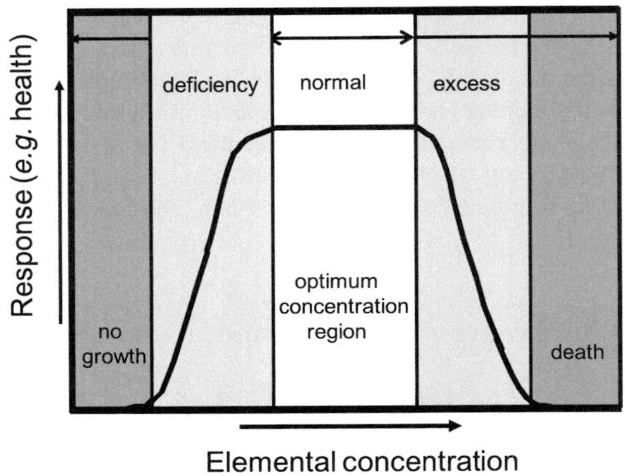

Fig. 1.9 Response curve of the biological system to the dose of the elements as the nutrients

biological systems (humans, animals, plants, and microorganisms) are generally examined as the response curve of the biological systems to the dose of the elements (generally contained in nutrients, foods, chemicals, and so forth). As an example, the dose-response curve is shown in Fig. 1.9, where the concentration of the element as a nutrient is taken on the horizontal axis and the biological response (e.g., health condition) is taken on the vertical axis. In the figure, the health condition is taken as the indicator of the response to the dose of the elements. As the biological response, body height or weight, cell growth, cell proliferation, medical indexes, and so forth are also taken into account for the response curve, where the upper position of the vertical axis in the curve indicates the better healthy condition (more normal).

It is well known that most of biological systems show the response curve in a trapezoid-shaped curve, as shown in Fig. 1.9. In such a biological response curve in the trapezoid shape, there are three regions such as deficiency region, normal region (or optimum concentration region), and excessive region, when the amount of element administered is increased from the low to higher concentration (from the left to the right). In the normal region, our health is physiologically or functionally maintained to be normal without any disease or dysfunctions. When the nutrient concentration becomes lower than the lower limit of the normal region, some specific symptom or disease is caused, depending on the kind of the element, for example, iron deficiency causes anemia. In the extreme case, no growth of the living systems is observed; a microorganism such as *Escherichia coli* (*E. coli*) does not grow completely, when the zinc concentration is below 1 ppb in the culture medium. On other hand, various symptoms are observed in the excessive region, where an element is administered more than the upper limit of the normal region. Especially, heavy metals such as Hg, Cd, Pb, and Sn cause serious damages or dysfunctions in the relatively low concentration, and in the extreme cases, such symptoms become fatal. Thus, such heavy metals are classified to toxic or hazardous metals. Most of environmental issues due to metals are caused as serious adverse effects by such toxic heavy metals. It should be pointed out here that any elements (and their compounds) cause some toxic effects on the biological systems if the living organisms are exposed to the excessive conditions by overdose of the elements, as in the right-hand side of Fig. 1.9. If an element causes some adverse effect or dysfunction by the smaller amount of dose, it is said that the toxicity of that element is greater.

According to definition, an element which is indispensable in our life system is called "essential element." In general, if the element satisfies with one of the following two cases, that element is classified to "essential element." The first case is that some abnormal diseases or physiological dysfunctions are observed because of the deficiency of a specific element, and the life system is recovered to normal condition by the dietary supplementation of that specific element. The second case is that a specific element is contained in biomaterials (mostly metalloproteins and/or metalloenzymes) with an important biological function for our life system, for example, Fe in hemoglobin, Zn in carboxypeptidase, and Se in glutathione peroxidase.

In Table 1.1, the essential elements contained in humans are indicated by the marks *, **, and ***. It can be seen from Table 1.1 that 11 elements such as H, C, N, O, Na, Mg, P, S, Cl, K, and Ca are essential major and minor elements. The cumulative amount of these 11 elements is 99.4% of the total, and they are mostly used to construct the whole-body structures of the living or biological systems. The cumulative amount of the rest of the elements is only 0.4% of the total, and all of them belong to trace elements, because their concentrations are below 100 ppm. Among those trace elements, eight elements (Cr, Mn, Fe, Co, Cu, Zn, Se, Mo, I) are essential for humans, and seven elements (F, Si, V, Ni, As, Sr, Sn, Pb) are essential for experimental mammals [26]. It is noted here that B is essential for plants (*see* Table 1.3), but not for animals.

In recent years, some people say that the trace elements such as Li, Rb, Cs, Ba, Al, Ge, Br, Sb, W, Ag, Au, Cd, and Hg might be essential for humans and animals, in which the toxic or hazardous elements such as Cd and Hg are included. As expected from the All-Element Present Theory described in Sect. 1.3, almost all elements can be determined or detected in the biological samples (*see* Table 1.3). These facts suggest that any elements are ubiquitously distributed in the biological systems including humans, so that some scientists are now thinking that even the hazardous or toxic elements might be necessary for maintenance of our life systems at the concentration level as low as the background concentration in nature. This is the reason why it is considered that other elements except for the well-established essential elements are possibly essential to the biological systems. However, since biological essentiality of such elements has not well elucidated so far, further extensive research on their essentiality as biological trace elements as well as their physiological functions should be carefully carried out in future research.

1.8.3 A Simplified Model of the Biological System

The biological cells containing various cell organs (organelles) are composed of numerous internal structures, so that the cell structures in the biological system (either prokaryotes or eukaryotes) are of complex assemblies, but well organized. At present, such cell structures composed of various organelles have been revealed by the advanced analytical technologies, especially electron microscopy. Taking into account of such cell structure, a simplified model of biological system is schematically illustrated in order to gain an insight into the scientific aspects of "metallomics" in Fig. 1.10 [1, 7], where the dotted line (inside) and continuous line (outside) indicate a biological cell unit and an organ/whole body, respectively, and biological fluids (e.g., blood) circulate between the cell membrane and organ. Some biological substances and their functions in the biological system are also indicated in Fig. 1.10.

On the left-hand side of the simplified model in Fig. 1.10, the scientific terms such as genomics, proteomics, glycomics, and metabolomics are shown along with metallomics to indicate their research areas in the biological system. Such a simplified model is helpful to understand the relationship of metallomics with genomics, proteomics, metabolomics, and glycomics. As is well known, genomics deals with the genetic information of DNAs and RNAs encoded as the sequences of nucleobases. The entirety of DNAs and RNAs called "genome" contains the information for synthesis of proteins and also for regulation of protein structures/

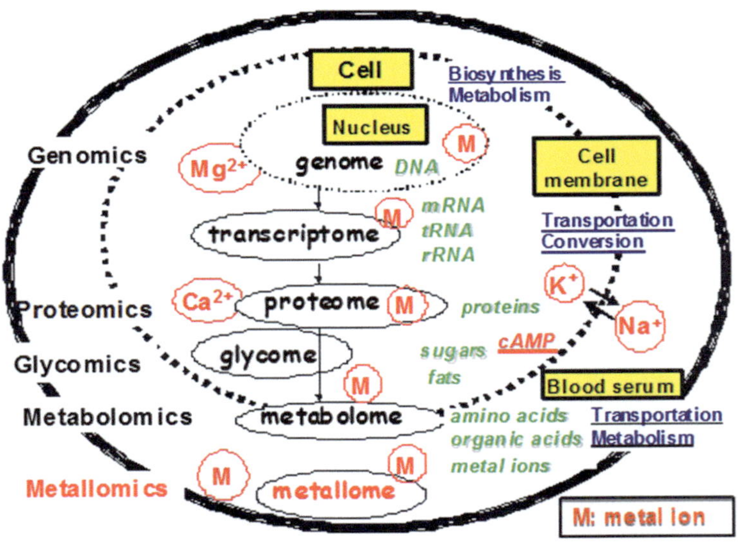

Fig. 1.10 Simplified model of biological system, showing the relationship of *omics* sciences [1, 4]. The *outer line* and the *inner dotted line* indicate organ (or whole body) and biological cell, respectively

functions. A large number of proteins are distributed inside and outside the cell as well as in the membranes, and they work as enzymes for synthesis and metabolism of various biological substances inside the cell. It is well known, for example, that DNAs and RNAs are synthesized by DNA polymerase and RNA polymerase, both of which are zinc enzymes. Since various proteins play essential roles to regulate and maintain the life system through the biosyntheses and metabolisms, "proteomics" as protein science has been receiving great attention as post-genome science linked with genomics.

In addition, many biological substances as well as metal ions are transported inside and outside the cell through the membrane. In general, since material conversion is actively occurring inside the cells and also often in the cell membranes, the scientific field for material conversion and transportation/exclusion processes are now called "metabolomics." Biological substances such as amino acids, organic acids, and other organic substances produced by metabolism are defined "metabolome" in a similar manner to genome in genomics [55].

As is seen from Fig. 1.10, metal ions are ubiquitously distributed inside and outside the cell to assist the physiological functions of genome, transcriptome, proteome, glycome, and metabolome, maybe, due to strong interactions and/or weak interactions. Since the limited space is available for the chapter, the detailed discussion about the roles of biometals is impossible, and then it is recommended to refer to the literatures [1, 26, 27, 56] in order to understand the functions of biometals in the biological systems.

1.8.4 Scientific Fields in Trace Metal Science

Biosciences concerned with metallic elements have been studied independently in many scientific fields such as physics, chemistry, biology, medicine, pharmacy, agriculture, environmental science, and other applied scientific fields. Such situations of trace metal studies are illustrated in Fig. 1.11, where the various scientific fields are shown in three groups: (i) basic science such as physics, chemistry, biology, and so on; (ii) applied science such as toxicology, food science, nutritional science, public hygiene, and others; and (iii) future science consisting of health science and environmental/green science. Since health science and environmental/ green science are very important fields for sustainability of humans and nature, they are shown apart from basic and applied sciences in Fig. 1.11. It is seen in Fig. 1.11 that metallomics is located at the center as the integrated biometal science and linked with all of basic and applied sciences. Then, this figure suggests that metallomics should be appreciated as multidisciplinary science in wide diversity.

As already mentioned, all scientific fields shown in Fig. 1.11 have complementary relationship with each other. Therefore, it is desirable that biometal science, that is, metallomics, will be promoted with the cooperative work of the metal-related scientific fields in the diverse communities. This is the reason why the present author proposed "metallomics as integrated biometal science." The most

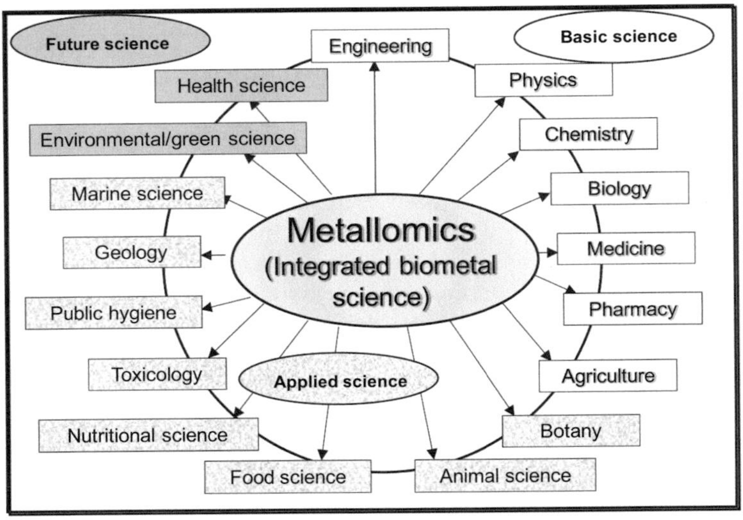

Fig. 1.11 Various scientific fields as studied in relation with metallomics

important research target in metallomics is the elucidation of the physiological roles and functions of biomolecules binding with metal ions for regulation of the life systems, so that metallomics may be considered, in another words, as "metal-assisted function bioscience".

Short comments about the difference between metallomics and metalloproteomics are added here briefly. In the literatures, the term of metalloproteomics is seen quite often. In fact, most of transition metal ions bind with proteins to form metalloproteins, which express the important biological or physiological functions in biological synthesis and metabolism. However, there are many other metal-binding compounds except for metalloproteins, for example, vitamin B12(Co) and chlorophylls(Mg), which also work as important bioactive substances. In addition, metal ions such as alkali and alkaline earth metal ions also play many important roles to regulate the physiological conditions in the biological systems. Thus, it is preferable that metals (metallic ions) not only in metalloproteins but also in nonprotein biomolecules should be included as metallome in metallomics. In that sense, metallomics, rather than metalloproteomics, may be more proper term as the academic terminology to represent biometal science.

1.8.5 Research Subjects in Metallomics

As an example, the research subjects to be performed in metallomics study are listed in Table 1.7 [1], which was summarized with taking into consideration the biological roles of biometals in life science. It can be seen from Table 1.7 that a

Table 1.7 Research subjects in metallomics [1]

(1)	Distributions of the elements in the biological fluids, cell, organs *etc*
(2)	Chemical speciation of the elements in the biological samples
(3)	Structural analysis of metallome (metal-binding molecules)
(4)	Elucidation of reaction mechanisms of metallome using model metal complexes (bioinorganic chemistry)
(5)	Identification of metalloproteins and metalloenzymes
(6)	Metabolisms of biological molecules and metals (metabolome and metabolites)
(7)	Medical diagnosis of health and disease related to trace metals on the multielement basis
(8)	Design of inorganic drugs for chemotherapy
(9)	Chemical evolution of the living systems and organisms on the earth
(10)	Other metal-assisted function biosciences in medicine, environmental science, food science, agriculture, toxicology, biogeochemistry *etc*

variety of research subjects related to biometals are involved in metallomics from the fundamental to the applied fields. Among the subjects listed in Table 1.7, Subject 1 "the distributions of the elements in the biological fluids, cell, organs, etc." and Subject 2 "chemical speciation of metals and metalloids (metallome) in the biological samples," about which the detailed description was given in the earlier part of this chapter, are important to obtain the fundamental information about biological species in the biological systems. It is really desirable that some regularity (general principle) of the distributions of metals (and metalloids) in the biological systems, in analogy to DNA sequences in genes and amino acid sequences in proteins, are found from the analytical data for further progress of metallomics. Of course, it is a difficult task, but such data may be helpful in the study especially on chemical evolution of the living system not only on the earth but also on other planets.

Subjects 3–5 in Table 1.7 have been extensively performed in the field of biological inorganic chemistry, and many interesting articles can be found in the *Journal of Biological Inorganic Chemistry (JBIC)* published from Springer. As for Subject 6, great progress has been achieved by applying the analytical technology for the profiling analysis of metabolites (metabolome) such as GC-MS and LC-MC to analysis of urine and blood serum. The results are also used for medical diagnosis of health and disease in Project 7.

Subject 8, i.e., the design of inorganic drugs for chemotherapy, is one of the most active research areas in metallomics. In order to understand such situations, the top 10 ranked articles (citation ranking) are listed in Table 1.8, which were selected from *Metallomics* by the literature survey using the keyword "metallomics" in the Web of Science (Thomson Reuters). In the citation ranking, the articles are arranged in the order of the times of citation referred to the Web of Science. It can be seen from the citation ranking that the researches related with drug design such as anticancer drugs, antitumor compounds, therapeutic agents, and biomedicine are actively performed in metallomics. That means the development of metal-containing drug design and chemotherapy is the most interested research area in

Table 1.8 Top-10 citation ranking of the articles published in *Metallomics*

The titles of articles[a]	Citation[b]	References
Recent developments in ruthenium anticancer drugs	263	[57]
Inhibition of transcription by platinum antitumor compounds	187	[58]
Epigenetics in metal carcinogenesis: nickel, arsenic, chromium and cadmium	132	[59]
Cytosolic zinc buffering and muffling: Their role in intracellular zinc homeostasis	129	[60]
Gold compounds as therapeutic agents for human diseases	123	[61]
Advances in metal-carbene complexes as potent anticancer agents	104	[62]
Role of metal dyshomeostasis in Alzheimer's disease	93	[63]
Interactions of Zn(II) and Cu(II) ions with Alzheimer's amyloid-beta peptide. Metal ion binding, contribution to fibrillization and toxicity	87	[64]
Trace metal imaging with high spatial resolution: Applications in biomedicine	76	[65]
Molecular and genetic features of zinc transporters in physiology and pathogenesis	68	[66]

[a]Total of publication 1020 on the date of February 29, 2016
[b]Times of citation were referred to the Web of Science of Thomson Reuters

metallomics now. Of course, other metal-assisted function biosciences in Subject 10 should be carried out in cooperation with Subjects 1–9.

1.9 Summary

The historical aspects of metallomics before and after the proposal in 2004 were described in order to understand how metallomics came into the academic research field and how it is going on and where it will go. Since metallomics is the interdisciplinary academic research field, it should be promoted as the integrated research field for biometal science, in cooperation with other omics sciences such as genomics, proteomics, and metabolomics. In genomics, all DNA sequences in a variety of biological species have been surveyed, and in proteomics, all amino acid sequence analysis of proteins in the biological systems has been promoted to understand their biological roles and functions in our life system. In that sense, it is really hoped that in metallomics the existence of all elements in single biological cell is proved from the viewpoint of the All-Element Present Theory to further understand our life systems.

References

1. Haraguchi H (2004) Metallomics as integrated biometal science. J Anal At Spectrom 19 (1):5–14
2. Caruso JA, O'Connor N (2009) Metallomics: integrating research related to biometals – a journal for an emerging community. Metallomics 1(1):14–16
3. Haraguchi H (2005) New development of chemical speciation analysis for metallomics research. Biomed Res Trace Elements (*in Japanese*) 16(3):217–232
4. Koppenaal DW, Hieftje GM (2007) Metallomics—the future of atomic spectroscopy? J Anal At Spectrom 22(2):111
5. Koppenaal DW, Hieftje GM (2007) Metallomics – an interdisciplinary and evolving field. J Anal At Spectrom 22(8):855
6. Mounicou S, Szpunar J, Lobinski R (2009) Metallomics: the concept and methodology. Chem Soc Rev 38(4):1119–1138
7. Lobinski R, Becker JS, Haraguchi H et al (2010) Metallomics: guidelines for terminology and critical evaluation of analytical chemistry approaches (IUPAC Technical Report). Pure Appl Chem 82(2):493–504
8. The Editorial Board of Metallomics (RSC) (2016) The scope of *Metallomics*. Metallomics 8 (1):8
9. Haraguchi H (1999) Multielement profiling analyses of biological, geochemical, and environmental samples as studied by analytical atomic spectrometry. Bull Chem Soc Jpn 72 (6):1163–1186
10. Vandecasteele C, Block CB (1991) Modern methods for trace element determination. Wiley, Chichester
11. Montaser A (ed) (1998) Inductively coupled plasma mass spectrometry. Wiley, New York
12. Haraguchi H, Sawatari H (1990) The analytical methods for trace elements (*In Japanese*). Mod Med (*Saishin Igaku*) 45(4):816–821
13. Templeton DM, Ariese F, Cornelis R et al (2000) Guidelines for terms related to chemical speciation and fractionation of elements. Definitions, structural aspects, and methodological approaches (IUPAC Recommendations 2000). Pure Appl Chem 72(8):1453–1470
14. Szpunar J (2000) Bio-inorganic speciation analysis by hyphenated techniques. Analyst 125 (5):963–988
15. Szpunar J, Lobinski R, Prange A (2003) Hyphenated techniques for elemental speciation in biological systems. Appl Spectrosc 57(3):102A–111A
16. Szpunar J (2005) Advances in analytical methodology for bioinorganic speciation analysis: metallomics, metalloproteomics and heteroatom-tagged proteomics and metabolomics. Analyst 130(4):442–465
17. International Human Genome Sequencing Consortium (2001) Initial sequencing and analysis of the human genome. Nature 409(6822):860–921
18. Venter JC, Adams MD, Myers EW et al (2001) The sequence of the human genome. Science 291(5507):1304–1351
19. James P (1997) Protein identification in the post-genome era: the rapid rise of proteomics. Q Rev Biophys 30(4):279–331
20. Fiehn O, Kopka J, Dormam P et al (2000) Metabolite profiling for plant functional genomics. Nat Biotechnol 18(11):1157–1161
21. Heumann KG, Gallus SM, Radlinger G et al (1998) Precision and accuracy in isotope ratio measurements by plasma source mass spectrometry. J Anal At Spectrom 13(9):1001–1008
22. Rodriguez-Gonzalez P, Marchante-Gayon JM, Alonso JIG et al (2005) Isotope dilution analysis for elemental speciation: a tutorial review. Spectrochim Acta B Atom Spectrosc 60 (2):151–207
23. Hasegawa T (2006) Ph. D. thesis, Nagoya University, Unpublished data
24. Noddack I (1936) Concerning the ubiquitous nature of the chemical elements. Angew Chem 47:835

25. Kuroda P (1982) The origin of the chemical elements and the Oklo phenomenon. Springer, Berlin
26. Sakurai H, Tanaka H (eds) (1996) Bio-trace elements. Nankodo, Tokyo (*in Japanese*), pp 1–11
27. Bowen HJM (1973) Trace elements in biochemistry. Academic, New York
28. Calvin M (1969) Chemical evolution–molecular evolution towards the origin of living systems on the earth and elsewhere. Oxford University Press, Oxford
29. Inagaki K, Haraguchi H (2000) Determination of rare earth elements in human blood serum by inductively coupled plasma mass spectrometry after chelating resin preconcentration. Analyst 125(1):191–196
30. Katsuki F, Hokura A, Iwahata D et al (1998) Multielement determination of major-to-ultratrace elements in cherry samples by ICP-MS and ICP-AES after acid digestion. Bunseki Kagaku 47(11):835–844
31. Haraguchi H, Ishii A, Hasegawa T et al (2008) Metallomics study on all-elements analysis of salmon egg cell and fractionation analysis of metals in cell cytoplasm. Pure Appl Chem 80 (12):2595–2608
32. Haraguchi H (2010) Metallomics research related to arsenic. In: Hongzhe S (ed) Biological chemistry of arsenic, antimony and bismuth. Wiley, London, pp 83–112
33. Williams RJP (2001) Chemical selection of elements by cells. Coord Chem Rev 216:583–595
34. Mounicou S, Lobinski R (2008) Challenges to metallomics and analytical chemistry solutions. Pure Appl Chem 80(12):2565–2575
35. Umemura T, Kitaguchi R, Haraguchi H (1998) Counterionic detection by ICP-AES for determination of inorganic anions in water elution ion chromatography using zwitterionic stationary phase. Anal Chem 70(5):936–942
36. Hasegawa T, Asano M, Takatani K et al (2005) Speciation of mercury in salmon egg cell cytoplasm in relation with metallomics research. Talanta 68(2):465–469
37. Inagaki K, Umemura T, Matsuura H et al (2000) Speciation of trace elements, binding and non-binding with proteins in human blood serum, by surfactant-mediated HPLC with element-selective detection by ICP-MS. Anal Sci 16(8):787–788
38. Kanwal R, Hua N (2012) Arsenic metabolism and thioarsenicals. Metallomics 4(9):881–892
39. Chen B, Liu Q, Popowich A et al (2015) Therapeutic and analytical applications of arsenic binding to proteins. Metallomics 7(1):39–55
40. Fergusonsmith AC, Ruddle FH (1988) The genomics of human homeobox-containing loci. Pathol Immunopathol Res 7(1–2):119–126
41. Zamir D, Tanksley SD (1988) Tomato genome is comprised largely of fast-evolving, low copy-number sequences. Mol Gen Genet 213(2–3):254–261
42. Gygi SP, Rist B, Gerber SA et al (1999) Quantitative analysis of complex protein mixtures using isotope-coded affinity tags. Nat Biotechnol 17(10):994–999
43. Nicholson JK, Lindon JC (2008) Systems biology: metabolomics. Nature 455 (7216):1054–1056
44. Haraguchi H (2002) A challenge to pico-world and metallomics: a new frontier of trace element chemistry. The invited lecture in the Tokushima Seminar on Chemical Engineering (The English abstract cited was translated from the Japanese one)
45. Haraguchi H, Matsuura H (2003) Trace element speciation for metallomics. In: Enomoto S, Seko Y (eds) Proceedings of the international symposium on Bio-trace elements 2002 (BITREL 2002). The Institute of Physical and Chemical Research (RIKEN), Wako, pp 3–8
46. Jakubowski N, Lobinski R, Moens L (2004) Metallobiomolecules. The basis of life, the challenge of atomic spectroscopy. J Anal At Spectrom 18:1–4
47. Szpunar J (2004) Metallomics: a new frontier in analytical chemistry. Anal Bioanal Chem 378 (1):54–56
48. Haraguchi H (2008) Preface; international symposium on metallomics 2007 (ISM 2007). Pure Appl Chem 80(12):iv
49. Haraguchi H (ed) (2007) Proceedings of the international symposium on metallomics 2007 (ISM 2007). Pure Appl Chem 80(12):2565–2750

50. Caruso JA (2010) 2009 international symposium on metallomics. Metallomics 2(2):103
51. Sperling M (2011) The third international symposium on metallomics 2011. Metallomics 3 (12):1263–1264
52. Montes-Bayon M, Bettmer J (2014) 4th international symposium on metallomics, 2013. Metallomics 6(2):187–188
53. Haraguchi H (2011) Metallomics in Japan. Metallomics 3(7):648–649
54. Ogra Y, Himeno S (2013) Metallomics in Japan. Metallomics 5(5):415–416
55. Dunn WB, Ellis DI (2005) Metabolomics: current analytical platforms and methodologies. TrAC Trends Anal Chem 24(4):285–294
56. Williams RJP, Frausto da Silva JJR (1996) The natural selection of the chemical elements. Clarendon, Oxford
57. Levina A, Mitra A, Lay PA (2009) Recent developments in ruthenium anticancer drugs. Metallomics 1(6):458–470
58. Todd RC, Lippard SJ (2009) Inhibition of transcription by platinum antitumor compounds. Metallomics 1(4):280–291
59. Arita A, Costa M (2009) Epigenetics in metal carcinogenesis: nickel, arsenic, chromium and cadmium. Metallomics 1(3):222–228
60. Colvin RA, Holmes WR, Fontaine CP et al (2010) Cytosolic zinc buffering and muffling: their role in intracellular zinc homeostasis. Metallomics 2(5):306–317
61. Berners-Price SJ, Filipovska A (2011) Gold compounds as therapeutic agents for human diseases. Metallomics 3(9):863–873
62. Gautier A, Cisnetti F (2012) Advances in metal-carbene complexes as potent anticancer agents. Metallomics 4(1):23–32
63. Bonda DJ, Lee HG, Blair JA et al (2011) Role of metal dyshomeostasis in Alzheimer's disease. Metallomics 3(3):267–270
64. Tougu V, Tiiman A, Palumaa P (2011) Interactions of Zn(II) and Cu(II) ions with Alzheimer's amyloid-beta peptide. Metal ion binding, contribution to fibrillization and toxicity. Metallomics 3(3):250–261
65. Qin Z, Caruso JA, Lai B et al (2011) Trace metal imaging with high spatial resolution: applications in biomedicine. Metallomics 3(1):28–37
66. Fukada T, Kambe T (2011) Molecular and genetic features of zinc transporters in physiology and pathogenesis. Metallomics 3(7):662–674

Part II
Analytical Techniques and Strategies Surrounding Metallomics

Chapter 2
Speciation and Identification of Chalcogen-Containing Metabolites

Yasumitsu Ogra

Abstract Selenium (Se) and tellurium (Te) are chalcogen elements belonging to group 16 in the periodic table. These elements have unique physical, chemical, and biological properties. Se and Te form Se- or Te-containing compounds having Se- or Te-carbon covalent bond(s), i.e., selenometabolites or tellurometabolites, respectively, in their metabolic pathways. In this chapter, the speciation and identification of selenometabolites and tellurometabolites in animal and plant samples are highlighted. First, the instruments required for analyses are overviewed. In particular, hyphenated techniques consisting of high-performance liquid chromatography and inductively coupled plasma mass spectrometry or electrospray ionization (tandem) mass spectrometry are focused on. Then, newly identified metabolites are introduced. The identification of selenosugars in urine, selenohomolanthionine in Se-accumulating plants, selenoneine in marine animals, trimethyltelluronium ion in urine, *Te*-methyltellurocysteine in Se-accumulating plants, and selenocyanate in cultured cells is discussed.

Keywords Selenium • Tellurium • Speciation • Metabolome • LC-ICP-MS • LC-ESI-MS • Selenosugar • Selenohomolanthionine • Selenoneine • Selenocyanate

2.1 Introduction

Selenium (Se) is an element belonging to the same group 16 in the periodic table as oxygen, sulfur (S), and tellurium (Te). Se has chemically and biologically ambivalent characteristics; it has similar chemical properties to S, a typical nonmetal, yet possesses chemical properties characteristic of a metal. Se is an essential element in animals but can become highly toxic when ingested in an amount that exceeds the nutritional level. In addition, the difference between nutritionally deficient and excessive levels is very small [1, 2], being one order of magnitude (0.1–1.0 μg/g diet or mL drinking water) in experimental animals. Se is required as the active

Y. Ogra, Ph.D. (✉)
Laboratory of Toxicology and Environmental Health, Graduate School of Pharmaceutical Sciences, Chiba University, 1-8-1 Inohana, Chuo, Chiba 260-8675, Japan
e-mail: ogra@chiba-u.jp

Y. Ogra, T. Hirata (eds.), *Metallomics*, DOI 10.1007/978-4-431-56463-8_2

center of selenoproteins that function as an antioxidant and participate in thyroid hormone production, DNA synthesis, and fertilization [3–6]. It acts as an active center in the formation of a selenol group (−SeH) on a selenocysteinyl (SeCys) residue in a selenoprotein sequence [7, 8]. Thus, animals have a very unique translational machinery for SeCys, which is called the “21st amino acid” [9, 10]. In contrast, Se is not essential in plants and exists as a “bystander” mineral. Nevertheless, its beneficial effects on plant growth have been reported [11, 12]. It is known that certain plants, including *Allium* plants belonging to family Liliaceae, e.g., garlic, onion, wild leek, and shallot, and *Brassica* plants, e.g., Indian mustard, broccoli, and radish, are able to accumulate Se [13–15]. Unique Se-containing compounds having Se-carbon covalent bond(s) (selenometabolites), such as *Se*-methylselenocysteine (MeSeCys), γ-glutamyl-*Se*-methylselenocysteine (GluMeSeCys), and selenomethionine (SeMet), are biosynthesized in some plants [16, 17] and fungi [18, 19].

Te is also a chalcogen element classified under group 16 of the periodic table. In contrast to Se, Te is not an essential element in animals and plants. Te is called a metalloid because it possesses properties that are intermediate between those of metals and nonmetals. Because of its unique chemical and physical properties as a metalloid, Te is widely used in industry [20]. For instance, Te is employed as an alloy for phase-change optical magnetic disks, solar panels, and Peltier devices. This suggests that the risk of exposure to Te exists ubiquitously in everyday life. In addition, the accident at the Fukushima Daiichi Nuclear Power Plant led to the discharge of artificial radionuclides, including ^{129m}Te, ^{132}Te, ^{131}I, ^{132}I, ^{134}Cs, ^{136}Cs, and ^{137}Cs, into the environment [21]. Although Te is considered nonessential, non-beneficial, and harmful to animals and plants, its biological and toxicological effects are little understood [22, 23]. However, plants are expected to detoxify Te because Te is presumed to share the same metabolic pathway with S and Se. Although Se is also a nonessential element in plants, some plants, such as the Indian mustard, broccoli, and garlic, are Se accumulators. Indeed, it was reported that plants that accumulated Se were able to accumulate Te as well, although their ability to accumulate Te was lesser than their ability to accumulate Se [24]. This indicates that human and animals face the risk of Te exposure by ingesting edible Te-accumulating plants through the food web.

In this chapter, the speciation and identification of selenometabolites and tellurometabolites are addressed to understand the entire metabolic pathway of Se and Te in animals and plants. First, the techniques used for the speciation and identification of selenometabolites and tellurometabolites are overviewed. Then, the verification of biological, pharmacological, nutritional, and toxicological significance of selenometabolites and tellurometabolites by speciation and identification is discussed.

2.2 Hyphenated Techniques

The hyphenated technique is the most frequently used analytical technique for speciation analysis. The term "speciation" of an element (metal or metalloid) is defined as the distribution of an element among defined chemical species in a system, whereas speciation analysis is defined as the analytical activities undertaken to identify and/or measure the quantities of one or more chemical species in a sample [25]. The hyphenated technique for metal/metalloid speciation consists of two analytical techniques: the separation technique and the detection technique [26]. As the separation technique, chromatography [27], capillary electrophoresis [28], and gel electrophoresis [29] are generally used. The separation techniques for speciation have been reviewed [30–32]. As the detection technique, inductively coupled plasma mass spectrometry (ICP-MS) and electrospray ionization mass spectrometry (ESI-MS) are frequently used.

The inductively coupled plasma (ICP) is a high-energy ionization source, i.e., the plasma normally operates at 5000–10,000 K at atmospheric pressure. Thus, ICP is a hard ionization source for the generation of ions of elemental atoms, and the generated ions are sampled by a mass spectrometer. Single quadrupole (Q), triple quadrupole (QQQ), time of flight (TOF), and sector field (SF) mass spectrometers have been employed with ICP. ICP equipped with a single-quadrupole mass spectrometer (hereinafter, ICP-MS) is the most widely used among the ICP-mass spectrometers. ICP-MS is a powerful tool for elemental detection because of its multielement detection capability, low detection limit, ability to measure isotope ratio, and robustness to sample matrix. Hence, ICP-MS is used for not only elemental detection but also speciation analysis. Because of easy hyphenation with HPLC, LC-ICP-MS is the technique of choice for speciation of biological samples. However, LC-ICP-MS has one crucial disadvantage: as ICP-MS provides only elemental information of metabolites, the identification by LC-ICP-MS is effective only when the retention times of samples could be matched with those of certified or authentic species (standard compounds). In other words, if the authentic Se compound of an unknown metabolite is not available, the unknown metabolite would not be identifiable by LC-ICP-MS. To overcome this disadvantage, other mass spectrometric techniques are used.

As an alternative to LC-ICP-MS, electrospray ionization (ESI) or atmospheric pressure chemical ionization (APCI) mass spectrometry is used. As ESI and APCI are softer ionization sources than ICP, they can provide molecular information [33]. Those mass spectrometric techniques are often called molecular mass spectrometry or organic mass spectrometry, in contrast to ICP-MS, which is called elemental/atomic mass spectrometry or inorganic mass spectrometry. In addition, tandem mass spectrometry (MS-MS) and multistage cascade of mass spectrometry (MS^n) enable structure elucidation. The fact that Se and Te have characteristic clusters of signals that reflect their isotope patterns, i.e., for Se, ^{74}Se, 0.89 %; ^{76}Se, 9.36 %; ^{77}Se, 7.63 %; ^{78}Se, 23.8 %; ^{80}Se, 49.6 %; and ^{82}Se, 8.73 % and for Te, ^{120}Te, 0.09 %; ^{122}Te, 2.55 %; ^{123}Te, 0.89 %; ^{124}Te, 4.74 %; ^{125}Te, 7.07 %; ^{126}Te, 18.84 %;

^{128}Te, 31.74 %; and ^{130}Te, 34.08 %, respectively, has facilitated the scanning of Se- or Te-containing species by mass spectrometry. Because of this, Se and Te are the preferred targets of organic mass spectrometry. Such organic mass spectrometric techniques as ESI-MS-MS and ESI-MS^n are available for the identification of unknown seleno- or tellurometabolites. However, organic mass spectrometry has two weak points compared with ICP-MS. First, the detection limit of organic mass spectrometry for Se- or Te-containing compounds is inferior to that of ICP-MS [34]. A Fourier transform orbital ion trap mass spectrometer has been developed for the identification of selenometabolites [35]. Although the instrument features multistage fragmentation, accurate mass determination, and a large intrascan dynamic range with high sensitivity compared to that of ICP-MS, it still has to gain widespread use in the analysis of seleno- and tellurometabolites. Second, organic mass spectrometry is severely affected by the sample matrix. More elaborate pretreatment is required for the analysis by ESI-MS than that by ICP-MS. Thus, the complementary use of inorganic and organic mass spectrometry is the best approach to the identification of unknown seleno- and tellurometabolites (Fig. 2.1). In some cases, ESI-MS cannot distinguish enantiomers and diastereomers. NMR spectroscopy, in contrast, can rigorously define the structure and/or configuration of a chemical species, so it is regarded as the most valuable technique for identification and can be utilized for the precise identification of metabolites. However, NMR spectroscopy requires more stringent purification methods and a higher concentration of sample than ESI-MS. Thus, this technique seems to be not always applicable to the identification of trace seleno- or tellurometabolites in animal and plant samples.

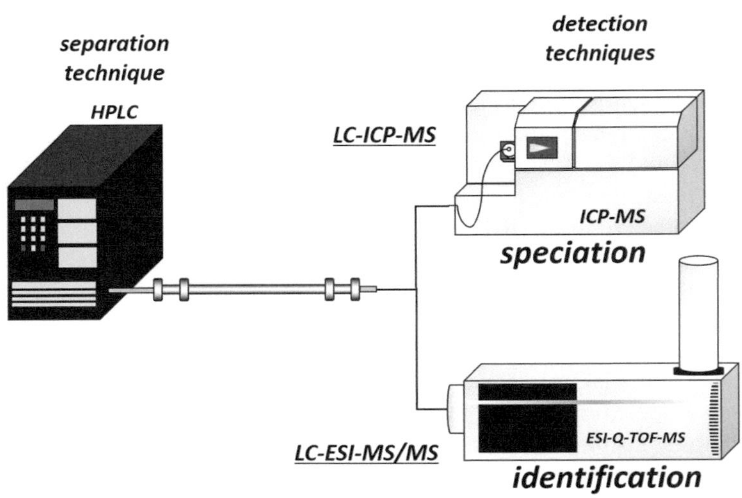

Fig. 2.1 Schematic diagram of complementary use of LC-ICP-MS and LC-ESI-MS-MS for the speciation and identification of metal/metalloid-containing metabolome

2.3 Identification of Chalcogen-Containing Metabolites

2.3.1 Selenosugars

Se is mostly excreted in urine. Within the nutritional level and the low toxicity level, the major urinary metabolite is *Se*-methylseleno-*N*-acetylgalactosamine (MeSeGalNAc or selenosugar 1) as identified by LC-ICP-MS, ESI-MS-MS, and NMR [36]. The mass spectra of the fragment ions of MeSeGalNAc obtained by ESI-MS-MS are indicated in Fig. 2.2. Three precursor ions, *m*/*z* 298, 300, and 302, representing Se isotopes, ^{78}Se, ^{80}Se, and ^{82}Se, respectively, were extracted with the first quadrupole MS and then introduced into a collision cell to obtain their fragment ions. Thereafter, the fragment ions were detected with the second quadrupole MS. At low collision energies, all the parent ions produced the same major product ion at *m*/*z* 204, suggesting that a Se-containing moiety was cleaved off from each of the parent ions (Fig. 2.2a, b, and c). Because the difference in *m*/*z* between the precursor ion and the fragment ion was 96, the common fragment ion was assumed to be a result of the removal of CH_3SeH from the precursor ion, and the

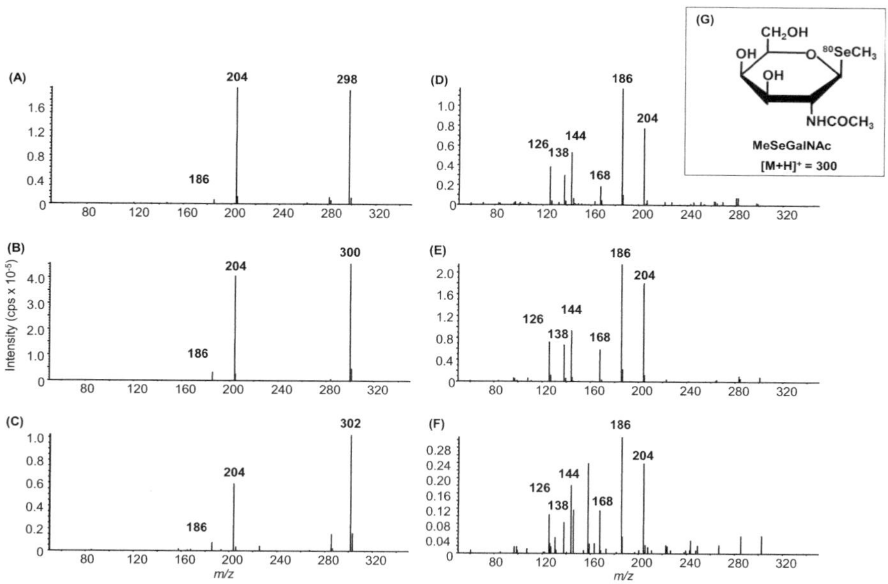

Fig. 2.2 Collision-induced dissociation mass spectra (ESI-MS-MS) of the major urinary selenometabolite (MeSeGalNAc) in urine of Se-administered rat. The dissociation of each Se-containing molecular ion, i.e., *m*/*z* 298 (^{78}Se, **a** and **d**), 300 (^{80}Se, **b** and **e**), or 302 (^{82}Se **c** and **f**), was induced in the collision cell with 10 (**a**, **b**, and **c**) and 20 eV (**d**, **e**, and **f**) collision energies, and then, each fragment ion was detected with the second mass spectrometer. The structure of MeSeGalNAc is shown in panel **g**

presence of a methylselenyl group (CH_3Se–) in MeSeGalNAc was suggested. A minor product ion appearing at *m*/*z* 186 could be easily assigned to the fragment ion that appeared after the loss of H_2O from the major product ion at *m*/*z* 204. At high collision energies, common fragment ions originating in the precursor ions appeared at *m*/*z* 186, 144, 138, and 126. Three of the four fragment ions, *m*/*z* 186, 144, and 126, were assigned to $[204\text{-}H_2O]^+$, $[204\text{-}CH_3COOH]^+$, and $[144\text{-}H_2O]^+$, respectively (Fig. 2.2d, e, and f). Consequently, the major urinary selenometabolite could be deconvoluted, as shown in Fig. 2.2g.

In addition to the major urinary selenosugar, MeSeGalNAc, minor selenosugars, such as *Se*-methylseleno-*N*-acetylglucosamine (MeSeGlcNAc or selenosugar 2) and *Se*-methylselenogalactosamine ($MeSeGalNH_2$ or selenosugar 3), were identified in urine by ESI-MS-MS [37, 38]. *Se*-Glutathionylseleno-*N*-acetylgalactosamine (GSSeGalNAc) was also identified in the liver of experimental animals by ESI-MS-MS [36]. This selenosugar is recognized as the precursor of the major urinary selenosugar, MeSeGalNAc.

The second major urinary metabolite is trimethylselenonium ion (TMSe), which is simply the methylated compound of selenide. Initially, it was thought that TMSe appeared in urine when Se exceeding the nutritional level was ingested [39–42]. It was reported recently that there is a specific group of people who constantly excrete TMSe, in addition to the major selenosugar, in urine [43–45].

2.3.2 Selenohomolanthionine

Selenohomolanthionine (SeHLan) is the common name of 4,4′-selenobis [2-aminobutanoic acid] or *Se*-(3-amino-3-carboxypropyl)-homocysteine. This unique selenoamino acid was first identified in Se-fortified Japanese pungent radish [46]. The mass spectra of SeHLan are shown in Fig. 2.3. Signals composing the isotope pattern of Se were obtained, and the peak at *m*/*z* 285 was assigned to a ^{80}Se-containing molecular ion, $[M+H]^+$. The ion at *m*/*z* 285 obtained from Se-fortified Japanese pungent radish was extracted and subjected to MS-MS analysis as a precursor ion. All peaks observed at low (20 eV, Fig. 2.3a) and high (30 eV, Fig. 2.3b) collision energies were assignable to fragments of the novel Se compound depicted in Fig. 2.3c.

SeHLan is less toxic than inorganic Se, and the tissue distribution of SeHLan differs from that of selenomethionine (SeMet) in experimental animals. In addition, it was reported that SeHLan exhibited an antiseptic effect in mouse [47]. The metabolism of SeHLan in plants and animals as well as its toxicological and pharmacological effects is explained in a review [48].

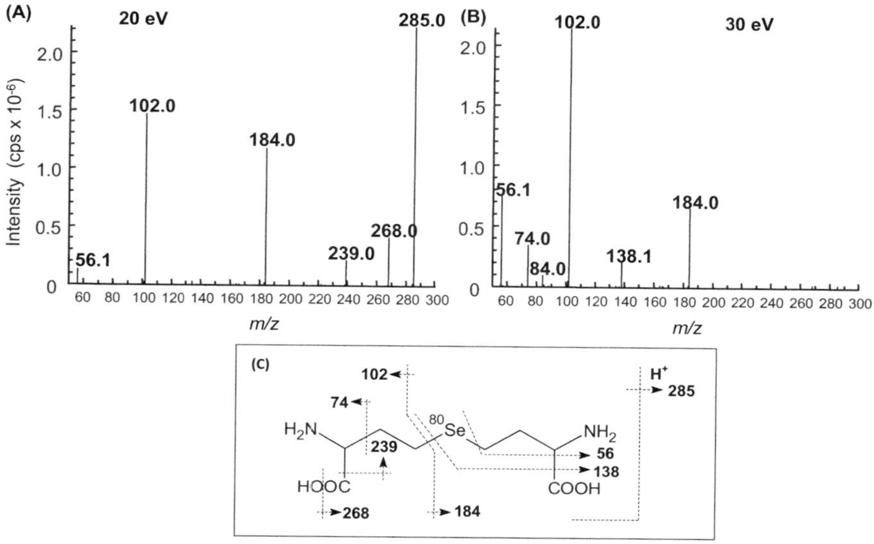

Fig. 2.3 Collision-induced dissociation mass spectra (ESI-MS-MS) of Se-containing positive molecular ions in a selenometabolite purified from Se-fortified Japanese pungent radish. The dissociation of each Se-containing molecular ion of the selenometabolite at *m*/*z* 285 was induced with 20 eV (**a**) and 30 eV (**b**) collision energies, and fragment ions were detected with the second mass spectrometer. The assignments of the fragments obtained are shown in panel C

2.3.3 Selenoneine

Selenoneine (2-selenyl-N^α,N^α,N^α-trimethyl-L-histidine) was first identified in bluefin tuna and other fish species [49]. It was also reported that sea turtles accumulate Se in their livers in spite of the low hepatic Hg concentration and that the majority of Se in the liver exists as selenoneine [50]. ESI-MS-MS has contributed significantly to the identification of selenoneine (Fig. 2.4).

Selenoneine is specifically detected in marine biota. Although selenoneine is detected in substantial amounts in the livers of sea turtles, it is below the detection limit in the livers of terrestrial turtles [50], suggesting that selenoneine is transferred among marine organisms via the food web to result in its ubiquitous distribution in the marine ecosystem. However, the origin of selenoneine is still unclear.

Selenoneine is also detected in blood samples from a fish-eating population [51, 52]. *Se*-Methylselenoneine is identified in a fish-eating population as well, suggesting that the redox activity of selenoneine is likely reduced in human, i.e., the toxicity of selenoneine is masked by the replacement of the selenyl/selone group with a methyl group [51].

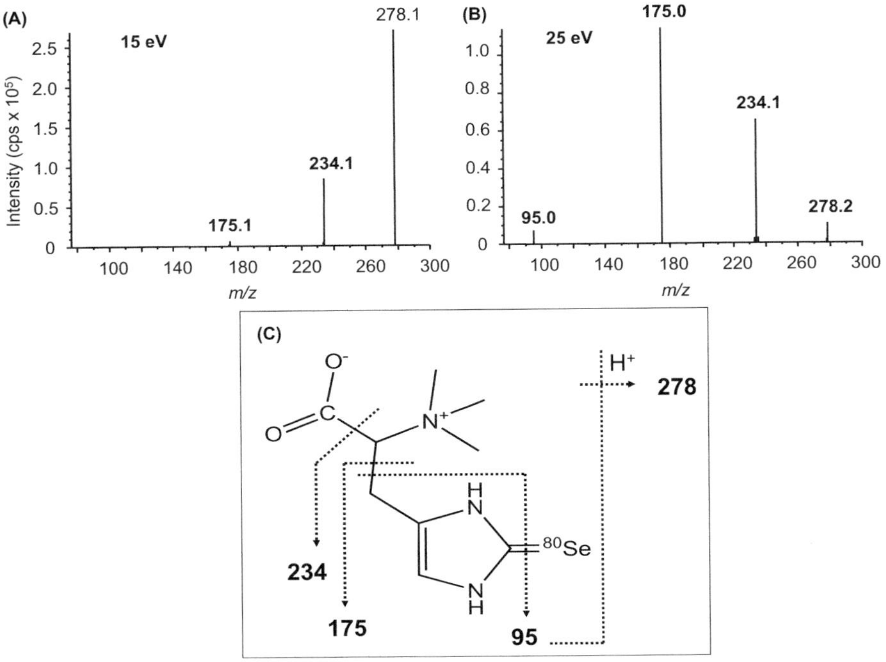

Fig. 2.4 Collision-induced dissociation mass spectra (ESI-MS-MS) of Se-containing positive molecular ions in a selenometabolite purified from a sea turtle. The dissociation of each Se-containing molecular ion of the selenometabolite at *m/z* 278 was induced with 15 eV (**a**) and 25 eV (**b**) collision energies, and fragment ions were detected with the second mass spectrometer. The assignments of the fragments obtained are shown in panel **c**

2.3.4 Selenocyanate

We recently identified selenocyanate in cultured mammalian cell lines as a selenometabolite by ESI quadrupole time-of-flight mass spectrometry (ESI-Q-TOF-MS) [53]. Signals composing the isotope pattern of Se were obtained, and the peak at *m/z* 106 was assigned to a ^{80}Se-containing molecular ion, as observed in the negative ion mode by ESI-MS. As the molecular mass of 106 seemed to be too small for MS-MS analysis as a precursor ion, we used ESI-Q-TOF-MS to determine the elemental composition of the novel Se metabolite based on its exact molecular mass. We speculated that the negative ion at *m/z* 106 was $[^{80}SeCN]^-$, i.e., selenocyanate. The observed mass containing ^{80}Se in the sample was 105.92055, and the theoretical mass of $^{80}SeCN^-$ is 105.91959 (Fig. 2.5). The difference between the observed and theoretical masses was 9.06 ppm. This indicated that the novel selenometabolite found in mammalian cultured cells was selenocyanate.

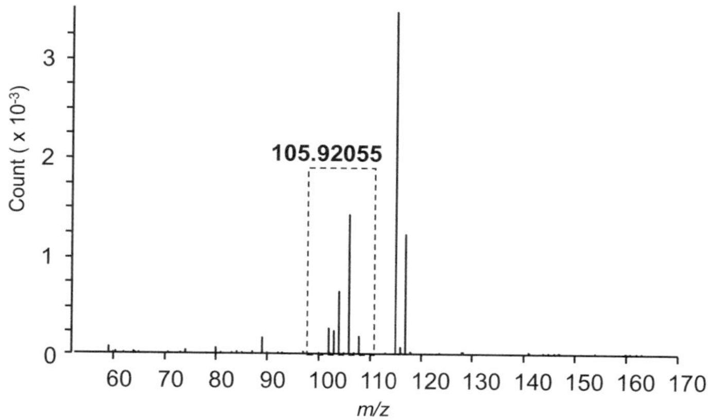

observed	105.92055	mass containing ^{80}Se	Δ9.06 ppm
theoretical	105.91959	$^{80}Se^{12}C^{14}N^-$	

Fig. 2.5 ESI-Q-TOF-MS spectrum of the supernatant of human hepatoma cell line, HepG2 cells, incubated with sodium selenite and glutathione. A 5.0 μL aliquot of sample was continuously introduced with 0.3 % ammonia into the ion source of ESI-Q-TOF-MS by a syringe pump at the flow rate of 0.1 mL/min. The ions were detected in the negative ion mode. Signals composing a Se-isotope pattern are highlighted by *broken lines*. Table shows the difference between observed and theoretical masses

Selenocyanate was assimilated into selenoproteins and selenometabolites in rats in the same manner as selenite [53]. Thus, selenocyanate can be utilized as a nutritional source of Se.

2.3.5 Trimethyltelluronium Ion

Te is a widely used metalloid in industry because of its unique chemical and physical properties. However, information about the biological and toxicological activities of Te in plants and animals remains limited. Although Te is expected to be metabolized in organisms via the same pathway as S and Se, no precise metabolic pathways are known. In this section, the identification of the major urinary Te metabolite (UTM) is introduced.

We synthesized trimethyltelluronium ion (TMTe) and utilized it as a standard compound. To confirm its structure, synthesized TMTe was subjected to ESI-MS. As mentioned above, Te consists of eight isotopes. Signals composing the isotope pattern of Te were observed in the mass spectra of synthesized TMTe, and the signal at *m*/*z* 173 was assigned to a ^{128}Te-containing molecular ion. Hence, the molecular ion at *m*/*z* 173 was assumed to correspond to trimethyltelluronium ion, $(CH_3)_3{}^{128}Te^+$, as $[M]^+$ (Fig. 2.6c). It was found that the major urinary Te

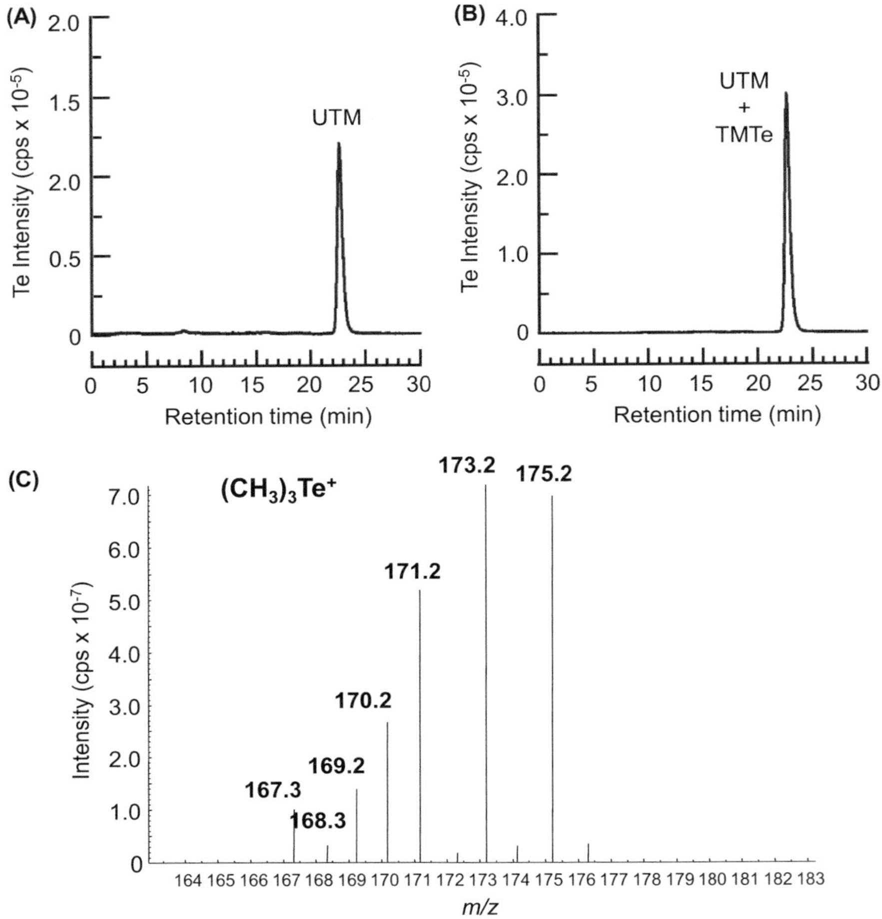

Fig. 2.6 Identification of major urinary Te metabolite (UTM) by LC-ICP-MS. Urine sample collected on day 7 was applied to a GS-320HQ column, which is a multimode gel filtration column. The sample was spiked (**b**) or not spiked (**a**) with synthesized TMTe. The eluate was directly injected into the ICP-MS, and Te was detected at *m/z* 128. The mass spectrum of synthesized TMTe was obtained with ESI-MS (**c**)

metabolite (UTM) was eluted at the retention time of 22.6 min on a GS-320HQ column (Fig. 2.6a). The retention time of synthesized TMTe used as standard was consistent with that of UTM on the column (Fig. 2.6b). Therefore, it was concluded that the UTM is TMTe. In this case, ESI-MS was indirectly used for the identification of unknown metabolite, i.e., the identification was done by LC-ICP-MS. In the original paper for the identification of UTM, two types of column, a multimode gel filtration column and an anion exchange column, were used [54]. At least two types of column are required for the unambiguous identification by LC-ICP-MS.

Interestingly, no Te-containing sugar metabolite(s), i.e., tellurosugar, was detected in the urine of rat-administered inorganic Te compounds. This indicated that the enzyme(s) participating in the biosynthesis of selenosugar could distinguish Se from Te although methyltransferase could not do so.

2.3.6 Te-*Methyltellurocysteine*

To reveal the metabolic pathway of Te in plants, garlic, a well-known Se accumulator was chosen as the model plant. Garlic was hydroponically cultivated and exposed to sodium tellurate, and Te-containing metabolites in the water extract of garlic leaves were identified by LC-ICP-MS or LC-ESI-MS-MS. At least three unknown Te-containing metabolites were detected by LC-ICP-MS, and two of them were subjected to LC-ESI-MS-MS for identification. The identification of one of the three unknown Te metabolites in garlic is introduced.

The ^{130}Te-containing molecular ions at m/z 246, 248, and 250 were extracted and subjected to MS-MS analysis as precursor ions. All the peaks observed at low (8 eV, Fig. 2.7a–c) and high (15 eV, Fig. 2.7d–f) collision energies were assignable to the fragments of the unknown tellurometabolite in Te-exposed garlic, as depicted in Fig. 2.7, i.e., *Te*-methyltellurocysteine telluroxide was identified [55]. In addition, the exact molecular mass of the unknown tellurometabolite was determined by HPLC-Orbitrap MS. Orbitrap MS can give us the exact mass of ions. The experimental masses of the unknown tellurometabolite in the positive and negative ion modes were 249.97184 and 247.95699, respectively. The theoretical masses of *Te*-methyltellurocysteine telluroxide in the positive and negative ion modes were 249.97176 and 247.95721, respectively. The differences between the experimental and theoretical masses were 0.3 and −0.9 ppm in the positive and negative ion modes, respectively. These results confirmed that the unknown tellurometabolite was *Te*-methyltellurocysteine telluroxide. However, the CH_3-Te moiety was rapidly oxidized to telluroxide form under ambient conditions. Thus, we speculated that *Te*-methyltellurocysteine telluroxide is the artificially oxidized form of *Te*-methyltellurocysteine.

2.4 Conclusion

Se- or Te-containing compounds having Se- or Te-carbon covalent bond(s), i.e., selenometabolites or tellurometabolites, are formed in animals and plants. Hence, the identification of selenometabolites and tellurometabolites is a means to determine the metabolic pathways of Se and Te in animals and plants. The metabolic pathway of Se in animals is depicted in Fig. 2.8, and those of Se and Te in plants are shown in Figs. 2.9 and 2.10, respectively. Because there are some unknown seleno- and tellurometabolites in animals and plants which can be detected by LC-ICP-MS

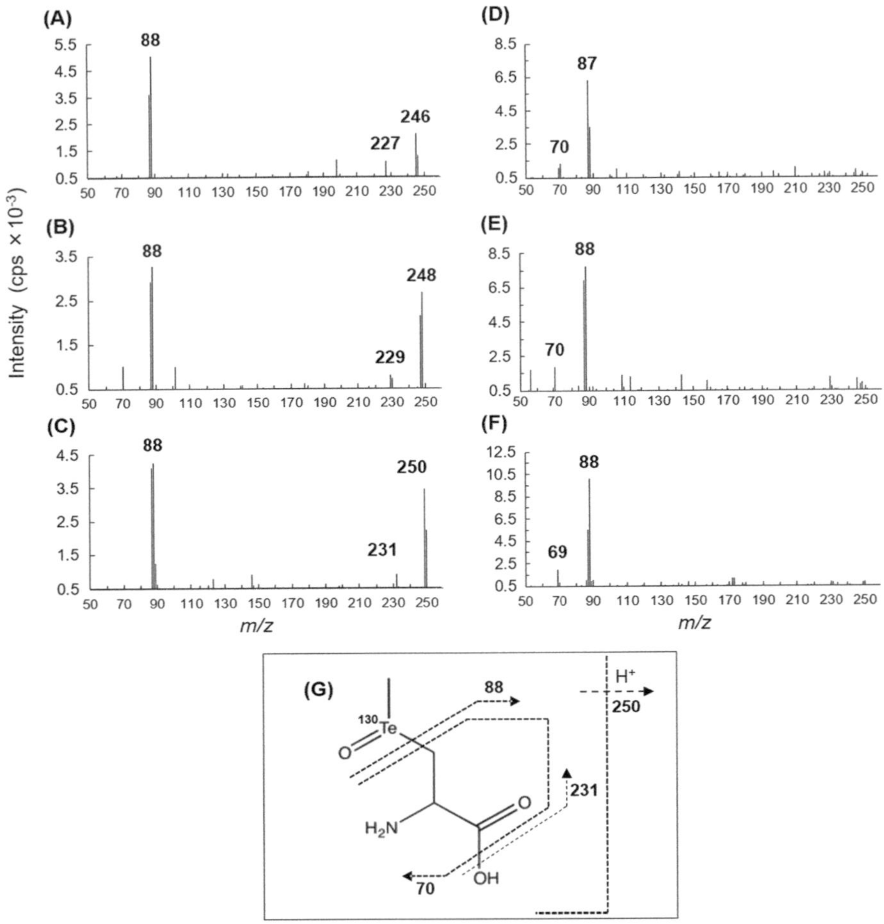

Fig. 2.7 Collision-induced dissociation mass spectra of unknown tellurometabolite in Te-exposed garlic measured by LC-ESI-MS-MS in the positive ion mode. The collision-induced dissociation mass spectra of precursor ions at *m/z* 246 (^{126}Te, **a** and **d**), 248 (^{128}Te, **b** and **e**), and 250 (^{130}Te, **c** and **f**) were obtained. The dissociation was induced with collision energies of 8 (**a**–**c**) and 15 (**c**–**f**) eV. The assignments of the fragments obtained and the structure of *Te*-methyltellurocysteine oxide are shown in panel **g**

but not LC-ESI-MS, further improvement of the sensitivity of organic mass spectrometry is expected to lead to the complete identification of seleno- and tellurometabolites, thereby paving the way to revealing the entire metabolic pathway of Se and Te.

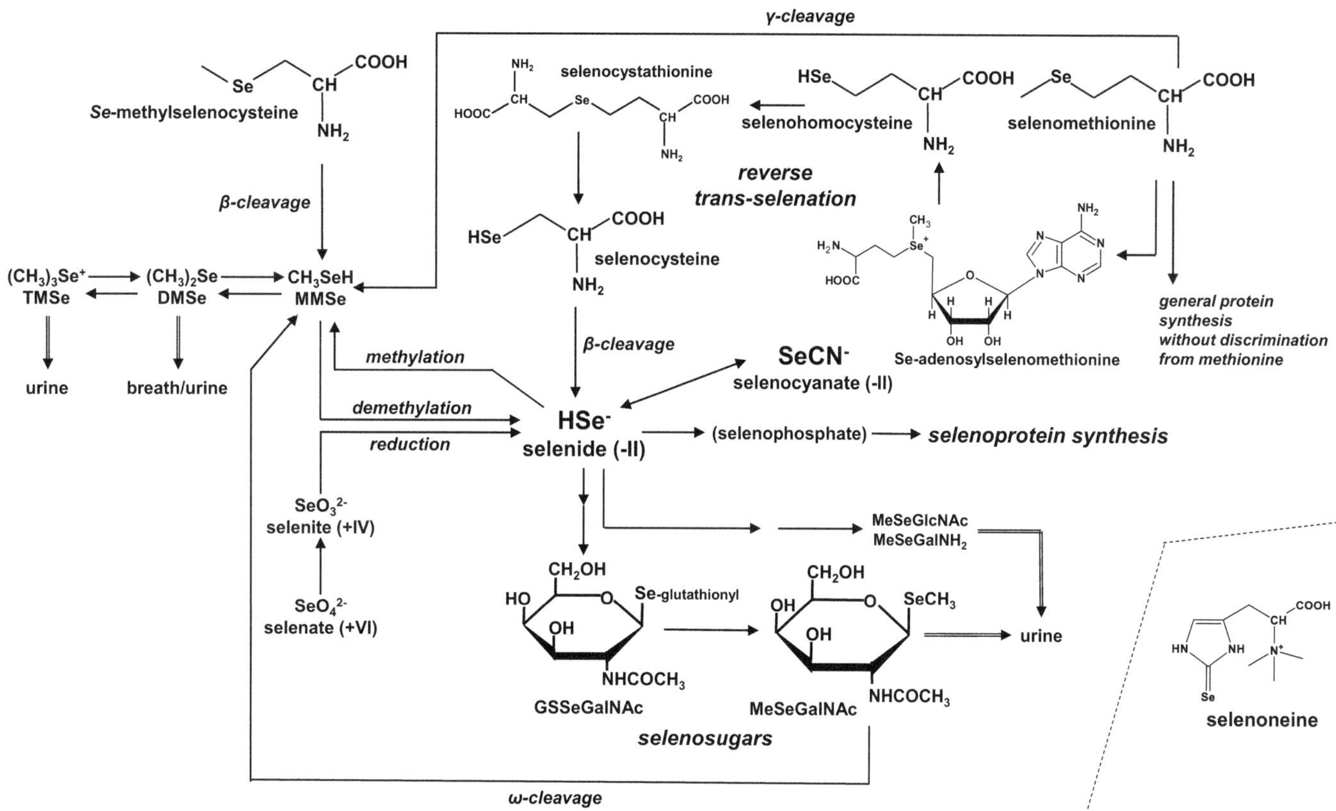

Fig. 2.8 Proposed metabolic pathway of Se in animals. Selenoneine is specifically observed in marine animals, and its metabolic pathway has not been identified yet

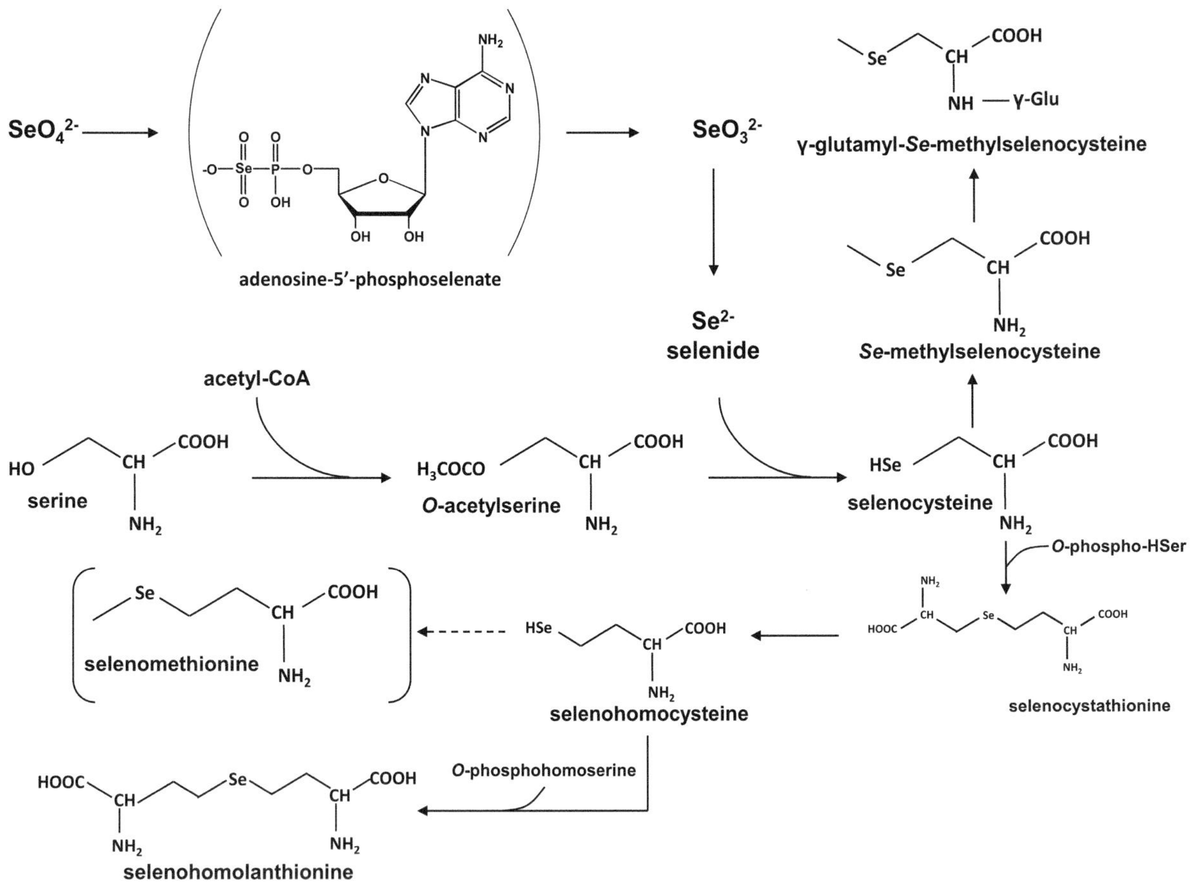

Fig. 2.9 Proposed metabolic pathway of Se in plants. The structure in parentheses has not been identified yet

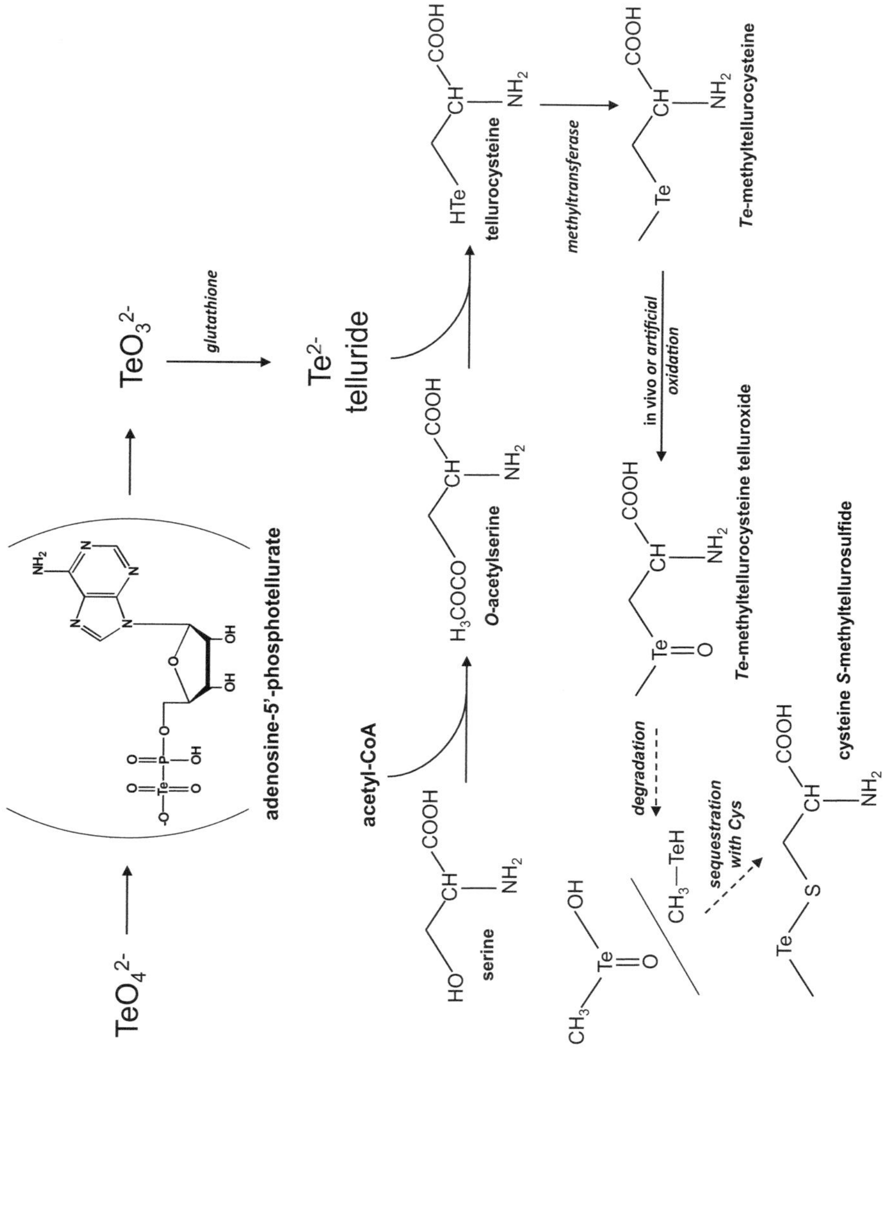

Fig. 2.10 Proposed metabolic pathway of Te in plants. The structure in parentheses has not been identified yet

Acknowledgments This study was supported by JSPS KAKENHI Grant Numbers 24659022, 26293030, 15K14991, and 16H05812, and a grant from the Food Safety Commission, Cabinet Office, Government of Japan (Research Program for Risk Assessment Study on Food Safety, No 1601).

References

1. Szpunar J, Łobiński R, Prange A (2003) Hyphenated techniques for elemental speciation in biological systems. Appl Spectrosc 57(3):102A–112A. doi:10.1366/000370203321558128
2. Whanger PD (2002) Selenocompounds in plants and animals and their biological significance. J Am Coll Nutr 21:223–232
3. Suzuki KT, Ogra Y (2002) Metabolic pathway for selenium in the body: speciation by HPLC-ICP MS with enriched Se. Food Addit Contam 19 (10):974–983. doi:10.1080/02652030210153578
4. Ogra Y (2008) Integrated strategies for identification of selenometabolites in animal and plant samples. Anal Bioanal Chem 390:1685–1689
5. Yamashita Y, Yamashita M (2010) Identification of a novel selenium-containing compound, selenoneine, as the predominant chemical form of organic selenium in the blood of bluefin tuna. J Biol Chem 285(24):18134–18138. doi:10.1074/jbc.C110.106377
6. Kobayashi Y, Ogra Y, Ishiwata K, Takayama H, Aimi N, Suzuki KT (2002) Selenosugars are key and urinary metabolites for selenium excretion within the required to low-toxic range. Proc Natl Acad Sci U S A 99(25):15932–15936
7. Ogra Y (2015) Selenohomolanthionine and its potential use in medicine and nutrition. In: Preedy VR (ed) Food and nutritional components in focus No. 9, Selenium: chemistry, analysis, function and effects. The Royal Society of Chemistry, London, pp 354–362
8. Anan Y, Mikami T, Tsuji Y, Ogra Y (2011) Distribution and metabolism of selenohomolanthionine labeled with a stable isotope. Anal Bioanal Chem 399 (5):1765–1772. doi:10.1007/s00216-010-4143-4
9. Anan Y, Nakajima G, Ogra Y (2015) Complementary use of LC-ICP-MS and LC-ESI-Q-TOF-MS for selenium speciation. Anal Sci 31(6):561–564. doi:10.2116/analsci.31.561
10. Ogra Y, Kobayashi R, Ishiwata K, Suzuki KT (2007) Identification of urinary tellurium metabolite in rats administered sodium tellurite. J Anal At Spectrom 22:153–157
11. Lu J, Holmgren A (2009) Selenoproteins. J Biol Chem 284:723–727
12. Kinoshita N, Sueki K, Sasa K, Kitagawa J, Ikarashi S, Nishimura T, Wong YS, Satou Y, Handa K, Takahashi T, Sato M, Yamagata T (2011) Assessment of individual radionuclide distributions from the Fukushima nuclear accident covering central-east Japan. Proc Natl Acad Sci U S A 108(49):19526–19529. doi:10.1073/pnas.1111724108
13. Caruso J, Montes-Bayón M (2003) Elemental speciation studies – new directions for trace metal analysis. Ecotoxicol Environ Safety 56:148–163
14. Gammelgaard B, Grimstrup Madsen K, Bjerrum J, Bendahl L, Jøns O, Olsen J, Sidenius U (2003) Separation, purification and identification of the major selenium metabolite from human urine by multi-dimensional HPLC-ICP-MS and APCI-MS. J Anal At Spectrom 18:65–70
15. Francesconi KA, Pannier F (2004) Selenium metabolites in urine: a critical overview of past work and current status. Clin Chem 50(12):2240–2253
16. Anan Y, Ishiwata K, Suzuki N, Tanabe S, Ogra Y (2011) Speciation and identification of low molecular weight selenium compounds in the liver of sea turtles. J Anal At Spectrom 26 (1):80–85
17. Ogra Y, Kobayashi R, Ishiwata K, Suzuki KT (2008) Comparison of distribution and metabolism between tellurium and selenium in rats. J Inorg Biochem 102(7):1507–1513. doi:S0162-0134(08)00041-X [pii] 10.1016/j.jinorgbio.2008.01.012

18. Ogra Y, Ishiwata K, Ruiz-Encinar J, Łobinski R, Suzuki KT (2004) Speciation of selenium in selenium-enriched shiitake mushroom, *Lentinula edodes*. Anal Bioanal Chem 379:861–866
19. Cowgill UM (1988) The tellurium content of vegetation. Biol Trace Elem Res 17:43–67
20. Ogra Y, Kitaguchi T, Ishiwata K, Suzuki N, Iwashita Y, Suzuki KT (2007) Identification of selenohomolanthionine in selenium-enriched Japanese pungent radish. J Anal At Spectrom 22:1390–1396
21. Djanaguiraman M, Durga Devi D, Shanker AK, Annie Sheeba J, Bangarusamy U (2005) Selenium – an antioxidative protectant in soybean during senescence. Plant Soil 272:77–86
22. Rayman MP (2000) The importance of selenium to human health. Lancet 356:233–241
23. Prange A, Pröfrock D (2005) Application of CE–ICP–MS and CE–ESI–MS in metalloproteomics: challenges, developments, and limitations. Anal Bioanal Chem 383:372–389
24. Ip C, Birringer M, Block E, Kotrebai M, Tyson JF, Uden PC, Lisk DJ (2000) Chemical speciation influences comparative activity of selenium-enriched garlic and yeast in mammary cancer prevention. J Agric Food Chem 48:2062–2070
25. Taylor A (1996) Biochemistry of tellurium. Biol Trace Elem Res 55:231–239
26. Ogra Y, Ishiwata K, Iwashita Y, Suzuki KT (2005) Simultaneous speciation of selenium and sulfur species in selenized odorless garlic (*Allium sativum* L. Shiro) and shallot (*Allium ascalonicum*) by HPLC–inductively coupled plasma-(octopole reaction system)-mass spectrometry and electrospray ionization-tandem mass spectrometry. J Chromatogr A 1093:118–125
27. Ravn-Haren G, Krath BN, Overvad K, Cold S, Moesgaard S, Larsen EH, Dragsted LO (2008) Effect of long-term selenium yeast intervention on activity and gene expression of antioxidant and xenobiotic metabolising enzymes in healthy elderly volunteers from the Danish Prevention of Cancer by Intervention by Selenium (PRECISE) pilot study. Br J Nutr 99 (6):1190–1198. doi:S0007114507882948 [pii] 10.1017/S0007114507882948
28. Reeves MA, Hoffmann PR (2009) The human selenoproteome: recent insights into functions and regulation. Cell Mol Life Sci 66:2457–2478
29. Reid ME, Duffield-Lillico AJ, Slate E, Natarajan N, Turnbull B, Jacobs E, Combs GF Jr, Alberts DS, Clark LC, Marshall JR (2008) The nutritional prevention of cancer: 400 mcg per day selenium treatment. Nutr Cancer 60(2):155–163, doi:791706363 [pii] 10.1080/01635580701684856
30. Bierla K, Szpunar J, Yiannikouris A, Lobinski R (2012) Comprehensive speciation of selenium in selenium-enriched yeast. Trends Anal Chem 41:122–132
31. Burk RF, Hill KE (2005) Selenoprotein P: an extracellular protein with unique physical characteristics and a role in selenium homeostasis. Annu Rev Nutr 25:215–235. doi:10.1146/annurev.nutr.24.012003.132120
32. Foster LH, Sumar S (1997) Selenium in health and disease: a review. Crit Rev Food Sci Nutr 37(3):211–228
33. Mehta A, Rebsch CM, Kinzy SA, Fletcher JE, Copeland PR (2004) Efficiency of mammalian selenocysteine incorporation. J Biol Chem 279(36):37852–37859. doi:10.1074/jbc.M404639200 [doi] M404639200 [pii]
34. Mounicou S, Szpunar J, Łobiński R (2009) Metallomics: the concept and technology. Chem Soc Rev 38:1119–1138
35. Szpunar J (2005) Advances in analytical methodology for bioinorganic speciation analysis: metallomics, metalloproteomics and heteroatom-tagged proteomics and metabolomics. Analyst 130:442–465
36. Francesconi KA, Pannier F (2004) Selenium metabolites in urine: a critical overview of past work and current status. Clin Chem 50:2240–2253
37. Freeman JL, Lindblom SD, Quinn CF, Fakra S, Marcus MA, Pilon-Smits EA (2007) Selenium accumulation protects plants from herbivory by Orthoptera via toxicity and deterrence. New Phytol 175:490–500

38. Squires JE, Berry MJ (2008) Eukaryotic selenoprotein synthesis: mechanistic insight incorporating new factors and new functions for old factors. IUBMB Life 60(4):232–235. doi:10.1002/iub.38
39. Jager T, Drexler H, Goen T (2016) Human metabolism and renal excretion of selenium compounds after oral ingestion of sodium selenite and selenized yeast dependent on the trimethylselenium ion (TMSe) status. Arch Toxicol 90(5):1069–1080. doi:10.1007/s00204-015-1548-z
40. Yamashita M, Yamashita Y, Ando T, Wakamiya J, Akiba S (2013) Identification and determination of selenoneine, 2-selenyl-N^a, N^a, N^a-trimethyl-L-histidine, as the major organic selenium in blood cells in a fish-eating population on remote Japanese Islands. Biol Trace Elem Res 156(1–3):36–44. doi:10.1007/s12011-013-9846-x
41. Kuehnelt D, Engstrom K, Skroder H, Kokarnig S, Schlebusch C, Kippler M, Alhamdow A, Nermell B, Francesconi K, Broberg K, Vahter M (2015) Selenium metabolism to the trimethylselenonium ion (TMSe) varies markedly because of polymorphisms in the indolethylamine N-methyltransferase gene. Am J Clin Nutr 102(6):1406–1415. doi:10.3945/ajcn.115.114157
42. Fairhill LT (1969) Tellurium, Industrial toxicology. Hafner Publishing Co., New York
43. Anan Y, Kimura M, Hayashi M, Koike R, Ogra Y (2015) Detoxification of selenite to form selenocyanate in mammalian cells. Chem Res Toxicol 28(9):1803–1814. doi:10.1021/acs.chemrestox.5b00254
44. Bock A, Flohe L, Kohrle J (2007) Selenoproteins – biochemistry and clinical relevance. Biol Chem 388(10):985–986. doi:10.1515/BC.2007.148
45. Falandysz J (2008) Selenium in edible mushrooms. J Environ Sci Health C Environ Carcinog Ecotoxicol Rev 26:256–299
46. Byard JL (1969) Trimethyl selenide. A urinary metabolite of selenite. Arch Biochem Biophys 130:556–560
47. Anan Y, Yoshida M, Hasegawa S, Katai R, Tokumoto M, Ouerdane L, Lobinski R, Ogra Y (2013) Speciation and identification of tellurium-containing metabolites in garlic, Allium sativum. Metallomics Integr Biometal Sci 5(9):1215–1224. doi:10.1039/c3mt00108c
48. Kuehnelt D, Juresa D, Kienzl N, Francesconi KA (2006) Marked individual variability in the levels of trimethylselenonium ion in human urine determined by HPLC/ICPMS and HPLC/vapor generation/ICPMS. Anal Bioanal Chem 386:2207–2212
49. Kraus RJ, Foster SJ, Ganther HE (1985) Analysis of trimethylselenonium ion in urine by high-performance liquid chromatography. Anal Biochem 147(2):432–436, doi:0003-2697(85)90293-3 [pii]
50. Montes-Bayón M, Grant TD, Meija J, Caruso JA (2002) Selenium in plants by mass spectrometric techniques: developments in bio-analytical methods. J Anal At Spectrom 17:1015–1023
51. Janghorbani M, Xia Y, Ha P, Whanger PD, Butler JA, Olesik JW, Daniels L (1999) Quantitative significance of measuring trimethylselenonium in urine for assessing chronically high intakes of selenium in human subjects. Br J Nutr 82(4):291–297, doi:S000711459900149X [pii]
52. Klein M, Ouerdane L, Bueno M, Pannier F (2011) Identification in human urine and blood of a novel selenium metabolite, Se-methylselenoneine, a potential biomarker of metabolization in mammals of the naturally occurring selenoneine, by HPLC coupled to electrospray hybrid linear ion trap-orbital ion trap MS. Metallomics Integr Biometal Sci 3(5):513–520. doi:10.1039/c0mt00060d
53. Becker JS, Jakubowski N (2009) The synergy of elemental and biomolecular mass spectrometry: new analytical strategies in life sciences. Chem Soc Rev 38:1969–1983
54. Templeton DM, Ariese F, Cornelis R, Danielsson L-G, Muntau H, van Leeuwen HP, Łobiński R (2000) Guidelines for terms related to chemical speciation and fractionation of elements.

Definitions, structural aspects, and methodological approaches (IUPAC Recommendations 2000). Pure Appl Chem 72:1453–1470
55. Szpunar J, Łobiński R (2002) Multidimensional approaches in biochemical speciation analysis. Anal Bioanal Chem 373:404–411

Chapter 3
Visualization of Intracellular Elements Using Scanning X-Ray Fluorescence Microscopy

Mari Shimura, Lukasz Szyrwiel, Satoshi Matsuyama, and Kazuto Yamauchi

Abstract Recent technological developments have enabled the imaging of chemical elements in cells, although quantitative analyses, such as by inductively coupled plasma mass spectrometry, were developed previously. Applications allowing high-resolution imaging at the single-cell level are anticipated in cell biology and medicine, where the roles of elements, especially in relation to intracellular molecules such as proteins, nucleic acids, lipids, and sugars, are essential for understanding cellular functions. The expression of proteins and genes varies depending on cellular function, and multiple elements are likely to be associated with biological molecules in the functioning of cell proliferation, differentiation, aging, and stress responses. In this review, we describe a scanning X-ray fluorescence microscopy system, which can reliably determine the cellular distribution of multiple elements by a sub-100-nm focusing approach, together with its applications. Visualizing intracellular elements and understanding their dynamics at the single-cell level may provide great insight into their behaviors.

Keywords X-ray microscopy • Single-cell level • Intracellular element • Labeling • Synchrotron • X-ray fluorescence analysis

---

M. Shimura (✉)
Department of Intractable Diseases, Research Institute, National Center for Global Health and Medicine, Shinjuku, Tokyo 162-8655, Japan

RIKEN SPring-8 Center, 1-1-1 Kouto, Sayo, Hyogo 679-5148, Japan
e-mail: mshimura@ri.ncgm.go.jp

L. Szyrwiel
RIKEN SPring-8 Center, 1-1-1 Kouto, Sayo, Hyogo 679-5148, Japan

CNRS/UPPA, LCABIE, UMR5254, F-64053 Pau, France

Department of Chemistry of Drugs, Wrocław Medical University, ul. Borowska 211, 50-552 Wrocław, Poland

S. Matsuyama • K. Yamauchi
Department of Precision Science and Technology, Graduate School of Engineering, Osaka University, Suita, Osaka 565-0871, Japan

Y. Ogra, T. Hirata (eds.), *Metallomics*, DOI 10.1007/978-4-431-56463-8_3

3.1 Imaging Intracellular Elements at the Single-Cell Level Provides New Insights

Imaging provides information, such as on the pattern and distribution of molecules, which is essential to understand the characteristics of objects that cannot be determined by measuring element concentrations. Light microscopy, fluorescence microscopy, and electron microscopy have been widely used to reveal cellular functions. Recently, technologies for imaging elements in cells have been established, such as laser ablation inductively coupled plasma mass spectrometry (LA-ICP-MS), microparticle-induced X-ray emission (PIXE), and X-ray fluorescence microscopy (XRF) [1–5]. Their applications in biology and medicine have often been reported, but a high resolution is required to observe the distribution of elements intracellularly at the single-cell level. Previous findings on proteins and nucleic acids indicate that cells are not a homogeneous population. For example, cell growth proceeds via different stages of the cell cycle (Fig. 3.1, cell cycle). Cell division also occurs, which is a dynamic process in which proteins and nucleic acids are produced and exhibit complex changes in their formations. Previous findings suggested that intracellular elements are related to cell proliferation [6, 7] and may undergo changes in their intracellular distribution. In addition, cancer cells often

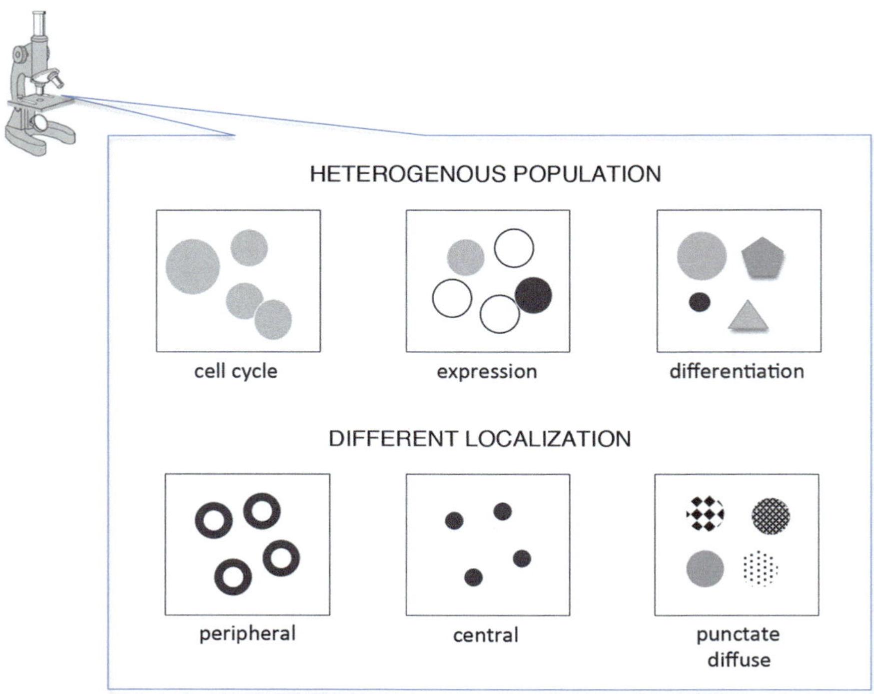

Fig. 3.1 Imaging intracellular elements at the single-cell level provides new insights

exhibit genetic variation, showing different protein expression levels within a population, which has been observed by flow cytometry (Fig. 3.1, expression). Normal cell populations are more quiescent, but they also differentiate and age (Fig. 3.1, differentiation). In these various processes, molecules in cells may exhibit different distributions, such as peripheral, focal, punctate, or diffuse (Fig. 3.1, different localizations). These different distributions suggest different binding partners, which is probably similar in the case of elements, since some essential elements are associated with proteins or nucleic acids [8, 9]. Imaging technology is also useful in the medical field. Imaging a series of sections using different staining procedures can be informative despite the small volume studied. Imaging of multiple intracellular elements may have the potential to clarify unknown disease mechanisms. In this chapter, we introduce our developed method of scanning X-ray fluorescence microscopy (SXFM), which has been especially established for imaging cells at the single-cell level.

3.2 Scanning X-Ray Fluorescence Microscopy

SXFM (also abbreviated as μXRF) is a very powerful tool that can quantitatively and nondestructively measure the distributions of elements with high sensitivity. As a result, it is widely applied in many industries and scientific fields and even in the fields of art and archeology [10]. There are a range of SXFM apparatuses, from compact systems that are commercially available to large systems installed at special facilities. A schematic of a typical apparatus is shown in Fig. 3.2. A typical

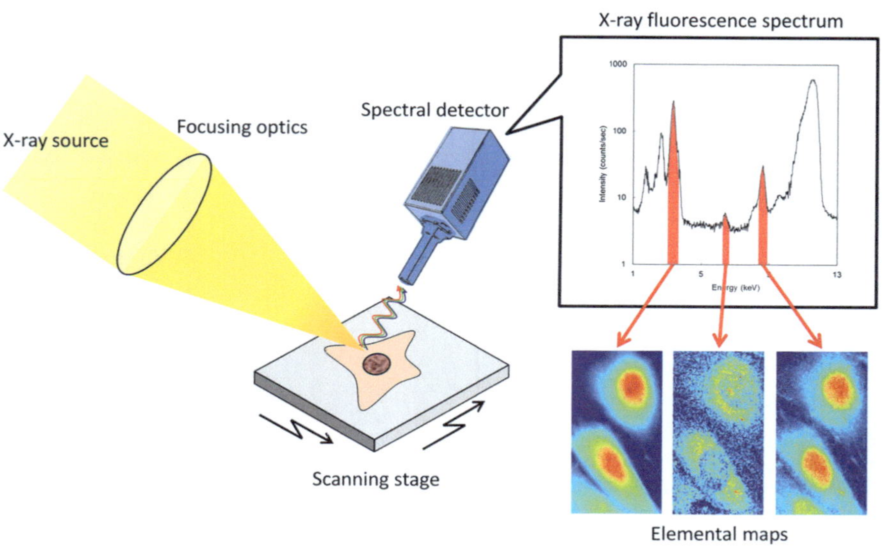

Fig. 3.2 Schematic of a scanning X-ray fluorescence microscope (SXFM)

system consists of an X-ray source, focusing optics, a sample scanner, and a detector to acquire X-ray spectra. In this context, the following question arises: What are the requirements for single-cell imaging? The most important features are an X-ray source with intense brightness and high-quality X-ray focusing optics.

3.2.1 Synchrotron Radiation X-Ray Source

An X-ray tube is a very common X-ray generator at the laboratory scale that can produce X-rays via collisions between accelerated electrons of several tens of kilovolts and a material, usually tungsten, molybdenum, or copper. However, it is too weak and has too large a source size and divergence for SXFM at the single-cell level. A synchrotron radiation (SR) X-ray source (Fig. 3.3, left) was developed to overcome these issues. Electrons traveling almost at the speed of light (99.9999998 % of light speed when an electron has energy of 8 GeV) can produce X-rays when they are bent by a magnetic field (Fig. 3.3, right). The produced X-rays are over a billion times as intense as an X-ray tube. This is advantageous because generally the sensitivity is proportional to the intensity of input X-rays. In addition, the X-ray source size and divergence are very small, which leads to the formation of an intense focused beam because the demagnification of practical focusing optics is limited. The typical values expressed as the standard deviation are 300 μm (H) × 7 μm (V) and 13 μrad (H) × 4 μrad (V) in the SPring-8 (Hyogo, Japan) standard undulator X-ray source [11].

3.2.2 X-Ray Focusing Optics

One of the difficulties for scanning-type X-ray microscopes is the development of X-ray focusing optics. This is due to the tolerance of fabrication error being very limited because the acceptable size of fabrication errors decreases with a decrease

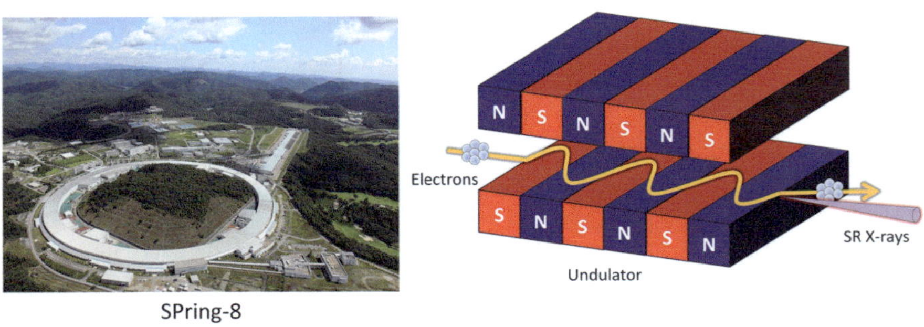

Fig. 3.3 *Left*, photograph of SPring-8; *right*, X-ray generation at a synchrotron radiation facility

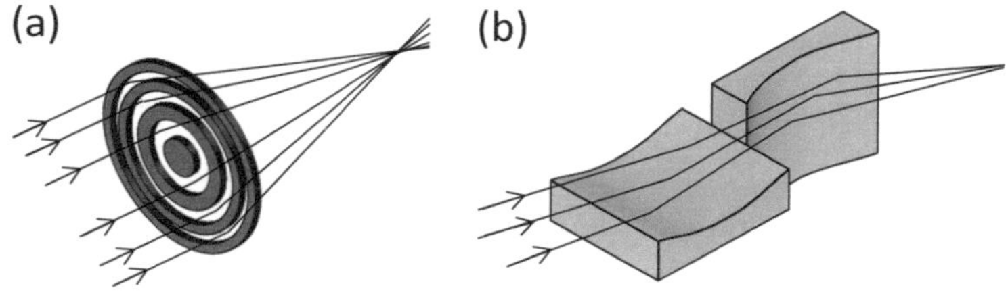

Fig. 3.4 Schematic of (**a**) Fresnel zone plate and (**b**) Kirkpatrick-Baez mirrors

in the wavelength of light. In addition, we cannot utilize a conventional lens, often used in the visible light region, as X-ray focusing optics because the refractive index in the X-ray region is less than but very close to 1 (e.g., refractive index of gold at 10 keV = 0.99997). Specialized focusing optics for X-rays is essential. Currently, Fresnel zone plates and Kirkpatrick-Baez (KB) mirrors are often used as practical and high-performance X-ray focusing optics (Fig. 3.4). A Fresnel zone plate is a circular diffraction grating with unequally spaced periods. It has major advantages in terms of usability and quality. However, its throughput is not very high because only the first-order diffracted light is used for focusing. In addition, diffraction significantly depends on the wavelength, which results in achromatic aberration. On the other hand, KB mirrors are focusing optics that utilize the phenomenon of total reflection. They consist of two elliptical mirrors arranged perpendicular to each other to focus X-rays two-dimensionally. Owing to the separation of the mirrors in the vertical and horizontal directions, almost planar mirrors that can be fabricated with extreme precision can be used as focusing mirrors. The benefits of this approach are high throughput, large acceptance, and no chromatic aberration. However, it has the disadvantages of difficult mirror fabrication, high cost, and difficult mirror adjustment. The former can be explained using an equation that can calculate acceptable figure errors on an X-ray mirror to reach the theoretically minimum spot (diffraction limit):

$$d = \frac{\lambda}{8 \sin \theta}$$

where λ is wavelength and θ is grazing-incidence angle. Typical d at 10 keV is ~4 nm. This accuracy is very severe, even using state-of-the-art precision machining. To realize SXFM with high resolution and high sensitivity, we developed high-quality KB mirror optics [12]. The mirrors were superpolished by the elastic emission machining (EEM) technique [13], which can realize nanometer-level shape accuracy and an atomically smooth surface. The achieved shape accuracy is 2 nm (Fig. 3.5), which is sufficiently small to satisfy the accuracy requirements. Focusing tests conducted at SPring-8, a synchrotron radiation facility in Japan (Fig. 3.3, left), revealed that our focusing system could focus X-rays down to an almost diffraction-limited spot size of 48 (H) nm × 36 (V) nm at X-ray energy of 15 keV (Fig. 3.5) [14]. In addition, owing to the smooth surface, loss of reflectivity

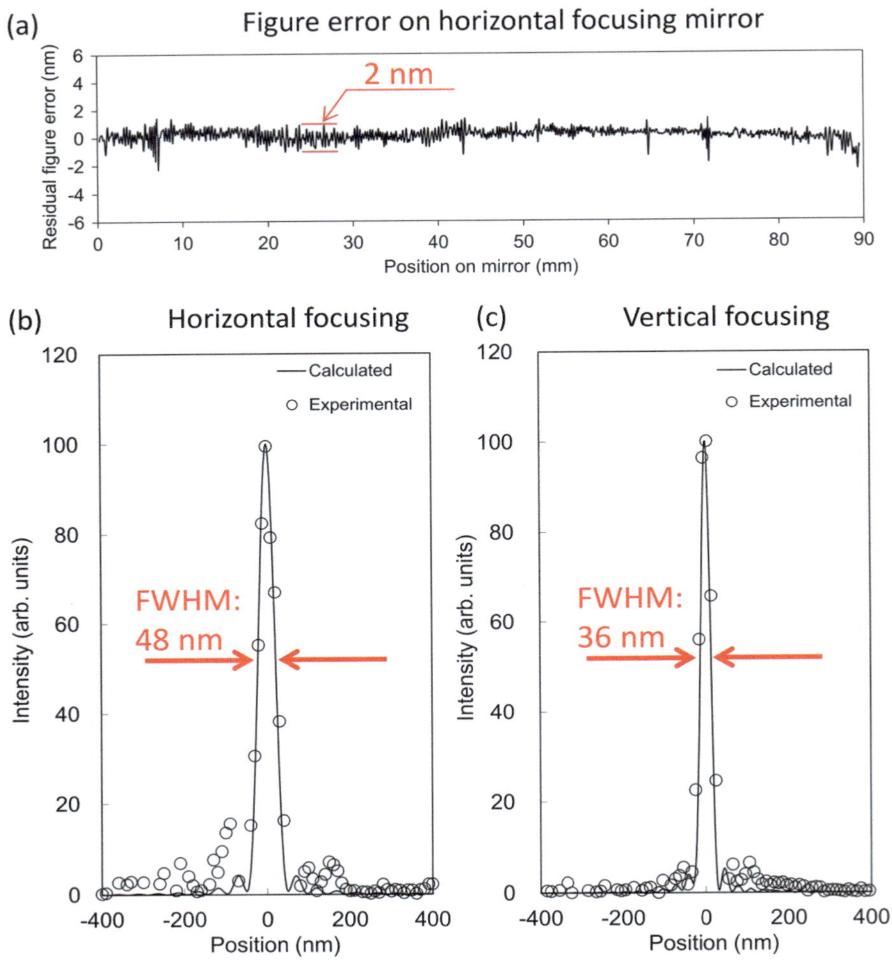

Fig. 3.5 Performance of fabricated Kirkpatrick-Baez mirrors. (**a**) Figure error on the horizontal focusing mirror used. (**b**, **c**) Measured beam profiles at the focus

was almost zero. Thus, we could successfully develop a focusing system that is highly suitable for SXFM.

3.2.3 Demonstration of SXFM Performance

The prototype for our SXFM was developed at BL29XUL of SPring-8 (Fig. 3.6). The system consisted of the KB focusing system, a detector system including a silicon drift detector (SDD), a fast multichannel analyzer (MCA), and a fine stage

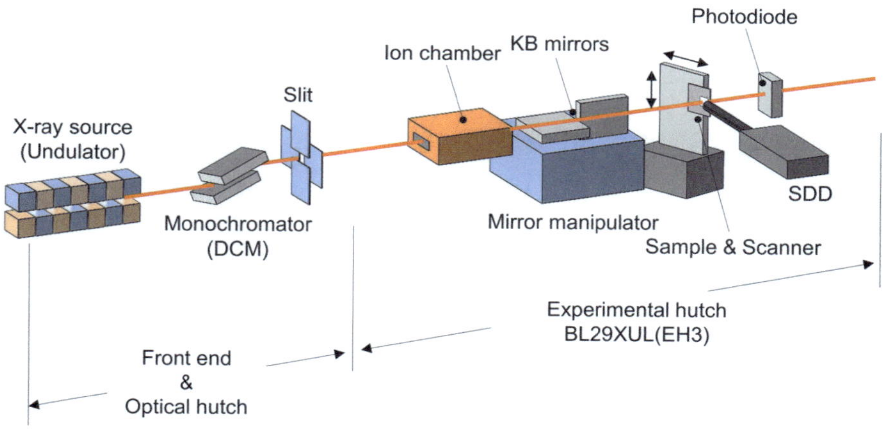

Fig. 3.6 Schematic of SXFM system

for sample scanning. The spectra formed by the MCA are recorded for every pixel over the measurement area to select chemical elements to be visualized after measurement and to process noise reduction and peak fitting. To demonstrate the performance, a fine test pattern, fabricated with a focused ion beam figuring system, was observed [12]. To find regions of interest effectively, first, a wide beam with a width of 1000 nm and then a 40-nm beam were used. This beam broadening could be adjusted using a slit placed downstream of the undulator X-ray source, in which the beam size is controllable between the diffraction limit (30×50 nm^2) and ~1000 nm. The distributions of elements were gradually magnified, and finally the target area was found. As can be seen in the gallium maps in Fig. 3.7, a 50-nm line width was successfully visualized.

Next, as a test of practical cell imaging, cultured mouse cells with mitochondria labeled with fluorescein isothiocyanate and gold colloids were observed [15]. The cells supported on a thin polymer film were chemically fixed with paraformaldehyde and dehydrated with ethanol. The obtained element maps are shown in Fig. 3.8. The right panel is a magnified image of the region of interest shown by the white square in the left panel in this figure. The color scale shows the semi-quantitatively determined levels of elements in the irradiated area. A calibration curve for the semiquantitative analysis was made using the fluorescence signals from platinum films, in which the element levels were determined in advance. The SXFM provided various element maps, such as for phosphorus, sulfur, chlorine, calcium, iron, copper, zinc, and gold. The gold maps indicated the locations of mitochondria. Interestingly, zinc, and probably copper, seemed to be distributed in the mitochondria and nucleus. The obtained distributions of elements suggest that the SXFM system can visualize the distributions of multiple trace elements at the single-cell level. The detection limits were estimated from various observation results. Detection limits tend to increase with a decrease of atomic number because of the decreases of absorption and fluorescence yield. Typical detection limits of

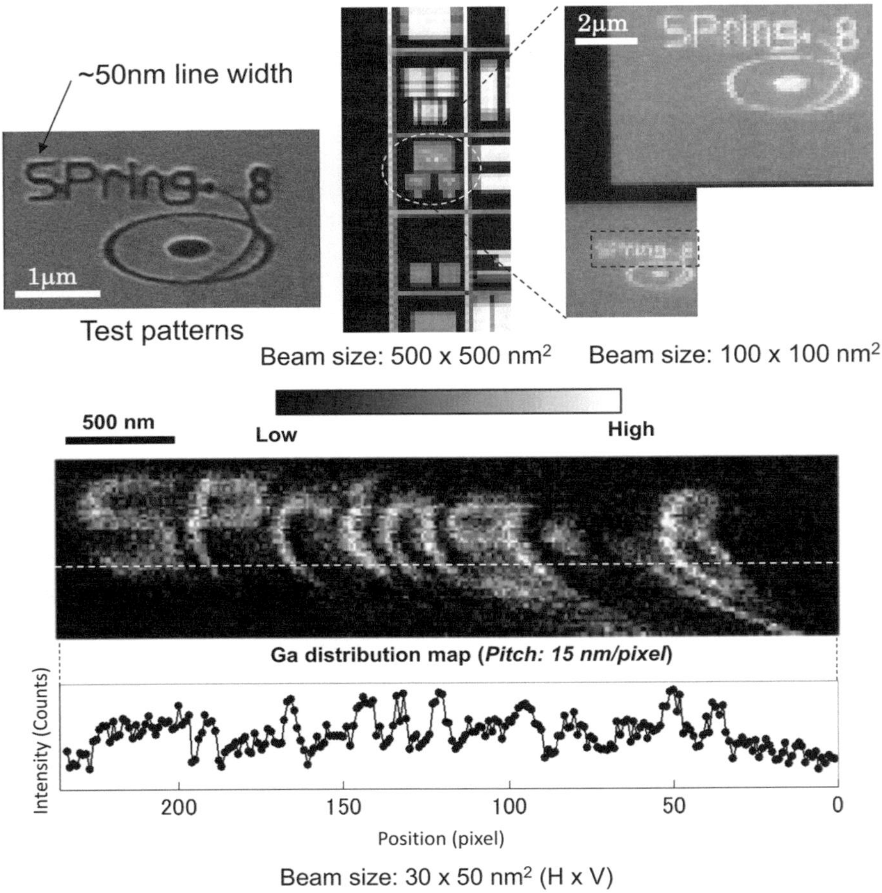

Fig. 3.7 High-resolution gallium maps of a test pattern. X-ray energy = 15 keV (The figure is taken from Matsuyama et al. 2006, with permission [12])

our SXFM are as follows, which depends on sample preparation, sample thickness, supporting membrane, and coexisting elements within a sample: Si (1 fg), P (0.4 fg), S (0.2 fg), Cl (0.3 fg), Ca (0.03 fg), Fe (0.01 fg), Cu (0.004 fg), Zn (0.005 fg), Ga (0.009 fg), W (0.01 fg), and Au (0.009 fg) [16]. The smallest spatial resolution was 30 nm, although there is a trade-off between spatial resolution and sensitivity. This is because the achievable minimum beam size can be obtained only when the X-ray source size is limited to 10 μm by a slit. In this case, beam intensity decreases with a decrease in the area of the source. Even current SR X-ray sources are not sufficiently small for such applications, although they are much better than an X-ray tube. Improvements of the sensitivity and spatial resolution are thus problems that remain to be resolved. An approach for achieving this is described in "Future SXFM."

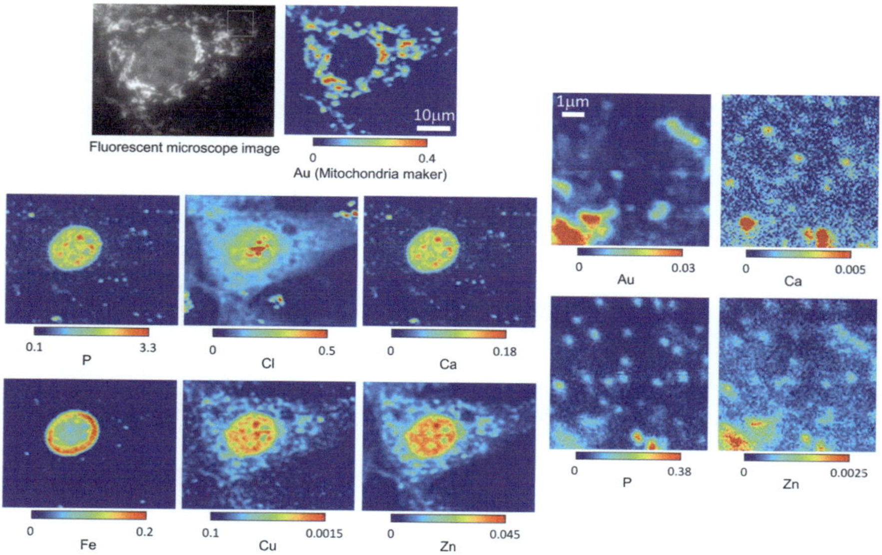

Fig. 3.8 Intracellular element maps of a whole cell (*left*) and the high-resolution element maps (*right*). Parameters of the *left* images: exposure = 1 s/pixel, scanning step = 500 nm/pixel, and X-ray energy = 15 keV. Parameters of the *right* images: exposure = 2 s/pixel, scanning step = 70 nm/pixel, and X-ray energy = 15 keV. *Color bars* indicate the elemental contents (fg) in an irradiation area (The figure is taken from Matsuyama et al. 2009, with permission [15])

3.3 Applications in Biology and Medicine

3.3.1 SXFM Is Now User-Friendly

To effectively apply SXFM to practical experiments, it was improved to allow biologists and medical scientists to use it easily. We developed a practical version of SXFM by adding a sample changer, an optical differential interference contrast (DIC) microscope (NA = 0.25, Mag = ×10, reflection type), and a GUI-based software (Fig. 3.9). The sample changer consisting of a motorized θ stage and a multiple-sample holder can hold 12 samples, so that users can continuously observe samples for a long time without interruption caused by sample introduction. This also contributes to stable observation because the temperature is disturbed by sample introduction. The DIC microscope was placed just downstream of the sample without disturbing the X-ray path. It enables users to observe cells while measuring the distributions of elements with the SXFM. The software can control the sample changer and the DIC microscope, as well as the X-ray focusing system, the detector system, and the scanning stage. Users can easily measure the area they desire by making the corresponding rectangle on the DIC microscopic image in the software. Only the desired cells can be selected and effectively observed. In addition, it is important to maintain the temperature approximately constant (within

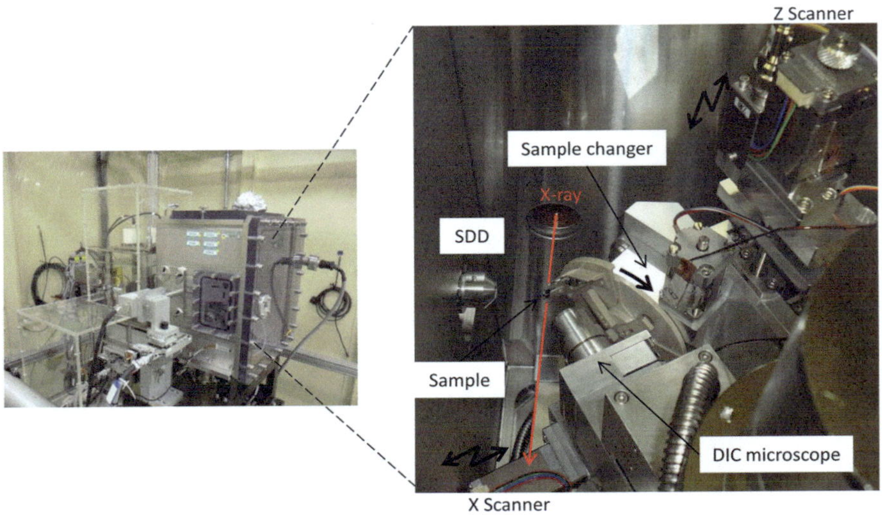

Fig. 3.9 Photograph of user-friendly SXFM. The whole system including the KB mirrors and the sample chamber (*left*) and inside the sample chamber (*right*)

~0.1 K) for user operation because beam broadening occurs upon a change in the temperature and readjustment of the focusing system is laborious. The room temperature control system consisting of platinum resistance temperature detectors and panel heaters was developed in collaboration with the optics group of the Japan Synchrotron Radiation Research Institute (JASRI). It is very suitable for KB mirror optics because it provides no vibration to the KB mirror system, which is sensitive to vibration. SXFM observation without adjustment of the focusing system could be performed over ~5 days.

3.3.2 Sample Preparation

The imaging of living cells without fixation and staining is the best way to observe their native state; however, living cells are not available for SXFM due to cell damage by the synchrotron X-rays via radicals. Fixation by washing and drying has often been applied for SXFM, but this alters the status of elements (Fig. 3.10). Free ions such as potassium and calcium were shown to be almost completely lost after fixation and washing, while zinc and copper were well maintained. These findings suggested that elements that bind strongly to cellular molecules are resistant to fixation and washing. These findings suggest the reliability of data for samples subjected to fixation and washing, as far as the data can be reproducibly obtained with appropriate controls, although we recommend freeze-drying or flash-freezing for SXFM (cryo-SXFM, Fig. 3.11), which are widely used to maintain the native

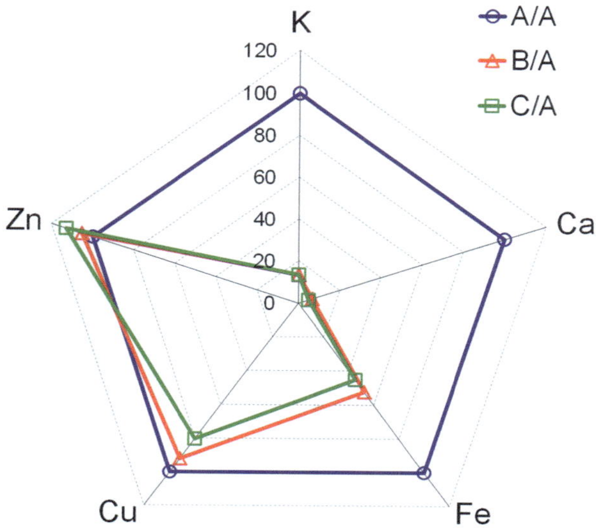

Fig. 3.10 Leakage of intracellular elements due to cell preparation. The intracellular potassium, calcium, and iron levels were markedly influenced by the cell preparation, whereas zinc and copper were unaffected. HeLa cells (10^7 cells) were centrifuged at 300 × g for 3 min (**Sample A**). The cells were then washed with phosphate-buffered saline (PBS) and fixed with 4 % paraformaldehyde in PBS for 10 min (**Sample B**). They were then permeabilized for 5 min with 0.2 % Triton X-100 in PBS and subsequently washed with PBS (**Sample C**). Average element levels (potassium, calcium, iron, copper, and zinc) of the samples A, B, and C were measured by inductively coupled plasma mass spectrometry (ICP-MS). The axis in the chart shows the ratios of B and C to A [18]

state in other forms of microscopy, such as electron microscopy, LA-ICP-MS, and NanoSIMS [17]. The fixation of clinical samples thus appears to be a practical approach.

3.3.3 Basement for Cells

Selecting a basement for cells is very important for imaging intracellular trace elements, since the basement influences the image background. We selected a commercially produced SiN basement. SiN has advantages in terms of its relative purity and thin layer (200 nm), which reduces elastic scattering. Its hydrophilicity also promotes cell adhesion and growth. In addition, the flatness of the membrane enables easy detection by DIC microscopy. On the other hand, SiN is associated with an intense silica peak, disturbing the detection of phosphorus and sulfur, which are essential for cellular function. We thus used a high-polymer membrane to avoid an intense Si peak as a background and applied carbon deposition to increase the hydrophilicity, resulting in sufficient cell growth (Fig. 3.12b). We call this

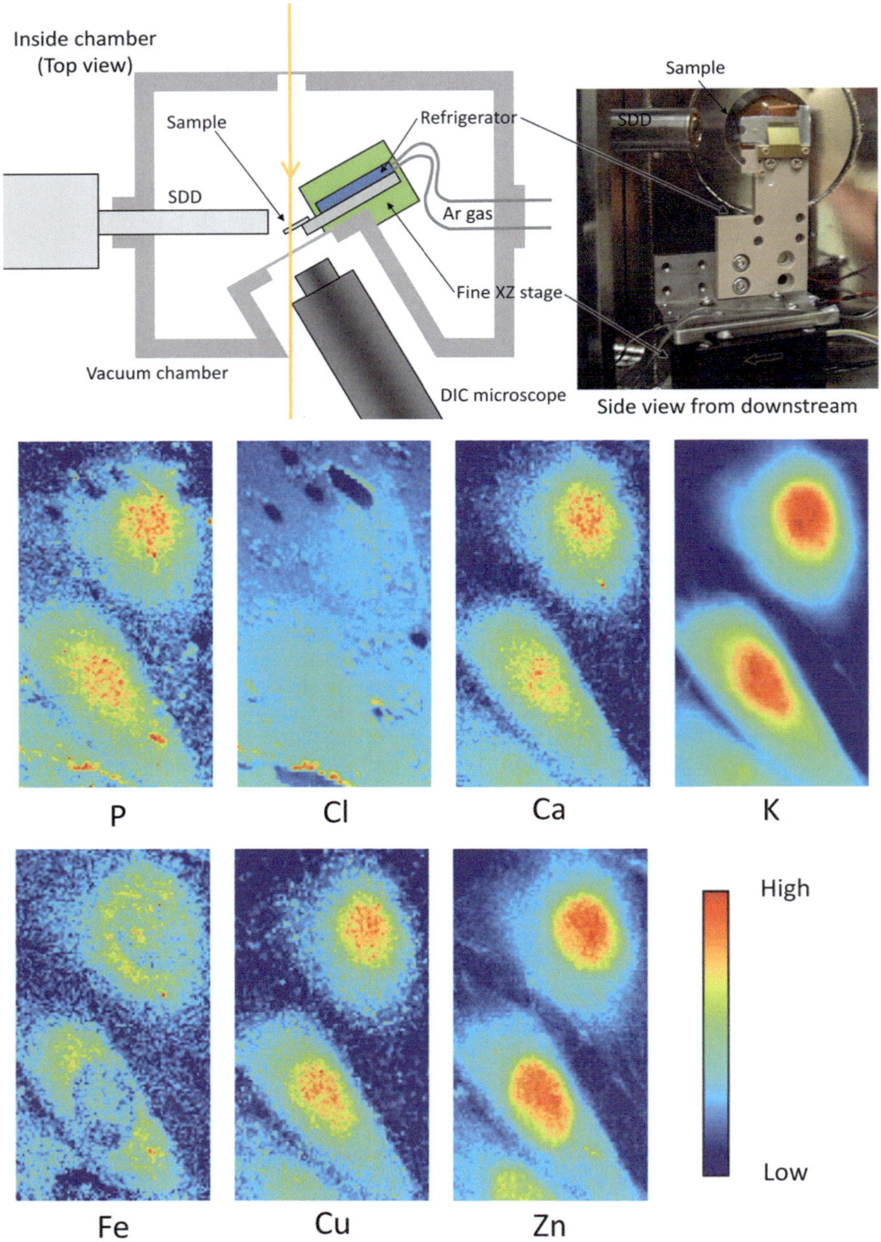

Fig. 3.11 Schematic and photograph of cryo-SXFM system (*upper*) and element mapping of frozen-hydrated cells (*lower*) (The figure is taken from Matsuyama et al. 2010, with permission [18])

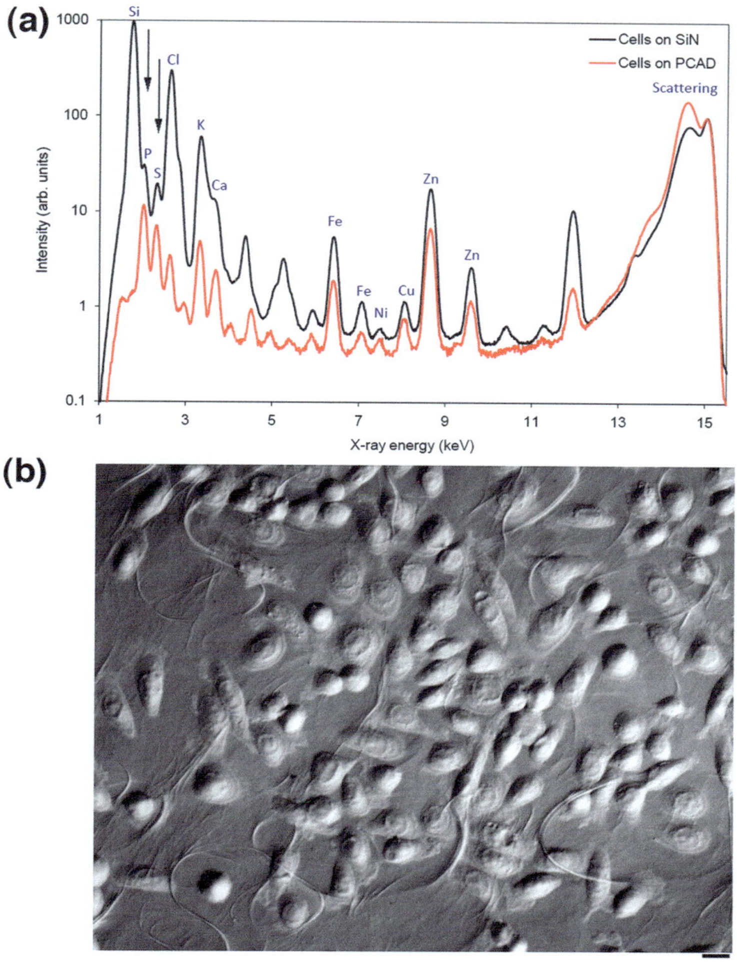

Fig. 3.12 PCAD basement. (**a**) Typical X-ray fluorescence spectra of cells on a SiN basement and polymer film with carbon deposition (PCAD) basement. *Arrows*, PCAD showed better peaks of phosphorus and sulfur than SiN. (**b**) Differential interference contrast (DIC) images of cells on a PCAD basement. Cells were well attached and growing. Scale bar, 10 μm (The figure is from Shimura and Matsuyama JSR 2010 via a copyright agreement)

membrane polymer film with carbon deposition (PCAD). PCAD allows the appearance of clear sulfur and phosphorus peaks and sufficient cellular adhesion (Fig. 3.12a).

3.3.4 Element Array

The system used for array analysis based on information on multiple elements from a single cell has been termed an element array [4]. For example, an element array can clarify the zinc dynamics of cells sensitive and resistant to anticancer drugs (Fig. 3.13). We also found that reduced glutathione was highly correlated with zinc regulation. Glutathione is a zinc-related protein that functions in excretion. We hypothesized that glutathione pumps out the drug in drug-resistant cells. In fact, a zinc-chelating agent combined with drug treatment for 5 consecutive days suppressed drug excretion and increased drug uptake and effects [4]. Array analysis thus suggested a new solution to drug resistance. High-resolution mapping images of the intracellular distribution are not always required for array analysis; however, array analysis is effective to evaluate the cell features with regard to the metallome and also in relation to cellular molecules.

3.3.5 A Cell Line Subjected to Sample Imaging at the Single-Cell Level

After a user-friendly version of SXFM was established, we could obtain more images at the single-cell level to reveal cellular function. Representative images obtained by SXFM are shown. Images of a 500-nm cryo-section of multiple myeloma (MM, blood cancer cell lines) were taken using the PCAD basement. The PCAD made it possible to image phosphorus clearly at the nucleus (Fig. 3.14), where there were some phosphorus-enriched regions. Zinc did not overlap with the phosphorus-enriched regions. In contrast, iron colocalized with the phosphorus-enriched regions, but showed a punctate distribution. These different localizations and patterns suggested that each element has different binding partners and roles. These elements may play roles in cancer cell proliferation. In further experiments, it may be possible to upregulate or suppress gene function to influence the localization of these elements. Information on the distribution of elements at the single-cell level has the potential to reveal novel nuclear functions.

3.3.6 Clinical Samples

Most clinical samples had already been fixed in formaldehyde solution at a pathology department. However, even fixed samples are informative if the data are reproducible and supported by control samples. We applied clinical samples to SXFM using cryo-sections (Fig. 3.15). Cryo-sections of bone marrow cells from MM cases and healthy volunteers were prepared by the Tokuyasu method, which is widely used for electron microscopy [19]. There was greater accumulation of

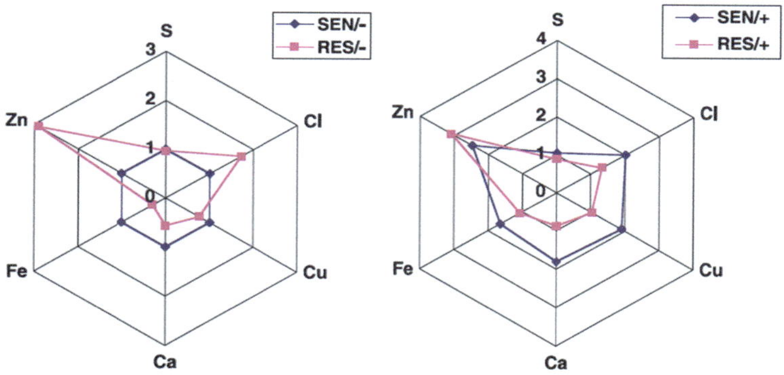

Fig. 3.13 Element array analysis based on mapping data from SXFM. The mean signal intensity of each element obtained by SXFM was calculated, and the fold increase of elements in drug-resistant (RES) cells is depicted using the mean signal intensity in drug-sensitive (SEN) cells as a standard (*left*). SEN and RES cells were treated with (+) or without (−) 1 μM cisplatin, a platinum-based anticancer drug. Notably, the fold increase of zinc in RES cells was greater than that in SEN cells without treatment (*left, red framed*). The fold increase of zinc in SEN cells was increased by cisplatin uptake, while the level in RES cells was stable (*right*). *SEN* cisplatin-sensitive cells, *RES* cisplatin-resistant cells, *S* sulfur signal, *Zn* zinc signal, *Fe* iron signal, *Ca* calcium signal, *Cu* copper signal, and *Cl* chlorine signal (The figure is taken from Shimura et al. 2005 [4])

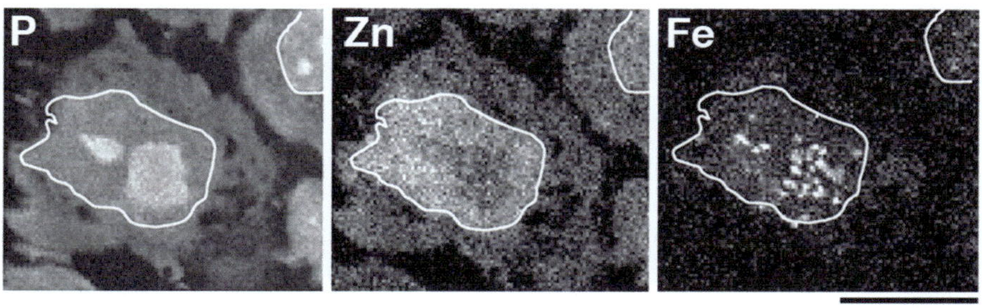

Fig. 3.14 Images of intracellular phosphorus, zinc, and iron mapping using a multiple myeloma (MM) cell line. Cryo-sections (500 nm thick) of cells were imaged with the following parameters: exposure = 1 s/pixel, scanning step = 250 nm/pixel, and X-ray energy = 15 keV. Bar, 10 μm. A *white* border indicates the nucleus (The figure is from Shimura and Matsuyama JSR 2010 via a copyright agreement)

copper and zinc in the MM cytoplasm than in that of healthy volunteers. There was also marked accumulation of nickel in MM bone marrow cells, compared with that in healthy volunteers, which we had not observed in the MM cell line (Fig. 3.14). Nickel was well detected in MM patient serum and was suspected of being a contaminant due to environmental pollution [20], although the mechanism by which it exerts carcinogenic effects is still unknown. Our imaging data suggested that nickel influences bone marrow cells. In fact, cancer was thought to emerge at bone marrow cells during the process of cell maturation [21]. Since nickel induced

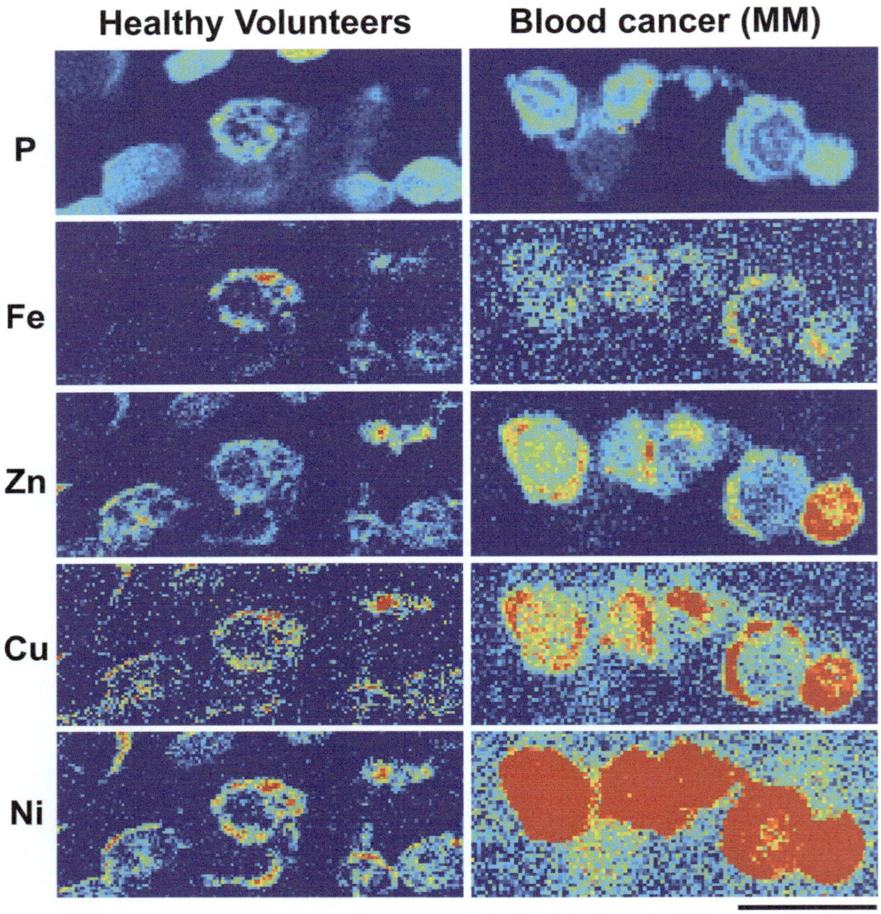

Fig. 3.15 Images of intracellular phosphorus, iron, zinc, copper, and nickel mapping in bone marrow cells derived from a clinical case of MM (*right*) and healthy volunteers (*left*). Cryo-sections (3 μm) of each sample were taken using the following parameters: exposure = 1 s/pixel, scanning step = 600 nm/pixel, and X-ray energy = 15 keV. Bar, 10 μm. Samples were obtained after approval from the Committee of the NCGM (#NCGM-G-00623-02), in accordance with the Declaration of Helsinki

DNA damage [22] (see also the metal targeting system in the next section), we cannot rule out the possibility of a relationship between nickel contamination and MM. Our data suggest that the clinical sample reflected the background of the donor, such as their living environment and diet, in the past and present, which is distinct from culture cells in the laboratory. Element imaging in clinical samples may thus shed light on the etiology and pathology and might contribute to diagnosis and prognosis in future medical applications.

3.3.7 Application in Metal Transport with Targeting

Previous findings suggested that the carcinogenic activity of Ni^{2+} depends on how this ion enters the cell and reaches the chromatin in the nucleus [23]. As the intracellular location of Ni^{2+} is related to cytotoxicity [24], genotoxicity [22], and the induction of oxidative stress [25], it would be desirable to characterize such effects in relation to intracellular metal localization. To understand the effects and cytotoxicity of metals in each part of the cell, it is necessary to have appropriate technology in terms of both the targeting system and the imaging for localization at the single-cell level. For this purpose, we applied our new metal transport system combined with SXFM.

3.3.8 Transporter

Our main goal for targeting metal transporters was to introduce metal ions (Ni^{2+}) at the selected cellular organelle without altering their natural activity. Such ions are delivered to the target organ and then released by a transporter (Fig. 3.16a). Note that the process of metal dissociation from a transporter differs from that for probes with a radionuclide semi-covalently bound to the chelators. In contrast, metals may be located diffusely or absorbed by membranes, for example, in the endoplasmic reticulum or mitochondria, or accumulate due to endocytosis by natural pathways (Fig. 3.16b). Two major conditions were taken into account for designing transporters. First, they should have a selective targeting domain. Ideal candidates have to show high specificity for the targeted organ, such as transactivator of transcription (TAT) (see the section on cell-penetrating peptides). Second, the metal-binding domain has to be constructed carefully, as metal-binding affinity is adjusted to hold the metal during transport, but not to chelate the metal completely so that it can be released at the cellular target. Metal dissociation may be controlled by different processes, for example, by changes in the redox potential at the organelle, different levels of intracellular ligands, which may compete with the transporter in metal binding, or transporter proteolysis, which potentially influences the metal-binding stability. Other issues also have to be taken into account, such as the cellular response or chemical kinetic features of metal ion binding. Stabilization of metal ion binding with various modifications in the structure of the three arms of 2,3-diaminopropionic acid (Dap)-branched peptides and the increased cell penetration and improved selectivity properties of branched "shuttles" are good candidates for such transporting platforms.

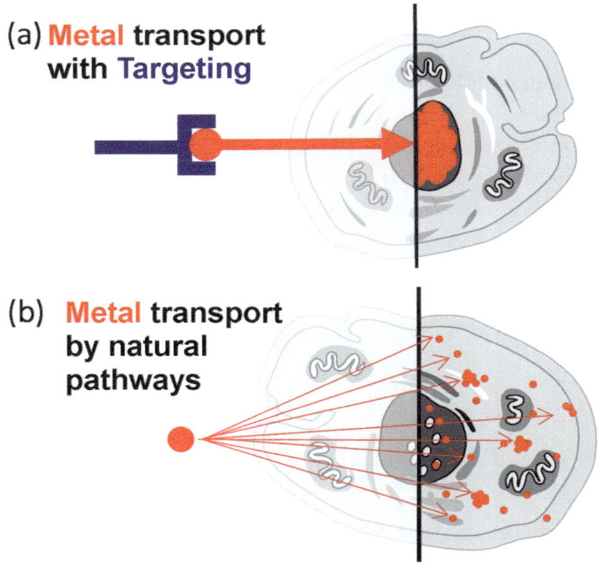

Fig. 3.16 Schematic presentation of metal transport with targeting (**a**) in contrast to metal transport via natural pathways (**b**)

3.3.9 Cell-Penetrating Peptides (CPPs)

The application of CPPs for targeting organelles is a well-known approach [26]. Such peptides are usually developed based on the natural sequences of proteins or synthetically developed sequences. They are applied to the transfer of nucleotides [27, 28], small molecules [26, 29], and tightly bound metal chelator radionuclides [30, 31] in both in vitro and in vivo systems. One of the most common applications of CPPs is as peptides from the TAT protein of the human immunodeficiency virus [32]. Recent developments in peptide branching have allowed us to increase their penetration properties [33], which enhance multiple effects associated with cell uptake [34]. Moreover, branching with Dap can stabilize metal ion-binding activity in monomeric complexes [35, 36]. It was shown that CPP branching increased the permeation [37] and molecular recognition of biomolecules [38]. The TAT fragments were shown to be compatible for branching to increase cell targeting. This potential to exhibit a targeting function when in branched structures has been reported in the literature. We chose the fragment TAT_{47-57} to design an artificial Ni^{2+}-branched peptide transporter by adding the Dap residue to the N-terminal. This allows further extension of peptide N-terminals with histidine donors to create a metal-binding junction. Finally, the H-HisDap (H-His)-Tyr^{47}-Gly-Arg-Lys-Lys-Arg-Arg-Gln-Arg-Arg-Arg-OH(A) peptide was synthesized (Fig. 3.17). An electrospray ionization MS study indicated that nickel is bound in an equimolar complex, and hot electron capture dissociation (HECD) MS fragmentation data indicate that nickel is bound in the designed fork (Fig. 3.17). To obtain information on metal-binding affinity, we performed

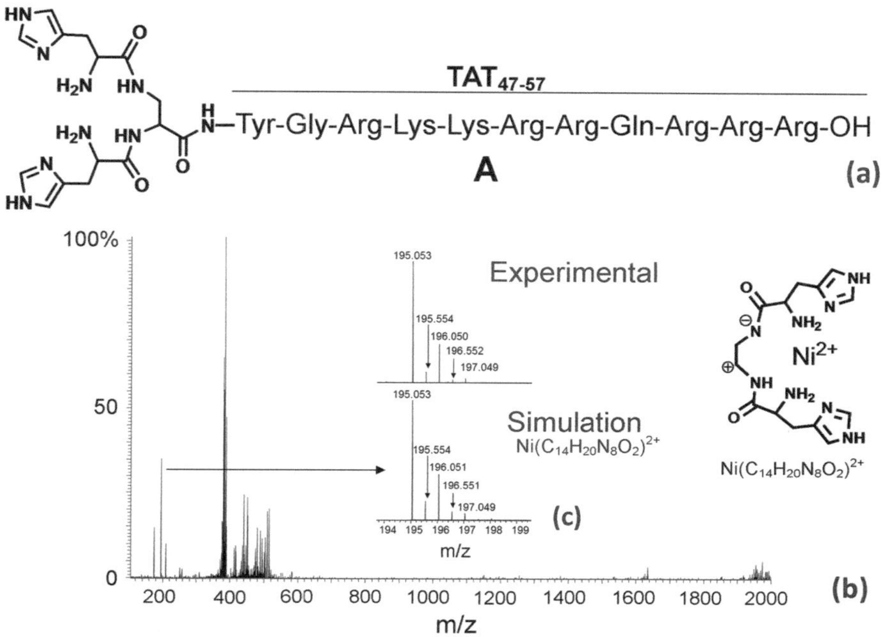

Fig. 3.17 The electrospray ionization dual MS (ESI-MS/MS) measurements were performed using an Orbitrap Velos mass spectrometer in positive ion full scan mode. The concentration and molecular ratio of apopeptide A were 6×10^{-5} M and 1 Ni:1 A, respectively. (**a**) Structure of the branched peptide. (**b**) The higher-energy collision dissociation spectra. (**c**) The signal corresponding to the $Ni(C_{14}H_{20}N_8O_2)^{2+}$ species. Data acquisition and treatment were performed using Xcalibur 2.1 software (The figure is taken from Szyrwiel et al. [5])

potentiometric experiments with the fragment responsible for metal ions binding H-His-Dap-(H-His)-Tyr-NH_2. Potentiometric studies have indicated that Ni^{2+} ions bind with $pNi^{2+} = 5.9$ at physiological pH (where $pNi^{2+} = -\log[Ni^{2+}_{\mathrm{frre}}]$)[5]. Our transport with a targeting system allowed Ni^{2+} transport into the nucleus, which was imaged by SXFM (Fig. 3.18, NiA) [5]. On the other hand, effective metal uptake above the background level was not seen in the cells treated with an equivalent number of moles of nickel ions. The level of nickel-induced DNA damage was ten times greater for NiA than for nickel ions (Fig. 3.19). These findings imply that our transport actually moved nickel ions into the nucleus and induced DNA damage [5].

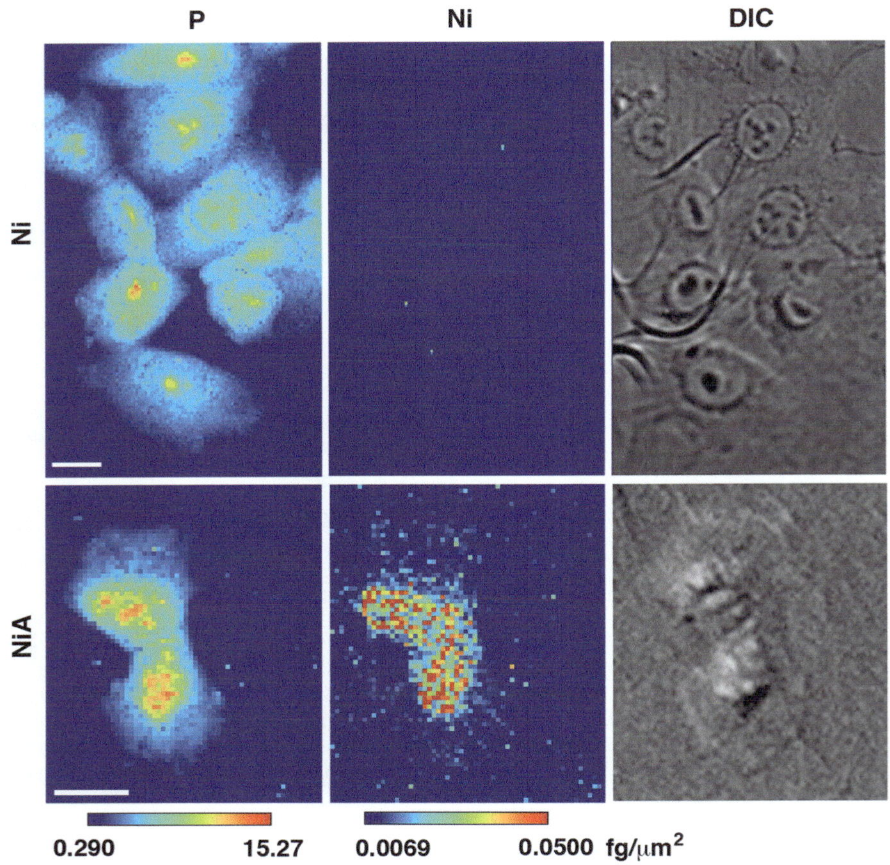

Fig. 3.18 Images of intracellular nickel mapping in cells treated with Ni or NiA. Cells were treated with Ni ion (*top*) or NiA (*bottom*) for 20 min. DIC, differential interference contrast images. A brighter color indicates higher signal intensity. Color bar, fg/μm^2; bar, 10 μm (The figure is taken from Szyrwiel et al. 2015 [5])

3.4 Future SXFM

3.4.1 Improvement of X-Ray Sources

Innovative X-ray sources suitable for X-ray nano-spectroscopy including SXFM are now strongly anticipated. One of them is the diffraction-limited storage ring (DLSR), using an ultralow-emittance storage ring with an undulator [39]. It is capable of generating a small and intense X-ray source leading to the formation of a very intense nanobeam with focusing optics. As mentioned above, the current SR X-ray source is much larger than that required to focus X-rays down to a few

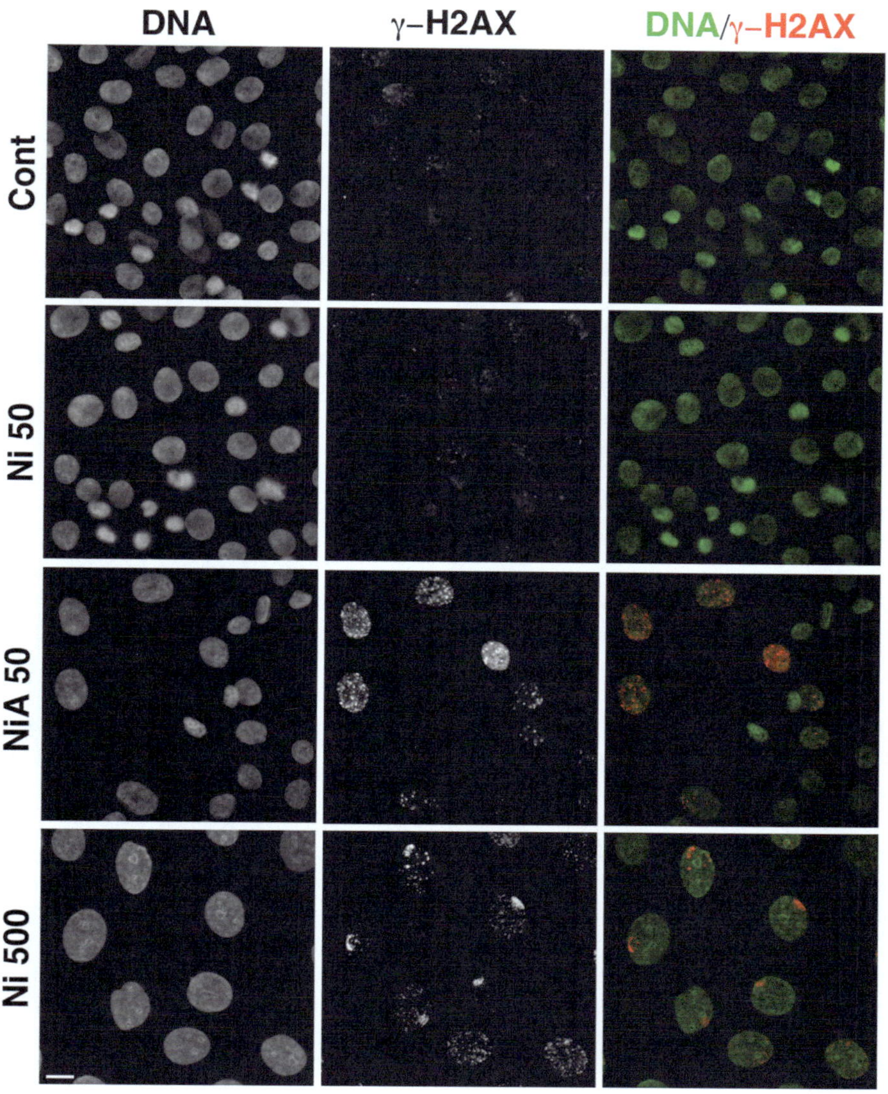

Fig. 3.19 DNA damage in NiA- and Ni ion-treated cells. DNA damage was estimated based on focus formation by immunostaining with an anti-γ-H2AX antibody. Note that NiA50 cells showed more significant focus formation (DNA damage) than Ni50 cells (*second* and *third panels*). Cont, untreated control cells; Ni50, cells treated with 50 μM Ni ions; NiA50, cells treated with 50 μM NiA for 48 h; and Ni500, cells treated with 500 μM Ni ions (The figure is taken from Szyrwiel et al. 2015 [5])

tens of nanometers without loss of the generated X-rays. Therefore, we have to reluctantly block X-rays with a slit near the light source. However, if DLSRs become available, it will be possible to collect almost all X-rays at the focus with existing focusing optics. A paper introducing the DLSR of SPring-8 [40] describes that very intense nanobeams with photon flux of ~10^{14} photons/s and beam size of 230 (H) nm × 120 (V) nm can be achieved with existing techniques. This is ~1000 times as many photons as our formed nanobeam. SXFM with both high resolution and high sensitivity will enable the detection of trace elements present at very low levels in a single mitochondrion within a short time. Fortunately, plans to construct DLSRs are being implemented all over the world, such as NSLS-II in the USA (started operation in 2015), MAX IV in Sweden, Sirius in Brazil, ESRF Upgrade in France, SPring-8 II in Japan, and APS Upgrade in the USA. The development of beamline optics, focusing optics, and X-ray microscopes has also been initiated, with the aim of realizing optics and systems specialized for the new X-ray sources.

3.4.2 High-Resolution SXFM

The current goal in developing scanning-type X-ray microscopes including SXFM is to achieve high resolution better than 10 nm. Possible techniques to achieve this are the multilayer Laue lens (MLL) [41] and the multilayer focusing mirror [42, 43] (Fig. 3.20a, b). An MLL, which is a kind of diffraction lens, can overcome the difficulty in fabricating diffraction lenses, in that the fabrication of very fine and thick patterns with a minimum period of less than 10 nm and device thickness of larger than a few microns is very challenging. An MLL fabricated with a magnetron sputtering deposition technique could focus 12-keV X-rays to an 11.2-nm (full width at half maximum, FWHM) spot with 15 % efficiency [41]. In addition, a multilayer focusing mirror can overcome the limitation that the grazing-incidence angle, proportional to the numerical aperture, of a total-reflection focusing mirror cannot be increased. Multilayer focusing mirrors can reflect and focus X-rays by using Bragg reflection, even when the grazing-incidence angle is large, leading to a very small focus size. FWHM of 7 nm has now been achieved using a multilayer KB system consisting of two multilayer focusing mirrors at X-ray energy of 20 keV (Fig. 3.21) [42, 43]. However, because there are no MLLs and multilayer mirrors to completely satisfy the requirements of spectroscopy users with regard to throughput, stability, and so on, few practical experiments have been implemented. Future improvements together with DLSR developments are anticipated. In the text below, other possible upgrades are briefly described. For more detail, please see the corresponding references, **deformable mirror** [44]. A deformable mirror (Fig. 3.22a) is a very powerful tool to change the optical system adaptively and moreover to directly compensate for wave front errors. For example, we can change the focal length depending on the experiment, the sample size, and the temporary apparatus used, such as a cooler or a heater. We can also produce an ideally focused beam even when wave front aberration is caused by beamline optics, such as a

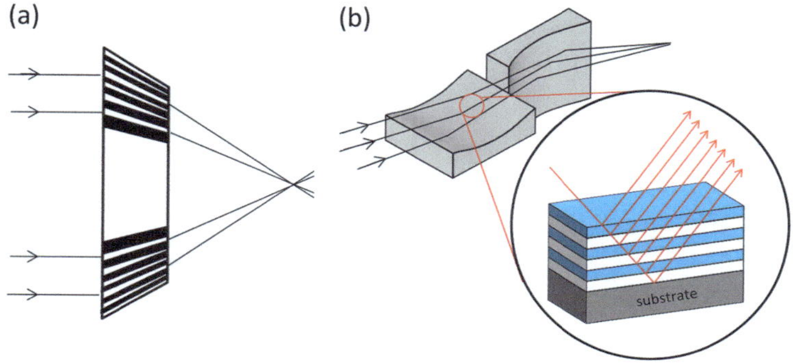

Fig. 3.20 High-resolution focusing optics. (**a**) Multilayer Laue lens. (**b**) Multilayer Kirkpatrick-Baez mirrors

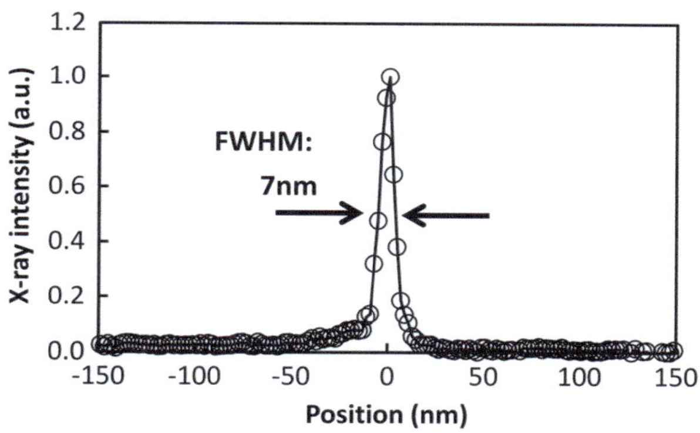

Fig. 3.21 X-ray focused beam with full width at half maximum of 7 nm (The figure is taken from Yamauchi et al. 2011, with permission [43])

monochromator, mirrors, or a vacuum window. Very precise adaptive mirrors and related techniques have been successfully developed, **ellipsoidal mirror** [45]. An ellipsoidal mirror (Fig. 3.22b) can produce a two-dimensional focusing beam with a single mirror, unlike KB mirror optics consisting of two mirrors. It has many advantages with regard to focusing efficiency, simple alignment, and stability of the focusing beam size and position because of the reduction in the number of mirrors. However, the fabrication process is very difficult because the shape involves a radius of curvature of a few millimeters in the short axis direction. Substantial efforts have been made to develop ultraprecise machining and measurement methods, **imaging mirror as focusing optics** [46]. SXFM usually uses focusing mirror optics such as KB mirrors, not imaging mirror optics such as advanced KB mirrors (Fig. 3.22c) and Wolter mirrors. Although KB mirror optics can focus X-rays, it cannot form clear images because it suffers from fatal comatic aberration leading to an unclear image, except in the central region. If imaging

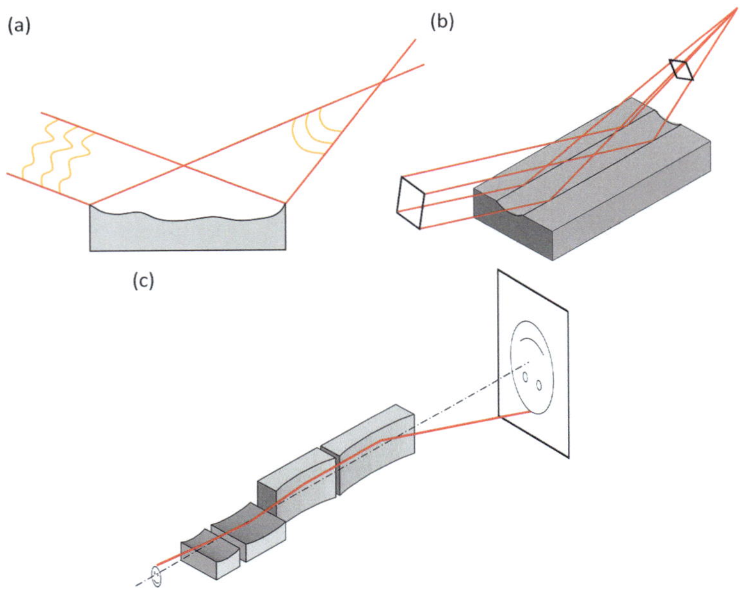

Fig. 3.22 Advanced focusing optics. (**a**) Deformable mirror. (**b**) Ellipsoidal mirror. (**c**) Advanced Kirkpatrick-Baez mirrors

mirror optics at the demagnification geometry is used, X-rays can invariably be focused even when the mirror inclines due to thermal drift or misalignment. This is because imaging mirrors can focus X-rays at off-axis conditions as well as at on-axis ones. Valuable beam time at the SR facilities can thus be saved by reducing the adjustment time for focusing optics of SXFM.

3.4.3 Possible Labeled Probes

Single-element labeling combined with SXFM may allow the imaging of nucleotides, proteins, secondary antibodies, and synthetic biomolecules, such as metabolites, transmitters, or organic drugs. In fact, SXFM successfully imaged a mitochondrion, an organelle with a diameter of less than 1 μm, using a metal-conjugated secondary antibody, as well as fluorescence imaging (Fig. 3.8). Data also suggested the possibility of linking SXFM to other -omics, such as metallomics, metabolomics, genomics, and proteomics (Fig. 3.23). Intracellular small molecules, such as fatty acids, sugars, and their derivatives, which are difficult to label for imaging due to their structural alterations, might be among the candidates for imaging using SXFM. If a single element is used to label these molecules at appropriate positions, SXFM has the potential to reveal their intracellular location during their metabolism. In Fig. 3.24a, elements suitable for the

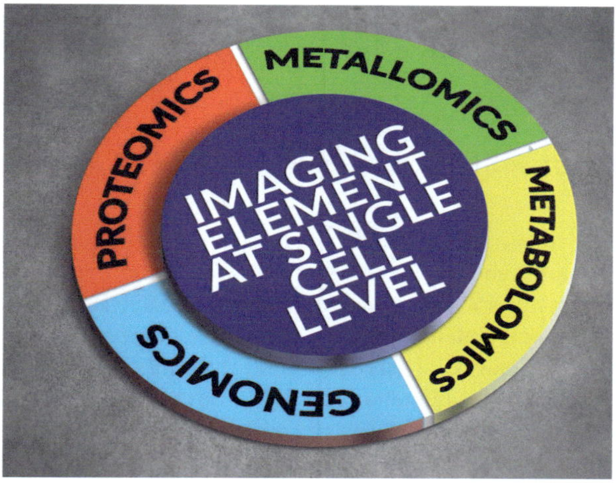

Fig. 3.23 The linkage between imaging elements (biomolecules) at the single-cell level and other -omics fields

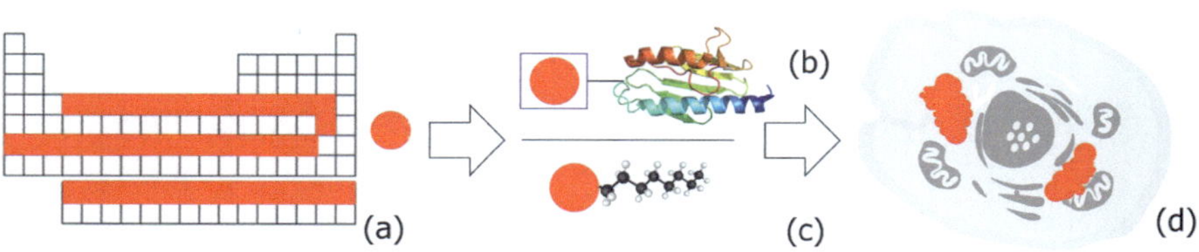

Fig. 3.24 Schematic presentation of biomolecules labeled with chemical elements for SXFM: (**a**) Chemical elements detectable by the current SXFM are indicated in *red*. (**b**) Labeling biomolecules with binding groups or nanoparticles attached to high-molecular-weight molecules. (**c**) Biomolecules directly labeled with elements for low-molecular-weight molecules. (**d**) Imaging of intracellular elements used to label biomolecules by SXFM

current SXFM are marked in red. It is notable that many elements are available for labeling molecules for detection by SXFM; however, major criteria for selecting elements for labeling depend on the actual cell status, as follows: (1) The signal intensity of the element label has to be higher than the cellular background level, for example, cells have a high level of zinc in the nucleus, so zinc labeling is not suitable for nuclear detection by SXFM. (2) The labeled probe with elements also has to exhibit thermodynamic and chemical stability to avoid intracellular exchange with cell components; consideration of the intracellular thiol compounds is particularly necessary. (3) Finally, it is also necessary for the labeled compound not to alter biological functions. Labeling biomolecules are mainly categorized as follows: (i) those with binding groups or nanoparticles attached to high-molecular-weight molecules (Fig. 3.24b) and (ii) biomolecules directly labeled with elements for low-molecular-weight molecules (Fig. 3.24c). If the element labels the complex appropriately, SXFM has the potential to reveal its intracellular location (Fig. 3.24d).

3.4.4 Application of Medical Probes to SXFM

Potential SXFM probes are suitable for other medical imaging methodologies, such as positron emission tomography, single-photon emission computed tomography, computer tomography, and magnetic resonance imaging. By introducing an element suitable for selected methodologies, such as gadolinium or bromine, into their structure, compounds can be observed within the entire body by medical imaging and within the cell by SXFM upon a biopsy. Moreover, cellular imaging of labeled drugs or radiopharmaceuticals at the micro-organelle level could reveal their uptake status and the processes affecting them in vitro, which would be interesting for elements such as gadolinium, bromine, technetium, and other reference isotopes for contrast media. This technology may open up new opportunities and challenges for the application of SXFM, improving our knowledge of cellular functions and disease processes, as well as helping in new drug development.

3.4.5 SXFM Application for Different Types of Microscopy

To understand cellular functions, a range of information, such as on cellular morphology, element distribution, protein distribution, and other cell biological and biochemical quantitative parameters, has to be collated. In particular, it is useful to combine information from multiple images collected from the same cell using various microscopes. As mentioned above, labeling proteins of interest with a metal or a fluorescent dye and observing them using a visible-light fluorescence microscope and SXFM is one possible approach (Fig. 3.8). It may be possible to link the obtained information with other -omics, for example, metallomics, metabolomics, genomics, and proteomics (Fig. 3.23). Moreover, the addition of electron microscopic data is very helpful to understand the relationship between ultrafine structures and the distributions of elements. X-ray absorption fine structure (XAFS) analysis in combination with a scanning technique, XAFS imaging, is also compatible with SXFM.

3.5 Collaboration on Platforms in Different Fields

Platforms based on many fields, such as biology, medicine, chemistry, engineering, and physics, are essential for innovative studies and the future development of element imaging at the single-cell level (Fig. 3.25). Element imaging at the single-

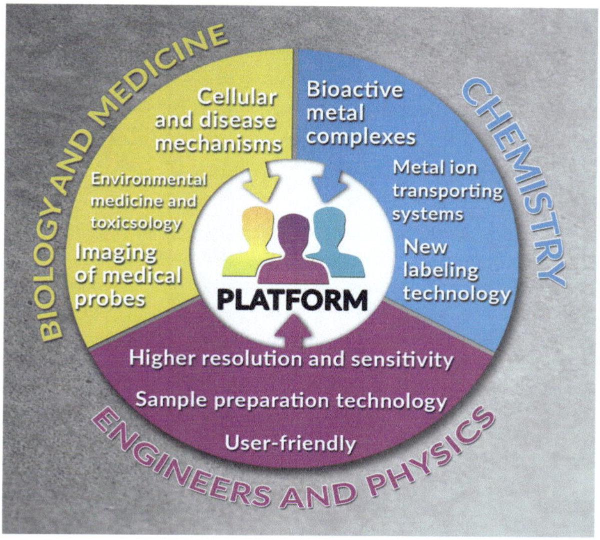

Fig. 3.25 A platform based on many fields, such as biology, medicine, chemistry, engineering, and physics, for innovative studies and future development of imaging at the single-cell level

cell level is beneficial for studies in biology and basic medicine by enabling the exploration of general cellular function and disease mechanisms. It is also important to apply such imaging to clinical samples that are expected to be affected by environmental pollution, diet, and toxicology, but for which it is difficult to draw definitive conclusions by conventional medical laboratory tests. Additionally, element imaging at the single-cell level can be applied to imaging drugs, contrast media, and radiopharmaceuticals labeled with elements. In the field of chemistry, a new technology of labeling molecules with elements should lead to the development of multiple new applications. Chemists may also propose new bioactive metal complexes and metal-ion-transporting systems. Moreover, collaboration among physicists, engineers, biologists, and chemists should enhance the development of other physicochemical methodologies that focus on intracellular elements, such as Q-dot [47], and should improve detection limits and sensitivity, including sample preparation technology, and be user-friendly. Collaboration among different fields may also lead to the imaging of other organisms such as viruses that are less than 100 nm in size at the electron microscopic level. While this may produce novel insights and products in various fields, integration of knowledge arising in diverse fields may initially be challenging.

3.6 Conclusion

The SXFM system described in this paper can reliably determine the cellular distribution of multiple elements using a sub-100-nm focusing approach. Labeling with a single element and imaging by SXFM should contribute to the understanding of interactions among intracellular elements (the metallome) and other biological molecules such as proteins, nucleic acids, lipids, and sugars (the other -omics). Future advanced mirror systems and more intense photon beams at synchrotron facilities should enable us to obtain higher-resolution images at higher speed than in current imaging. Visualizing intracellular elements at the single-cell level may provide great insights into the effects of such elements on cellular dynamics.

Acknowledgment We would like to thank Tetsuya Ishikawa at RIKEN for providing advice and encouragement during this study. We also acknowledge the help of Akihiro Matsunaga at the NCGM for the measurements and analyses of images, Shotaro Hagiwara at the NCGM hospital for clinical studies, and Yoshinori Nishino at Hokkaido University and Yoshiki Kohmura at RIKEN for assistance with the beamline adjustment. This study was supported by CREST from the Japan Science and Technology Agency (MS, SM, LS); MS was supported by a Grant-in-Aid for Research on Advanced Medical Technology, Ministry of Health, and Labor and Welfare of Japan; LS was supported by a Marie Curie Intra-European Fellowship from the European Union (PIEF-GA-2012-329969).

References

1. Crone B, Aschner M, Schwerdtle T et al (2015) Elemental bioimaging of Cisplatin in Caenorhabditis elegans by LA-ICP-MS. Metallomics 7(7):1189–1195
2. Godinho RM, Cabrita MT, Alves LC et al (2014) Imaging of intracellular metal partitioning in marine diatoms exposed to metal pollution: consequences to cellular toxicity and metal fate in the environment. Metallomics 6(9):1626–1631
3. Thompson CM, Seiter J, Chappell MA et al (2015) Synchrotron-based imaging of chromium and gamma-H2AX immunostaining in the duodenum following repeated exposure to Cr(VI) in drinking water. Toxicol Sci 143(1):16–25
4. Shimura M, Saito A, Matsuyama S et al (2005) Element array by scanning X-ray fluorescence microscopy after cis-diamminedichloro-platinum(II) treatment. Cancer Res 65(12):4998–5002
5. Szyrwiel L, Shimura M, Shirataki J et al (2015) A novel branched TAT(47–57) peptide for selective Ni(2+) introduction into the human fibrosarcoma cell nucleus. Metallomics 7 (7):1155–1162
6. Blackiston DJ, McLaughlin KA, Levin M (2009) Bioelectric controls of cell proliferation: ion channels, membrane voltage and the cell cycle. Cell Cycle 8(21):3527–3536
7. Zhang J, Wei J, He Q et al (2015) SKF95365 induces apoptosis and cell-cycle arrest by disturbing oncogenic Ca(2+) signaling in nasopharyngeal carcinoma cells. Onco Targets Ther 8:3123–3133
8. Jourdan E, Marie Jeanne R, Regine S et al (2004) Zinc-metallothionein genoprotective effect is independent of the glutathione depletion in HaCaT keratinocytes after solar light irradiation. J Cell Biochem 92(3):631–640
9. Klug A (2010) The discovery of zinc fingers and their development for practical applications in gene regulation and genome manipulation. Q Rev Biophys 43(1):1–21
10. Mantler M, Schreiner M (2000) X-ray fluorescence spectrometry in art and archaeology. X-Ray Spectrom 29(1):3–17
11. Tanaka T SPECTRA synchrotron radiation calculation code. SPring-8 Center, Hyogo, pp 679–5148. http://radiant.harima.riken.go.jp/spectra/index.html
12. Matsuyama S, Mimura H, Yumoto H et al (2006) Development of scanning x-ray fluorescence microscope with spatial resolution of 30nm using Kirkpatrick-Baez mirror optics. Rev Sci Instrum 77(10):103102

13. Yamauchi K, Mimura H, Inagaki K et al (2002) Figuring with subnanometer-level accuracy by numerically controlled elastic emission machining. Rev Sci Instrum 73(11):4028–4033
14. Matsuyama S, Mimura H, Yumoto H et al (2006) Development of mirror manipulator for hard-x-ray nanofocusing at sub-50-nm level. Rev Sci Instrum 77(9):093107
15. Matsuyama S, Shimura M, Mirnura H et al (2009) Trace element mapping of a single cell using a hard x-ray nanobeam focused by a Kirkpatrick-Baez mirror system. X-Ray Spectrom 38 (2):89–94
16. Matsuyama S, Mimura H, Katagishi K et al (2008) Trace element mapping using a high-resolution scanning X-ray fluorescence microscope equipped with a Kirkpatrick-Baez mirror system. Surf Interface Anal 40(6–7):1042–1045
17. Walther P, Studer D, McDonald K (2007) High pressure freezing tutorial. Microsc Microanal 13(S02):440–441
18. Matsuyama S, Shimura M, Fujii M et al (2010) Elemental mapping of frozen-hydrated cells with cryo-scanning X-ray fluorescence microscopy. X-Ray Spectrom 39(4):260–266
19. Griffiths G, SLOT JW, Webster P (2015) Kiyoteru Tokuyasu: a pioneer of cryo-ultramicrotomy. J Microsc 260(3):235–237
20. Egedahl R, Coppock E, Homik R (1991) Mortality experience at a hydrometallurgical nickel refinery in Fort Saskatchewan, Alberta between 1954 and 1984. Occup Med 41(1):29–33
21. Gentry SN, Jackson TL (2013) A mathematical model of cancer stem cell driven tumor initiation: implications of niche size and loss of homeostatic regulatory mechanisms. PLoS One 8:e71128
22. Cameron KS, Buchner V, Tchounwou PB (2011) Exploring the molecular mechanisms of nickel-induced genotoxicity and carcinogenicity: a literature review. Rev Environ Health 26 (2):81–92
23. Costa M, Klein CB (1999) Nickel carcinogenesis, mutation, epigenetics, or selection. Environ Health Perspect 107(9):A438
24. Salnikow K, Kasprzak KS (2007) In: Sigel A, Sigel H, Sigel RK (eds) Nickel and its surprising impact in nature: metal ions in life sciences, vol 2. Wiley, Chichester, pp 581–618
25. Das KK, Büchner V (2007) Effect of nickel exposure on peripheral tissues: role of oxidative stress in toxicity and possible protection by ascorbic acid. Rev Environ Health 22(2):157–173
26. Milletti F (2012) Cell-penetrating peptides: classes, origin, and current landscape. Drug Discov Today 17(15):850–860
27. Welser K, Campbell F, Kudsiova L et al (2012) Gene delivery using ternary lipopolyplexes incorporating branched cationic peptides: the role of peptide sequence and branching. Mol Pharm 10(1):127–141
28. Liu Y, Kim YJ, Ji M et al (2014) Enhancing gene delivery of adeno-associated viruses by cell-permeable peptides. Mol Ther Methods Clin Dev 1:1–12
29. Sakhrani NM, Padh H (2013) Organelle targeting: third level of drug targeting. Drug Des Dev Ther 7:585
30. Polyakov V, Sharma V, Dahlheimer JL et al (2000) Novel Tat-peptide chelates for direct transduction of technetium-99m and rhenium into human cells for imaging and radiotherapy. Bioconjug Chem 11(6):762–771
31. Bullok KE, Dyszlewski M, Prior JL et al (2002) Characterization of novel Histidine-tagged tat-peptide complexes dual-labeled with 99mTc-tricarbonyl and fluorescein for scintigraphy and fluorescence microscopy. Bioconjug Chem 13(6):1226–1237
32. Zhao M, Weissleder R (2004) Intracellular cargo delivery using tat peptide and derivatives. Med Res Rev 24(1):1–12
33. Eggimann GA, Blattes E, Buschor S et al (2014) Designed cell penetrating peptide dendrimers efficiently internalize cargo into cells. Chem Commun 50(55):7254–7257
34. Eggimann GA, Buschor S, Darbre T et al (2013) Convergent synthesis and cellular uptake of multivalent cell penetrating peptides derived from Tat, Antp, pVEC, TP10 and SAP. Org Biomol Chem 11(39):6717–6733

35. Szyrwiel Ł, Szczukowski Ł, Pap JS et al (2014) The Cu2+ binding properties of branched peptides based on l-2, 3-diaminopropionic acid. Inorg Chem 53(15):7951–7959
36. Szyrwiel Ł, Pap JS, Szczukowski Ł et al (2015) Branched peptide with three histidines for the promotion of Cu II binding in a wide pH range–complementary potentiometric, spectroscopic and electrochemical studies. RSC Adv 5(70):56922–56931
37. AbrreyáMonreal I (2015) Branched dimerization of Tat peptide improves permeability to HeLa and hippocampal neuronal cells. Chem Commun 51(25):5463–5466
38. Wynn JE, Santos WL (2015) HIV-1 drug discovery: targeting folded RNA structures with branched peptides. Org Biomol Chem 13(21):5848–5858
39. Eriksson M, van der Veen JF, Quitmann C (2014) Diffraction-limited storage rings-a window to the science of tomorrow. J Synchrotron Radiat 21(5):837–842
40. Yabashi M, Tono K, Mimura H et al (2014) Optics for coherent X-ray applications. J Synchrotron Radiat 21(5):976–985
41. Huang X, Yan H, Nazaretski E et al (2013) 11 nm hard X-ray focus from a large-aperture multilayer Laue lens. Sci Rep 3:3562
42. Mimura H, Handa S, Kimura T et al (2010) Breaking the 10 nm barrier in hard-X-ray focusing. Nat Phys 6(2):122–125
43. Yamauchi K, Mimura H, Kimura T et al (2011) Single-nanometer focusing of hard x-rays by Kirkpatrick–Baez mirrors. J Phys Condens Matter 23(39):394206
44. Goto T, Nakamori H, Kimura T et al (2015) Hard X-ray nanofocusing using adaptive focusing optics based on piezoelectric deformable mirrors. Rev Sci Instrum 86(4):043102
45. Hirokatsu Y, Takahisa K, Satoshi M, Kazuto Y, Haruhiko O (2016) Stitching interferometry for ellipsoidal x-ray mirrors. Rev Sci Instrum 87:051905
46. Matsuyama S, Kidani N, Mimura H et al (2012) Hard-X-ray imaging optics based on four aspherical mirrors with 50 nm resolution. Opt Express 20(9):10310–10319
47. Rubin M, Medeiros-Ribeiro G, O'shea J et al (1996) Imaging and spectroscopy of single InAs self-assembled quantum dots using ballistic electron emission microscopy. Phys Rev Lett 77 (26):5268

Chapter 4
Quantitative Elemental Bioimaging Protocol Using Femtosecond-Laser Ablation-ICP-Mass Spectrometry Coupled with Glass Standard Reference Material

Yoshiki Makino, Seiya Ohara, Marina Yamada, Sho Mukoyama, Kentaro Hattori, Shuhei Sakata, Yuki Tanaka, Toshihiro Suzuki, Atsuko Shinohara, Takehisa Matsukawa, Kazuhito Yokoyama, and Takafumi Hirata

Y. Makino • S. Ohara • M. Yamada • S. Mukoyama • K. Hattori • Y. Tanaka
Division of Earth and Planetary Sciences, Graduate School of Science, Kyoto University, Kitashirakawa, Sakyo-ku, Kyoto 606-8502, Japan

S. Sakata
Division of Earth and Planetary Sciences, Graduate School of Science, Kyoto University, Kitashirakawa, Sakyo-ku, Kyoto 606-8502, Japan

Present Address: Department of Chemistry, Gakushuin University, Mejiro 1-5-1, Toshima-ku, Tokyo 171-8588, Japan

T. Suzuki
Department of Earth and Planetary Sciences, Tokyo Institute of Technology, Ookayama 2-12-1, Meguro-ku, Tokyo 152-8551, Japan

A. Shinohara
Department of Epidemiology and Environmental Health, Faculty of Medicine, Juntendo University, Hongo 2-1-1, Bunkyo-ku, Tokyo 113-8421, Japan

Present Address: Research Institute for Cultural Studies, Seisen University, Higashi Gotanda 3-16-21, Shinagawa-ku, Tokyo 141-8642, Japan

T. Matsukawa • K. Yokoyama
Department of Epidemiology and Environmental Health, Faculty of Medicine, Juntendo University, Hongo 2-1-1, Bunkyo-ku, Tokyo 113-8421, Japan

T. Hirata (✉)
Division of Earth and Planetary Sciences, Graduate School of Science, Kyoto University, Kitashirakawa, Sakyo-ku, Kyoto 606-8502, Japan

Present Address: Geochemical Research Center, The University of Tokyo, Hongo 7-3-1, Tokyo 113-0033, Japan
e-mail: hrt1@eqchem.s.u-tokyo.ac.jp

Y. Ogra, T. Hirata (eds.), *Metallomics*, DOI 10.1007/978-4-431-56463-8_4

Abstract We have developed new analytical technique for the quantitative elemental imaging for trace elements using a femtosecond-laser ablation-ICP-mass spectrometry (fs-LA-ICP-MS). With the laser ablation under the finely controlled fluence, sliced alveolar tissue samples (thickness 1 μm), placed on the glass slide, were totally ablated, and the laser-induced sample aerosols were introduced into the ICP. Hence, no laser ablation was achieved on the glass substrate because the laser fluence was lower than the energy threshold for the laser ablation phenomena of glass materials. Under the total ablation conditions, the ablated volume of the sample could be well defined, and therefore, concentrations of the analytes can be calibrated based on the signal intensity data and the volume of the sample and glass standard reference material. Combination of the LA-ICP-MS technique and the preferential and total ablation achieved by the soft ablation protocol can become a powerful method for studying the trace elements and pharmacokinetics containing metal elements through the quantitative imaging. Analytical spatial resolution of the imaging analysis as well as the detection limit of the present analytical technique will be discussed in this study.

Keywords LA-ICP-MS • Elemental imaging • Quantitative imaging • Soft ablation • Calibration with non-matrix-matched standards • Preferential ablation • Trace elements • Alveolus tissue

4.1 Introduction

Inorganic nutrients play an important role in the various biochemical reactions. Both the excessive intake and deficiency of the inorganic nutrients can cause a declination of the biochemical reactions or a serious disease, and therefore, the evaluation of the nutritional status as well as dietary consumption and metabolism of inorganic nutrients are still key issues [1–3]. Imaging (elemental mapping) data for trace elements provides piercing information concerning the biological functions of the elements or the biological interaction with the organic molecular [4, 5]. Moreover, quantitative imaging data can be applied to understand the rate of circulation, excretion, or turnover time of the elements, and therefore, reliable quantification technique is highly desired for the further studies for the elemental metabolism.

Several analytical techniques have been applied to obtain imaging data for trace elements from biological samples. The X-ray fluorescence (XRF) and autoradiography have been widely used for the mapping analysis from biological tissue samples [6]. However, the sensitivity of the XRF technique was not high enough to detect the trace elements (e.g., <100 μg g^{-1}). Secondary ion mass spectrometry (SIMS) also provides both the high sensitivity and high spatial resolution for the element analysis [7, 8]. Despite the high spatial resolution (<5 μm) of the SIMS technique, the resultant abundance data for the elements could be erroneous because of the large contribution of the matrix effect. With the SIMS technique, preparation of matrix-matched standard and careful sample preparation procedures are highly desired to obtain reliable abundance values for the trace elements.

Recently, combination of laser ablation sampling technique and ICP-mass spectrometry (LA-ICP-MS) has been developed, and the resulting spatial resolution and elemental sensitivity are now comparable to those achieved by the SIMS and XRF techniques. The LA-ICP-MS technique is sensitive enough to determine the abundance of the trace elements [9–13]. One of the great advantages to use the LA-ICP-MS technique is that the sample is located under the atmospheric pressure, and neither evacuation of the sample housing nor coating with conductive materials is required, obviating both the loss and contamination of analytes through sample preparation procedures. Moreover, because of both the minimum preparation procedures and the high repeatability of measurement, the LA-ICP-MS technique becomes a fast and accurate method of choice for the quantitative imaging technique for the trace elements of biological samples [14–22].

For the quantitative imaging by LA-ICP-MS, calibration with an internal standardization correction has been widely employed. Hence, reliable and accurate abundance data for analytes depends on the choice of suitable standard reference materials. Because of the contribution of the matrix effect (non-mass spectrometric interferences) mainly due to difference in color or hardness between the standard and sample, the resulting abundance data could be erroneous when the non-matrix-matched standard was used. Meanwhile, quantitative imaging for several trace elements (Cu, Zn, Cd, Hg, and Pb) was achieved by the calibration using the matrix-matched standard [16, 17, 19], isotope dilution, or standard addition [5, 23]. However, these methods have risk of elemental contaminations of the standard materials originating from making the standard; in addition, preparing characteristic (color, hardness, density, etc.)-matched standard is difficult. Hence, another approach based on standardization using non-matrix-matched standard is worth testing because it allows us to prepare reliable calibration standard sample more easily and greatly reduces the cost about quantitative imaging of trace elements in biological samples.

To improve the accuracy of the abundance values calibrated using non-matrix-matched standard, correction of sample weight or volume becomes a critical problem. In general, the total amount of biological samples ablated would be quite difficult to estimate because the amount can change due to the heterogeneity in color or hardness of the samples. To overcome this, internal standardization using ^{13}C was widely used to calibrate the abundances for trace elements from biological samples [24, 25]. The main drawback of the internal standardization using the ^{13}C is that carbon would be present as at least two different phases or chemical forms: solid particulates (aerosols) or gaseous phase (CO_2) [26]. Some of the carbon from organic matter can be converted into the gaseous CO_2 through the laser-induced oxidation reactions at the ablation pit. Since the total energy required for evaporation, dissociation/atomization, and ionization of carbon would be different between the gaseous carbon (CO_2) and carbon present as particulates, the resulting signal intensity of carbon can vary with the ratio of gaseous carbon and particulate carbon. This suggests that the changes in the amount of sample ablated could not be accurately corrected by the internal standardization using the carbon.

Another possible source of the analytical error under the laser ablation with high fluence would be the changes in the signal intensities of analytes, mainly due to the laser ablation of glass substrate (slide glass) and the mass loading effect (Fig. 4.1a). Moreover, the total amount of sample ablated would vary with the color and hardness of the biological samples (Fig. 4.1b). This suggests that the sampling depth could be kept constant (1 μm) despite the differences in color or hardness of the sample, and therefore, reliable calibration procedure can be made based on the signal intensity data and comparison of the sampling volumes between standard and samples. With the laser ablation using the low fluence, the sliced sample (1 μm thickness) can be preferentially and totally ablated without any laser ablation of

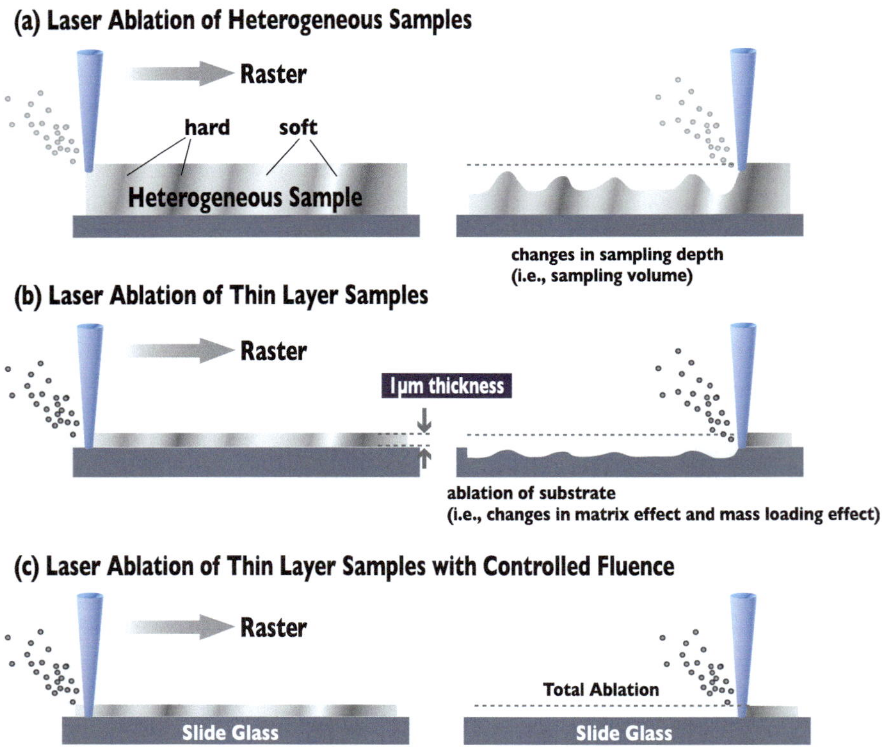

Fig. 4.1 Schematic diagram of sampling depth after the laser ablation. The resulting sampling depth can vary when the heterogeneous sample was ablated (**a**). With the conventional laser ablation, not only biological samples but also glass substrate was ablated (**b**). With the laser beam of reduced fluence (soft ablation protocol), only the sliced biological tissue samples could be preferentially ablated (**c**). This is very important to control the weight (volume) of sample ablated and also to minimize the contamination from glass substrate

substrate (glass slide). Hence, the ablation depth (i.e., volume) can be regarded as constant for the sliced sample despite the difference hardness or color within the sample, and the reliable correction for the ablated volume or weight could be made (Fig. 4.1c).

In this study, to obtain reliable abundance data for trace elements, we have developed new calibration and quantification protocol using the preferential and total ablation. It is potentially recognized that the energy threshold for laser ablation of biological samples is generally lower than those required for the glass materials, and therefore, only the biological samples can be ablated with the laser beam when the fluence was lower than the energy threshold for laser ablation of glass materials. Comparison of the energy threshold will be discussed in the experiment section. In addition, the femtosecond-laser ablation-ICP-mass spectrometry (fs-LA-ICP-MS) was applied to obtain the elemental mapping of trace elements in the biological sample for using convenient glass standard (non-matrix-matched standard). Using femtosecond laser, particle size distribution became smaller among materials; thus the matrix effect in ablation and ionization can be reduced between sample and glass standard reference material [27–33]. Among the elements, imaging analysis for Cu and Zn were made in this study. Main reasons to choose the elements are that both Cu and Zn were one of the most important trace inorganic nutrients, and also these were ubiquitously present in the most of tissues. Hence, evaluation of the reliability of the present quantitative imaging technique using the glass standard reference material was rigorously tested through a comparison with the previously reported values [34–36].

4.2 Experimental Protocols

4.2.1 Instrumentations

The laser ablation system used in this study was an in-house CyberProbe UV laser ablation system utilizing a titanium-sapphire (Ti:S) femtosecond laser, operating under the wavelength of 260 nm with pulse duration of 227 femtosecond [37, 38]. In this study, ArF Excimer laser ablation system (NWR193, ESI New Wave Research, Oregon, USA) was also used to measure the energy threshold for the laser ablation. The ICP-MS instrument used here was a double-focusing ICP-MS (AttoM, Nu Instruments, Wrexham, UK). Instrumentation, analytical conditions, and the procedures for data reductions were summarized in Table 4.1.

Table 4.1 Instrumentation and operational setting

1. ICP-mass spectrometer	
Instrument	AttoM (Nu Instruments, Wrexham, UK)
ICP RF incident power	1300 W
Plasma gas flow rate	13.0 L min^{-1}
Auxiliary gas	0.75 L min^{-1}
Mass resolution	300 (m/Δm)
He flow rate	600 mL min^{-1}
Ar makeup flow rate	900 mL min^{-1}
Scan mode	Electrostatic scan (E-scan)
Monitored isotope	^{63}Cu, ^{64}Zn
Date acquisition mode	Time-resolved analysis (TRA)
Dwell time	1 ms
No. of sweeps	150
Total integration time	100 ms per slice
Detector	Pulse counting
2. Laser ablation system	
Instrument	CyberProbe UV
Laser	Ti:S femtosecond laser
Pulse duration	227 fs
Wavelength	260 nm (THG)
Objective lens	f-theta lens (f = 100 mm)
Raster speed	2 μm s^{-1}
Repetition rate	20 Hz
Fluence	4 J cm^{-2} for tissue samples
	8 J cm^{-2} for glass standard (NIST SRM610)
Ablation pit size	18 μm
Stabilizer	Baffle type [45]
3. Standardization	
Analytical protocol	Preferential and total ablation
Calibration standard	Glass standard reference material (NIST SRM610)
Cu concentration	430.3 μg g^{-1} [44]
Zn concentration	456.3 μg g^{-1} [44]

4.2.2 *Sample Preparations*

All animal experiments conducted in this study were approved by the Institutional Animal Care and Use Committee at Juntendo University School of Medicine (No. 250142). Male ICR/crj mice were purchased at 5 weeks of age (Sankyo Labo Service Corporation, INC., Tokyo, Japan) and housed in a plastic cage at room temperature (22–24°C) with 12 h light and dark cycle. The mice were allowed free access to food (CA-1, CLEA Japan INC., Tokyo, Japan) and sterile tap water. After the 5 weeks of acclimation, the mice were sacrificed by cervical dislocation, and the thorax was opened. After the injection of 10 % buffered formalin (1 ml) into the lungs via tracheas, the lungs were removed and fixed with 10 % buffered formalin for 24 h or more. The formalin-fixed lungs were embedded in paraffin. Then, the thin sections of the lung specimens were sliced with an ultramicrotome (EM UC7, Leica Microsystems GmbH, Wetzlar, Germany) to a thickness of 1 μm and mounted on glass microscope slides using standard histopathological techniques. Conventional freeze-drying procedures were not applied in this study in order to maintain the original structure of the alveolus.

4.2.3 Experiment Procedures

The experiment parameters of LA-ICP-MS were optimized to obtain maximum intensity of ^{63}Cu and ^{64}Zn obtained by the laser ablation of NIST SRM610 glass standard reference material.

Two-dimensional mappings (imaging) for Cu and Zn were obtained by repeated line-profiling analysis across the sample with raster speed of 2 μm s^{-1}. Ablation pit size of 18 μm and the repetition rate of 20 Hz were employed to obtain imaging data. The distance between the lines was 20 μm. The resulting line-profile data were visualized using the in-house iQuant2 software, specially designed for elemental/isotopic mapping analysis.

Energy threshold for the laser ablation phenomena is dependent upon various physicochemical properties such as chemical composition, crystal structure, color, and hardness of the sample and also dependent upon laser type (wavelength and pulse duration). In the laser ablation using the ArF Excimer laser, the olivine phase can be ablated under the low fluence (<1 J cm^{-2}), whereas the fluence required for the laser ablation of both the glass materials (NIST SRM610) and metallic samples is significantly higher (>2 J cm^{-2}) (Fig. 4.2a). In contrast, with the laser ablation using the Ti:S femtosecond laser, the fluence required for the laser ablation of metallic samples (2 J cm^{-2}) was significantly lower than those required for olivine and glass materials (3 J cm^{-2} for olivine and 7 J cm^{-2} for glass materials) (Fig. 4.2b). Moreover, the fluence required for laser ablation of biological samples (tissue) was about 40 % lower than those for laser ablation of glass materials, indicating that the biological samples could be exclusively ablated with specific range of fluence.

The soft ablation protocol adopted in this study and the optimized fluence allowed for the preferential and total ablation of only the biological sample on the glass substrate (Fig. 4.1c). This suggests that the ablation depth could be kept constant (1 μm) despite the difference in color or hardness of the sample, and therefore, a reliable calibration procedure could be designed on the basis of the signal intensity data and the comparison of sampling volume between standard and sample. The sampling volume was calculated by the diameter and depth of the crater measured at the vertical edge of standard material by digital microscope.

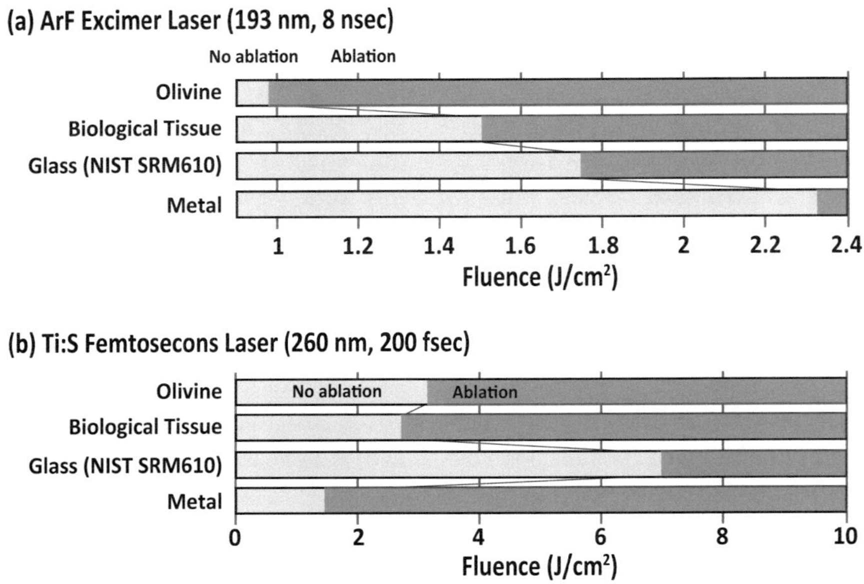

Fig. 4.2 Comparison of energy threshold for the laser ablation for ArF Excimer laser (**a**) and Ti:S femtosecond laser (**b**). Energy threshold for the laser ablation can vary significantly with sample composition and laser types. This was well demonstrated by the signal intensity of ^{57}Fe obtained from olivine, glass materials (NIST SRM610), and metallic Fe-Ni alloy (iron meteorite) under the various energy fluences and laser types: (**a**) ArF Excimer 193 nm laser and (**b**) Ti:S Femtosecond THG 260 nm laser

4.3 Results and Discussion

4.3.1 Calibration Strategy

With the preferential and total ablation protocol using the soft ablation technique, analyte concentration can be calibrated based on the signal intensity data, the volume, and the density of sample ablated using Eq. (4.1):

$$C_{sam} = C_{std} \times \frac{I_{sam}}{I_{std}} \times \frac{V_{std}}{V_{sam}} \times \frac{\rho_{std}}{\rho_{sam}} \tag{4.1}$$

where C_{sam} and C_{std} represent the concentration of analyte in sample and standard, respectively, and I_{sam} and I_{std} denote the signal intensities for sample and standard. The V_{sam} and V_{std} denote the ablated volume of the sample and the standard by the laser. The weight of sample and standard can be calculated by the volume and density (ρ_{std} and ρ_{sam}) for sample and standard material. Hence, the glass standard reference material (NIST SRM610: $\rho_{std} = 2.6$ g cm^{-3} [39]) was used as a calibration standard.

Density of the biological sample (ρ_{sam}) can vary significantly by reflecting the chemical composition, crystallinity, as well as the water contents of the tissue samples: 1.067 g cm^{-3} for muscle, 0.902 g cm^{-3} for fat, and 1.0506 g cm^{-3} for blood [40, 41]. Moreover, the cell density, as well as cell size and volume, is one of the fundamental indicators to evaluate the cell cycle progression, apoptosis, differentiation, disease state, and malignant transformation. Recent approach to measure the cell density using microchannel resonator revealed that the cell densities would be about 1.05–1.06 g cm^{-3}, suggestive of very limited deviation from those for overall tissue samples 1.05 g cm^{-3} [42]. Another important parameter to calibrate the abundance values is the density of the paraffin. As mentioned before, the sample was paraffinized prior to the slicing treatment. The density of paraffin ($\rho_{paraffin} = 0.9$ g cm^{-3} [43]) would be slightly lower than those for the tissue samples, suggesting that the overall density for the mixture of biological tissue and paraffin would be intermediate ($\rho_{paraffin} < \rho_{sam} < \rho_{tissue}$). Thus we take a practical assumption that the density of the sample (ρ_{sam}) is "1 g cm^{-3}." Potential error in the ρ_{sam} would be smaller than 10 %, and thus the uncertainties in the $\rho_{paraffin}$ values would be smaller than the analytical uncertainties achieved in this study.

The standard reference material used in this study was NIST SRM610 (National Institute of Standard and Technology, Maryland, USA). The concentration values (C_{std}) for Cu and Zn previously reported were 430.3 μg g^{-1} and 456.3 μg g^{-1}, respectively [44].

4.3.2 Ablated Volume

The volume of the sample ablated was calculated from the diameter of ablation pit (18 μm) and the thickness of the sample slice (1 μm), and the volume of standard ablated (V_{std}) was calculated from the diameter and measured depth of the ablation pit based on the assumption that cross section of the ablation pit after the line scanning had a triangle shape. Figure 4.3 illustrates the cross-sectional view of the ablation pits on NIST SRM610 obtained with various fluences. For the standard reference material

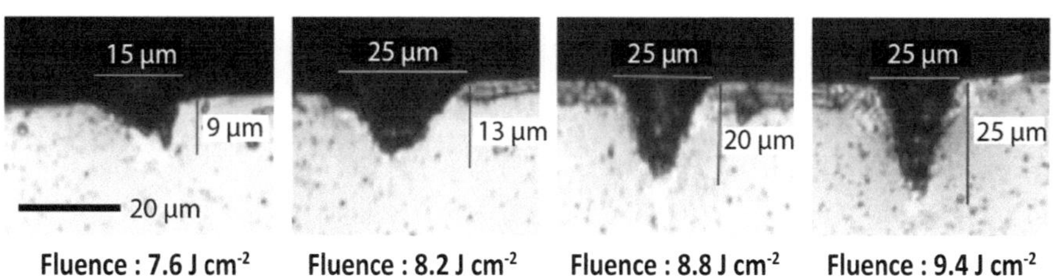

Fig. 4.3 Cross-sectional photograph for ablation crater of glass standard reference material (NIST SRM610) by femtosecond laser. Photographic image for the cross-sectional area of the ablation pits on glass standard. The volume of the standard ablated (Vstd) was calculated from the diameter and measured depth of the ablation pit based on the assumption that cross section of the ablation pit after the line scanning had a triangle shape

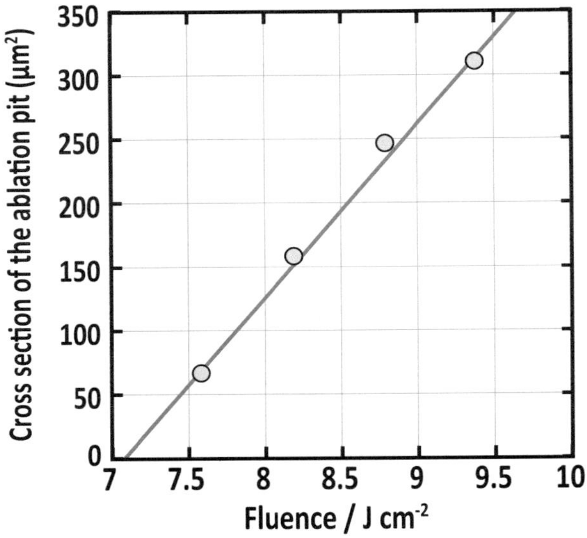

Fig. 4.4 The pit size dependence on fluence. The pit size was calculated from fluence. The volume of the sample ablated was also calculated from the diameter of ablation pit (18 µm) and the thickness of the sample slice (1 µm)

(NIST SRM610), the ablated volume of the sample was calculated based on the cross-sectional area of the ablation pits and the length of line scanning. Figure 4.3 also illustrates the dependence of the cross-sectional area of the ablation pits upon the energy fluence, which is demonstrative of linear correlation (Fig. 4.4). Once the fluence was defined, the volume of the standard ablated was calculated based on the cross-sectional area and the length of line scanning obtained here.

In the present technique, neither matrix-matched standard nor homogeneously spiked film (e.g., PMMA film) was required. More importantly, the traceability of the resultant abundance data could be maintained because the standard reference material (e.g., NIST SRM610) was employed for the calibrations.

4.3.3 Quantitative Imaging

For quantitative imaging, detection limits can be defined in Eq. (4.2):

$$DL_E = 3 \times \frac{\sqrt{\sigma_E}}{I_E} \tag{4.2}$$

where DL_E denotes the detection limit for the element E, σ_E is the intensity of background counts for element E, and I_E denotes the signal intensity (count per second) per unit concentration multiplied volume (e.g., μg g^{-1} μm^{-3}) as expressed as cps/(μg g^{-1} μm^{-3}). Because of high background counts due to memory effect on our ICP-MS instruments, the intensities of background counts for Cu (σ_{Cu}) and Zn (σ_{Zn}) were about 200 cps and 450 cps, respectively. Typical signal intensity per unit concentrations was 2000 cps/(μg g^{-1} μm^{-3}) for ^{63}Cu and 1800 cps/(μg g^{-1} μm^{-3}) for ^{64}Zn. Based on the signal intensity and background data, the detection limits for Cu and Zn achieved in this study were about 0.40 μg g^{-1} and 0.66 μg g^{-1}.

Microscopic image was shown in Fig. 4.5a (220 μm x 400 μm area). A capillary vessel surrounded by an alveolar wall could be observed. The resulting Cu and Zn mapping images for the mice alveolar tissue were given in Fig. 4.5c, d. With the soft ablation, only the biological tissue sample was preferentially and totally ablated, whereas no laser ablation was achieved on the glass slide (Fig. 4.5b), suggestive of very limited elemental contamination from the substrate. The abundance values for the Cu and Zn were calibrated using the NIST SRM610 based on the present calculation protocol defined in Eq. (4.1). For Cu, concentration varies from 10 to 70 μg g^{-1}, and Cu was basically distributed in the wall of the alveolus and the capillary vessel (Fig. 4.5c). For Zn, concentration range was about 5 μg g^{-1}, which was almost one order of magnitude lower than that for Cu. Our result is consistent with the concentrations of Cu and Zn in lung tissue reported by previous research [31–33]. Figure 4.5d revealed that the majority of Zn was distributed along the capillary vessel, demonstrating that the Zn was present in the blood. With the soft ablation protocol, elemental abundances can be calibrated.

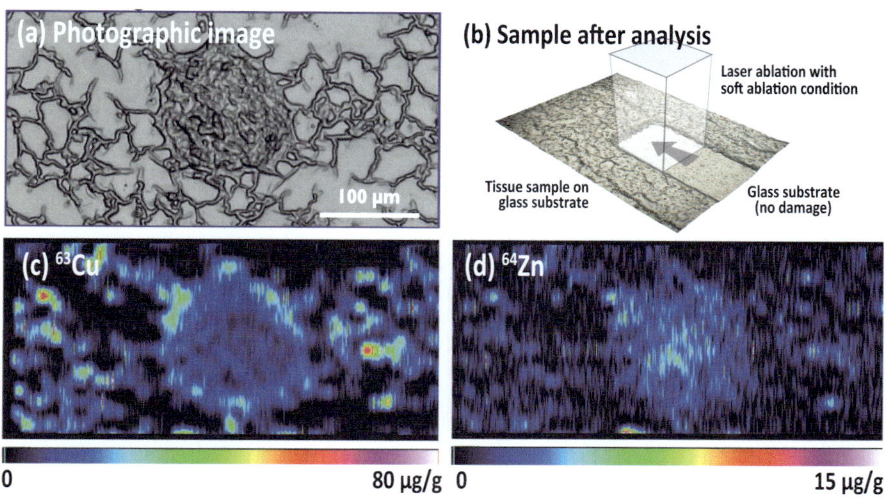

Fig. 4.5 Quantitative elemental images for Cu and Zn, obtained from the mouse alveolar tissue sample. Photographic image for the mouse alveolar tissue sample (**a**) and schematic diagram for preferential laser ablation of only biological samples (**b**). The resulting quantitative elemental images for Cu (**c**) and Zn (**d**) were shown

4.4 Conclusion

In this study, quantitative imaging analysis could be achieved for Cu and Zn in the biological tissue by means of laser ablation-ICP-mass spectrometry technique combined with preferential and total ablation protocol using femtosecond laser. Calibrations for concentration values were achieved by comparing both the signal intensities of the analytes between sample and standard and the ablated volume of sample and standard.

The variation in the volume of the biological tissue samples ablated was minimized by the total ablation of sliced tissue samples (1 μm thickness). Under the laser ablation with the controlled fluence, only the tissue samples were totally ablated, whereas no laser ablation was made for glass substrate. Concentration values for the analytes could be calibrated by the present calibration protocol, and therefore, metabolism or excretion rates for the trace elements could be estimated from the series of imaging data for the sample of various time sequences.

Another analytical feature achieved by present calibration protocol is that no biological or matrix-matched standard was required. Hence, glass standard reference material (NIST SRM610) was used to calibrate the concentration of analytes in samples. This is very promising to determine the concentration of analytes for which non-matrix-matched standards were available.

Acknowledgments We are grateful to express my gratitude to Mr. M. Tanaka, Mr. Y. Higashi, Mr. H. Obayashi (Kyoto Univ., Japan), and Dr. T. Shimamura (Kitasato Univ., Japan) for providing valuable suggestions and scientific insights concerning the biological activities of trace metals. This work was partly supported by a Grant-in-Aid for Scientific Research A26247094 for TH and C24590760 for AT from the Ministry of Education, Culture, Sports, Science and Technology, Japan.

References

1. Walczyk T, Blanckenburg FV (2005) Deciphering the iron isotope message of the human body. Int J Mass Spectrom 242:117–134
2. Bullen TD, Walczyk T (2009) Environmental and biomedical applications of natural metal stable isotope variations. Elements 5:381–385
3. Aramendia M, Rello L, Resano M, Vanhaecke F (2013) Isotopic analysis of Cu in serum samples for diagnosis of Wilson's disease: a pilot study. J Anal At Spectrom 28:675. doi:10.1039/c3ja30349g
4. Zoriy MV, Dehnhardt M, Reifenberger G, Zilles K, Becker JS (2006) Imaging of Cu, Zn, Pb and U in human brain tumor resections by laser ablation inductively coupled plasma mass spectrometry. Int J Mass Spectrom 257:27–33
5. Hare D, Austin C, Doble P (2012) Quantification strategies for elemental imaging of biological samples using laser ablation-inductively coupled plasma-mass spectrometry. Analyst 137:1527–1537
6. Takeda A, Tamano H, Enomoto S, Oku N (2001) Zinc-65 imaging of rat brain tumors. Cancer Res 61:5065–5069

7. Chandra S, Morrison GH (1995) Imaging ion and molecular transport at subcellular resolution by secondary ion mass spectrometry. Int J Mass Spectrom 143:161–176
8. McDonnell LA, Heeren RMA (2007) Imaging mass spectrometry. Mass Spectrom Rev 26:606–643
9. Günther D, Frischknecht R, Heinrich C, Kahlert HJ (1997) Capabilities of an argon fluoride 193 nm excimer laser for laser ablation inductively coupled plasma mass spectrometry microanalysis of geological mateirals. J Anal At Spectrom 12:939–944
10. Ghazi AM, Wataha JC, O'Dell NL, Singh BB, Simmons R, Shuttleworth S (2002) Quantitative concentration profiling of nickel in tissues around metal implants: a new biomedical application of laser ablation sector field ICP-MS. J Anal At Spectrom 17:1295–1299
11. Durrant SF (1999) Laser ablation inductively coupled plasma mass spectrometry: achievements, problems, prospects. J Anal At Spectrom 14:1385–1403
12. Russo RE, Mao X, Gonzalez JJ, Zorba V, Yoo J (2013) Laser ablation in analytical chemistry. Anal Chem 85:6162–6177
13. Limbeck A, Galler P, Bonta M, Bauer G, Nischkauer W, Vanhaecke F (2015) Recent advances in quantitative LA-ICP-MS analysis: challenges and solutions in the life sciences and environmental chemistry. Anal Bioanal Chem 407:6593–6617
14. Jackson B, Harper S, Smith L, Flinn J (2006) Elemental mapping and quantitative analysis of Cu, Zn, and Fe in rat brain sections by laser ablation ICP-MS. Anal Bioanal Chem 384:951–957
15. Lobinski R, Moulin C, Ortega R (2006) Imaging and speciation of trace elements in biological environment. Biochimie 88:1591–1604
16. Zoriy MV, Becker JS (2007) Imaging of elements in thin cross sections of human brain samples by LA-ICP-MS: a study on reproducibility. Int J Mass Spectrom 264:175–180
17. Becker JS, Zoriy M, Becker JS, Dobrowolska J, Matusch A (2007) Laser ablation inductively coupled plasma mass spectrometry (LA-ICP-MS) in elemental imaging of biological tissues and in proteomics. J Anal At Spectrom 22:736–744
18. Santos MC, Wagner M, Wu B, Scheider J, Oehlmann J, Cadore S, Becker JS (2009) Biomonitoring of metal contamination in a marine prosobranch snail (Nassarius reticulatus) by imaging laser ablation inductively coupled plasma mass spectrometry (LA-ICP-MS). Talanta 80:428–433
19. Becker JS, Breuer U, Hsieh HF, Osterholt T, Kumtabtim U, Wu B, Matusch A, Caruso JA, Qin Z (2010) Bioimaging of metals and biomolecules in mouse heart by laser ablation inductively coupled plasma mass spectrometry and secondary ion mass spectrometry. Anal Chem 82:9528–9533
20. Hare D, Austin C, Doble P, Arora M (2011) Elemental bio-imaging of trace elements in teeth using laser ablation-inductively coupled plasma-mass spectrometry. J Dent 39:397–403
21. Gordalíza EM, Giesen C, Lázaro A, Fernández DE, Humanes B, Cañas B, Panne U, Tejedor A, Jakubowski N, Gómez-Gómez MM (2011) Elemental bioimaging in kidney by LA-ICP-MS as a tool to study nephrotoxicity and renal protective strategies in cisplatin therapies. Anal Chem 83:7933–7940
22. Konz I, Fernandez B, Fernandez ML, Pereiro R, Sanz-Medel A (2011) Absolute quantification of human serum transferrin by species-specific isotope dilution laser ablation ICP-MS. Anal Chem 83:5353–5360
23. Sela H, Karpas Z, Zoriy M, Pickhardt C, Becker JS (2007) Biomonitoring of hair samples by laser ablation inductively coupled plasma mass spectrometry (LA-ICP-MS). Int J Mass Spectrom 261:199–207
24. Hoffmann E, Ludke C, Skole J, Stephanowitz H, Ullrich E, Colditz D (2000) Spatial determination of elements in green leaves of oak trees (Quercus robur) by laser ablation-ICP-MS. Fresenius J Anal Chem 367:579–585
25. Wu B, Zoriy M, Chen Y, Becker JS (2009) Imaging of nutrient elements in the leaves of Elsholtzia splendens by laser ablation inductively coupled plasma mass spectrometry (LA-ICP-MS). Talanta 78:132–137

26. Frick DA, Günther D (2012) Fundamental studies on the ablation behaviour of carbon in LA-ICP-MS with respect to the suitability as internal standard. J Anal At Spectrom 27:1294–1303
27. Eggins SM, Kinsley LPJ, Shelly JMG (1998) Deposition and element fractionation processes during atmospheric pressure laser sampling for analysis by ICP-MS. Appl Surf Sci 127–129:278–286
28. Günther D, Heinrich K (1999) Enhanced sensitivity in laser ablation-ICP mass spectrometry using helium-argon mixtures as aerosol carrier. J Anal At Spectrom 14:1363–1368
29. Guilong M, Günther D (2002) Effect of particle size distribution on ICP induced elemental fractionation in laser ablation inductively coupled plasma mass spectrometry. J Anal At Spectrom 17:831–837
30. Kuhn H, Günther D (2004) Laser ablation-ICP-MS: particle size dependent elemental composition studies on filter-collected and online measured aerosols from glass. J Anal At Spectrom 19:1158–1164
31. Koch J, Günther D (2006) Femtosecond laser ablation inductively coupled plasma mass spectrometry: achievements and remaining problems. Anal Bioanal Chem 387:149–153
32. Pisonero J, Günther D (2008) Femtosecond laser ablation inductively coupled plasma mass spectrometry: fundamentals and capabilities for depth profiling analysis. Mass Spectrom Rev 27:609–623
33. Flamigni L, Koch J, Günther D (2014) The effect of carrier gas humidity on the vaporization of laser-produced aerosols in inductively coupled plasmas. J Anal At Spectrom 29:280–286
34. Morita A, Kimura M, Itokawa Y (1994) The effect of aging on the mineral status of female mice. Biol Trace Elem Res 42:165–177
35. Chen Z, Meng H, Yuan H, Xing G, Chen C, Zhao F, Wang Y, Zhang C, Zhao Y (2007) Identification of target organs of copper nanoparticles with ICP-MS technique. J Radioanal Nucl Chem 272:599–603
36. Fujihara J, Tongu M, Hashimoto H, Yamada T, Kimura-Kataoka K, Yasuda T, Fujita Y, Takeshita H (2015) Distribution and toxicity evaluation of ZnO dispersion nanoparticles in single intravenously exposed mice. J Med Invest 62:45–50
37. Hirata T, Kon Y (2008) Evaluation of the analytical capability of NIR femtosecond laser ablation-inductively coupled plasma mass spectrometry. Anal Sci 24:345–353
38. Yokoyama TD, Suzuki T, Kon Y, Hirata T (2011) Determinations of REE abundance and U-Pb age of zircons using multispot laser ablation-ICP-mass spectrometry. Anal Chem 83:8892–8899
39. Kuhn HR, Günther D (2006) A quantification strategy in laser ablation ICP-MS based on the transported aerosol particle volume determined by optical particle size measurement. J Anal At Spectrom 21:1209–1213
40. Huang HK, Wu SC (1976) The evaluation of mass densities of the human body in vivo from CT scans. Comput Biol Med 6:337–343
41. Trundnowski RJ, Rico RC (1974) Specific gravity of blood and plasma at 4 and 37 °C. Clin Chem 20:615–616
42. Bryan AK, Hechet VC, Shen W, Payer K, Grover WH, Manalis SR (2014) Measuring single cell mass, volume, and density with dual suspended microchannel resonator. Lab Chip 14:569–576
43. Seyer WF, Patterson RF, Keays JL (1944) The density and transition points of the n-paraffin hydrocarbons. J Am Chem Soc 66:179–182
44. Pearce NJG, Perkins WT, Westgate JA, Gorton MP, Jackson SE, Neal CR, Chenery SP (1997) A compilation of new and published major and trace element data for NIST SRM610 and NIST SRM612 glass reference materials. Geostand Newslett 21:115–144
45. Tunheng A, Hirata T (2004) Development of signal smoothing device for precise elemental analysis using laser ablation-ICP-mass spectrometry. J Anal At Spectrom 19:932–934

Chapter 5
Single Cell Analysis by Using ICP-MS

Shin-ichi Miyashita, Shin-ichiro Fujii, Kaori Shigeta, and Kazumi Inagaki

Abstract Single cell analysis by using inductively coupled plasma mass spectrometry (ICP-MS) has been attracting attention for investigating a cell-to-cell variation of content of elements and biomolecules. ICP-MS has a unique feature that can be applied to a highly sensitive elemental analysis for evaluating a cell-to-cell variance or cellular uptake of elements, drugs, and nanoparticles and a multiparametric analysis for characterizing a cell population based on multiparameters, like a flow cytometry (so-called mass cytometry). In addition, subcellular localization analysis including multiparametric imaging can be performed with the coupling of a laser ablation system. In this chapter, we describe an overview of single cell analysis by using ICP-MS and its future prospects, focusing on cytometric analysis and imaging analysis.

Keywords Single cells • ICP-MS • Mass cytometry • Time-resolved measurement • Imaging mass spectrometry • Laser ablation

5.1 Introduction

In the field of metallomics, inductively coupled plasma mass spectrometry (ICP-MS) has been an essential technique for detecting and quantifying trace elements and biomolecules [1–3]. ICP-MS has an excellent detection capability for many elements; the detection limits are ng L^{-1} to sub-ng L^{-1} levels for almost the elements in the periodic table, and the dynamic range in a measurement is 5–9 orders of magnitudes with a good measurement stability and reproducibility [4, 5]. In addition, an ICP-MS instrument can easily be coupled with flow

S.-i. Miyashita • S.-i. Fujii • K. Inagaki (✉)
National Metrology Institute of Japan (NMIJ), National Institute of Advanced Industrial Science and Technology (AIST), Tsukuba Central 3-9, Umezono1-1-1, Tsukuba 305-8563, Japan
e-mail: k-inagaki@aist.go.jp

K. Shigeta
Environmental Management Research Institute (EMRI), National Institute of Advanced Industrial Science and Technology (AIST), Tsukuba West, Onogawa16-1, Tsukuba, Ibaraki 305-8569, Japan

Y. Ogra, T. Hirata (eds.), *Metallomics*, DOI 10.1007/978-4-431-56463-8_5

separation device such as gas and liquid chromatography and capillary electrophoresis device and also can be coupled with a solid sampling device such as a laser ablation system [6, 7]. The applications of these coupling systems have been spreading for the detection and quantification of not only trace elements and metal-containing biomolecules but also amino acids, peptides, proteins, and nucleic acids by measuring heteroatoms such as phosphorus and sulfur or by tracking element tags such as metal-labeled antigens and nanoparticles. Now single cell analysis by using ICP-MS has been attracting attention for investigating a cell-to-cell variation of content of elements and biomolecules [8–12].

Single cell analysis targeting individual cells or cellular components (organelles, molecules, and ions) has been attracting attention in various research fields [13, 14]. Biological cells are heterogeneous even though they proliferate from a same parent cell with a same cultivate condition. The heterogeneity will give differences of not only visible phenotype but also content of cellular components including messenger RNA, proteins, metabolites, and ions. Single cell analysis can be applied for investigating such cell-to-cell variation in contrast to conventional bulk analysis (which provides average information from a large cell population) and thus would be expected to contribute, for example, to elucidating basic phenomena such as cell development, adaptation, and differentiation in the biological field and to understanding antibiotic resistances and functions of immune systems in the medical and pharmaceutical fields.

Various single cell analysis methods have been developed, and most of them are based on fluorometry or mass spectrometry [10, 13–15]. Fluorometric methods such as fluorescence microscopy and flow cytometry (FC) are noninvasive and nondestructive methods and are superior in specificity and sensitivity for detecting target molecules if molecule-specific fluorescence labels can be used. For example, fluorescence microscopy has been applied to in situ observation of biomolecules in living cells. Flow cytometer has been used to characterize the properties of each individual cell in a large population of cells and to sort cells based on properties of interest. However, these methods have some technical problems to be solved for applying multiparametric analysis, i.e., limitation on the number of labels used simultaneously due to spectral overlapping, quenching of fluorescence, and autofluorescence from cells themselves. On the other hand, mass spectrometric methods are destructive methods and are superior for profiling and imaging cells by measuring biological molecules and elements. Various types of mass spectrometry, such as electrospray ionization (ESI)-MS, laser desorption ionization (LDI)-MS, matrix-assisted laser desorption ionization (MALDI)-MS, secondary ion MS (SIMS), and ICP-MS, have been applied for profiling and imaging analysis of cells. Among them, ICP-MS has a unique feature that can be applied to multiparametric analysis of elements and biomolecules, like a flow cytometry (often called mass cytometry). In mass cytometry, much number of labels can be used simultaneously, compared with FC, and has no quenching and autofluorescence problems in measurement [10]. In addition, subcellular localization analysis including multiparametric imaging can be performed with the coupling of a laser ablation system. Thus, ICP-MS has been attracting attention for profiling and imaging analysis of cells, although the number of applications of ICP-MS to single cell analysis is fewer than those of other types of mass spectrometry.

In this chapter, we describe an overview of single cell analysis by using ICP-MS and its future prospects, focusing on cytometric analysis and imaging analysis.

5.2 Overview of Cytometric Analysis by Using ICP-MS

Cytometric analysis by using ICP-MS is based on a time-resolved measurement, along with direct introduction of cells into the plasma [16–20]. Time-resolved measurement is a technique for measuring very short transient signals or spike signals with a high time resolution. The time resolution in corresponds to a measurement time at one data point, including not only acquisition time (i.e., dwell time on mass-over-charge ratio *m*/*z*) but also a settling time for *m*/*z* or duty cycle time (the details are described later).

The schematic image of cytometric analysis by time-resolved ICP-MS is shown in Fig. 5.1 [18, 19]. In the cytometric analysis, cells are introduced into the plasma by nebulizing or dispensing a suspension of cells. Cells are readily decomposed, and ion plumes corresponding to individual cells are generated during passing through the plasma. The ion plumes generated can be detected as very short transient signals or spike signals (here after referred to cell events or event signals) without overlapping each other, if the time resolution is high enough for detecting the ion plume and the introduction interval of individual cells is longer than the duration of the ion plume generation.

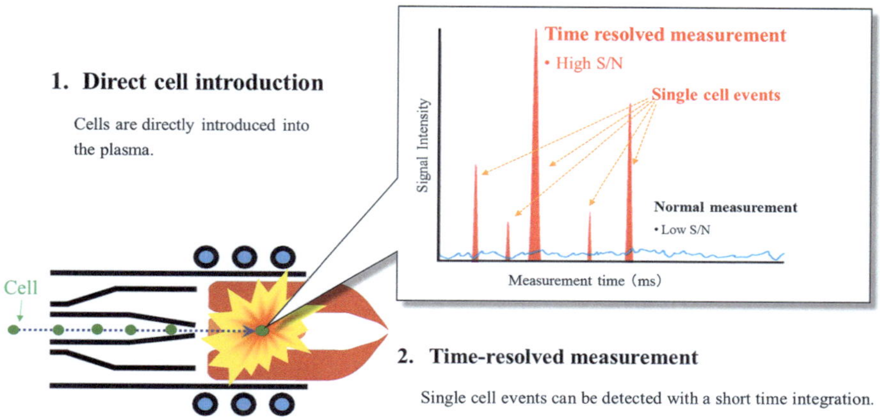

Fig. 5.1 Schematic image of cytometric analysis by time-resolved ICP-MS (With permission from Anal Sci)

The durations of the ion plume generation corresponding to cells having a diameter of 1–10 μm are typical in the range of 100–500 μs [16, 21] and depend on the size of cells, the central gas flow rate (i.e., the sum of the carrier gas and other support gas flow), and the on-axis sampling position in the plasma. The central gas flow rate controls the residence time of cells inside the plasma for completely decomposing the cells to atoms. The on-axis sampling position should be set at the position where the cell is completely atomized and then ionized and the ions less diffused inside the plasma, since ions in a small region of the plasma can only be sampled via the sampling interface.

The intensity of cell events can be directly related to the amount of elements in individual cells. The amount of elements natively contained in a single cell is small, but the concentrations of some elements are enough for detecting by ICP-MS. Therefore, the ions of these elements in a single cell can be measured by the time-resolved measurement, because the signals corresponding to the ions are not averaged with background signals. The amounts of biomolecules such as amino acids, peptides, proteins, and nucleic acids can also be measured by the time-resolved measurement with element tagging techniques, if the tag contains much amounts of elements for detecting, like nanoparticles [16, 17].

The sensitivity of the measurement strongly depends on the time resolution of the measurement [19–21]. Figure 5.2 shows an example of time-resolved ICP-MS profiles of phosphorous in yeast cells (*Saccharomyces cerevisiae*) with different

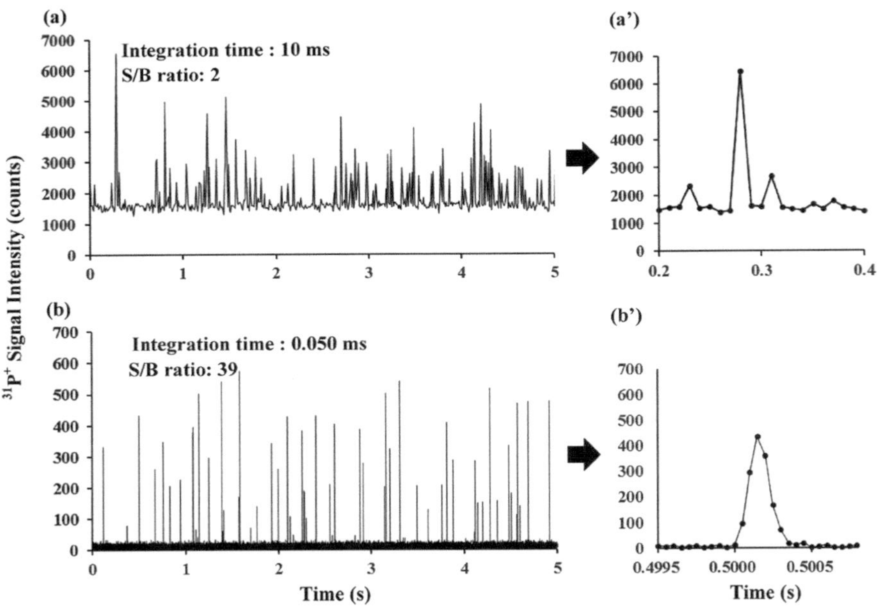

Fig. 5.2 Example of time-resolved ICP-MS profiles of phosphorous in yeast cells (*S. cerevisiae*) with different time resolution settings (**a**) $\Delta t = 10$ ms, (**b**) $\Delta t = 0.05$ ms (With permission from Anal Sci)

time resolution settings (Δt) from 0.05 and 10 ms [19]. On the time resolution of 10 ms, which is longer than the ion plume generation in the plasma, the signals corresponding to cell events are measured as spike signals, averaged with the background, resulting in low signal-to-background (S/B) ratio. In contrast, on the time resolution of 0.05 ms, which is much shorter than the duration of a cell being in the plasma, cell events are measured as transient signals. The S/B ratio of cell events is improved by increasing the time resolution due to minimizing the averaging of the event signals with the background signals. The maximum S/B ratio would be obtained at a time resolution less than 0.01 ms.

The cell throughput in measurement also depends on the time resolution. The cell throughput can be increased by increasing the time resolution, but the maximum cell throughput is limited by the duration of ion plume generation. This is because cell events can be individually measured when the interval of the cell introduction into the plasma is longer than the time resolution. Olesik et al. calculated the probabilities of detecting signals from one particle or more than one within an acquisition time, when the duration of ion plume generation is ca. 400 μs [21]. The probability of detecting signals from more than one particle can be reduced by increasing the time resolution (decreasing the acquisition time), but the number of particles entering the plasma must be limited to reduce the probability of the overlapping of the signals to acceptably low. The theoretical limitation of cell throughput is ca. 500–2000 cells per second when the duration is 100–500 μs [21], although it would vary with the size of cells and the operating conditions of ICP-MS, which can change the duration of ion plume generation.

5.3 ICP Mass Spectrometers for Cytometric Analysis

For cytometric analysis, the following three types of ICP-MS instruments have been applied: quadrupole ICP-MS (ICP-QMS) [18–20, 22–26], sector field ICP-MS (ICP-SFMS), [27–29] and time-of-flight ICP-MS (ICP-TOFMS) [10–12, 16–18, 30].

Among them, only ICP-TOFMS can be applied for the multiparametric analysis of single cells, because it can quasi-simultaneously measure a wide range of *m*/*z* in time-resolved measurement due to its fast scanning speed. One drawback of ICP-TOFMS is a low sensitivity in measurement. The sensitivity is one or two orders of magnitude lower than that of ICP-QMS or ICP-SFMS, respectively. This is mainly due to the duty cycle of TOF mass analyzer, which is usually defined as the fraction of extracted ions that make it into the mass analyzer. In the ICP-TOFMS measurement, an ion packet is extracted from the ions focused by ion optics in several microseconds and then acquired by sweeping a wide range of *m*/*z* around 30 μs. Therefore, a substantial amount of ions per m/z at the respective acquisition time is lower than in the case of ICP-QMS and ICP-SFMS.

Tanner et al. developed a specialized ICP-TOFMS system for multiparametric analysis of single cells in 2008 [16, 17] (Fig. 5.3). The ICP-TOFMS system was designed for utilizing multiplex antibodies tagged with enriched lanthanide

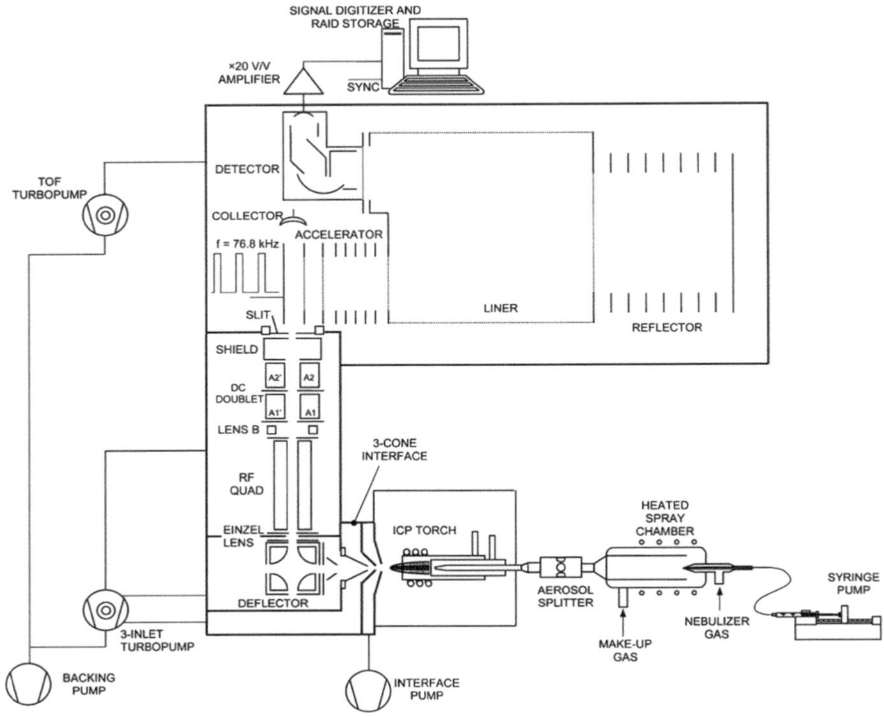

Fig. 5.3 Schematic diagram of CyTOF, which is a specially customized ICP-TOFMS for utilizing multiplex antibodies tagged with enriched lanthanide isotopes (Adapted from Bandura et al. [17] with permission from Anal Chem)

isotopes and has a customized sample introduction system for improving a cell introduction efficiency into the plasma. The system was launched as "CyTOF" by DVS Science, and third-generation instrument "Helios," which has the improved sensitivity of measurement and extended mass range, has been launched by Fluidigm in 2015.

ICP-QMS and ICP-SFMS can measure only one *m*/*z* at one time in the time-resolved measurement, because settling and hopping times for each *m*/*z* in measurement are over several hundred microseconds, which are almost the same as the duration of ion plume generation. However, their sensitivities are superior to that of ICP-TOFMS.

For cytometric analysis by using ICP-QMS or ICP-SFMS, a suitable time resolution setting is around 100 μs, considering the sensitivity and the cell throughput in measurement. Some ICP-QMS and ICP-SFMS instruments still suffer from limitation on their settable time resolution (typically 1–10 ms) owing to the control unit and operating software, although new instruments launched around 2015 have a sufficient time resolution of 100 μs. For improving the time resolution in time-resolved ICP-MS measurement, some external current reading systems have been developed for reading pulse or analog signals directly from a secondary electron multiplier (SEM) [20, 31–33]. Gschwind et al. [31] and Iwai et al. [32] used a digital storage oscilloscope for directly reading analog signals from a detector in order to acquire shorter dwell times (below 10 μs). Miyashita et al. [20] used a pulse counting unit being connected to a SEM via a function generator, for reading pulse signals with 50-μs dwell time. Strenge and Engelhard [33] built a data acquisition unit for reading pulse signals from a SEM with 5-μs dwell time. The reading process with the abovementioned systems has a shorter and negligible reading dead time (below 10 ns), but it still suffers from the detector dead time when highly intense cell events are measured.

Before applying such external current reading system for time-resolved measurement, the adequacy of parameter settings for data acquisition should be checked because in some instruments, a settling time of a mass filter and gate closing time of a detector between acquisitions of a data point might be preset, resulting in overlooking cell events. In such cases, both parameters of the settling time and the gate closing time should be made invalid.

Absolute detection limits (ADLs) of time-resolved ICP-QMS and ICP-SFMS for elements are roughly sub-femtogram (fg), when the time resolution is set below 100 μs. In the case of an ICP-QMS instrument, about four orders of magnitude number of ions are lost in the process of ion sampling and passing through a mass filter, and thus hundreds to a few thousands ions can be detected if 1 fg of a target element is contained in a cell introduced into the plasma (e.g., in the case of uranium, ca. 250 atoms finally detected from the starting 2,500,000 atoms). ADLs in ICP-TOFMS are one or two orders of magnitude worse than those in ICP-QMS, due to lower sensitivity.

5.4 Cell Introduction Devices for Cytometric Analysis

An efficient cell introduction into the plasma is desirable for reducing sample consumption in cytometric analysis. Conventional sample introduction system of an ICP-MS instrument, consisting of a nebulizer and spray chamber, has poor cell introduction efficiency into the plasma (i.e., less than 10 %) [34]. This is because aerosols generated with conventional nebulizers are coarse, thereby around 90 % of the aerosols were cut off by using a spray chamber before introducing into the plasma. Therefore, some special sample introduction systems have been developed for improving the cell introduction efficiency.

The CyTOF has a special sample introduction system consisting of a concentric nebulizer and a small-volume on-axis spray chamber with a heater, as shown in Fig. 5.3 [17]. A cell suspension is introduced into the nebulizer at a flow rate of ca. 50 μL min^{-1}, and aerosols are dried inside the spray chamber heated to a temperature of ca. 200 °C. The cell introduction efficiency reported was 30–40 %, but it can vary depending on cell sizes.

Groombridge et al. [18] developed a high efficiency cell introduction system (HECIS) consisting of a modified triple-tube concentric nebulizer, called high performance concentric nebulizer (HPCN), and a custom-made small-volume on-axis cylinder chamber with a sheath gas flow, as is shown in Fig. 5.4. The HPCN was originally designed for total consumption sample introduction in ICP-MS, and its unique structure makes a flow-focusing effect at the tip of the nozzle, by which a fine aerosol generation and a high tolerance for salts can be realized at less than 10 μL min^{-1} liquid flow rate [35, 36]. For accepting wide range of different-sized cells (from 2.0 to 6.4 μm in mean diameter), they modified the HPCN to have a large-bore center capillary tube (LB-HPCN) and changed a design of the spray chamber with longer total length and shorter inner tube length for reducing an impact loss of cells to walls inside the spray chamber [20]. By using LB-HPCN at 10 μL min^{-1} liquid flow rate, over 90 % (volume per volume) of aerosols are below 10 μm diameter in size, which is critical size of an aerosol for complete decomposition in the plasma; thereby most of the aerosols can be introduced into the plasma. Reproducible cell introduction efficiency of 80 % (for the unicellular green alga *Chlamydomonas reinhardtii* CC-125 with a mean cell diameter of 6.4 μm) to ca. 100 % (for the yeast *S. cerevisiae*, the unicellular cyanobacterium *Synechocystis* sp. PCC 6803 and the unicellular red algae *Cyanidioschyzon merolae* 10D and *Galdieria sulphuraria* with mean cell diameters of 2.0–3.0 μm) was achieved.

A microdroplet generator (μDG) using a piezo actuator has been applied for cell introduction into the plasma [28, 29, 31, 37, 38]. Commercially available μDGs can generate uniformly sized microdroplets, but the droplet sizes were quite larger than the size of aerosols generated by nebulizers, in a range of 30–75 μm depending on a capillary diameter of μDG. Therefore, the size of droplets reduced to acceptable ones for being completely decomposed in the plasma. Shigeta et al. [28] achieved a uniformly dispensing of 23 μm droplets with a triple-pulse mode of the μDG, and the droplets were downsized to <1 μm by using a low-temperature desolvation system (Fig. 5.5), which is almost the same system produced by Günther's research group [31]. In the desolvation system, helium was used as a drying gas without any heating, and all the droplets generated could be completely introduced into the

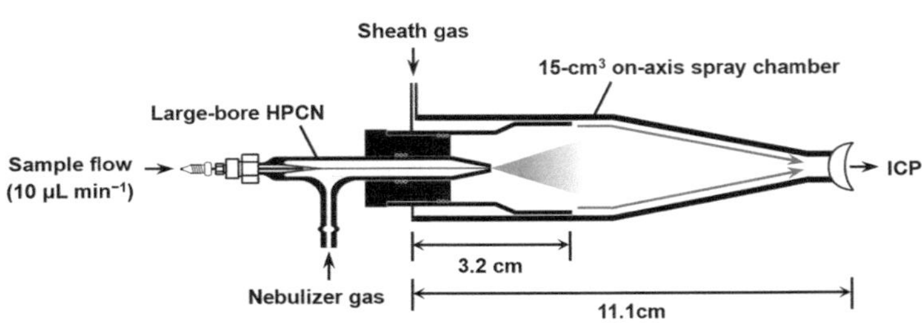

Fig. 5.4 Schematic diagram of the high efficiency cell introduction system (HECIS), consisting of a modified high performance concentric nebulizer (HPCN) and a custom-made on-axis small-volume cylinder chamber (With permission from J Anal At Spectrom)

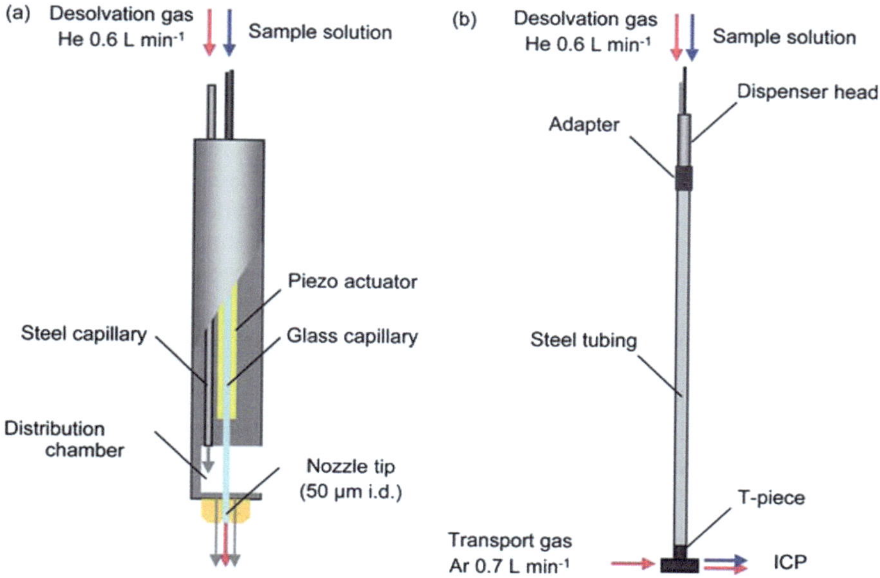

Fig. 5.5 Schematic diagram of the low-temperature desolvation system used for achieving a uniformly dispensing of 23-μm droplets with a triple-pulse mode (With permission from J Anal At Spectrom)

plasma. They demonstrated the performance of the system by applying it to single cell analysis of selenized yeast cells (*S. cerevisiae*). Okino's research group [37, 38] modified the desolvation system by making a heating and cooling zone in a drying pathway to reduce water vapor loading into the plasma and successfully improved the sensitivity of the elements in single algal cells in time-resolved ICP-OES measurement. Even though μDGs are very efficient for introducing cells into the plasma, they have the following drawbacks: they are prone to clogging and difficult to clean, and the dispensing performance is strongly influenced by physical properties of liquid such as pH, velocity, and salt content.

Verboket et al. [39] developed a new microfluidics-based droplet generator, which is a liquid-assisted droplet ejection (LADE) chip made entirely of poly (dimethylsiloxane) (PDMS). The chip produced monodisperse aqueous droplets with a diameter range of 40–60 μm in a stream of a highly volatile carrier phase perfluorohexane (PFH). They demonstrated that the chip is robust to changes of liquid sample properties and applicable to single cell analysis of bovine red blood cells by interfacing the chip with ICP-MS via a custom-built transport system including a cartridge heater and a membrane desolvator, which permits the PFH vapor removal.

5.5 Application of Cytometric Analysis by Using ICP-MS

Owing to the abovementioned unique features, time-resolved ICP-MS has been applied as the following two cytometric analysis techniques: (1) a highly sensitive elemental analysis technique of elements, drugs, and nanoparticles in single cells, which has been applied for evaluating a cell-to-cell variance or cellular uptake of the analytes, and (2) a multiparametric analysis technique (so-called mass cytometry), which has been applied for characterizing a cell population based on multiparameters. More details regarding the two techniques are described as follows.

5.5.1 *Highly Sensitive Elemental Analysis by Time-Resolved ICP-MS*

Time-resolved ICP-MS can be applied as a highly sensitive elemental analysis technique in single cells. The following various applications have been reported: investigation of distribution patterns of native elements in cells [18–20, 22, 23, 28], exposure test of hazardous elements [22, 27], tracking evaluation of metallo-drugs [24, 25], and cytotoxicity assessment of nanoparticles including quantum dots [26].

Li et al. [27] applied time-resolved ICP-MS to investigate the behavior of individual uranium-exposed bacterial cells in the plasma. The cells were introduced into an ICP-SFMS by a microconcentric nebulizer through a desolvation system utilized to dry the bacterial aerosol before injection into the plasma. They successfully observed U^+ spike signals only when the bacterial aerosol (regardless of wet or dry conditions) was introduced into the plasma.

Ho et al. [22] reported single cell analysis of a unicellular alga *Chlorella vulgaris* by using ICP-QMS. They investigated a quantification of elements in single algal cells by using metal oxide particles for a response calibration in the measurement. They demonstrated a kinetic study of sorption of Cr^{3+} onto the cells without separation of the cells from suspension culture. They also applied this method for tracking bismuth antiulcer drug uptake in *Helicobacter pylori* cells [24]. They demonstrated that the changes in bismuth content in the cells can be rapidly monitored by time-resolved ICP-MS analysis, providing a snapshot of the bismuth uptake in the cells.

Zheng et al. applied time-resolved ICP-MS for quantitative evaluation of cellular uptake of antitumor agents, $Gd@C_{82}(OH)_{22}$ and cisplatin [25], and quantum dots [26]. They proposed a new quantification method of the concentrations of the antitumor agents and quantum dots in human cancer cells (HeLa) and normal cells (16HBE) and validated the accuracy of the method on comparing with a bulk analysis by ICP-MS with an acid digestion. The validation results indicate that the proposed method can provide quantitative information on the concentrations of the antitumor agents and quantum dots taken up by the cells, at a single-cell level.

Wang et al. [23] applied time-resolved ICP-MS for analyzing the contents and distribution patterns of essential mineral elements such as Fe, Cu, Zn, Mn, P, and S in two types of cancer cells (HeLa and A549) and one type of normal cells (16HBE). Based on the analytical results of the cells at single-cell level, obvious differences in the contents and distribution patterns of the elements were observed among the three types of cells.

Shigeta et al. developed a μDG sample introduction system into the plasma [29] and applied it for analyzing selenized yeast cells [28]. Uniformly sized droplets including the cells were dispensed by the μDG operated in a triple-pulse mode with a repetition frequency of 50 Hz and, thereafter, completely introduced into the plasma through a desolvation system. They successfully obtained the frequency distributions of Cu, Zn, and Se in individual cells, and, with the use of the distributions, they determined the absolute amounts of these elements in a single cell: Cu (0.54 fg), Zn (1.2 fg), and Se (72 fg).

Groombridge et al. [18–20] applied their originally developed cell introduction systems, high efficiency cell introduction system (HECIS), and a pulse counting unit for analyzing microbial cells with 50-μs dwell time. They obtained time-resolved measurement profiles of biologically important elements such as C, Mg, Al, P, S, K, Ca, Cr, Mn, Fe, and Zn for yeast (*S. cerevisiae*), cyanobacterium (*Synechocystis* sp. PCC 6803), red algae (*C. merolae* 10D and *G. sulphuraria*), and green alga (*C. reinhardtii* CC-125) [20]. They also conducted a preliminary investigation into the potential for multielement correlation analysis of individual yeast cells by time-resolved ICP-TOFMS, and a relatively strong correlation was observed for the time-resolved measurement profiles between P and Zn (correlation factor: 0.69), between P and Mg (0.63), and between Mg and Zn (0.63) [18].

5.5.2 Mass Cytometry

Mass cytometry is a method that enables high-throughput, qualitative, and quantitative multiparametric analysis of single cells (such as a real-time immunochemical analysis of multiple antigens in single cells), as is in Fig. 5.6. It likes fluorescence-based flow cytometry, but ICP-TOFMS in combination with multielement tagging techniques is used for detecting multiple targets simultaneously in a single cell [10–12, 16, 17].

Fluorescence-based polychromatic flow cytometry (PFC) is the most advanced technique for multiparametric analysis of single cells, since it is superior in specificity and sensitivity. It has the advantage of analyzing over 20 parameters simultaneously but still has some technical problems to be solved, such as spectrum interference that limits the number of fluorescent colors used simultaneously, quenching of fluorescence, and autofluorescence from cells themselves. On the other hand, mass cytometry has no similar problems like those mentioned above,

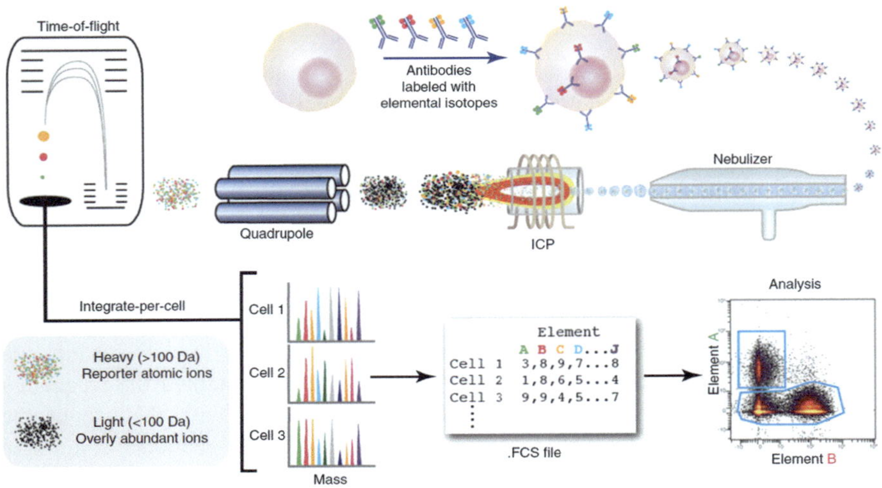

Fig. 5.6 Schematic image of mass cytometry. An affinity product (e.g., antibody) tagged with a specific element binds to the cellular epitope. Signals corresponding to each elemental tag are correlated with the presence of the respective marker (Adapted from Bendall et al. [10] with permission from Trends Immunol)

because monoisotopic lanthanide-labeled antibody and DNA intercalator (i.e., isotopic reporters) are used as element tags for detection. The current state-of-the-art mass cytometer can measure 100 different stable isotopes of metals, but around 40 isotopic reporters are currently available; therefore, around 40 parameters including DNA content, cell survival, and relative cell size can be analyzed simultaneously by mass cytometry [12]. Meanwhile, the cell throughput of mass cytometry is theoretically limited to about 1000 cells s^{-1}, although a commercially available PFC has better cell throughput of over 20,000 cells s^{-1}. Moreover, the cell introduction efficiency of mass cytometry is low (i.e., about 30–40 % as mentioned before) in contrast to PFC, where almost all the cells can be introduced into the flow cytometer [10]. Since these drawbacks of mass cytometry (i.e., limited cell throughput and low cell introduction efficiency) have been recently overcome to some extent, mass cytometry is becoming an alternative method for multiparametric analysis of cells. Various applications of mass cytometry have been reported and summarized in excellent reviews for several years [40–44].

Typical applications of mass cytometry are briefly introduced below. Bendall et al. [30, 45] applied mass cytometry to the study of hematopoiesis in bone marrow, and they found that hematopoietic stem cells produce various blood cell components in several differentiation pathways. The surface proteins of the individual cells were quantified by using mass cytometry, and the cells were clustered by chemical characteristics by spanning-tree progression analysis of density-normalized events (SPADE). This technique allowed step-by-step tracking of cell maturation through differentiation processes. Thus, the combination of single cell analysis by mass cytometry and SPADE analysis is a powerful tool for characterizing cells being in a healthy state or disease state and cells after drug intervention. This technique has been already used for various studies such as characterization of CD8-positive T-cell populations [46], identification of bacterial species [47], and validation of the cancer stem cell hypothesis [48]. Moreover, based on this technique, a cell-based multiplexing technique, called mass-tag cellular barcoding (MCB), has been developed for simultaneous characterization of multiple samples. By using the MCB, comprehensive analyses of the differences among the human peripheral blood mononuclear cells provided by several donors and the differences of the effects of various inhibitors on the cells were carried out [49].

5.6 Single Cell Analysis by Imaging Mass Spectrometry

Imaging mass spectrometry is a technique that provides information on the spatial distribution of chemical compositions (e.g., elements, compounds, biomarkers, metabolites, peptides, and proteins) within or across the surface of a single cell. This technique has an advantage that by using an ionization probe that is significantly smaller than the size of an individual cell, multiple mass spectra can be obtained for a single cell. The image obtained by this technique is created based on the relative intensity of the signal of each target substance. Therefore, although the localization of the same substance between different cells is comparable to each other, that of the different substances between different cells is not simply comparable to each other because the ionization efficiencies of different substances are different, unless the signal intensity of each target substance is corrected with an internal standard. Many discussions on the methodology of such internal standard correction have been made so far [50]. In recent years, the spatial resolution of laser-based ionization approaches has been improved by the following: (1) several technological and methodological advances, (2) a new sampling technique called oversampling (where a laser is irradiated at smaller step intervals than the laser spot size), (3) improved laser optics (i.e., a smart beam shaping), and (4) improved matrix application techniques. Thus, imaging mass spectrometry is becoming a powerful single cell analysis platform [14].

Laser ablation ICP-MS (LA-ICP-MS) is a technique that allows the in situ analysis of elements in solid samples without sample preparation. This technique has been often used for micromapping of elements and nanoparticles (or their aggregate forms) in single cells [9]. LA-ICP-MS is becoming a common bioimaging tool in the single cell analysis field but requires further technological development especially on LA technology for increasing the spatial resolution (i.e., making smaller a laser spot size and step intervals). It is important to note, however, that a smaller laser spot size is not necessarily suitable for LA-ICP-MS analysis because decrease in the laser spot size leads to reduction in the amount of gas-phase particles produced in a single ablation, resulting in low S/B ratios. Thus, it is necessary to appropriately determine the laser spot size depending on the sample characteristics and purpose of analysis.

Jakubowski's research group [51, 52] has successfully used LA-ICP-MS in a differential scanning mode for micromapping of nanoparticles in single cells. They optimized the ablation/scanning parameters in order to obtain a certain laser spot size of 4 μm or 8 μm and then irradiated a laser as the spots are overlapped each other. Consequently, the lateral resolution in the scanning direction was achieved to be much smaller than the laser spot size, which enabled to obtain elemental mapping images of single cells with several hundred pixels. In addition, the counting of nanoparticles incorporated into single cells was also achieved based on a matrix-matched calibration using nitrocellulose membranes doped with nanoparticle suspension. However, the obtainable lateral resolution was still limited by washout time of the LA cell and the time resolution of the ICP-MS instrument.

In recent years, Günther's research group [53] has developed a novel tubular LA cell in combination with an excimer laser and achieved fast washout time of 30 ms (compared with roughly 1 s in commercial devices) for high spatial resolution of ~1 μm. Moreover, they connected the developed tubular LA cell with CyTOF and successfully performed immuno-imaging of 32 proteins and their modifications in formalin-fixed, paraffin-embedded breast cancer tissue at subcellular resolution [54]. This pioneering study showed that the designing of new LA cell with a fast washout time is a breakthrough in this field to break the limitation by the conventional spatial resolution and then realize advanced multiparameter immunoimaging at subcellular resolution. The combination technique of LA and mass cytometry (such as CyTOF) is now often called "imaging mass cytometry," and it is complementing existing imaging techniques. Although this technique still has challenges to obtain faster scanning speeds, higher sample throughput, and higher sensitivity, it is becoming a powerful tool for basic studies of tissue heterogeneity and function. Giesen et al. [54] expected that by substantially boosting sensitivity (e.g., reduction of aerosol dispersion, use of multiple antibody clones against a target, and an increase in the number of metal atoms bound to each antibody through the use of metal nanoparticles), quantitative analysis on ~100 biomarkers on a given tissue with submicrometer resolution in less than 1 h will be possible in the future.

5.7 Conclusion and Future Prospect

In this chapter, an overview of single cell analysis by using ICP-MS has been presented, focusing on cytometric analysis and imaging analysis.

The current situations of the technological development relevant to the single cell analysis by using ICP-MS still have a lot of challenges for the future. For example, cell sample preparation and cell manipulation techniques suitable for subsequent single cell analysis by using ICP-MS are left to be improved. Microfluidics is one of the most promising techniques for such purposes because it has been applied so far in the field of single cell analysis with mass spectrometry (especially ESI-MS for protein or peptide analysis [55–57]) mainly for cell manipulation before mass spectrometric analysis. For another example, LA-ICP-MS techniques need to be improved for achieving higher spatial resolution as mentioned in the main text. By progressing the technological improvement and development relevant to the single cell analysis by using ICP-MS, it would become possible to acquire more detailed information on individual cells.

References

1. Szpunar J (2004) Metallomics: a new frontier in analytical chemistry. Anal Bioanal Chem 378:54–56
2. Sanz-Medel A, Montes-Bayon M, Bettmer J, Fernandez-Sanchez M, Encinar J (2012) ICP-MS for absolute quantification of proteins for heteroatom-tagged, targeted proteomics. Trac Trends Anal Chem 40:52–63
3. Sanz-Medel A, Montes-Bayon M, de la Campa M, Encinar J, Bettmer J (2008) Elemental mass spectrometry for quantitative proteomics. Anal Bioanal Chem 390:3–16
4. Montaser A (1998) Inductively coupled plasma mass spectrometry. Wiley-VCH, New York
5. Tomas R (2013) Practical guide to ICP-MS: a tutorial for beginners, 3rd edn. CRC Press, Boca Raton
6. Pröfrock D, Prange A (2012) Inductively coupled plasma-mass spectrometry (ICP-MS) for quantitative analysis in environmental and life sciences: a review of challenges, solutions, and trends. Appl Spectrosc 66:843–868
7. Engelhard C (2011) Inductively coupled plasma mass spectrometry: recent trends and developments. Anal Bioanal Chem 399:213–219
8. Dittrich P, Jakubowski N (2014) Current trends in single cell analysis. Anal Bioanal Chem 406:6957–6961
9. Mueller L, Traub H, Jakubowski N, Drescher D, Baranov VI, Kneipp J (2014) Trends in single-cell analysis by use of ICP-MS. Anal Bioanal Chem 406:6963–6977
10. Bendall SC, Nolan GP, Roederer M, Chattopadhyay PK (2012) A deep profiler's guide to cytometry. Trends Immunol 33:323–332
11. Ornatsky O, Bandura D, Baranov V, Nitz M, Winnik MA, Tanner S (2010) Highly multiparametric analysis by mass cytometry. J Immunol Methods 361:1–20
12. Atkuri KR, Stevens JC, Neubert H (2015) Mass cytometry: a highly multiplexed single-cell technology for advancing drug development. Drug Metab Dispos 43:227–233
13. Schmid A, Kortmann H, Dittrich PS, Blank LM (2010) Chemical and biological single cell analysis. Curr Opin Biotechnol 21:12–20
14. Trouillon R, Passarelli MK, Wang J, Kurczy ME, Ewing AG (2013) Chemical analysis of single cells. Anal Chem 85:522–542
15. Haselgrubler T, Haider M, Ji BZ, Juhasz K, Sonnleitner A, Balogi Z, Hesse J (2014) High-throughput, multiparameter analysis of single cells. Anal Bioanal Chem 406:3279–3296
16. Tanner SD, Bandura DR, Ornatsky O, Baranov VI, Nitz M, Winnik MA (2008) Flow cytometer with mass spectrometer detection for massively multiplexed single-cell biomarker assay. Pure Appl Chem 80:2627–2641
17. Bandura DR, Baranov VI, Ornatsky OI, Antonov A, Kinach R, Lou XD, Pavlov S, Vorobiev S, Dick JE, Tanner SD (2009) Mass cytometry: technique for real time single cell multitarget immunoassay based on inductively coupled plasma time-of-flight mass spectrometry. Anal Chem 81:6813–6822

18. Groombridge AS, Miyashita S, Fujii S, Nagasawa K, Okahashi T, Ohata M, Umemura T, Takatsu A, Inagaki K, Chiba K (2013) High sensitive elemental analysis of single yeast cells (*Saccharomyces cerevisiae*) by time-resolved inductively-coupled plasma mass spectrometry using a high efficiency cell introduction system. Anal Sci 29:597–603
19. Miyashita S, Groombridge AS, Fujii S, Takatsu A, Chiba K, Inagaki K (2014) Time-resolved ICP-MS measurement: a new method for elemental and multiparametric analysis of single cells. Anal Sci 30:219–224
20. Miyashita S, Groombridge AS, Fujii S, Minoda A, Takatsu A, Hioki A, Chiba K, Inagaki K (2014) Highly efficient single-cell analysis of microbial cells by time-resolved inductively coupled plasma mass spectrometry. J Anal At Spectrom 29:1598–1606
21. Olesik JW, Gray PJ (2012) Considerations for measurement of individual nanoparticles or microparticles by ICP-MS: determination of the number of particles and the analyte mass in each particle. J Anal At Spectrom 27:1143–1155
22. Ho KS, Chan WT (2010) Time-resolved ICP-MS measurement for single-cell analysis and on-line cytometry. J Anal At Spectrom 25:1114–1122
23. Wang HL, Wang B, Wang M, Zheng LN, Chen HQ, Chai ZF, Zhao YL, Feng WY (2015) Time-resolved ICP-MS analysis of mineral element contents and distribution patterns in single cells. Analyst 140:523–531
24. Tsang CN, Ho KS, Sun HZ, Chan WT (2011) Tracking bismuth antiulcer drug uptake in single *Helicobacter pylori* cells. J Am Chem Soc 133:7355–7357
25. Zheng LN, Wang M, Zhao LC, Sun BY, Wang B, Chen HQ, Zhao YL, Chai ZF, Feng WY (2015) Quantitative analysis of $Gd@C_{82}(OH)_{22}$ and cisplatin uptake in single cells by inductively coupled plasma mass spectrometry. Anal Bioanal Chem 407:2383–2391
26. Zheng LN, Wang M, Wang B, Chen HQ, Ouyang H, Zhao YL, Chai ZF, Feng WY (2013) Determination of quantum dots in single cells by inductively coupled plasma mass spectrometry. Talanta 116:782–787
27. Li F, Armstrong DW, Houk RS (2005) Behavior of bacteria in the inductively coupled plasma: atomization and production of atomic ions for mass spectrometry. Anal Chem 77:1407–1413
28. Shigeta K, Koellensperger G, Rampler E, Traub H, Rottmann L, Panne U, Okino A, Jakubowski N (2013) Sample introduction of single selenized yeast cells (*Saccharomyces cerevisiae*) by micro droplet generation into an ICP-sector field mass spectrometer for label-free detection of trace elements. J Anal At Spectrom 28:637–645
29. Shigeta K, Traub H, Panne U, Okino A, Rottmann L, Jakubowski N (2013) Application of a micro-droplet generator for an ICP-sector field mass spectrometer – optimization and analytical characterization. J Anal At Spectrom 28:646–656
30. Bendall SC, Simonds EF, Qiu P, Amir EAD, Krutzik PO, Finck R, Bruggner RV, Melamed R, Trejo A, Ornatsky OI, Balderas RS, Plevritis SK, Sachs K, Pe'er D, Tanner SD, Nolan GP (2011) Single-cell mass cytometry of differential immune and drug responses across a human hematopoietic continuum. Science 332:687–696
31. Gschwind S, Flamigni L, Koch J, Borovinskaya O, Groh S, Niemax K, Günther D (2011) Capabilities of inductively coupled plasma mass spectrometry for the detection of nanoparticles carried by monodisperse microdroplets. J Anal At Spectrom 26:1166–1174
32. Iwai T, Shigeta K, Aida M, Ishihara Y, Miyahara H, Okino A (2015) A transient signal acquisition and processing method for micro-droplet injection system inductively coupled plasma mass spectrometry (M-DIS-ICP-MS). J Anal At Spectrom 30:1617–1622
33. Strenge I, Engelhard C (2016) Capabilities of fast data acquisition with microsecond time resolution in inductively coupled plasma mass spectrometry and identification of signal artifacts from millisecond dwell times during detection of single gold nanoparticles. J Anal At Spectrom 31:135–144
34. Todoli JL, Mermet JM (2006) Sample introduction systems for the analysis of liquid microsamples by ICP-AES and ICP-MS. Spectrochim Acta B 61:239–283
35. Inagaki K, Fujii S, Takatsu A, Chiba K (2011) High performance concentric nebulizer for low-flow rate liquid sample introduction to ICP-MS. J Anal At Spectrom 26:623–630

36. Groombridge AS, Inagaki K, Fujii S, Nagasawa K, Okahashi T, Takatsu A, Chiba K (2012) Modified high performance concentric nebulizer for inductively coupled plasma optical emission spectrometry. J Anal At Spectrom 27:1787–1793
37. Kaburaki Y, Nomura A, Ishihara Y, Iwai T, Miyahara H, Okino A (2013) Development of injection gas heating system for introducing large droplets to inductively coupled plasma. Anal Sci 29:1147–1151
38. Ishihara Y, Aida M, Nomura A, Miyahara H, Hokura A, Okino A (2015) Development of desolvation system for single-cell analysis using droplet injection inductively coupled plasma atomic emission spectroscopy. Anal Sci 31:781–785
39. Verboket PE, Borovinskaya O, Meyer N, Günther D, Dittrich PS (2014) A new microfluidics-based droplet dispenser for ICPMS. Anal Chem 86:6012–6018
40. Tanner SD, Baranov VI, Ornatsky OI, Bandura DR, George TC (2013) An introduction to mass cytometry: fundamentals and applications. Cancer Immunol Immunother 62:955–965
41. Bjornson ZB, Nolan GP, Fantl WJ (2013) Single-cell mass cytometry for analysis of immune system functional states. Curr Opin Immunol 25:484–494
42. Di Palma S, Bodenmiller B (2015) Unraveling cell populations in tumors by single-cell mass cytometry. Curr Opin Biotechnol 31:122–129
43. Nair N, Mei HE, Chen SY, Hale M, Nolan GP, Maecker HT, Genovese M, Fathman CG, Whiting CC (2015) Mass cytometry as a platform for the discovery of cellular biomarkers to guide effective rheumatic disease therapy. Arthritis Res Ther 17:127–135
44. Proserpio V, Lönnberg T (2016) Single-cell technologies are revolutionizing the approach to rare cells. Immunol Cell Biol 94:225–229
45. Qiu P, Simonds EF, Bendall SC, Gibbs KD Jr, Bruggner RV, Linderman MD, Sachs K, Nolan GP, Plevritis SK (2011) Extracting a cellular hierarchy from high-dimensional cytometry data with SPADE. Nat Biotechnol 29:886–891
46. Newell EW, Sigal N, Bendall SC, Nolan GP, Davis MM (2012) Cytometry by time-of-flight shows combinatorial cytokine expression and virus-specific cell niches within a continuum of $CD8^+$ T cell phenotypes. Immunity 36:142–152
47. Leipold MD, Ornatsky O, Baranov V, Whitfield C, Nitz M (2011) Development of mass cytometry methods for bacterial discrimination. Anal Biochem 419:1–8
48. Gibbs KD Jr, Jager A, Crespo O, Goltsev Y, Trejo A, Richard CE, Nolan GP (2012) Decoupling of tumor-initiating activity from stable immunophenotype in HoxA9-Meis1-driven AML. Cell Stem Cell 10:210–217
49. Bodenmiller B, Zunder ER, Finck R, Chen TJ, Savig ES, Bruggner RV, Simonds EF, Bendall SC, Sachs K, Krutzik PO, Nolan GP (2012) Multiplexed mass cytometry profiling of cellular states perturbed by small-molecule regulators. Nat Biotechnol 30:858–867
50. Hare D, Austin C, Doble P (2012) Quantification strategies for elemental imaging of biological samples using laser ablation-inductively coupled plasma-mass spectrometry. Analyst 137:1527–1537
51. Drescher D, Giesen C, Traub H, Panne U, Kneipp J, Jakubowski N (2012) Quantitative imaging of gold and silver nanoparticles in single eukaryotic cells by laser ablation ICP-MS. Anal Chem 84:9684–9688
52. Drescher D, Zeise I, Traub H, Guttmann P, Seifert S, Büchner T, Jakubowski N, Schneider G, Kneipp J (2014) In situ characterization of SiO_2 nanoparticle biointeractions using BrightSilica. Adv Funct Mater 24:3765–3775
53. Wang HAO, Grolimund D, Giesen C, Borca CN, Shaw-Stewart JRH, Bodenmiller B, Günther D (2013) Fast chemical imaging at high spatial resolution by laser ablation inductively coupled plasma mass spectrometry. Anal Chem 85:10107–10116
54. Giesen C, Wang HAO, Schapiro D, Zivanovic N, Jacobs A, Hattendorf B, Schüffler PJ, Grolimund D, Buhmann JM, Brandt S, Varga Z, Wild PJ, Günther D, Bodenmiller B (2014) Highly multiplexed imaging of tumor tissues with subcellular resolution by mass cytometry. Nat Methods 11:417–422

55. Mellors JS, Jorabchi K, Smith LM, Ramsey JM (2010) Integrated microfluidic device for automated single cell analysis using electrophoretic separation and electrospray ionization mass spectrometry. Anal Chem 82:967–973
56. Mao P, Wang HT, Yang P, Wang D (2011) Multinozzle emitter arrays for nanoelectrospray mass spectrometry. Anal Chem 83:6082–6089
57. Yin H, Marshall D (2012) Microfluidics for single cell analysis. Curr Opin Biotechnol 23:110–119

Chapter 6
Synchrotron Radiation X-Ray Analysis of Metal-Accumulating Plants

Akiko Hokura and Emiko Harada

Abstract X-ray microanalyses have been utilized to determine mechanisms of metal detoxification and homeostasis in plants and are powerful tools, especially for target plants unsuitable for genetic analyses. This chapter describes several synchrotron-based research methods to determine metal element distribution and speciation of metal-accumulating plants. The procedures are described, to show how micro-X-ray techniques are applicable to various biological materials, including those that are solid or liquid, soft or hard, or wet or dry. Combination of X-ray microanalysis with other methodologies, including TEM (transmission electron microscope), SEM (scanning electron microscope), and techniques of biochemistry and molecular biology, is described. Overall, this chapter attempts to encourage cross-disciplinary work in metallomics.

Keywords Metal-accumulating plant • Synchrotron radiation • XRF • XRD • EXAFS • XANES

6.1 Introduction

Synchrotron radiation X-ray analysis has been utilized to assess elucidating metal accumulation by plants, along with mechanisms of metal detoxification and homeostasis. Features of synchrotron radiation X-rays are well suited for research on metallomics.

This chapter provides an outline of synchrotron radiation (SR) X-ray techniques, including X-ray fluorescence (XRF), X-ray absorption spectrometry (XAS), and X-ray diffraction (XRD), as well as reviewing research work using these techniques on metal-accumulating plants, including methods of sample preparation. At the end of the chapter, prospective users of this technology should have information

A. Hokura (✉)
Tokyo Denki University, Adachi, Tokyo, Japan
e-mail: hokura@mail.dendai.ac.jp

E. Harada
The University of Shiga Prefecture, Hikone, Shiga, Japan
e-mail: harada.e@ses.usp.ac.jp

Y. Ogra, T. Hirata (eds.), *Metallomics*, DOI 10.1007/978-4-431-56463-8_6

enabling them to design proposal X-ray experiments using a beamline in an SR facility.

6.2 Synchrotron Radiation X-Rays

SR is an electromagnetic wave emitted from an electron traveling at almost the speed of light. The wave propagates tangential to the electron beam, with its path bent by a magnetic field (Fig. 6.1). Among the useful features of SR are its brightness, high directionality, and variable polarization [1]. Many SR facilities have been built throughout the world, with each facility having many beamlines designed for specific purposes. SR-based X-ray technology, including XRF, XRD, and μ-CT (computed tomography) [2], has been utilized for biological analyses.

6.3 X-Ray Analyses

6.3.1 Interactions of X-Rays with Matter

X-rays are one type of electromagnetic wave that interact with matter (Fig. 6.2). The interactions of X-rays with material objects can provide useful information on the elements present in that object, its crystal structure, and the chemical state of targeted elements. Because X-ray analysis is nondestructive, it can provide in vivo information on biological samples.

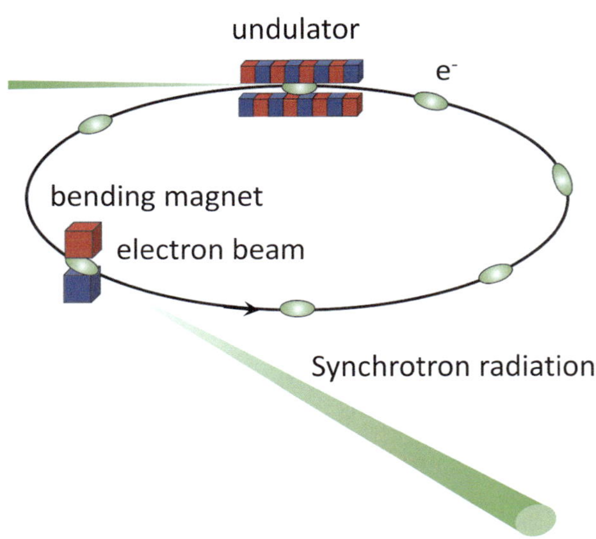

Fig. 6.1 Diagram illustrating synchrotron radiation

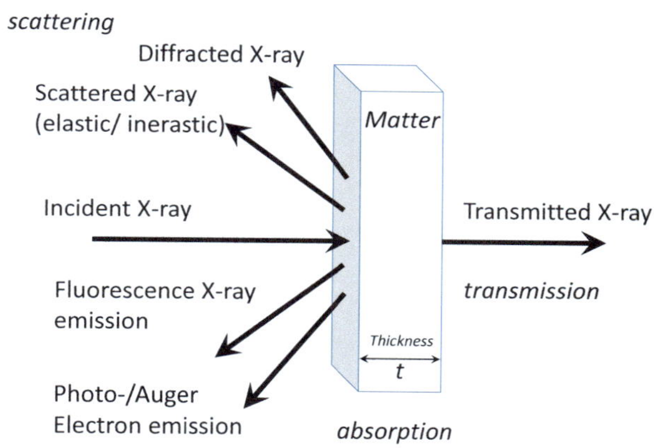

Fig. 6.2 Interactions of X-rays with matter

6.3.2 X-Ray Fluorescence

When materials are exposed to X-rays, their component atoms may become ionized. Exposure of an atom to radiation of an energy greater than its ionization potential may result in the ejection of one or more electrons from that atom. Figure 6.3 shows the electron transitions in atoms resulting from the interactions of X-rays with an object. Because each element has a unique set of energy levels, each element produces X-rays at a unique set of energies. This ionization causes the emission of both characteristic X-rays (fluorescence X-rays) and Auger electrons (Fig. 6.4), with the two effects competing with each other.

As shown in Fig. 6.4, the fluorescence yield for the K-line of heavy elements is approximately 70–90 %. This high fluorescence yield makes XRF analysis of heavy elements highly sensitive. Electron beams also cause the emission of X-rays characteristic of each element, but the energy of these electron probes is insufficient to excite the K-line of heavy elements. Rather, SEM (scanning electron microscope) equipped with EDS (energy-dispersive X-ray spectrometer) is suitable for detection of the K-line of light elements.

The construction of two- or three-dimensional images from one-dimensional analyses is called mapping or imaging. SR-based XRF imaging is comparable to state-of-the-art microscopy. The high intensity and transmissivity of SR-based X-rays enable analyses of samples under normal pressure in air. The resulting signal is the sum of the light through that path.

6.3.3 X-Ray Absorption Spectroscopy

X-ray absorption spectroscopy (XAS) is a widely used to probe the local electronic structure of materials [3, 4]. When a beam of monochromatic X-rays passes through

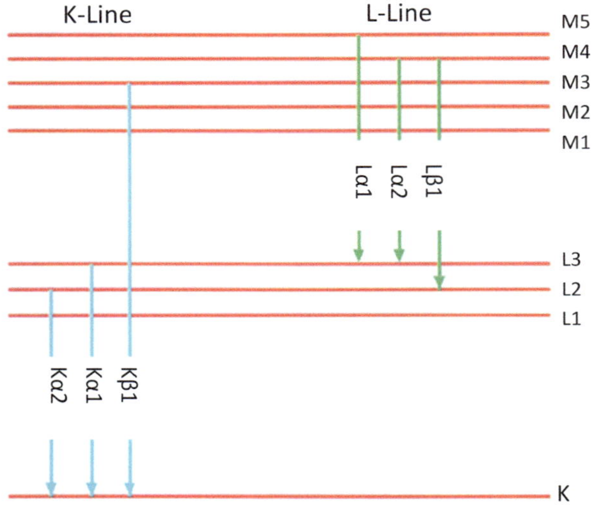

Fig. 6.3 Electron transitions and fluorescence X-rays

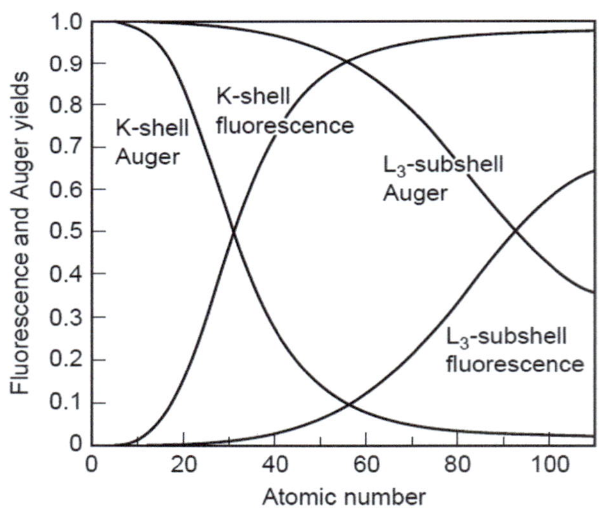

Fig. 6.4 Fluorescence and Auger emission yields $\omega + \alpha = 1$ Here, ω and α represent the fluorescence and the Auger emission yield, respectively. For excitation of the electron shell K of zinc (atomic number, Z = 30), $\omega = \alpha = 0.5$

an object, the beam loses intensity due to its interactions with the atoms in the object. The intensity of the transmitted X-ray can be described using the equation:

$$I = I_0 e^{-\mu t} \tag{6.1}$$

where I_0 is the X-ray intensity incident on a sample, μ is the absorption coefficient, t is the sample thickness, and I is the intensity transmitted through the sample.

Equation (6.1) can also be written as

$$\mu t = \ln \frac{I_0}{I} \tag{6.2}$$

The absorption coefficient μ decreases as X-ray energy increases, except for special photon energies. When X-ray energy is sufficient to overcome the binding energy of a core electron in an element, the absorption coefficient will increase sharply (Fig. 6.5). This energy is specific for every element, making XAS an element-selective method. XAS spectra can provide information about the composition of an object.

The region above the edge is normally divided into two subregions, with the first 50–100 eV above the edge called the X-ray absorption near-edge structure (XANES) and the region above the XANES region called the extended X-ray absorption fine structure (EXAFS). XANES spectra can be used to determine the average oxidation state of elements in a sample. XANES spectra are also sensitive to the coordination environment of the absorbing atom in a sample. Fingerprinting methods have been used to match the XANES spectra of an unknown sample to those of known "standards." Linear combination fitting of several standard spectra can provide estimates of the amount of each element corresponding to these standard spectra of an unknown sample.

Photoelectrons in the EXAFS region are characterized by high kinetic energy and single backscattering from nearest-neighbor atoms, providing information about the number and identity of these atoms, as well as their distance from the absorbing atom.

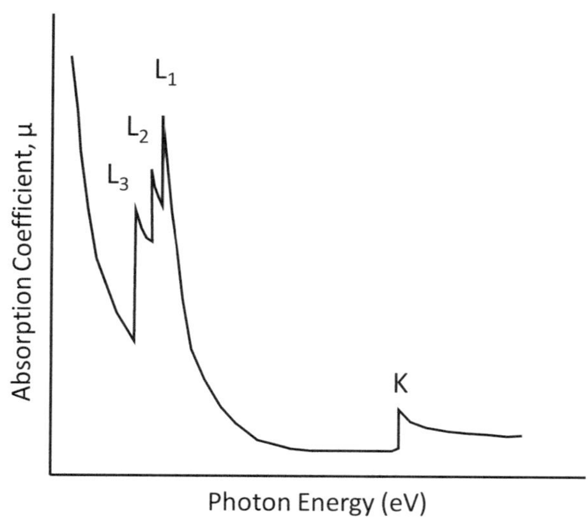

Fig. 6.5 Absorption spectrum. The absorption coefficient decreases smoothly at higher energy, except at special photon energies. When the X-ray energy is sufficient to overcome the binding energy of a core electron in an element, the absorption coefficient increases sharply

Among the primary mechanisms of metal detoxification in plants are the formation of metal-chelating molecules and the immobilization of metals inside cell or plant tissue. XANES and EXAFS have shown the storage forms of Mn in Mn-hyperaccumulating leaves ([5]; Fig. 6.6). Several standards must be prepared, to serve as candidate chemical species for interpretation of XAS spectra. If a standard chemical is not commercially available, it can be synthesized by chemical methods in the laboratory or estimated from its crystal structure determined in an SR facility. Samples containing several chemicals have spectra that appear as the sum of the spectra of all components. Several application programs are available for linear combination fitting of spectra; these software packages include ATHENA [6], available as a component of the free package IFEFFIT at (http://bruceravel.github.io/demeter/) and REX2000 (Rigaku Corporation, Tokyo) (Fig. 6.6). The additive property of component spectra enables the ratio of the components of an object to be estimated when the chemical status and spectrum of each component have been determined. The combination of XRF imaging and XANES/EXAFS analysis will be a powerful tool for determining the forms of metal storage in plants.

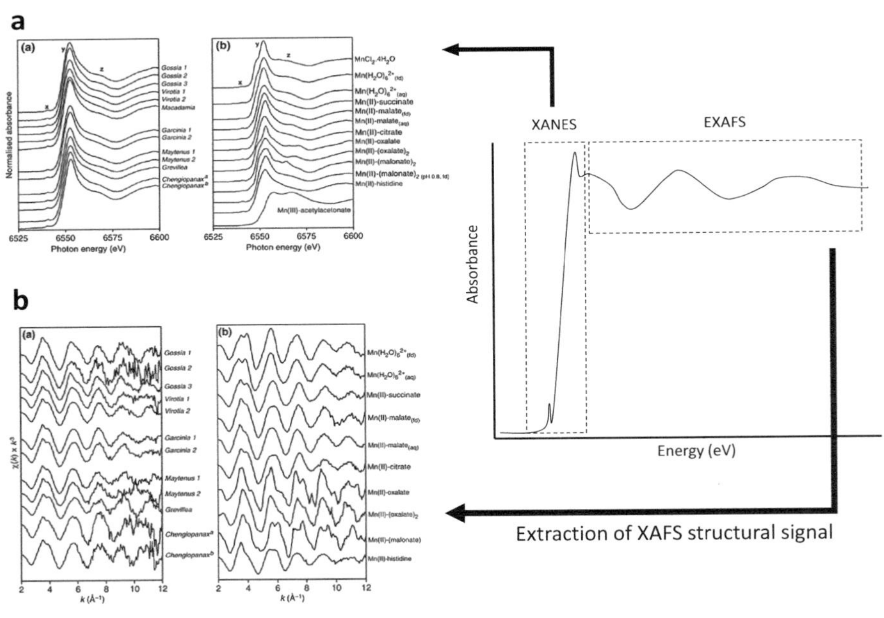

Fig. 6.6 Energy ranges of XANES (**a**) and EXAFS (**b**) spectra from Mn-accumulating leaves and reference chemicals for the speciation of endogenous Mn [5] X-ray absorption spectra generally have three regions, including the pre-edge; XANES, including the first 50–100 eV after the absorption edge; and EXAFS, which can be detected in suitable samples up to at least 1000 eV beyond the edge

6.3.4 X-Ray Diffraction

X-ray diffraction (XRD) is a rapid analytical technique that provides information on the structure of crystalline materials. This technique is used primarily to identify and characterize compounds based on their diffraction patterns. The interaction of incident X-rays with the sample produces constructive interference (and a diffracted ray) when conditions satisfy Bragg's law ($n\lambda = 2d \sin \theta$). This technique can also provide information on unit cell dimensions.

6.3.5 X-Ray Focusing Optics

X-ray focusing optics have recently been applied to multiple analyses by, for example, XRF, XAS, and XRD. Many synchrotron research techniques are attempting to focus on smaller-sized spots. Because the X-ray refractive index of most materials is extremely close to 1.0, the principal methods of manipulating X-rays are therefore reflection, diffraction, and interference [7]. X-ray focusing optics mainly used in X-ray analysis include Kirkpatrick-Baez (KB) mirrors, capillary optics, and Fresnel zone plates (FZP). Each of these has both advantages and disadvantages.

The KB mirror system consists of two elliptical, totally reflective mirrors with two focal points, one at the light source and the other at the focal point. One mirror is used for vertical focusing and the other for horizontal focusing. The KB system is achromatic, and the high reflectivity achieved with mirrors can enable X-rays to be focused simply and very efficiently over a wide energy range. This feature is suitable for XAS, making the KB mirror system the most powerful tool for multiple X-ray analyses [8].

Capillary optics consist of bundles of light-guiding pipes used to focus beams, while FZP uses diffraction effects to focus.

Figure 6.7 shows the multitechnique setup at the nanoimaging station ID22NI of the European Synchrotron Radiation Facility.

XRF enables the imaging (mapping) of the locations of elements in plant tissues. The KB mirror focusing system provides high-flux X-ray beams with diameters of a few micrometers, enabling the specimen set on the path to be scanned, resulting in fluorescent X-rays. Signals collected from each point were used to construct a two-dimensional distribution of each element. At present, the resolution has reached cellular and subcellular levels. In contrast to SEM-EDX imaging, which is based on a surface analytical technique, SR-based XRF sums all signals from a sample, because X-ray beams are passed through a thin (<100 μm) section or segment (Fig. 6.8). Metals are unevenly distributed throughout living organisms, with metal sinks sequestering toxic elements from metal sensitive parts. Localizing elements can help determine the detoxification mechanism in metal-accumulating

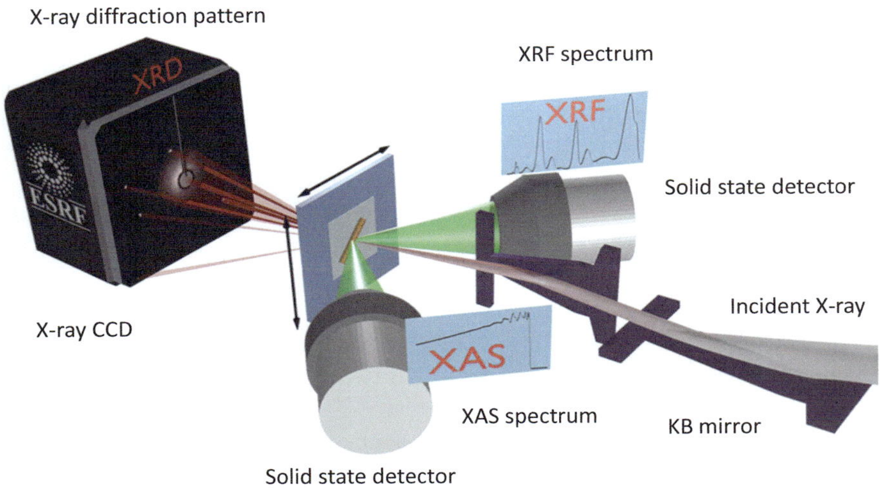

Fig. 6.7 Multitechnique setup at the nanoimaging station ID22NI of the European Synchrotron Radiation Facility [8]

plants. The large step size (50 μm) should be first, enabling monitoring of the candidate area at low resolution. After selecting the area of interest, high-resolution (smaller step size) analysis should be performed. Increasing the sampling time can enhance signal intensity and improve S/N ratio.

6.4 Preparation of Biological Materials for Synchrotron Radiation (SR)-Based X-Ray Analyses and Applications in Metallomics

6.4.1 What Are Heavy Metal-Accumulating Plants?

A number of terrestrial plants absorb and accumulate extraordinarily high concentrations of metals and metalloids in their shoots. To date, 500 species of metal hyperaccumulating plants have been identified throughout the world (Table 6.1). Several methods are available to determine the mechanism(s) for metal detoxification and accumulation, including methods involving genetics, molecular physiology, and biochemistry. Physicochemical analyses, including X-ray microanalyses, are also powerful tools, especially in target plants not suitable for genetic analyses. This method can also be used in plants that are not model plants or when genome information is not available, whether because the genome was too large in size or because polyploidy was high.

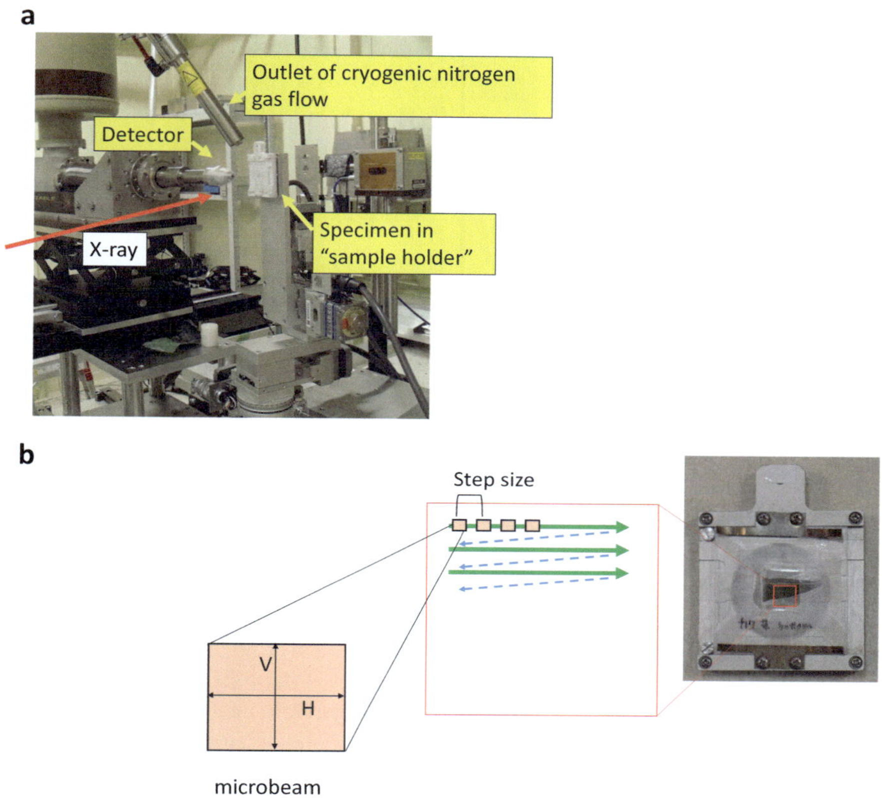

Fig. 6.8 Diagram of SR-based XRF imaging. (**a**) Detection unit installed on the BL37XU beamline at SPring-8. To analyze biological materials, cryogenic nitrogen gas flow is often sprayed on the specimen. The direction of the X-ray beam is shown as a *red arrow*. (**b**) A segment of a willow leaf mounted in "sample holder" and illustration of scanning by micrometer diameter of a μ-X-ray beam focused by the KB mirror. The target area (*red square*) of the specimen is scanned horizontally and the fluorescence X-ray signals are collected by the detector. Step size corresponds to the resolution of the image. A longer "measurement time" increases signal intensity. V, vertical; H, horizontal diameters of a microbeam

6.4.2 General Preparation of Biological Materials

Figure 6.9 illustrates the workflow for SR-based XRF, XANES, and EXAFS. Prior to working in the synchrotron facility, metal contents must be validated by quantitative analysis. Samples were digested with acids, such as HNO_3, H_2O_2, H_2SO_4, and $HClO_4$, and elemental concentrations were determined by AAS (Atomic Absorption Spectrometry), ICP-OES (Inductively Coupled Plasma Optical Emission Spectrometry), and ICP-MS (Inductively Coupled Plasma Mass Spectrometry). Optical and electron microscopy, such as SEM and TEM with or without EDX equipment, may also help estimate the size of the target.

Table 6.1 Criteria for hyperaccumulating concentrations of metals and metalloids by plants (μg g^{-1})

Element		Hyperaccumulation concentration (μg g^{-1})	
		Classical definition	Recently suggested
Antimony	Sb	1000	
Arsenic	As	1000	
Cadmium	Cd	100	
Cobalt	Co	1000	300
Copper	Cu	1000	300
Chromium	Cr	1000	300
Lead	Pb	1000	
Manganese	Mn	10,000	
Nickel	Ni	1000	
Selenium	Se	1000	100
Thallium	Tl	1000	100
Zinc	Zn	10,000	3000
Rare earth elements			3000

Baker and Brooks [9], Krämer [10], and van der Ent et al. [11]

Sample preparation techniques and tools for "classical" electron microscopy, such as TEM and SEM, may also be helpful. Care should be taken to maintain the thickness of a section or segment, because thickness may affect signal intensity. The samples should be fixed tightly, such that the sample cannot move, shrink, or melt during the analysis. Frozen or freeze-dried specimens are therefore preferred to fresh specimens, primarily because it is easier to monitor these samples.

If a tissue sample is composed of several spatial layers of cells, the X-ray signals of these signals will also overlap. This outcome may be avoided by preparing vertical or transverse sections of tissues. Normally, tissue thickness ranges between 30 and 100 μm. In analyzing macroscale localization, the microslicer has been found useful for preparing sections with the thickness of 40 μm to several mm in diameter. For example, SR-XRF imaging was used for tissue-specific localization of elements in rice grains [12]. Note the uneven thickness of sections directly affecting signal intensity.

The resulting frozen sections were directly mounted onto sample holders in cryomicrotome chambers and analyzed in a liquid nitrogen gas stream to prevent radiation damage. Frozen-hydrated sections were also available for these analyses. These samples were much easier to transport from their own institutions to the SR facility, as the drying procedure may change the distribution of metals. Sections may be prepared at individual institutions and transferred to that synchrotron facility. Alternatively, living organisms can be transferred to a facility, with the sample prepared in situ just before use (Fig. 6.10). Permission must be requested at each facility in advance.

For analyses, the specimens were mounted in sample holders (Fig. 6.11 in beamline 37XU in SPring-8). The samples were placed on a thin polymer film, such as Mylar (polyethylene) or Kapton (polyimide), stretched over an acrylic plate

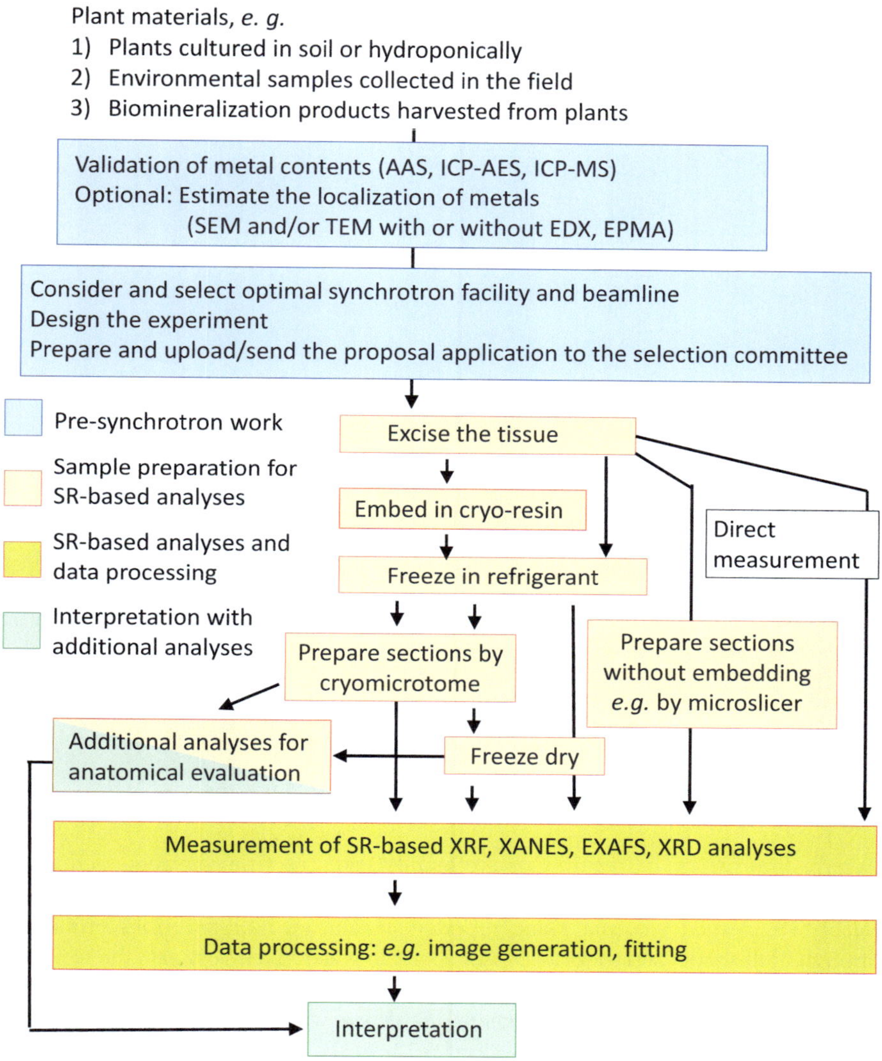

Fig. 6.9 General workflow of SR-based X-ray microanalyses including sample preparation

with a hole, and then held in the metallic holder. Acrylic plates (1 mm thick, 4×4 cm^2) were designed to adjust to the size of metallic sample holder, and a hole, 0.3–3 cm in diameter, was placed in the center of the plate. The samples were fixed in place with double-sided tape.

Fig. 6.10 Photograph of the "Green house," temporarily placed along a beamline. μ-XRF analysis of the willow plant requires maintenance of living plant materials in the synchrotron radiation facility, such that the are available for sample preparation. A wet cloth or spray will maintain moderate humidity. Plastic or aluminum lids on the pot cultures were designed to avoid the scattering of soil particles. The fluorescent lamps should be set carefully so that electricity does not leak

6.4.3 Example of Direct Analysis of Plant Tissues

In terrestrial plants, inorganic ions are mainly taken up from the roots and transferred to the shoots. *Egeria densa* (Brazilian waterweed, Hydrocharitaceae) is a freshly submerged vascular plant, reported to take up inorganic ions efficiently through both shoots and roots when immersed in water. To investigate the behavior of Cs in the aquatic ecosystem and to predict the ability of the radionuclide to remove this Cs from the environment, *E. densa* was cultured in a hydroponic medium supplemented with ^{133}Cs (Fig. 6.12, [13]). Because the leaves of these plants contain a single longitudinal vascular bundle and the blade consists of only two layers of cells, μ-XRF imaging with microfocused X-ray beams has shown the elemental distribution of K, Cs, Zn, Fe, Ca, and Mn in a leaf segment from an unprepared section of *E. densa* tissue. Relative fluorescence intensity and signals consistent with the strength of each sampling point are shown in successive color variations. A strong signal directly indicated the accumulation of the corresponding element. Nevertheless, more attention must be paid when comparing several elements. Because the radiation yield and the excitation efficiency of incident X-rays are dependent on the element, these features are directly related to the sensitivity on the detection of each element.

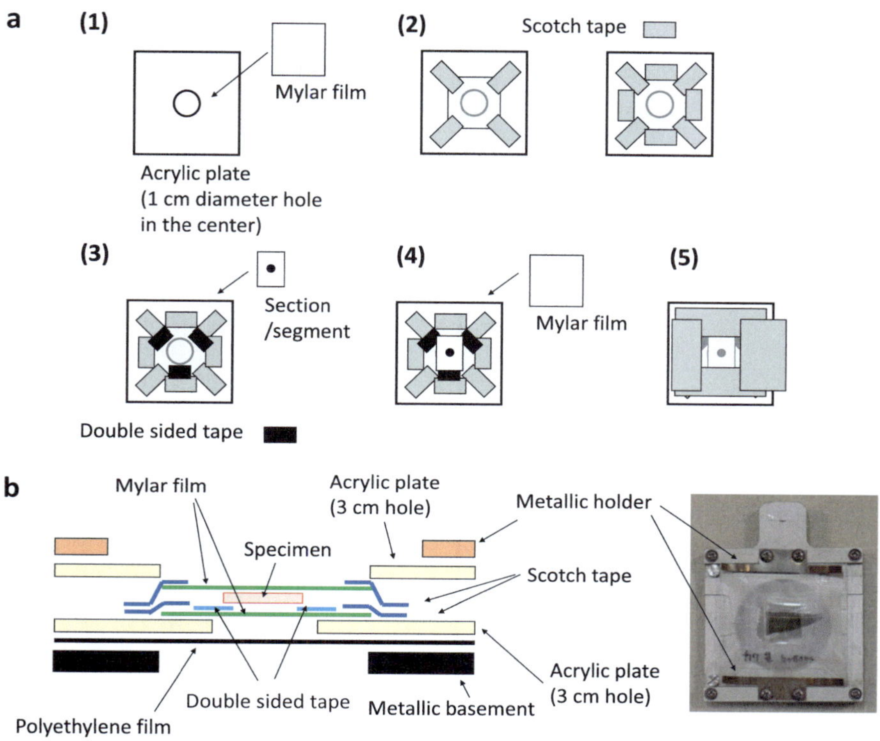

Fig. 6.11 Mounting of a section or segment for SR-μ-XRF analyses. (**a**) Procedure for mounting a section or segment on an acrylic holder (4 cm × 4 cm, 1 mm thick) with a hole (1 cm diameter) in BL37XU in SPring-8. (1) A piece of Mylar film is placed in the middle of an acryl holder. (2) The Mylar film is stretched and pasted with scotch tape. (3) A few pieces of double sided tape are placed on the Mylar film, followed by a cryosection or plant segment. (4) Another piece of Mylar film is placed on the sample and fixed with scotch tape, preventing the generation wrinkles. (5) The edges of the sample are sealed tightly to avoid drying. The extruding scotch tape is cut away. (**b**) Illustration of a cross section of a mounted specimen. This specimen was covered by another acrylic holder with 3 cm diameter hole and set in metal holders. Polyethylene film was placed over the specimen to avoid ice deposits

A leaf segment was removed from the plant, fixed using several pieces of double-sided tape and scotch tape (Fig. 6.5), and analyzed by SR-XRF. Figure 6.13 shows the distribution of metals in the leaves of a willow (*Salix miyabeana*) grown in metal-contaminated soils [14]. Before mounting onto the holder, the leaf segments were flattened, if necessary, by placing them between two heavy books, to keep them smooth. Leaf veins may also be removed in advance to maintain the stability of the segment. To maintain the humidity, a small piece of wet filter paper was mounted together with the segment. To determine the combined and long-term effects of multiple heavy metals on *S. miyabeana* grown in soil, variable-pressure scanning electron microscopy fitted with energy-dispersive X-ray analysis

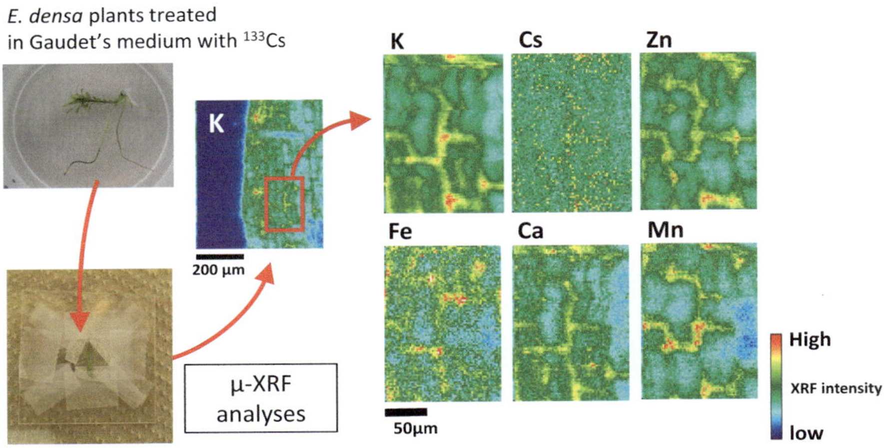

Fig. 6.12 μ-XRF imaging of a leaf of *E. densa* treated with CsCl in Gaudet's medium. The distribution showed the predominant localization of Cs, K as well as several divalent metals in the cell wall or apoplastic regions, suggesting the possible absorption mechanism of Cs in the plants. The leaf was excised and set over a 4×4 cm^2 acrylic plate with a 1.0 cm diameter hole and covered with a Mylar film. The high resolution imaging area is shown in the *red square*. The two-dimensional distribution of K, Cs, Zn, Fe, Ca, and Mn. The XRF signal intensity is shown as a color scale. Beam energy: 10 keV. Step size: 2.0 μm (H) × 2.0 μm (V). Exposure time: 0.5 s per point [13]

(VPSEM-EDX) and SR-XRF was used to characterize the serration of the leaves. The exudate was observed by VPSEM on the tip cells in serrations. In contrast, EDX showed that the exudate contained Ca and Sr, but not heavy metals, because SR-XRF analysis could detect the metal accumulation inside these tip cells.

6.4.4 *Preparation and Analyses of Frozen Tissue Sections*

Cryofixation involved the embedding of a tissue segment in a resin, such as an OCT (optimal cutting temperature) compound (Sakura Tissue-Tek, Tokyo, Japan), followed by freezing in liquid nitrogen (Fig. 6.14a). If the ice crystals grew slowly, thawing and refreezing of the tissue can disrupt cellular structure. Moreover, metals can dissolve from the surface of the segment if specimens are immersed in resin in room temperature for a long time.

Liquid nitrogen (melting point −210 °C, boiling point −195 °C) is generally used for flash freezing, stabilizing biological materials, halting metabolic activity, and preserving internal structure. However, freezing may be affected by the Leidenfrost effect, in which liquid nitrogen boils rapidly around a specimen and the cooling rate is relatively slow. Liquid propane (melting point −187.6 °C, boiling point −42.09 °C) or hexane (melting point −95 °C, boiling point 69 °C) cooled with liquid nitrogen is also available. During the freezing procedure, the

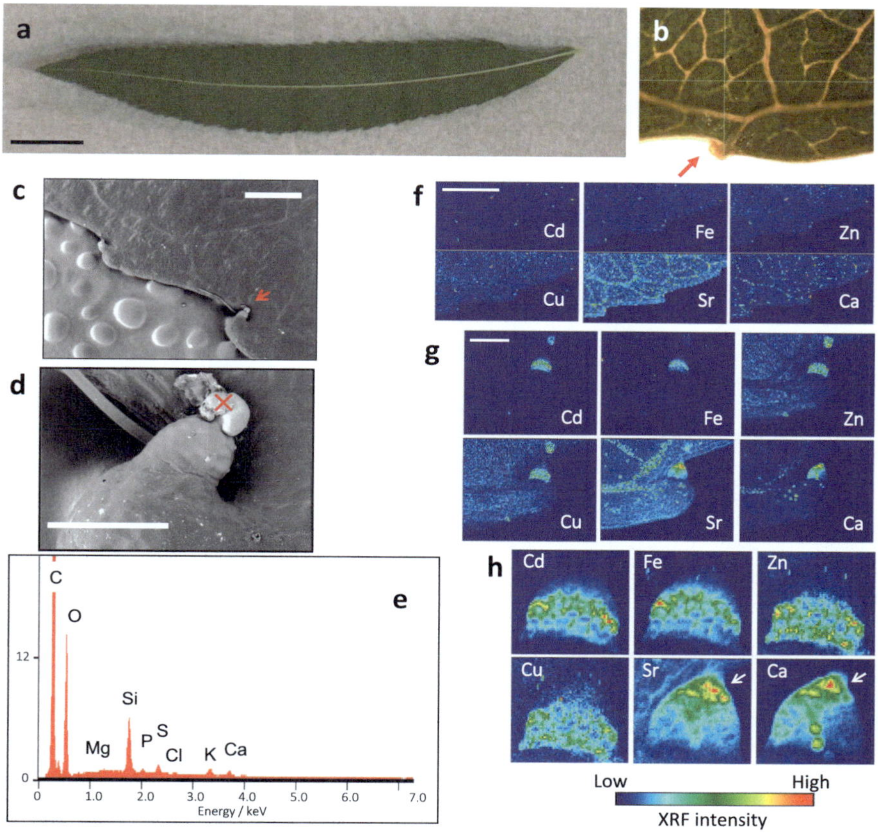

Fig. 6.13 Characterization of tip cells in the serrations of a willow (*S. miyabeana*) leaf by a combination of VPSEM-EDX and μ-XRF. Although these tip cells accumulated heavy metals, they were not responsible for their excretion in willow plants. (**a**) Photograph of an entire willow leaf. Bar = 1 cm. (**b**) A serration of a willow leaf (*red arrow*). (**c**) VPSEM observation of an intact leaf. Bar = 500 μm. (**d**) Secretion detected on the tip of a serration (an *arrow* in (**c**)). Bar = 100 μm. (**e**) Elemental analysis of the exudate in (**d**) by EDX. The exudate contained K, Ca, Si and several non-metallic elements, but the concentrations of heavy metals, such as Cd, Fe, Cu and Zn, were below the limit of detection. (**f**) μ-XRF imaging of a willow leaf segment. Step size: 50 μm, bar = 2000 μm. (**g**) Distribution of elements in a serration. Step size: 5 μm, bar = 200 μm. (**h**) The *white arrows* show that Ca and Sr were enriched in the tip of the serrations, but heavy metals were not. Step size: 2 μm; beam size: 1.8 μm × 1.8 μm; measurement time: 0.3 s per point in μ-XRF analysis (a-b: [15]; c-h: [14])

formation of bubbles in the resin should be avoided, such that the section is not disrupted. Several variations have been described including contact [5] and high-pressure [16] freezing methods.

Woody hyperaccumulators accumulate metals not only in leaves but in stems, especially in the bark (Fig. 6.14b; [14, 15]). To prepare cryosections from these blocks, each frozen block was mounted onto a cryomicrotome. Microtome blades for "thin cryosectioning" or "hard tissue block sectioning" are recommended for

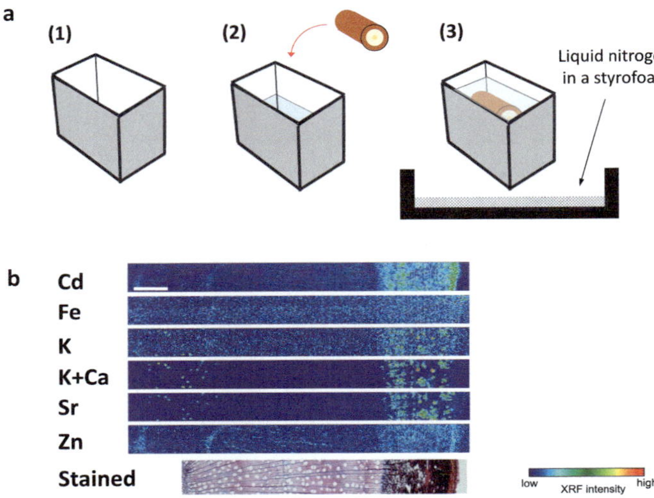

Fig. 6.14 Distribution of elements of a stem of a willow treated with Cd in hydroponic culture. (**a**) Preparation of a frozen block for cryosection. (1) A dish was formed of aluminum foil. (2) A small amount of cryoresin was poured into the bottom of the dish, and a segment of the stem (ca. 1 cm diameter) harvested from the Cd-treated willow was placed onto it. (3) Cryoresin was poured onto the plant segment. The bottom of the dish was placed onto the surface of liquid nitrogen to make a frozen block. (**b**) μ-XRF imaging of a stem of S. miyabeana. XRF signal intensity is shown in color. A double-stained image and two-dimensional distributions of Cd, Fe, K, K + Ca, Sr and Zn are shown in a 50 μm thick transverse segment of a stem. Step size: 5 μm × 5 μm. Bar = 300 μm. Beam size: 0.65 μm × 1.1 μm [15]

woody materials. Tungsten carbide blades recommended for especially hard bark, *e.g.*, cherry bark. A chamber temperature between −10 °C and −20 °C may also affect the quality of the cryosections.

X-ray imaging to determine the tissue and cellular distribution of metals involves visualization of a limited number of elements detectable under corresponding conditions. Staining of adjacent sections by other methods provides anatomical orientation and help in interpreting the results of SR-based imaging (Fig. 6.14b). Horizontal sections of the stem of a willow were anatomically identified by the staining of transverse sections with aqueous solutions of 1 % astra blue and 1 % safranine and visualization under an optical microscope [15]. Hydrated sections can be examined directly by optical microscopy [17].

6.4.5 *Liquid and Powder Samples*

The chemical state of elements can be determined by analysis of standard solutions. Liquid samples can be examined directly or allowed to infiltrate a piece of filter paper, which is sealed in a polyethylene bag.

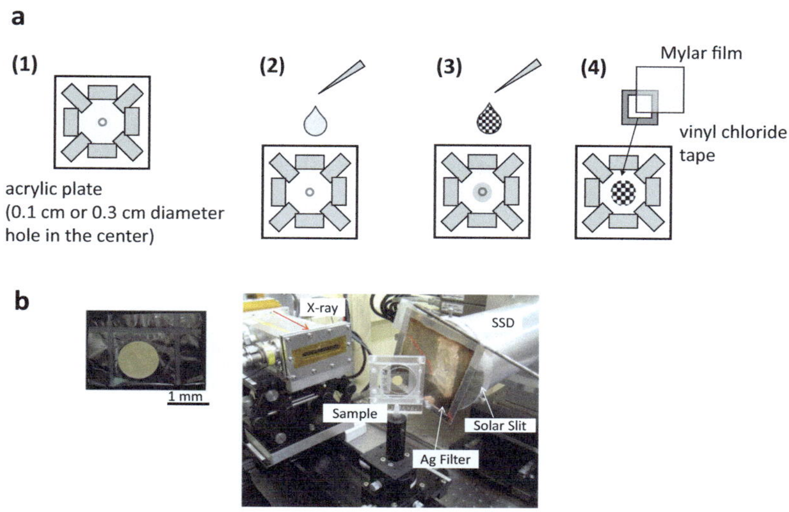

Fig. 6.15 System for metal distribution and speciation analysis in tobacco BY-2 cells using micro-XRF-XAFS analysis. (**a**) Mounting of a suspension culture of BY-2-cells. (1) A Mylar film was stretched and pasted in the center of the acrylic holder with a hole 1 mm or 3 mm in diameter. (2) α-Polylysine was dropped onto the Mylar film and allowed to spread evenly over its surface. (3) Cell suspension was then placed on the film. (4) The specimen was covered by another piece of Mylar film with the flame of a vinyl chloride tape (ca. 0.2 mm thickness), so that the Mylar film bag would not crush the cells, followed by freezing in liquid nitrogen. (**b**) Sample preparation for XAFS analysis in KEK PF-AR NW10A, Japan. BY-2 cells were freeze-dried for 48 h; tablets were formed and packed in a Mylar film bag

For example, bulk EXAFS was used to analyze Zn in the xylem sap of the Zn/Cd hyperaccumulator *Sedum alfredii* treated with 100 μM or 500 μM Zn [18]. To prevent ice crystal formation, 30 % glycerol was added to xylem sap, followed by freezing at −80 °C for EXAFS analysis. During the analysis, the samples were maintained at approximately 10 K in a liquid helium flow cryostat to minimize the loss of signal intensity. Chemically, most of the Zn in the xylem sap of *S. alfredii* is similar to aqueous Zn, with the remaining proportion associated with Zn citrate.

Metal distribution can also be assessed in cell suspension cultures (Fig. 6.15). A piece of Mylar film is initially coated with the positively charged polymer α-polylysine to improve cell adherence (Fig. 6.15a). Cells are harvested by centrifugation or filtration, freeze-dried, and subjected to speciation analyses (Fig. 6.15b).

Solid standards are ground in an agate mortar and pestle, followed by dilution, if necessary, with boron nitride powder. To obtain a satisfactory S/N ratio, the concentration of the standard chemical in boron nitrate should be calculated with the assistance of ATHENA software (described above), so that the absorbance can be adjusted to $\mu t = 1$. The powdered standard is placed in a polyethylene bag or dispersed on the sticky side of scotch tape.

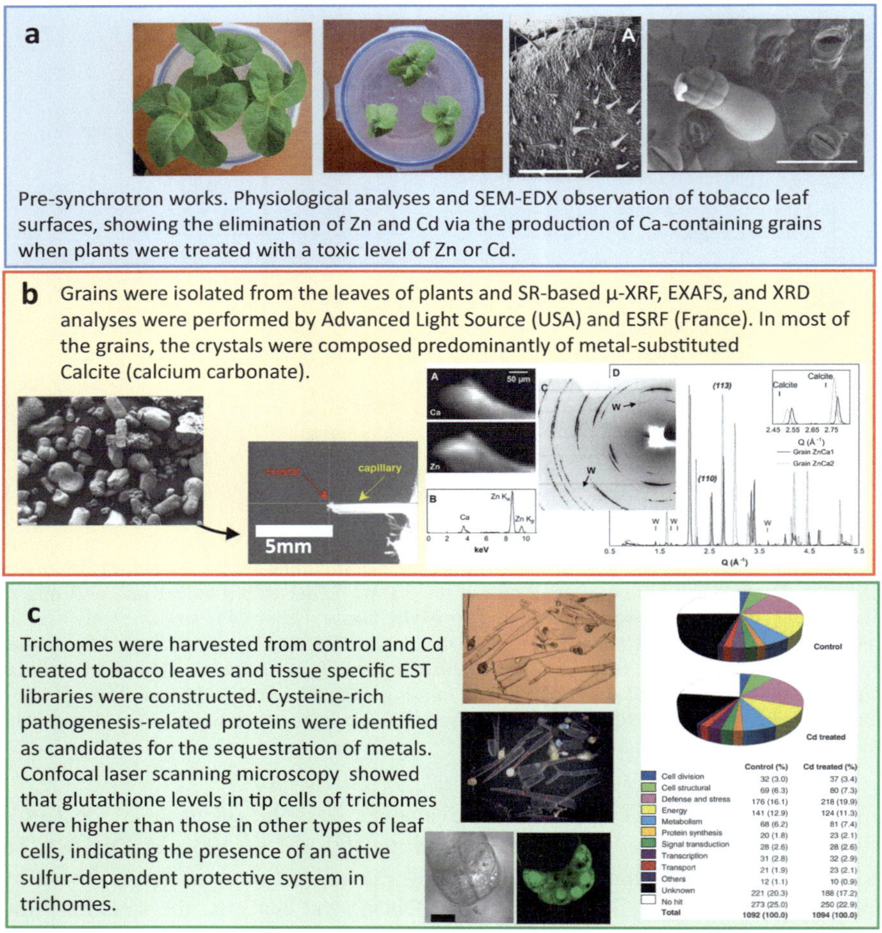

Fig. 6.16 Combination of SR-based X-ray microanalyses and plant physiology/molecular biology. Elucidation of metal exudation mechanism of tobacco (N. tabacum) leaf trichomes is shown as an example. (**a**) Physiological and electron microscopic analyses performed in advance of (**b**) SR-based analyses. (**c**) Advanced project based on the results of (**b**). (Pictures from [19–23])

6.4.6 Effects of Biomineralization on the Numbers of Grains Produced

Exposure of tobacco (*Nicotiana tabacum*) plants to toxic levels of Zn or Cd resulted in the production of Ca- and mineral-containing grains on the head cells of trichomes (Fig. 6.16a, [19–21]), indicating that tobacco plants actively exclude toxic Cd *via* these cells. To determine the mechanism of biomineralization, isolated grains were subjected to SR-based XRD, XRF, and EXAFS analyses.

To collect the grains, plants were immersed in 50-mL tubes containing pure water and vortexed for a few seconds. These grains had previously been shown to be insoluble in solutions between pH 4 and 12.5 but soluble under acidic conditions. Following centrifugation of the plant suspensions, the supernatants were carefully removed, and the grains were collected at the bottom with a pipette. A second isolation procedure consisted of careful collection of the grains using toothpicks under a stereomicroscope (Olympus SZX11-ST). Isolated tobacco grains were mounted on the top of a glass capillary tube, 0.2 mm in diameter, using adhesive glue containing no metal. The capillary tubes were fixed on a stage, and the grains were cooled to 150 K with a Cryostream cooler (Oxford 611) during measurements to minimize any potential beam damage (Fig. 6.16b). Figure 6.16c shows a schematic example of a combination of SR-XRF and a non-synchrotron-based method for assessing molecular biology.

6.5 Proposal Application for a Beamline

In proposing to use a beamline, the user must first obtain an ID number from the facility, because most synchrotron facilities accept online applications.

General application forms request the following information:

1. The design of the experiment, the method of measurement, the choice of SR facility and beamline, and the value and significance of the proposed research
2. Overall number of beamtimes (number of shifts) requested
3. Requested beamline equipment in the SR facility
4. Names of the project team members
5. Safety hazards of the items being brought to the facility
6. Value and significance of the proposed research
7. A list of publications by the project leader
8. Related pictures, if necessary

The targeted element and the spatial resolution of XRF mapping are important in selecting the synchrotron facility and beamline. For example, SPring-8 provides high-energy SR X-rays that excite the Cd K-line. The X-ray beam size defines the spatial resolution of XRF imaging. Individual proposing to use the microbeam (imaging cellular level) must add focusing optics (*e.g.*, KB mirror) to the list of equipment requested.

6.6 Prospective X-Ray Microanalyses in Metallomics

In this chapter, the authors attempt to encourage cross-disciplinary work to elucidate biological processes. Current knowledge of metal homeostasis, accumulation, and detoxification in plants is still limited. Recent progress in SR-based X-ray microanalyses has allowed determination of element distribution in single organelles and at the subcellular level. Combinations of physicochemical analyses and biological techniques, including those of molecular biology and biochemistry, may expand the frontier of plant biological science.

Acknowledgments The authors thank Ms. Saori Takada (Tokyo University of Science) and Mr. Fumihiro Masuyama (Tokyo Denki University) for their suggestions on preparing samples of biological materials.

References

1. Mobilio S, Boscherini F, Meneghini C (eds) (2014) Synchrotron radiation: basics, methods and applications. Springer, Berlin
2. Vijayan P, Willick IR, Lahlali R, Karunakaran C, Tanino KK (2015) Synchrotron radiation sheds fresh light on plant research: the use of powerful techniques to probe structure and composition of plants. Plant Cell Physiol 56:1252–1263. doi:10.1093/pcp/pcv080
3. Li YF, Chen C (2010) Chapter 6 X-ray absorption spectroscopy. In: Nuclear analytical techniques for metallomics and metalloproteomics. Royal Society of Chemistry, London
4. van Bokhoven JA, Lamberti C (2016) X-ray absorption and X-ray emission spectroscopy theory and applications. Wiley, Chichester
5. Fernando DR, Mizuno T, Woodrow IE, Baker AJ, Collins RN (2010) Characterization of foliar manganese (Mn) in Mn (hyper)accumulators using X-ray absorption spectroscopy. New Phytol 188:1014–1027. doi:10.1111/j.1469-8137.2010.03431.x
6. Ravel B, Newville M (2005) ATHENA, ARTEMIS, HEPHAESTUS: data analysis for X-ray absorption spectroscopy using IFEFFIT. J Synchrotron Radiat 12:537–541
7. Tsuji K, Injuk J, van Grieken R (2004) X-ray spectrometry: recent technological advances. Wiley, Chichester
8. Segura-Ruiz J, Martínez-Criado G, Denker C, Malindretos J, Rizzi A (2014) Phase separation in single In Ga N nanowires revealed through a hard X-ray synchrotron nanoprobe. Nano Lett 14:1300–1305
9. Baker AJM, Brooks RR (1989) Terrestrial higher plants which hyperaccumulate metallic elements –, a review of their distribution, ecology and phytochemistry. Biorecovery 1:81–126
10. Krämer U (2010) Metal hyperaccumulation in plants. Annu Rev Plant Biol 61:517–534
11. van der Ent A, Baker AJM, Reeves RD, Pollard AJ, Schat H (2013) Hyperaccumulators of metal and metalloid trace elements: facts and fiction. Plant Soil 362:319–334
12. Iwai T, Takahashi M, Oda K, Terada Y, Yoshida KT (2012) Dynamic changes in the distribution of minerals in relation to phytic acid accumulation during rice seed development. Plant Physiol 160:2007–2014. doi:10.1104/pp.112.206573
13. Kowata H, Nagakawa Y, Sakurai N, Hokura A, Terada Y, Hasegawa H, Harada E (2014) Radiocesium accumulation in Egeria densa, a submerged plant – possible mechanism of cesium absorption. J Anal At Spectrom 29:868–874. doi:10.1039/C3JA50346A
14. Harada E, Hokura A, Nakai I, Terada Y, Baba K, Yazaki K, Shiono M, Mizuno N, Mizuno T (2011) Assessment of willow (Salix sp.) as a woody heavy metal accumulator: field survey and in vivo X-ray analyses. Metallomics 3:1340–1346. doi:10.1039/C1MT00102G
15. Harada E, Hokura A, Takada S, Baba K, Terada Y, Nakai I, Yazaki K (2010a) Characterization of cadmium accumulation in willow as a woody metal accumulator using synchrotron radiation-based X-ray microanalyses. Plant Cell Physiol 51:848–853. doi:10.1093/pcp/pcq039
16. Manceau A, Simionovici A, Lanson M, Perrin J, Tucoulou R, Bohic S, Fakra SC, Marcus MA, Bedell JP, Nagy KL (2013) Thlaspi arvense binds Cu(II) as a bis-(L-histidinato) complex on root cell walls in an urban ecosystem. Metallomics 5:1674–1684. doi:10.1039/c3mt00215b

17. Yamaguchi N, Mori S, Baba K, Kaburagi-Yada S, Arao T, Kitajima N, Hokura A, Terada Y (2011) Cadmium distribution in the root tissues of solanaceous plants with contrasting root-to-shoot Cd translocation efficiencies. Environ Exp Bot 71:198–206. doi:10.1016/j.envexpbot.2010.12.002
18. Lu L, Tian S, Zhang J, Yang X, Labavitch JM, Webb SM, Latimer M, Brown PH (2013) Efficient xylem transport and phloem remobilization of Zn in the hyperaccumulator plant species Sedum alfredii. New Phytol 198:721–731. doi:10.1111/nph.12168
19. Choi YE, Harada E, Wada M, Tsuboi H, Morita Y, Kusano T, Sano H (2001) Detoxification of cadmium in tobacco plants: formation and active excretion of crystals containing cadmium and calcium through trichomes. Planta 213:45–50
20. Isaure MP, Sarret G, Harada E, Choi YE, Marcus MA, Fakra SC, Geoffroy N, Pairis S, Susini J, Clemens S, Manceau A (2010) Calcium promotes cadmium elimination as vaterite grains by tobacco trichomes. Geochim Cosmochim Acta 74:5817–5834
21. Sarret G, Harada E, Choi YE, Isaure MP, Geoffroy N, Fakra S, Marcus MA, Birschwilks M, Clemens S, Manceau A (2006) Trichomes of tobacco excrete zinc as zinc-substituted calcium carbonate and other zinc-containing compounds. Plant Physiol 141:1021–1034
22. Harada E, Choi YE (2008) Investigation of metal exudates from tobacco glandular trichomes under heavy metal stresses using a variable pressure scanning electron microscopy system. Plant Biotechnol 25:407–411. doi:10.5511/plantbiotechnology.25.407
23. Harada E, Kim JA, Meyer AJ, Hell R, Clemens S, Choi YE (2010b) Expression profiling of tobacco leaf trichomes identifies genes for biotic and abiotic stresses. Plant Cell Physiol 51:1627–1637. doi:10.1093/pcp/pcq118

Chapter 7
^{77}Se NMR Spectroscopy for Speciation Analysis of Selenium Compounds

Noriyuki Suzuki and Yasumitsu Ogra

Abstract Nuclear magnetic resonance (NMR) spectroscopy is widely used for the determination of the chemical structure of an organic compound. NMR spectroscopy is theoretically applicable to all metallic elements except cerium. In this chapter, we demonstrate selenium (Se) speciation in animal Se metabolites (selenometabolites) by NMR spectroscopy and review hitherto performed NMR spectroscopic studies of Se detection in biological samples. ^{77}Se, an NMR-active isotope of Se, was directly observed, but its NMR receptivity was lower than its sensitivity in mass spectrometry. However, the use of enriched stable isotope improved the receptivity. Each selenometabolite had its own chemical shift, suggesting that the chemical shift of ^{77}Se could be used as a fingerprint. Indirect measurement by heteronuclear multiple bond correlation (HMBC) spectroscopy with ^{1}H nuclides was also effective for the Se speciation.

Keywords NMR • Selenium • Speciation • HMBC • Stable isotope • Selenosugar • Selenomethionine • Selenohomolanthionine

7.1 Introduction

Nuclear magnetic resonance (NMR) spectroscopy is used for the investigation of typical elements, such as ^{1}H, ^{13}C, ^{19}F, and ^{31}P, and is, in principle, applicable to almost all of the metallic elements on the periodic table. All metals except cerium (Ce) have at least one NMR-active isotope, and therefore, NMR spectroscopy can be used for the direct detection of organometallic compounds and inorganic metals bound to a protein, peptide, or another biomolecule [1]. Differences in nuclear properties, such as resonance frequency and nuclear spin, among NMR-active nuclides can yield specific receptivity (sensitivity to NMR measurement), linewidth, and chemical shift (Table 7.1) [2]. However, few nuclides are practically observable due to their physicochemical properties.

N. Suzuki, Ph.D. (✉) • Y. Ogra
Laboratory of Toxicology and Environmental Health, Graduate School of Pharmaceutical Sciences, Chiba University, 1-8-1 Inohana, Chuo, Chiba 260-8675, Japan
e-mail: n-suzuki@chiba-u.jp

Y. Ogra, T. Hirata (eds.), *Metallomics*, DOI 10.1007/978-4-431-56463-8_7

Table 7.1 Basic properties of ^{77}Se for NMR spectroscopy

Spin	1/2
Natural abundance	7.63 %
Chemical shift range	from −1000 to 2000
Reference compound	Me_2Se
T1	~30 s
Receptivity (to ^{1}H at natural abundance)	5.37×10^{-4}
Receptivity (to ^{1}H when enriched)	7.04×10^{-3}

Selenium (Se) is an element belonging to the same group 16 on the periodic table as oxygen, sulfur, and tellurium. Se has chemically and biologically ambivalent characteristics; it has similar chemical properties to sulfur as a typical element, yet possesses chemical properties characteristic of a metal. Se is an essential element in animals but can become highly toxic when ingested in an amount that exceeds the nutritional level. Se is required as the active center of selenoproteins that function as an antioxidant and participate in thyroid hormone production, DNA synthesis, and fertilization [3, 4]. It acts as an active center in the formation of a selenol group (−SeH) on a selenocysteinyl (SeCys) residue in a selenoprotein sequence [5, 6]. Se ingested at physiological levels is excreted in urine as a selenosugar (*Se*-methylseleno-*N*-acetyl-galactosamine, selenosugar 1) and a trimethylselenonium ion (TMSe). In addition, *Se*-methylseleno-*N*-acetyl-glucosamine (selenosugar 2) and *Se*-methylseleno-galactosamine (selenosugar 3) are detected in urine as minor metabolites [7–9]. The ingestion of excess amounts of Se would yield volatile selenometabolites, such as dimethyldiselenide and dimethylselenide, in exhaled breath. On the other hand, Se is not essential in plants and is a bystander mineral for plant growth. Se-containing metabolites (selenometabolites), such as methylselenocysteine (MeSeCys), γ-glutamylmethylselenocysteine (GluMeSeCys), selenohomolanthionine (SeHLan), and selenomethionine (SeMet), are biosynthesized in certain plants [10, 11] and fungi [12].

^{77}Se is one of the nuclides detectable by NMR spectroscopy because of its relatively high NMR receptivity. Thus, ^{77}Se NMR spectroscopy has been utilized for the identification of Se compounds in organic chemistry [2]. Although ^{77}Se NMR spectroscopy is a reliable tool for the identification of organoselenium compounds and inorganic Se compounds, it has been rarely used in biological samples because the compounds are present in trace amounts.

Recent advances in NMR instrumentation and measurement techniques have enabled us to investigate Se compounds routinely without the need for special knowledge or equipment. Because Se is an ultra-trace element in organisms, it seems suitable to evaluate recent NMR techniques for metallomics research. In this chapter, we demonstrate the speciation of selenometabolites by ^{77}Se NMR spectroscopy.

7.2 Advantages of NMR Speciation Over Hyphenated Techniques

At the initial stage of Se analysis, much attention was given to the determination of total Se concentration in biological samples. Recent studies, however, have concentrated on the elucidation of the chemical structures of selenometabolites to clarify the physiological role of Se. In other words, both quantitative and qualitative analyses are required to understand the biological and toxicological roles of Se in biological samples. In this regard, the development of reliable techniques for Se speciation in biological samples is necessary to understand the absorption, distribution, metabolism, and excretion of Se. One of the most powerful methods for Se speciation in biological samples is high-performance liquid chromatography hyphenated with inductively coupled plasma mass spectrometry (LC-ICP-MS) because it has high sensitivity and selectivity for the determination of Se and is robust to biological matrices. Indeed, LC-ICP-MS has provided exceptional results [13–17]. However, although LC-ICP-MS has merits, it also has limitations for Se speciation in biological samples, namely, HPLC cannot detect unstable metabolic intermediates, and LC-ICP-MS is not suitable for the identification of unknown metabolites. In contrast, NMR speciation is applicable to these samples. An alternate hyphenated technique, HPLC coupled with electrospray ionization mass spectrometry, i.e., LC-ESI-MS, was developed for Se speciation [18]. In contrast to LC-ICP-MS, LC-ESI-MS can be used for the identification of unknown metabolites. However, exhaustive pretreatment is required to reduce the matrix in a biological sample. On the other hand, NMR spectroscopy is applicable to crude sample analysis without the need for any pretreatment or HPLC separation in the presence of a matrix because of its selectivity and specificity to each Se compound. In addition, NMR spectroscopy is a powerful method for directly probing the chemical structure and redox properties of Se atom. The X-ray absorption fine structure (XAFS) technique can also be used for Se speciation because it can provide the oxidation number and the coordination number of an element on the basis of the X-ray absorption near-edge structure (XANES) spectrum, as well as the structural information of neighboring atoms of an element, such as the number of ligand atoms and the distance between the element and each ligand atom, on the basis of the extended X-ray absorption fine structure (EXAFS) region [19]. In the same manner as NMR spectroscopy, the XAFS technique does not require any sample pretreatment; however, XAFS data are less informative than NMR data from the point of view of identification of Se compounds. Understandably, the major disadvantage of NMR spectroscopy compared to mass spectrometry is its low sensitivity, and because of this, it has not been used for Se speciation in biological samples in studies conducted to date. In the following sections, we highlight unique NMR techniques for Se speciation.

7.3 Observation of ^{77}Se Nuclide in Authentic Standards

7.3.1 Chemical Shifts of Bio-selenocompounds by Direct Detection

Due to the low natural abundance (7.6 %) of the NMR-active isotope ^{77}Se and the long relaxation time, Se was considered to be not an ideal element for NMR spectroscopy. In the direct detection of ^{77}Se nuclide, the low NMR receptivity of Se could be overcome by increasing the scan number. The relationships between the scan number and the signal-to-noise (S/N) ratio of ^{77}Se are shown in Fig. 7.1a. In this experiment, sodium selenite dissolved in sodium phosphate buffer (pH 7.4) at various concentrations was directly analyzed by an NMR instrument (ECZ600, JEOL, Tokyo, Japan), and the lowest concentration where ^{77}Se at natural abundance was detected was 50 mM under our analytical conditions. The data offer clues to knowing intuitively the NMR receptivity of Se. The S/N ratios were improved by enrichment (99.8 %) of ^{77}Se (Fig. 7.1b).

Similarly to ^{1}H and ^{13}C NMR chemical shifts, the ^{77}Se NMR chemical shifts are definitive for compounds containing ^{77}Se [20]. The chemical shifts of Se compounds having reduced Se (−II) are observed in the relatively high-field region, and those having oxidized Se (from + IV to + VI) are observed in the low-field region. This can be explained by the fact that nuclei in the high electron density region are more shielded from the applied field than those in the low electron density region. For the Se speciation by NMR spectroscopy, dimethylselenide is used as chemical shift reference ($\delta = 0$) [21]. The chemical structures of Se compounds in our library of selenometabolites are shown in Fig. 7.2. The chemical shifts measured by NMR ECZ600 are also shown in Fig. 7.2. Important selenometabolites, such as selenoamino acids, a selenosugar, and methylated Se metabolite, were observed in the narrow region from 0 to 300 ppm. The full width at half maximum of each ^{77}

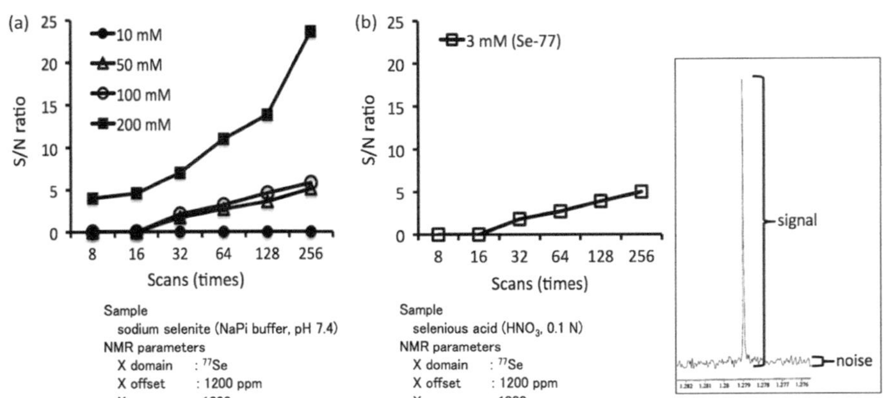

Fig. 7.1 Relationship between scan number and signal-to-noise (S/N) ratio of ^{77}Se

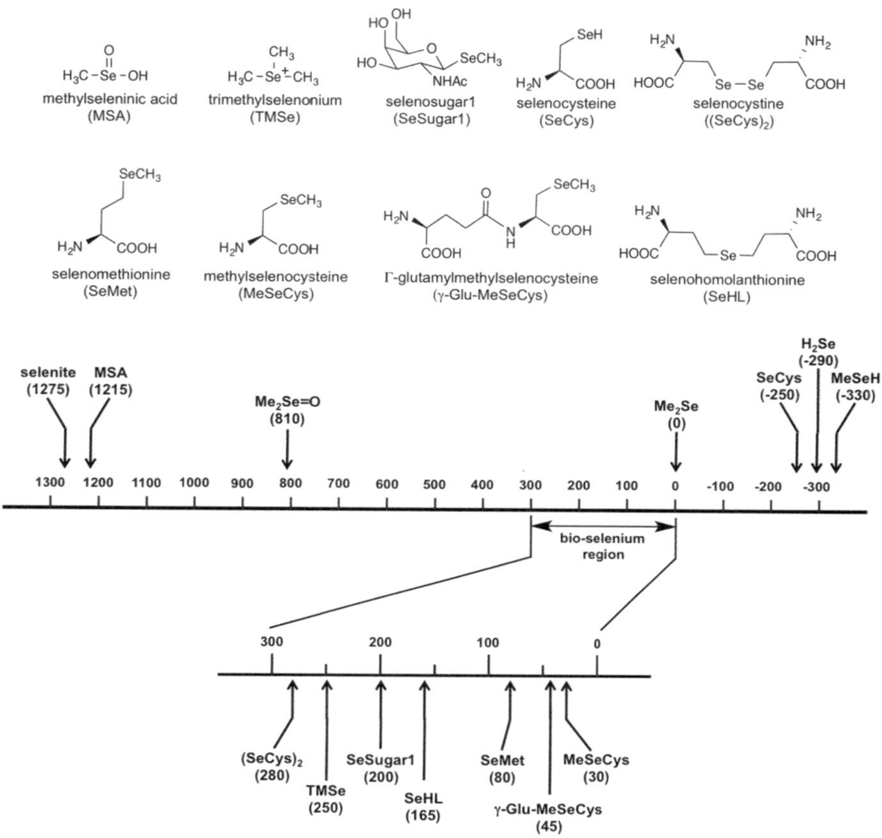

Fig. 7.2 Chemical structures and chemical shifts of Se compounds in our library of selenometabolites measured by NMR (ECZ600, JEOL)

Se signal was so narrow that each Se compound could be identified on the basis of its own chemical shift like a fingerprint. Hence, we propose that the region be named "bio-selenium region."

7.3.2 *Advanced Techniques for Detection of Bio-selenocompounds*

Because most selenometabolites contain a Se-C-H or a Se-C-C-H bond in their chemical structures, the S/N ratio would be reduced by geminal and vicinal ^{77}Se-^{1}H coupling when the detection is conducted in the single pulse mode without ^{1}H decoupling. Indeed, the S/N ratio was markedly increased in the complete ^{1}H decoupling mode by proton irradiation, although signal enhancement by the nuclear Overhauser effect was hardly observed.

Another method for enhancing ^{77}Se signal intensity is the indirect measurement by heteronuclear multiple bond correlation (HMBC) spectroscopy with ^{1}H nuclides [22–25]. As an example, the HMBC spectrum of selenosugar 1 is shown in Fig. 7.3. Because the two-dimensional correlation spectrum of ^{77}Se and ^{1}H could be obtained, this method would provide decisive information for chemical structure determination. Furthermore, because this technique is an inverse detection experiment using ^{1}H nuclei, the S/N ratio is markedly increased, namely, the sensitivity obtained at 50 mM would be the same as that obtained at 200 μM.

These advanced techniques for ^{77}Se detection by NMR spectroscopy are useful for Se speciation in biological samples.

7.4 Application of ^{77}Se NMR Spectroscopy to Se Speciation in Biological Samples

As an example of an application of ^{77}Se NMR spectroscopy to Se speciation, the HMBC spectrum of ^{77}Se and ^{1}H of Se compound(s) in a dietary Se supplement (“Yeast-Free” Selenium 200 mcg, Vitamin World®, New York, USA) is shown in Fig. 7.4. The procedures for sample preparation are summarized below.

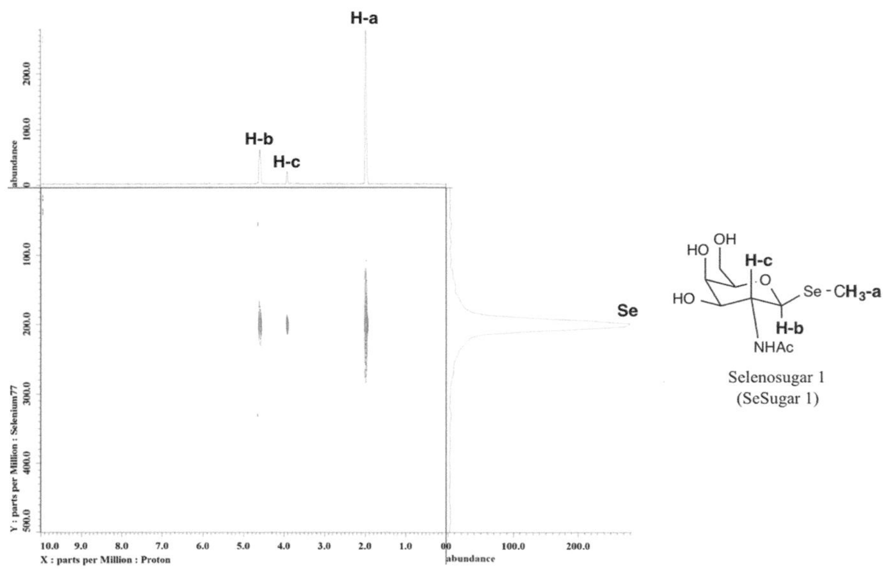

Fig. 7.3 Heteronuclear multiple bond correlation spectrum of selenosugar 1 (50 mM in D_2O)

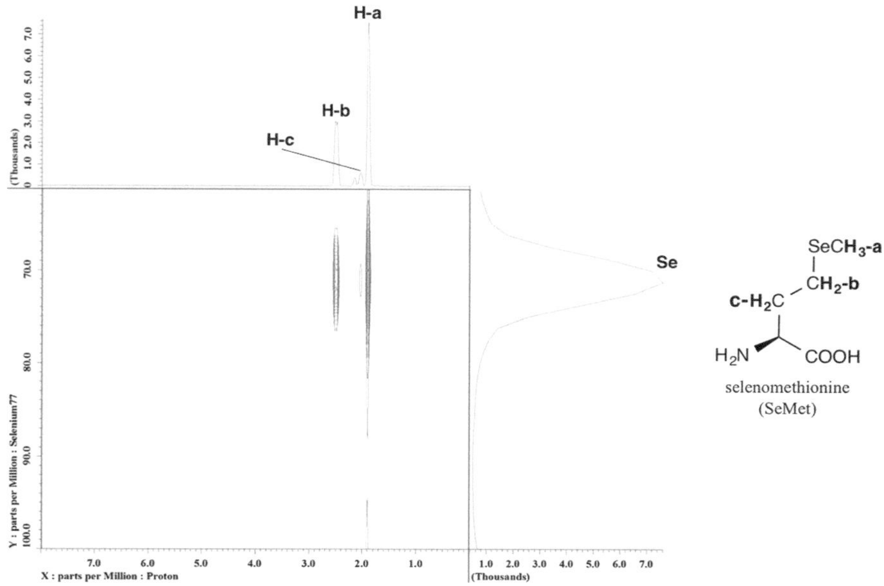

Fig. 7.4 Heteronuclear multiple bond correlation spectrum of Se compound in dietary supplement in D_2O

Five tablets of the dietary Se supplement were ground and extracted with a water and methanol mixture (1:1). The extract was lyophilized, and the lyophilizate was dissolved in D_2O. The HMBC spectrum of the lyophilizate corresponded to that of selenomethionine. The result coincided with the ingredient label showing that selenomethionine is the major Se species in the supplement (Fig. 7.4).

Recent studies have used ^{77}Se NMR spectroscopy for Se speciation in biological samples. Mobli et al. carried out ^{77}Se NMR measurements to determine the p*K*a values of selenocysteine residues incorporated into a bioactive peptide hormone and a neurotransmitter and demonstrated that the p*K*a values of selenocysteine residues in peptide were substantially lower than that of free selenocysteine [26]. Schaefer et al. employed the ^{77}Se NMR technique to probe the local electronic environment of reactive selenocysteine residues by preparing ^{77}Se-enriched proteins [27].

7.5 Conclusions

^{77}Se NMR spectroscopy provides concrete information of the chemical structures of selenometabolites, although the NMR receptivity of ^{77}Se in a biological sample is unsatisfactory compared with the sensitivity in mass spectrometry. The use of enriched stable isotopes is one of the ways to overcome the disadvantages of NMR spectroscopy. The complementary use of LC-ICP-MS, LC-ESI-MS(-MS), and NMR would be the best analytical approach for the identification of unknown selenometabolites. In addition, we expect that speciation analysis by NMR spectroscopy increasingly become an important technique in metallomics research of not only ^{77}Se but also other nuclides of metals/metalloids. Further development of the measurement technique and instrumentation is expected to lead to the widespread use of NMR spectroscopy in metallomics.

Acknowledgment This work was supported by JSPS KAKENHI Grants-in-Aid for Scientific Research (Grant Numbers 26460032 (N.S.), 26293030 (Y.O.), 15 K14991 (Y.O.), and 16H05812 (Y.O.)).

References

1. Bernhard M, Brinckman FE, Sadler PJ (eds) (1986) The importance of chemical "Speciation" in environmental processes. Springer, Heidelberg
2. Block E, Glass RS, Jacobsen NE, Johnson S, Kahakachchi C, Kamiński R, Skowrońska A, Boakye HT, Tyson JF, Uden PC (2004) Identification and synthesis of a novel selenium-sulfur amino acid found in selenized yeast: rapid indirect detection NMR methods for characterizing low-level organoselenium compounds in complex matrices. J Agric Food Chem 52:3761–3771
3. Suzuki KT, Doi C, Suzuki N (2008) Simultaneous tracing of multiple precursors each labeled with a different homo-elemental isotope by speciation analysis: distribution and metabolism of four parenteral selenium sources. Pure Appl Chem 80:2699–2713
4. Kövér KE, Kumar AA, Rusakov YY, Krivdin LB, Illyés T-Z, Szilágyi L (2011) Experimental and computational studies of $^{n}J(^{77}$Se, ^{1}H) selenium-proton couplings in selenoglycosides. Magn Reson Chem 49:190–194
5. Lardon M (1970) Selenium and proton nuclear magnetic resonance measurements on organic selenium compounds. J Am Chem Soc 92:5063–5066
6. Lu J, Holmgren A (2009) Selenoproteins. J Biol Chem 284:723–727
7. Szpunar J, Lobinski R (2003) Hyphenated techniques in speciation analysis. Royal Society of Chemistry, Cambridge
8. Böck A, Flohé L, Köhrle J (2007) Selenoproteins-biochemistry and clinical relevance. Biol Chem 38:985–986
9. Gammelgaard B, Bendahl L (2004) Selenium speciation in human urine samples by LC- and CE-ICP-MS-separation and identification of selenosugars. J Anal At Spectrom 19:135–142
10. Suzuki KT, Kurasaki K, Suzuki N (2007) Selenocysteine β-lyase and methylselenol demethylase in the metabolism of Se-methylated selenocompounds into selenide. Biochim Biophys Acta 1770:1053–1061
11. Weekley CM, Aitken JB, Finney L, Vogt S, Witting PK, Harris HH (2013) Selenium metabolism in cancer cells: the combined application of XAS and XFM techniques to the problem of selenium speciation in biological systems. Nutrients 5:1734–1756
12. Luthra NP, Dunlap RB, Odom JD (1983) The use of dimethyl selenide as a chemical shift reference in ^{77}Se NMR spectroscopy. J Magn Reson 52:318–322
13. Schroeder TB, Job C, Brown MF, Glass RS (1995) Indirect detection of selenium-77 in nuclear magnetic resonance spectra of organoselenium compounds. Magn Reson Chem 33:191–195
14. Mobli M, Morgenstern D, King GF, Alewood PF, Muttenthaler M (2011) Site-specific pK_a determination of selenocysteine residues in selenovasopressin by using 77Se NMR spectroscopy. Angew Chem Int Ed 50:11952–11955

15. Ogra Y, Kitaguchi T, Ishiwata K, Suzuki N, Iwashita Y, Suzuki KT (2007) Identification of selenohomolanthionine in selenium-enriched japanese pungent radish. J Anal At Spectrom 22:1390–1396
16. Kobayashi Y, Ogra Y, Ishiwata K, Takayama H, Aimi N, Suzuki KT (2002) Selenosugars are key and urinary metabolites for selenium excretion within the required to low-toxic range. Proc Natl Acad Sci U S A 9:15932–15936
17. Zhang M, Vogel HJ (1994) Two-dimensional NMR studies of selenomethionyl calmodulin. J Mol Biol 239:545–554
18. Tan K-S, Arnold AP, Rabenstein DL (1988) Selenium-77 nuclear magnetic resonance studies of selenols, diselenides, and selenenyl sulfides. Can J Chem 66:54–60
19. Foster SJ, Ganther HE (1984) Synthesis of [^{75}Se]trimethylselenonium iodide from [^{75}Se] selenocystine. Anal Biochem 137:205–209
20. Ogra Y (2008) Integrated strategies for identification of selenometabolites in animal and plant samples. Anal Bioanal Chem 390:1685–1689
21. Ogra Y, Kitaguchi T, Ishiwata K, Suzuki N, Toida T, Suzuki KT (2009) Speciation of selenomethionine metabolites in wheat germ extract. Metallomics 1:78–86
22. Ramadan SE, Razak AA, Yousseff YA, Sedky NM (1988) Selenium metabolism in a strain of Fusarium. Biol Trace Elem Res 18:161–170
23. Reeves MA, Hoffmann PR (2009) The human selenoproteome: recent insights into functions and regulation. Cell Mol Life Sci 66:2457–2478
24. Suzuki KT, Somekawa L, Kurasaki K, Suzuki N (2006) Simultaneous tracing of ^{76}Se-selenite and ^{77}Se-selenomethionine by absolute labeling and speciation. Toxicol Appl Pharmacol 217:43–50
25. Schaefer SA, Dong M, Rubenstein RP, Wilkie WA, Bahnson BJ, Thorpe C, Rozovsky S (2013) ^{77}Se enrichment of proteins expands the biological NMR toolbox. J Mol Biol 425:222–231
26. Suzuki KT, Ogra Y (2002) Metabolic pathway for selenium in the body: speciation by HPLC-ICP MS with enriched Se. Food Addit Contam 19:974–983
27. Tsuji Y, Suzuki N, Suzuki KT, Ogra Y (2009) Selenium metabolism in rats with long-term ingestion of Se-methylselenocysteine using enriched stable isotopes. J Toxicol Sci 34:191–200

Chapter 8
Protein Quantification and Quantitative Phosphorylation Analysis by the Determination of Hetero Atoms (S and P) by Means of nanoHPLC-ICPMS

Yoshinari Suzuki and Naoki Furuta

Abstract It becomes easier to identify proteins via determination of peptide mass or peptide sequence by ESI-MS or MALDI-MS. On the other hand, the quantification of proteins remains a major challenge in modern proteomics. Protein quantification was performed by the determination of sulfur (S) in trypsin-digested peptides by means of nanoHPLC-ICPMS under a flow of O_2 in an octopole reaction cell. Peptide concentrations were calculated as the ratio of the S concentration/the number of S atoms in the peptide/peak. Finally, protein concentration can be quantified through the peptide quantification, after taking into account tryptic digestion efficiency, via S quantification or phosphorus (P) quantification using ICPMS. This method was applied to human serum albumin and β-casein, and then the quantified protein concentrations showed a good agreement with the theoretical values, respectively. Degree of phosphorylation was also calculated as the P concentration divided by the peptide concentration. This quantitative phosphorylation analysis was applied to β-casein in cow milk.

Keywords Protein quantification • Phosphorylation • Sulfur • Phosphorus • NanoHPLC • ICPMS • Tryptic digestion • Peptides • Albumin • Casein

---

Y. Suzuki
Faculty of Life and Environmental Science, Department of Regional Environmental Science, Shimane University, 1060 Nishikawatsu-cho, Shimane 690-8504, Japan

N. Furuta (✉)
Faculty of Science and Engineering, Department of Applied Chemistry, Chuo University, 1-13-27 Kasuga, Bunkyo-ku, Tokyo 112-8551, Japan
e-mail: nfuruta@chem.chuo-u.ac.jp

Y. Ogra, T. Hirata (eds.), *Metallomics*, DOI 10.1007/978-4-431-56463-8_8

8.1 Introduction

Since the Human Genome Project was completed in 2003, the subject of bioscience has shifted to the next step of genomics, called "the post-genomic era," which attempts to explain biological phenomena by using genomic information. Among the many research fields of the post-genomic era, proteomics, which is the study of proteins synthesized based on genomic information, has drawn much concern because proteins play a central role in life activity. Proteome analysis has been greatly facilitated by the development of protein and peptide sequence databases by means of genomic analysis. In other words, it has become easier to identify proteins via the determination of peptide mass or peptide sequence by using electrospray ionization mass spectrometry (ESI-MS) or matrix-assisted laser desorption/ionization mass spectrometry (MALDI-MS) [1]. However, the answers to many biological questions require not only identification but also quantification of the proteins in a cell or tissue; therefore, the quantification of proteins remains a major challenge in modern proteomics. For protein quantification, various methods are already established. For example, the UV absorbance of aromatic amino acids at 280 nm, colorimetry methods such as the Bradford and Lowry assays, the bicinchoninic acid method, and gel staining with Coomassie Brilliant Blue R-250 or silver are well established and used for the quantification of crude proteins [2–7]. These are simple methods that are appropriate for many samples; however, they can have problems, such as fluctuations of the absorption depending on the protein and the presence of interfering substances.

In addition to the abovementioned methods, antigen-antibody reactions and mass spectrometry are used for protein quantification. The antigen-antibody reaction approach is based on the formation of an immune complex via the specific binding of an antibody to an antigen. Many methods have been developed that involve competitive or noncompetitive binding with a labeled substance, such as the radioimmunoassay, immunoradiometric assay, and enzyme-linked immunosorbent assay. Although these are highly specific methods, they are often time-consuming, and they always require an appropriate antibody. In other words, it is impossible to apply these methods to an unknown protein. With mass spectrometry, there can be many problems for quantification: proteins may not be completely tryptically digested, the ionization efficiencies of ESI-MS can fluctuate depending on the matrices and peptides, and the intensities of MALDI-MS signals can fluctuate depending on the area irradiated by the laser. Quantification of the relative abundances of peptides or proteins by using mass spectrometry can be achieved through the use of chemical labeling methods involving the derivatization of selected amino acid residues with isotopically (^{2}H, ^{13}C, ^{15}N, ^{18}O) modified reagents, such as ICAT (isotope-coded affinity tag) [8], ICPL (isotope-coded protein label) [9], and iTRAQ (isobaric tag for relative and absolute quantification) [10]. These methods require not only a specific antibody but also amino acid-specific labeling reagents. New protein quantification methods that do not require protein standards or specific reagents are needed.

Despite remarkable methodological and instrumental developments, especially within the field of mass spectrometry, the quantification of biomolecules is still not straightforward.

In addition to quantitative proteome analysis, analysis of posttranslational modification which alters protein functions has become important task. Especially, reversible phosphorylation is one of the most important posttranslational modifications and has attracted attention from a viewpoint of drug development.

Phosphorylation of protein could occur at serine (Ser), threonine (Thr), and tyrosine (Tyr) residues. Particularly, phosphorylation of Tyr involves cellular functions such as cell proliferation and differentiation, and abnormal control system of phosphorylation could be the main cause of disease such as cancer [11, 12]. Therefore, it is important to quantitatively evaluate degree of protein phosphorylation. Normally, key information of phosphorylation analysis using mass spectrometry is based on mass shift of +51 owing to additional phosphate group [12]. Hence, it is difficult to identify phosphorylated peptides from lots of peaks using ESI-MS/MS.

ICPMS, which is an element-specific detection method, does not provide any structural information about analyte molecules, because they are introduced into a high-temperature ion source and are broken down into atoms that are then ionized. In the case of HPLC-ICPMS, species are identified by comparing their retention times with those of standards. However, a key advance of ICPMS is a general detection method that allows species nonspecific element-selective quantification because ICPMS can generate singly charged ions effectively, and the responses are virtually compound- and matrix-independent. From the viewpoint of atomic mass spectrometry, ICPMS has been used to quantify metal proteins such as metallothionein and selenoproteins [13–18]. In addition to metallic and metalloid elements, nonmetal elements, such as sulfur (S) and phosphorus (P), are considered to be important for protein quantification. In particular, S is a key element for protein and peptide qualification because most proteins contain two S-containing amino acids, cysteine and methionine; cysteine is present in 96.6 % of human proteins, and methionine is present in 98.8 % [19]. As reported previously, it is easy to quantify a peptide if we know how many S atoms are contained in that peptide or protein [20–22]. But in the case of an unknown protein or peptide, it is impossible to determine how many S atoms are contained in a peptide by using ICPMS alone. As long as trypsin cleaves every arginine and lysine residue equally, the S concentration in a peptide should correlate with the number of S atoms contained in that peptide. Also, it is considered that the ratio of the S concentration/peptide concentration becomes an integral ratio that reflects the number of S atoms in the peptide. Also, to quantify the protein concentration from peptides, it is necessary to determine the efficiency of tryptic digestion. Moreover, quantitative phosphorylation analysis via accurate determination of P/S ratio is also possible.

For the determination of S and P using ICPMS, spectral interferences with major isotopes such as oxygen ions ($^{15}N^{16}O^+$, $^{16}O_2{}^+$, and $^{16}O^{18}O^+$) on $^{31}P^+$, $^{32}S^+$, and $^{34}S^+$ are big problems. Therefore, measurement of these isotopes can be performed with high-resolution ICPMS and more recently with reaction/collision cell ICPMS; for the latter case, Xe is used for S determination, and O_2 is used for both elements [23]. Polyatomic ions formed in plasma are suppressed by the addition of Xe into a collision cell. The first ionization potential of Xe (12.130 eV) is smaller than that of O (13.617 eV). Therefore, the charge of O is transferred to Xe, and then spectral interferences decrease due to the prevention of oxygen ionization. O_2 gas is used as a reaction gas for generation of oxide ions, especially for multi-element analysis. Due to the generation of oxidized ion, it is possible to generate a mass shift of plus 16, allowing simultaneous detection of phosphorus at 47 and sulfur at 48, which are less interference mass than 31 and 32.

Sample volumes from biopsy sample or cultured cell are often limited. Therefore, it is required that analytical system is able to cope with small-volume sample. For separation, nanoHPLC technique is one of the recent trends because of high separation efficiency, capability of introducing small amount of samples, low consumption of solvent, and so on, in analytical chemistry. NanoHPLC using 75 μm i.d. columns with eluent flow rates between 200 and 400 nL min^{-1} has become a key technique for high-efficiency separations of complex mixtures. If protein quantification methods, which enable to detect S-containing protein, by means of nanoHPLC-ICPMS, are established, it could analyze wide diverse (more than 300,000) proteome quantitatively.

In this study, we quantified peptides in a small amount of sample via S quantification using a nanoHPLC-ICPMS system equipped with an octopole reaction cell (ORC) and Xe or O_2 gas. By using this method of peptide quantification and taking into account the tryptic digestion efficiency, we were able to quantify human serum albumin in a test sample.

8.2 Experimental

8.2.1 Reagents

Standard solutions of S (10,000 mg L^{-1}, SPEX, Metuchen, NJ) and Ba (1000 mg L^{-1}, Kanto Chemical Co., Inc., Tokyo, Japan) were diluted and used for the optimization of the O_2 and Xe gas flow rates. The S standard was also used for S calibration. Three commercially available peptide standards (vasopressin, oxytocin, and amyloid β (human, 25–35)) were purchased from Peptide Institute Inc. (Osaka, Japan) and used for evaluating the nanoHPLC-ICPMS system. Vasopressin (CYFQNCPRG, M_W: 1084.2) and oxytocin (CYIQNCPLG, M_W: 1007.2) contain two cysteine residues and aromatic amino acids. Amyloid β (human, 25–35)

(GSNKGAIGLM, M_W: 1060.3) contains one methionine but does not contain any aromatic amino acids. A commercially available S-containing protein, human serum albumin (HSA; Sigma-Aldrich, Tokyo, Japan) and β-casein (from bovine milk, Sigma-Aldrich) were used for polyacrylamide gel electrophoresis (PAGE) and tryptic digestion.

N,N′-methylenebisacrylamide (electrophoresis grade: Nacalai Tesque, Inc., Kyoto, Japan), tris(hydroxymethyl)aminomethane (Sigma-Aldrich, Tokyo, Japan), ammonium persulfate (electrophoresis grade: Nacalai Tesque, Inc.), glycerol (99 %: Kanto Kagaku, Tokyo, Japan), Coomassie Brilliant Blue R-250 (electrophoresis grade: Nacalai Tesque, Inc.), methanol (99 %: Kanto Kagaku), and acetic acid (99 %, Kanto Kagaku) were used for electrophoresis. Ammonium bicarbonate (Kanto Kagaku), dithiothreitol (Sigma-Aldrich, Tokyo, Japan), iodoacetamide (Kanto Kagaku), and trypsin (from bovine pancreases, Sigma-Aldrich Co., St. Louis, MO) were used for tryptic digestion.

8.2.2 Tryptic Digestion Procedure

Native PAGE was performed according to a method of Laemmli [24] in slab gels (10 % gel) under nonreducing and non-denaturing conditions. After preparing of 36.4 μmol L^{-1} HSA, 20 μl of a sample (50 μg $lane^{-1}$) was applied to a lane. After electrophoresis, the proteins in the gel were stained for 30 min with 0.1 % Coomassie Brilliant Blue R-250 in 10 % acetic acid and 30 % methanol, and then the gel was destained in 10 % acetic acid and 30 % methanol. Gel bands containing albumin were excised from the gel, cut into pieces, and decolored with 50 % acetonitrile and 25 mmol L^{-1} ammonium bicarbonate (AB). The decolored sample was reduced with 10 mmol L^{-1} dithiothreitol and 25 mmol L^{-1} AB at 56 °C for 1 h and then alkylated with 55 mmol L^{-1} iodoacetamide and 25 mmol L^{-1} AB at room temperature for 45 min in a test tube. When the reduction and alkylation processes were finished, the excess solvent was removed, and the gel was carefully washed with 25 mmol L^{-1} AB. After that, 10 μg mL^{-1} trypsin and 25 mmol L^{-1} AB were added to the sample, which was then kept on ice for 30 min. Excess trypsin was then removed, and the sample was washed with 25 mmol L^{-1} AB and finally heated at 37 °C overnight. The trypsin-digested sample was extracted with 50 % aqueous acetonitrile containing 5 % trifluoroacetic acid (TFA) with shaking for 30 min. This procedure was repeated twice. The extract was then lyophilized and stored at −20 °C until analysis.

8.2.3 NanoHPLC-ICPMS

A systematic diagram of the nanoHPLC-ICPMS system is shown in Fig. 8.1. The capillary liquid chromatographic system used consisted of a degasser

(DG661, GL Sciences Inc., Tokyo, Japan), a capillary pump (MP710i, GL Sciences Inc.), and a sample injector (sample volume: 2.0 μL; Model C4-1344, Valco Instruments Co., Houston, TX) for introducing the sample and a make-up solution. A 2-μL aliquot was injected into the preconcentration cartridge (C_{18} PepMap 100, 300 μm i.d. × 5 mm, Nippon Dionex K.K., Osaka, Japan) at a flow rate of 10 μL min^{-1} with 0.05 % TFA in ultrapure water, for sample desalting and matrix elimination. The preconcentrated sample was eluted over a stepwise gradient, which is described later, by the nanoHPLC pump (Nippon Dionex K.K.) and introduced into the next separation column. The chromatographic separations were performed on a reverse-phase capillary column (C_{18} PepMap 100, 75 μm i.d. × 15 cm, 100 Å, Nippon Dionex K.K.) at a flow rate of 300 nL min^{-1}. The mobile phases A and B were 0.05 % TFA in 0.5 % acetonitrile in water and 0.05 % TFA in 99.5 % acetonitrile, respectively. S-containing peptides were separated on the nanoHPLC column by a stepwise gradient: 0–4 min 5 % B isocratic, 4–35 min 5–55 % B linear, and 35–45 min 95 % B isocratic conditions were used for the peptide standards; 0–5 min 5 % B isocratic, 5–35 min 5–22 % B linear, 35–40 min 22–55 % B linear, and 40–47 min 95 % B isocratic conditions were used for the trypsin-digested samples. Phosphorus-containing peptides were separated by the nanoHPLC column using a stepwise gradient of mobile phase A (0.1 % acetonitrile), B (5 % acetonitrile, 0.1 % TFA), and C (95 % acetonitrile, 0.1 % TFA): 0–15 min 100 % A isocratic, 15–25 min 100 % B, 25–48 min 100–60 % B and 0–40 % C linear, 48–63 min 60–45 % B and 40–55 % C linear, 63–71 min 45–5 % B and 55–95 % C linear, and 71–81 min 5 % B and 95 % C isocratic condition.

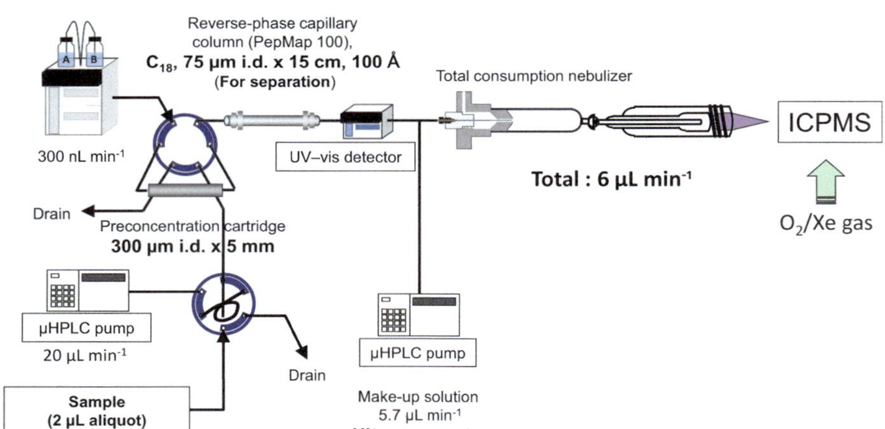

Fig. 8.1 Schematic diagram of the nanoHPLC-ICPMS system

Element measurements were performed on an ICPMS instrument (HP7500ce, Agilent, Tokyo, Japan) equipped with a total consumption nebulizer (CETAC CEI-100, Omaha, NE). Eluate of the nanoHPLC was introduced to the ICPMS after the addition of ultrapure water by using a capillary pump at a flow rate of 5.7 μL min^{-1}. In the case of phosphorylation, the eluate was made up with μHPLC with a stepwise gradient of make-up solution D (0.1 % TFA) and E (5 % acetonitrile, 0.1 % TFA): 0–15 min 100 % D isocratic, 15–25 min 100 % E isocratic, 25–48 min 0–40 % D and 100–60 % E linear, 48–63 min 40–55 % D and 60–45 % E linear, 63–71 min 55–90 % D and 45–10 % E linear, and 71–81 min 90 % D and 10 % E isocratic condition.

8.2.4 Screening of Phosphorylated Proteins in SDS-PAGE Gel Using LA-ICPMS

After a gel was dehydrated with methanol and dried with gel dryer, it was placed in a large format cell and then ablated with LA (UP-213, New Wave Research) under the following conditions: line-scanning mode; spot size, 100 μm; scan speed, 100 μm s^{-1}; laser fluence, 4.7 J cm^{-2}; and pulse frequency, 20 Hz. An ablated sample was directly introduced into ICPMS under He (1.0 ml min^{-1}) and Ar (1.0 ml min^{-1}) atmosphere, and the $^{31}P^{16}O$ signals were monitored.

8.2.5 Evaluation of the Tryptic Digestion Efficiency

A gel band of HSA separated in another lane was used for acid digestion. The decolored band was obtained as described above and digested in a microwave oven (MLS-1200 MEGA, Milestone General K.K., Kanagawa, Japan) after adding 400 μL of nitric acid and 200 μL of hydrogen peroxide. The acid-digested sample was diluted to 1000 μL with ultrapure water. A 2- μL aliquot was introduced to the ICPMS at a flow rate of 6 μL min^{-1} by using the capillary liquid chromatographic system described above. The total S concentrations in the acid- and trypsin-digested samples were then analyzed.

8.3 Results and Discussion

8.3.1 Optimization of the Octopole Reaction Gas Flow

To optimize the O_2 and Xe flows, we introduced a solution containing 20 μg L^{-1} Ba and 500 μg L^{-1} S to the ICPMS by using a capillary pump at a flow rate of

6.0 μL min^{-1} for the Xe and O_2 gas modes. For optimizing the O_2 flow, we monitored the oxide ion generation rate as the $^{138}Ba^{16}O/^{138}Ba$ ratio in accordance with the change in the O_2 flow. The optimum O_2 gas-flow rate was 0.38 mL min^{-1} because the maximum oxidized ion generation rate (12 %) was obtained under this condition. Thus, the intensities at $m/z = 48$ showed the maximum counts (1.2×10^4 cps), and the doubly charged ion generation rate (as 69/138) was less than 20 %. Under this condition, the detection limit (3σ, 3 replicates) for S as $^{32}S^{16}O$ was 0.7 μg L^{-1} in the chromatographic system. Taking into account the sample volume of 2 μL, this corresponds to an absolute detection limit of 1.4 pg (43 fmol).

When Xe was introduced to the ORC at a flow rate of 0.12–0.13 mL min^{-1}, the $^{34}S/^{32}S$ isotope ratio (4.5 ± 0.2) agreed with the theoretical isotope ratio (4.43). Therefore, these conditions were adequate for S determination. The detection limit (3σ, 3 replicates) was 10.9 μg L^{-1} for ^{32}S. The detection limit for S under the Xe gas mode was ten times that under the O_2 gas mode.

8.3.2 Hyphenation of nanoHPLC and ICPMS

A previous study reported that the S intensity decreased linearly with increasing acetonitrile concentration and was at background levels for 100 % acetonitrile under conditions in which Xe was introduced to the reaction cell [21]. Therefore, we checked the influence of acetonitrile on the ion intensities. We used 100 μg L^{-1} S-containing standards for the O_2 gas mode. When we used a make-up solution, the eluate from the nanoHPLC was 20 times diluted, and the final fluctuations of the ^{32}S and $^{32}S^{16}O$ intensities were within ±3 %, regardless of the acetonitrile concentration in the nanoHPLC eluent. This shows the advantage of using a make-up solution.

Figure 8.2 shows nanoHPLC-UV-ICPMS chromatograms of the standard peptides. NanoHPLC-ESI-MS/MS coupling is one of the principal techniques in today's proteomics that use nano columns because the concentration in the eluent depends on the section area of the column. Giusti et al. [25] developed a sheathless nano-nebulizer for ICPMS to analyze elements by parallel nanoHPLC-ICPMS and nanoHPLC-ESI-MS/MS. Compared with ESI-MS, which is a concentration-dependent detection method, ICPMS is a mass-dependent detection method. Therefore, the development of a nano-nebulizer that works at sample uptake rates of less than 500 nL min^{-1} is considered to be unnecessary if samples can be introduced to ICPMS while maintaining chromatogram separation after the addition of a make-up solution. Even if a make-up solution was added, the peak widths under the O_2 gas mode did not change compared with the UV chromatograms (within 5 %; compare Fig. 8.2a with 8.2b). On the other hand, the peak width under the Xe gas mode became 1.2–1.3 times as wide as that of the UV chromatogram, suggesting that S ions were diffused by the Xe in the ORC. The atomic weight of Xe is about four times that of S; therefore, the diffusion effect by Xe would also be bigger than that by O_2.

In this study, the $^{32}S^{16}O$ detection limits for vasopressin, oxytocin, and amyloid β were 1.0, 1.2, and 1.7 μg L^{-1}, respectively. The detection limits of ^{32}S under the Xe gas

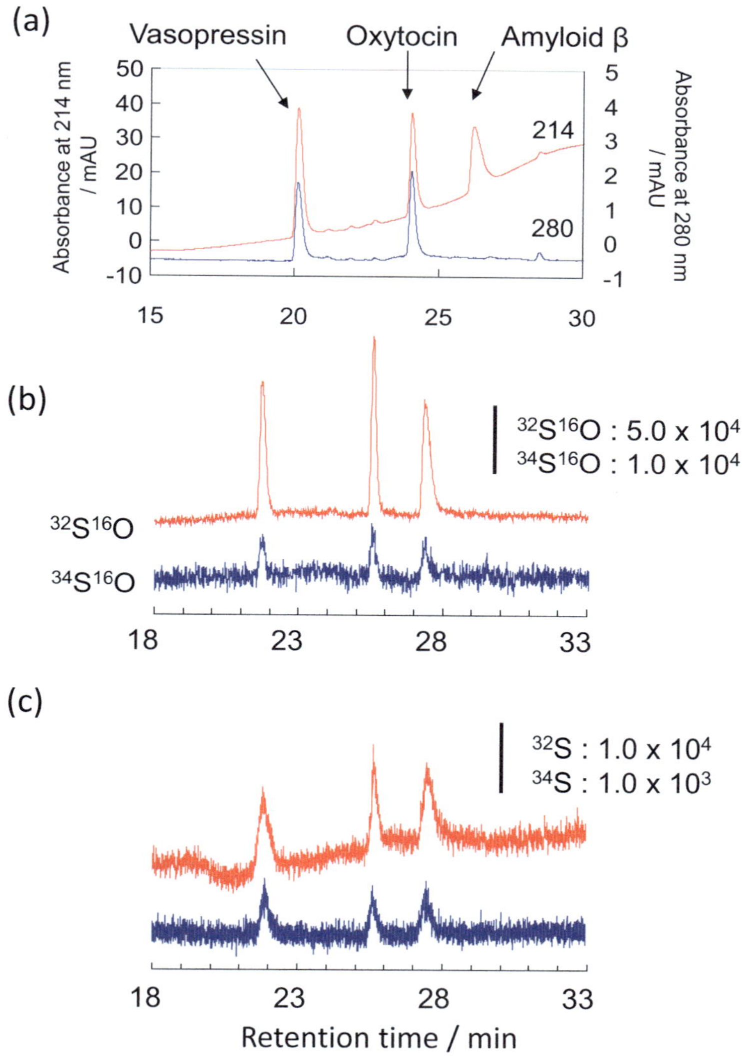

Fig. 8.2 NanoHPLC-ICPMS chromatograms for the commercially available peptides vasopressin, oxytocin, and amyloid β (human, 25–35). (**a**) UV absorbance at 214 and 280 nm, (**b**) $^{32}S^{16}O$ and $^{34}S^{16}O$ intensities under the O_2 gas mode, and (**c**) ^{32}S and ^{34}S intensities under the Xe gas mode

mode for vasopressin, oxytocin, and amyloid β were 31, 20, and 63 μg L^{-1}, respectively. The $^{32}S^{16}O$ detection limit was similar to the detection limits obtained by using other hyphenated ICPMS techniques for S detection [26–28]. Also, the detection limit of ^{32}S under the Xe gas mode was also similar to previously reported values [14, 20].

Taking into account the sample volume of 2.0 μL, the absolute detection limits for vasopressin, oxytocin, and amyloid β were 65, 77, and 110 fmol, respectively. In ICPMS under the Xe gas mode, the absolute detection limits for vasopressin,

Table 8.1 Quantification of sulfur in peptide standards

Peptide	Analytical value[a]/μg L^{-1}		Theoretical value/μg L^{-1}
	Xe mode	O_2 mode	
Vasopressin	596 ± 28	593 ± 15	591
Oxytocin	639 ± 24	638 ± 17	636
Amyloid β	620 ± 32	618 ± 22	608

[a]Calibrated with 0, 100, 200, 500, 1000 μg S L^{-1} of ammonium sulfate by flow injection analysis

oxytocin, and amyloid β were 2000, 1300, and 4000 fmol, respectively. The S detection limit as a concentration in this study was similar to, or slightly higher than, that reported in previous papers, but a lower absolute detection limit of S was achieved by the hyphenation of nanoHPLC and ICPMS [20–22, 27, 28].

To quantify peptides by using nanoHPLC-ICPMS, standard solutions containing 100, 200, 500, and 1000 μg g^{-1} of S were prepared from ammonium sulfate and used in an external standard method to determine the S concentration of the peptide by means of flow injection analysis with a 2 μL sample loop. The S concentration of each peptide was measured with relatively good accuracy (within 1.0 %) and precision (<5.0 % RSD) under the O_2 gas modes (Table 8.1). This result indicates that a peptide could be quantified by determining its S concentration by using ICPMS.

8.3.3 Assignment of S Atom Composition by the Use of nanoHPLC-ICPMS and Peptide Quantification

Albumin is a protein that is synthesized in hepatic parenchymal cells and accounts for 55 % of blood serum protein [29]. The molecular weight of albumin is 69 kDa; 585 amino acid residues and 41 S atoms are contained in its sequence. Trypsin cleaves exclusively on the C-terminal side of arginine and lysine residues. We thus expected that we could obtain 14 peptides containing 1 S atom, 8 peptides containing 2 S atoms, and 4 peptides containing 3 S atoms. Moreover, 2 peptides with the same amino acid sequence (CCK) containing 2 S atoms were also produced. Therefore, we expected to see a total of 25 peaks containing S atoms (Fig. 8.3).

NanoHPLC-UV-ICPMS chromatograms for trypsin-digested standard HSA are shown in Fig. 8.4. We eluted 23 peaks as quantifiable peaks. Similar nanoHPLC-ICPMS chromatograms were obtained for the other samples, except for peaks 10 and 14. We compared the nanoHPLC-UV-ICPMS chromatograms with predictable information from the amino acid sequence, such as the number of S-containing peptide, the number of S atoms of peptide, and the elution sequence of peptide. An ideal analytical strategy to assign the S atom composition in peptides involves parallel analyses of nanoHPLC-ICPMS with nanoHPLC-ESI-MS/MS. In the case of parallel analysis, proteins were identified by using a few detected peptides [25, 30, 31]. In this study, we did not identify the peptide sequence of each peak but used quantifiable S peaks as much as possible for further analysis, such as the

(a)

```
  1 DAHKSEVAHR FKDLGEENFK ALVLIAFAQY LQQCPFEDHV KLVNEVTEFA
 51 KTCVADESAE NCDKSLHTLF GDKLCTVATL RETYGEMADC CAKQEPERNE
101 CFLQHKDDNP NLPRLVRPEV DVMCTAFHDN EETFLKKYLY EIARRHPYFY
151 APELLFFAKR YKAAFTECCQ AADKAACLLP KLDELRDEGK ASSAKQRLKC
201 ASLQKFGERA FKAWAVARLS QRFPKAEFAE VSKLVTDLTK VHTECCHGDL
251 LECADDRADL AKYICENQDS ISSKLKECCE KPLLEKSHCI AEVENDEMPA
301 DLPSLAADFV ESKDVCKNYA EAKDVFLGMF LYEYARRHPD YSVVLLLRLA
351 KTYETTLEKC CAAADPHECY AKVFDEFKPL VEEPQNLIKQ NCELFEQLGE
401 YKFQNALLVR YTKKVPQVST PTLVEVSRNL GKVGSKCCKH PEAKRMPCAE
451 DYLSVVLNQL CVLHEKTPVS DRVTKCCTES LVNRRPCFSA LEVDETYVPK
501 EFNAETFTFH ADICTLSEKE RQIKKQTALV ELVKHKPKAT KEQLKAVMDD
551 FAAFVEKCCK ADDKETCFAE EGKKLVAASQ AALGL
```

(b)

Hydropathy index	Sequence no.	Amino acid sequence	No. of S atom
-1.18	277-281	E**CC**EK	2
-1.06	99-106	NE**C**FLQHK	1
-1.05	390-402	QN**C**ELFEQLGEYK	1
-0.93	565-573	ET**C**FAEEGK	1
-0.84	263-274	YI**C**ENQDSISSK	1
-0.78	52-64	T**C**VADESAEN**C**DK	2
-0.53	82-93	ETYGE**M**AD**CC**AK	3
-0.49	241-257	VHTE**CC**HGDLLE**C**ADDR	3
-0.37	118-136	PEVDV**MC**TAFHDNEETFLK	2
-0.25	501-519	EFNAETFTFHADI**C**TLSEK	1
-0.22	286-313	SH**C**IAEVENDE**M**PADLPSLAADFVESK	2
-0.20	314-317	DV**C**K	1
-0.07	486-500	P**C**FSALEVDETYVPK	1
-0.02	163-174	AAFTE**CC**QAADK	2
-0.01	200-205	**C**ASLQK	1
0.00	476-484	**CC**TESLVNR	2
0.18	360-372	**CC**AAADPHE**C**YAK	3
0.37	447-449	**CC**K	2
0.37	558-560	**CC**K	2
0.39	446-466	**M**P**C**AEDYLSVVLNQL**C**VLHEK	3
0.49	21-41	ALVLIAFAQYLQQ**C**PFEDHVK	1
0.51	324-336	DVFLG**M**FLYEYAR	1
0.58	546-557	AV**M**DDFAAFVEK	1
1.17	175-181	AA**C**LLPK	1
1.28	74-81	L**C**TVATLR	1
		total	41

*Hydropathy indices were calculated by using the ExPASy Proteomics Server (http://au.expasy.org/).

Fig. 8.3 Amino acid sequence of human serum albumin (Uniprot ID; P02768, aa 2-609). (**a**) Sequence positions where trypsin cleaves on the C-terminal side of arginine and lysine residues and (**b**) hydropathy indices of S-containing peptides. Boldface letters indicate S-containing amino acids

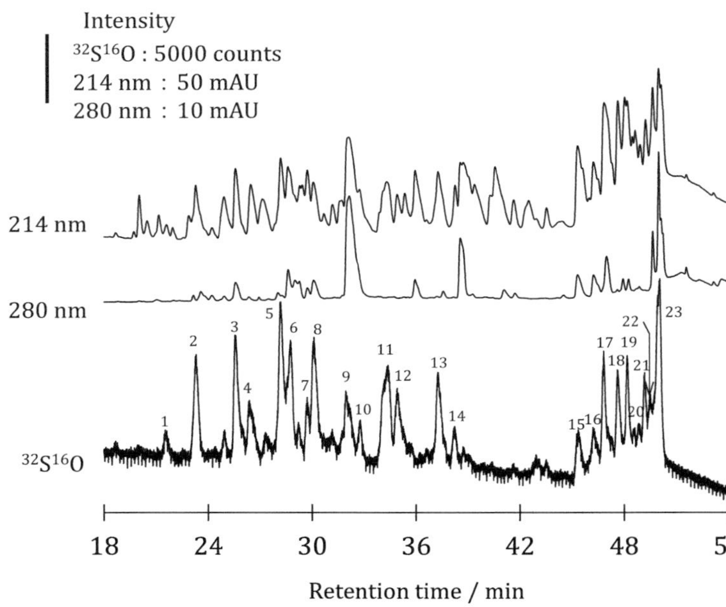

Fig. 8.4 NanoHPLC-UV-ICPMS chromatograms for trypsin-digested human serum albumin under O_2 gas introduced into an octopole reaction cell

assignment of the S atom number in each peak from the nanoHPLC-ICPMS chromatograms and protein quantification.

A schematic explanation of the algorithm that we used to assign the number of S atoms in the peaks is shown in Fig. 8.5. In the case of albumin, the maximum number of S atoms in any one peak is 4 (the same peptide sequences (CCK) are produced after tryptic digestion). Therefore, we assumed that 1, 2, 3, or 4 S atoms could be contained in a peak. Firstly, we assumed that peak n had k_n ($k_n = 1, 2, 3, 4$) S atom(s) in its sequence. Under this assumption, the number of S atom(s) ($A_{m/n}$) in the sequence of the peak m ($m \neq n$) could be calculated with the following equation:

$$A_{m/n} = \frac{\text{S concentration in peak } m}{\text{S concentration in peak } n} \times k_n \tag{8.1}$$

If peak m contained k_m ($k_m = 1, 2, 3, 4$) S atom(s), $A_{m/n}$ would be equivalent to k_m. This $A_{m/n}$ value is assumed to not exceed $k_m \pm 0.5$. Under the assumption of a triangular distribution, the type-B uncertainty is calculated as $\frac{0.5}{\sqrt{6}}$. Therefore, if a k_m that satisfies Eq. 8.2 exists, peak m would have k_m S atom(s) in its peptide:

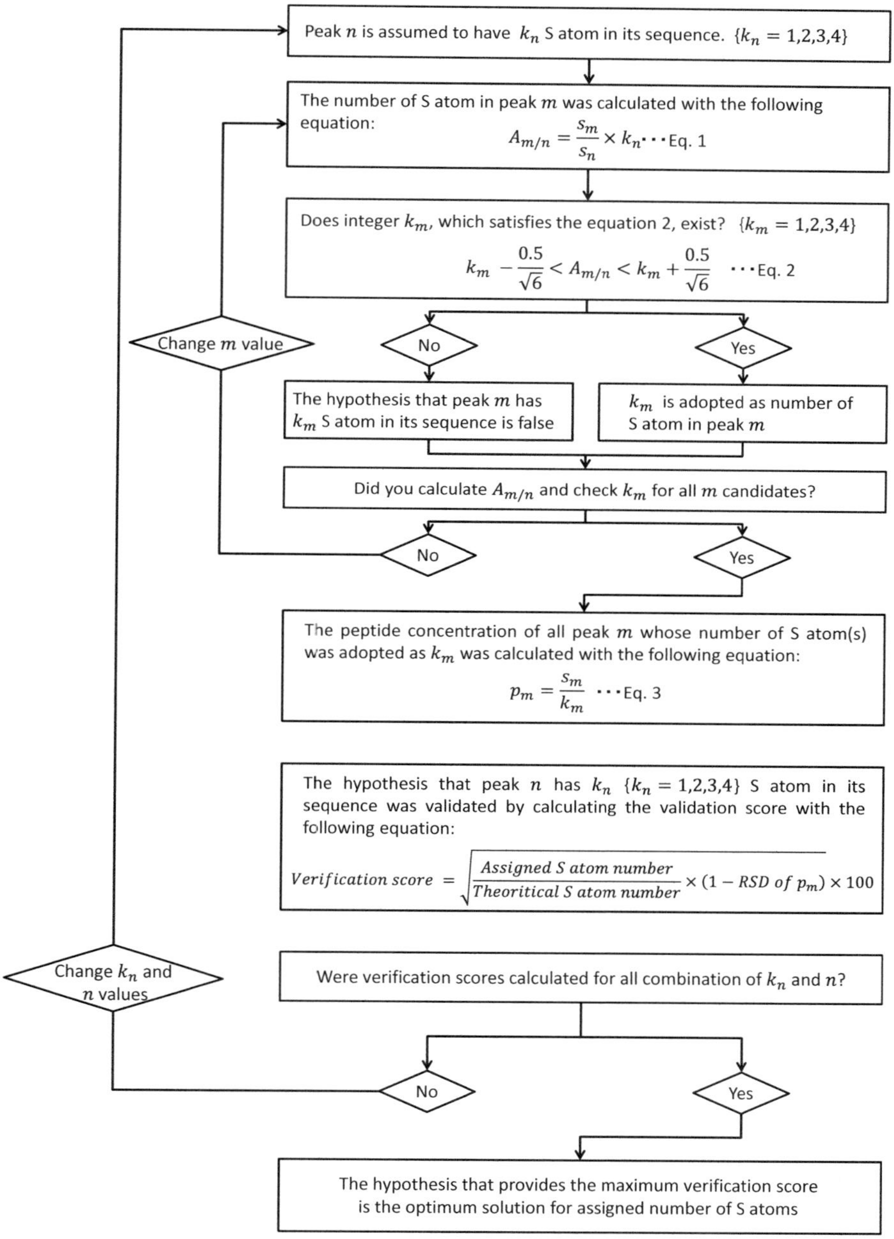

Fig. 8.5 Schematic of the algorithm used to assign the number of S atom(s) in the chromatographic peaks. s_n, S concentration in peak n; s_m, S concentration in peak m; $A_{m/n}$, number of S atom (s) in peak m; p_m, peptide concentration of peak m

$$k_m - \frac{0.5}{\sqrt{6}} < A_{\frac{m}{n}} < k_m + \frac{0.5}{\sqrt{6}} \quad (8.2)$$

If a k_m that satisfies Eq. 8.2 does not exist, the hypothesis that peak m contained k_m S atom(s) would be false.

After the number of S atoms in each peak was assigned, the peptide concentrations in peak m were calculated using

$$p_m = \frac{s_m}{k_m} \quad (8.3)$$

We then calculated the verification score as follows by using Eq. 8.4:

$$\text{Verification score} = \left\{ \left(\frac{\text{Assigned number of S atoms}}{\text{Theoritical number of S atoms}} \right) \times (1 - \text{RSD of } p_m) \right\}^{0.5} \times 100 \quad (8.4)$$

Finally, we compared the verification scores of 92 patterns (23 peaks × 4 patterns for S atoms; see Table 8.2). The hypothesis proves that the maximum verification score is the optimal solution for the assigned number of S atoms. In this case, the maximum verification score was 81.8, when peak 9 was assumed to contain 3 S atoms. Under this hypothesis, the %RSD of the peptide concentration was 7.1 %, and 29 S atoms were assigned (Table 8.3). The peptide concentration was calculated to be 6.66 ± 0.14 μmol L^{-1} (standard uncertainty of the mean, which is the experimental standard deviation divided by the positive square root of the sample size).

Actually, it could not be performed to distinguish peptides with two S atoms from two peptides with one S atom in their sequences. However, the %RSD of peptide concentration in each peak showed relatively good accuracy in other samples. In the case of other samples, the maximum verification scores (75.8 and 79.0) were observed when peak 19 contained 2 S atoms in sample 2 and peak 11 contained 4 S atoms in sample 3, respectively. Under our hypothesis, 25 and 27 S atoms, respectively, were recovered, and the peptide concentrations were 4.78 ± 0.09 and 6.63 ± 0.10 μmol L^{-1}, respectively.

8.3.4 Tryptic Digestion Efficiency and Protein Quantification

The tryptic digestion efficiency was calculated as the ratio of the total amount of S in the trypsin-digested sample to the total amount of S in the bands. The average amount of S in the albumin band before tryptic digestion was 31.2 ± 4.1 nmol. The total amount of S in the trypsin-digested sample was calculated as the sum of the S

Table 8.2 Comparison of verification scores for assigning number of S atom to each S peak obtained by nanoHPLC-ICPMS[a, b]

Peak n	Assigned number of S atom (k_n)			
	1	2	3	4
1	$(12.2 \times 97.7)^{0.5} = 34.5$	$(4.9 \times NC)^{0.5} = NC$	$(7.3 \times NC)^{0.5} = NC$	$(9.8 \times NC)^{0.5} = NC$
2	$(22.0 \times 89.3)^{0.5} = 44.3$	$(43.9 \times 92.5)^{0.5} = 63.7$	$(53.7 \times 94.2)^{0.5} = 71.1$	$(36.6 \times 96.1)^{0.5} = 59.3$
3	$(24.4 \times 89.2)^{0.5} = 46.7$	$(39.0 \times 89.6)^{0.5} = 59.1$	$(70.7 \times 91.2)^{0.5} = 80.3$	$(56.1 \times 95.1)^{0.5} = 73.0$
4	$(29.3 \times 92.7)^{0.5} = 52.1$	$(48.8 \times 91.7)^{0.5} = 66.9$	$(31.7 \times 94.5)^{0.5} = 54.7$	$(22.0 \times 97.7)^{0.5} = 46.3$
5	$(12.2 \times 90.8)^{0.5} = 33.3$	$(29.3 \times 91.9)^{0.5} = 51.9$	$(34.1 \times 91.9)^{0.5} = 55.8$	$(70.7 \times 91.2)^{0.5} = 80.3$
6	$(24.4 \times 89.2)^{0.5} = 34.5$	$(39.0 \times 89.6)^{0.5} = 59.1$	$(70.7 \times 91.2)^{0.5} = 80.3$	$(56.1 \times 95.1)^{0.5} = 73.0$
7	$(48.8 \times 91.7)^{0.5} = 66.9$	$(22.0 \times 97.7)^{0.5} = 46.3$	$(12.2 \times 97.1)^{0.5} = 34.4$	$(12.2 \times 97.7)^{0.5} = 34.5$
8	$(22.0 \times 86.8)^{0.5} = 43.7$	$(24.4 \times 93.6)^{0.5} = 47.8$	$(41.5 \times 91.9)^{0.5} = 61.7$	$(53.7 \times 91.5)^{0.5} = 70.1$
9	$(22.0 \times 90.4)^{0.5} = 44.6$	$(39.0 \times 89.6)^{0.5} = 59.1$	$\mathbf{(70.7 \times 92.9)^{0.5} = 81.1}$	$(63.4 \times 94.6)^{0.5} = 77.4$
10	$(68.3 \times 93.7)^{0.5} = 80.0$	$(24.4 \times 93.6)^{0.5} = 47.8$	$(7.3 \times NC)^{0.5} = NC$	$(19.5 \times 89.9)^{0.5} = 41.9$
11	$(12.2 \times 90.8)^{0.5} = 33.3$	$(29.3 \times 91.9)^{0.5} = 51.9$	$(34.1 \times 91.1)^{0.5} = 55.8$	$(70.7 \times 91.2)^{0.5} = 80.3$
12	$(19.5 \times 88.1)^{0.5} = 41.5$	$(34.1 \times 92.9)^{0.5} = 56.1$	$(34.1 \times 91.7)^{0.5} = 55.9$	$(48.8 \times 91.7)^{0.5} = 66.9$
13	$(22.0 \times 90.4)^{0.5} = 44.6$	$(36.6 \times 91.1)^{0.5} = 57.7$	$(68.3 \times 93.7)^{0.5} = 80.0$	$(63.4 \times 94.6)^{0.5} = 77.4$
14	$(63.4 \times 94.6)^{0.5} = 77.4$	$(4.9 \times NC)^{0.5} = NC$	$(19.5 \times 89.9)^{0.5} = 41.9$	$(19.5 \times 93.1)^{0.5} = 42.6$
15	$(24.4 \times 90.7)^{0.5} = 47.0$	$(34.1 \times 89.9)^{0.5} = 55.4$	$(26.8 \times 93.2)^{0.5} = 50.0$	$(17.1 \times 95.5)^{0.5} = 40.4$
16	$(39.0 \times 89.2)^{0.5} = 59.0$	$(43.9 \times 93.7)^{0.5} = 64.1$	$(34.1 \times 89.9)^{0.5} = 55.4$	$(26.8 \times 92.1)^{0.5} = 49.7$
17	$(22.0 \times 86.8)^{0.5} = 43.7$	$(29.3 \times 88.9)^{0.5} = 51.0$	$(36.6 \times 92.0)^{0.5} = 58.0$	$(31.7 \times 90.9)^{0.5} = 53.7$
18	$(31.7 \times 88.8)^{0.5} = 53.1$	$(70.7 \times 91.2)^{0.5} = 80.3$	$(26.8 \times 97.7)^{0.5} = 51.2$	$(34.1 \times 95.7)^{0.5} = 57.2$

(continued)

Table 8.2 (continued)

Peak n	Assigned number of S atom (k_n)			
	1	2	3	4
19	$(12.2 \times 83.7)^{0.5} = 31.9$	$(16.4 \times 98.7)^{0.5} = 38.0$	$(48.8 \times 91.7)^{0.5} = 66.9$	$(34.1 \times 89.9)^{0.5} = 55.4$
20	$(41.5 \times 94.8)^{0.5} = 62.7$	$(12.2 \times 97.1)^{0.5} = 34.4$	$(19.5 \times 93.1)^{0.5} = 42.6$	$(9.8 \times \mathrm{NC})^{0.5} = \mathrm{NC}$
21	$(26.8 \times 90.9)^{0.5} = 49.4$	$(41.5 \times 91.5)^{0.5} = 61.6$	$(39.0 \times 94.3)^{0.5} = 60.6$	$(41.5 \times 90.5)^{0.5} = 61.3$
22	$(41.5 \times 92.7)^{0.5} = 62.0$	$(56.1 \times 95.1)^{0.5} = 73.0$	$(24.4 \times 95.9)^{0.5} = 48.4$	$(26.8 \times 93.8)^{0.5} = 50.2$
23	$(2.4 \times \mathrm{NC})^{0.5} = \mathrm{NC}$	$(24.4 \times 89.2)^{0.5} = 46.7$	$(34.1 \times 93.8)^{0.5} = 56.6$	$(39.0 \times 89.6)^{0.5} = 59.1$

[a]Boldface type indicates that the best verification score was observed with this combination of n and k_n
[b]*NC* not calculated

Table 8.3 Assignment of number of S atoms in peptide sequence under hypothesis that peak 9 contained 3 S atoms in its sequence

Peak *n*	S concentration (μmol L^{-1})	Calculated number of S atom	Assigned number of S atom	Peptide concentration (μmol L^{-1})
1	1.37	0.20		
2	14.7	2.19	2	7.35
3	20.6	3.07	3	6.86
4	11.5	1.71		
5	27.6	4.11	4	6.89
6	20.7	3.08	3	6.89
7	5.68	0.85	1	5.68
8	23.1	3.44		
9	20.1	3.00	3	6.71
10	6.6	0.98		
11	27.5	4.10	4	6.87
12	22.8	3.41		
13	19.8	2.95	3	6.58
14	4.93	0.74		
15	8.24	1.23		
16	12.7	1.89	2	6.35
17	24.5	3.65		
18	14.1	2.10	2	7.05
19	16.7	2.50		
20	3.64	0.54		
21	12.1	1.81	2	6.05
22	10.4	1.55		
23	40.8	6.09		
				av. 6.66 ± 0.14

concentrations in the monopeptides determined by using nanoHPLC-ICPMS. In this case, we distinguished between monopeptides and polypeptides by calculating the number of S atoms ($A_{m/n}$). The $A_{m/n}$ values of peak 23 were higher than four for all samples (6.09, 6.85, and 6.79, respectively). Therefore, peak 23 was considered to be a polypeptide produced by incomplete tryptic digestion.

Taking into account the total sample volume of 40 μL, the total amounts of S in the trypsin-digested samples were 13.2, 8.9, and 12.3 nmol, respectively, and the tryptic digestion efficiencies were 42.2 ± 5.9, 28.4 ± 4.0, and 39.3 ± 5.5 %, respectively (Table 8.4).

Taking into account the dilution factor of 2, the protein concentration was calculated as twice the peptide concentration after it was divided by the tryptic digestion efficiency; the analytical results of the protein concentrations were 31.6 ± 4.5, 33.7 ± 4.8, and 33.6 ± 4.7 μmol L^{-1}, respectively (Table 8.4). The average concentration and the expanded uncertainty of measurement (combined standard uncertainty divided by the positive square root of *n* and multiplied by a

Table 8.4 Protein concentration (human serum albumin) derived from peptide concentration and tryptic digestion efficiency

	Recovered S amount (nmol)	Tryptic digestion efficiency[a](%)	Peptide concentration[b] (μmol L^{-1})	Protein concentration[c] (μmol L^{-1})
Sample 1	13.2	42.2 ± 5.9	6.66 ± 0.14	31.6 ± 4.5
Sample 2	8.9	28.4 ± 4.0	4.78 ± 0.09	33.7 ± 4.8
Sample 3	12.3	39.3 ± 5.5	6.63 ± 0.10	33.6 ± 4.7
				av. 33.0 ± 5.4

[a]Tryptic digestion efficiency was calculated as the sum of S concentration in the monopeptide divided by total S concentration (31.2 ± 4.4 nmol) in bands before digestion
[b]Average ± standard uncertainty which is calculated as standard deviation divided by $\sqrt{n}$
[c]Protein concentrations were calculated as peptide concentration divided by tryptic digestion efficiency and multiplied dilution factor of 2

coverage factor of 2) were 33.0 and 5.4 μmol L^{-1}, respectively, and the theoretical value (36.4 μmol L^{-1}) was within the range of the expanded uncertainty of measurement.

In the case of the elimination of peaks 10 and 14, which were detected in sample 1, the tryptic digestion efficiency of sample 1 was 40.7 %, and the protein concentration was 32.7 ± 4.7, which was close to the theoretical value. We might have overestimated the tryptic digestion efficiency because we could not tell whether the S in the chromatographic peak originated from HSA. If purified trypsin-digested samples were prepared in a less S contaminated environment, we would have been able to obtain more accurate tryptic digestion efficiency and protein quantification. Nevertheless, both the peptide concentration and the tryptic digestion efficiency could be determined by the same detection methodology, i.e., S detection using ICPMS.

8.3.5 Analysis of Degree of Phosphorylation

Radioisotope ^{32}P labeling, purification/concentrating of phosphorylated protein, and using phosphospecific antibody are used for screening of phosphorylated proteins [32–34]. In the case of screening using ^{32}P, usage of radioisotope ^{32}P was strictly limited of and its half-life is short as 14.3 days. Moreover, from the very beginning, the results may be affected by radiation; the results should be evaluated carefully. Therefore, using ^{32}P is one of the useful screening methods but has many limitations. In the case of purification/concentrating phosphorylated protein, phosphorylated proteins are concentrated using immobilized metal affinity chromatography (IMAC) [32], titania [33], or Phos-tag [34], and then tryptic digestion and mass spectrometry analysis are applied. These methods make analysis of phosphorylated protein easy. Since non-phosphorylated proteins are not target analyte for these methods, it is difficult to obtain information how much the protein has been phosphorylated (the degree of phosphorylation). In the case of analysis using antigen-antibody reaction, antibody for specific phosphorylated protein and anti-phosphorylated tyrosine antibody such as pY20 and 4G10 are known to be useful monoclonal antibody. These methods using antibody for specific phosphorylated protein are not adequate to screening. In addition, antigen-antibody reaction is sometimes time-consuming because of blocking, reaction, and washing process.

Now then LA-ICPMS analysis was adapted to direct detection of P in gel. LA-ICPMS is a method for the element-selective trace analysis of solid materials [35]. Sample surface is melted and evaporated by laser beam focused on solid sample. These aerosols are transported into ICP, and elements in aerosols are atomized, ionized, and measured. Besides, LA-ICPMS was adapted to β-casein standard (from bovine milk), and bovine milks were used as test samples. Casein account for 80 % of protein in bovine milk, and αS1, αS2, β, and κ-caseins are known as major form of casein. Although relative abundance of these four caseins varies according to various conditions, it is known that β-casein constitutes a large majority (35–45 %) of casein [36].

Phosphorus images about electrophoresis gel of β-casein standard (a and b) and bovine milk (c) were obtained using LA-ICPMS (Fig. 8.6). LA-ICPMS analysis enables to detect P-containing band by one line scan for analytical time of *ca.* 80 s. Furthermore, both 2D images of P signal showed good agreement with the CBB-stained image.

Subsequently, we examined whether the number of added phosphate group could be quantitatively analyzed or not using nanoHPLC-ICPMS. Chromatograms of $^{32}S^{16}O$, $^{31}P^{16}O$, and UV signals were shown in Fig. 8.7. As the same as HSA, peptide concentration in trypsin-digested β-casein could be determined by calculating S concentration is each peak/number of S atom ratio in each peptide. From amino acid sequence of β-casein, we expected that we could obtain 4 peptides containing 1 S atom and 1 peptide containing 2 S atoms after tryptic digestion

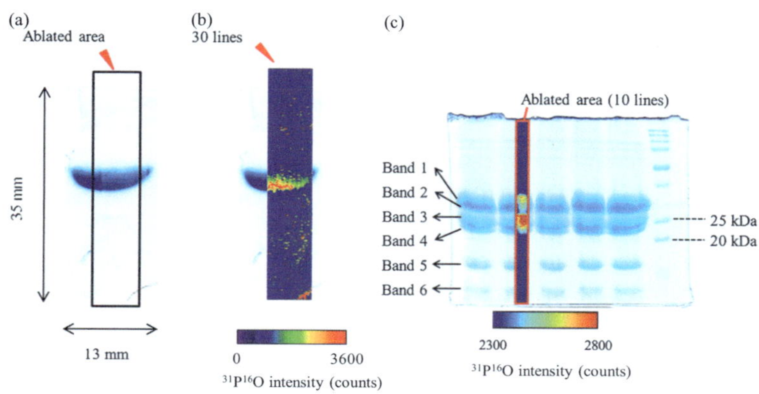

Fig. 8.6 Screening of phosphorylated protein after SDS-PAGE using LA-ICPMS. (**a**) CBB staining for *β*-casein standard form bovine milk, (**b**) 2D imaging result from 30 line scans for *β*-casein standard form bovine milk, (**c**) 2D imaging result from 10 line scans for milk sample

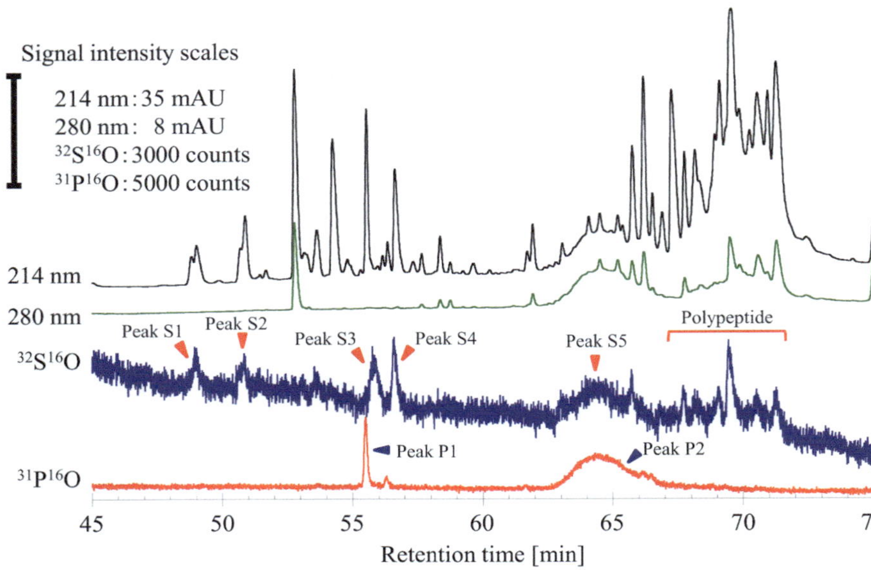

Fig. 8.7 NanoHPLC-UV-ICPMS chromatograms for tryptic digested β-casein from bovine milk under introduction of O_2 gas into octopole reaction cell

(Fig. 8.8). S concentration of $^{32}S^{16}O$ peak at 49.8 min (peak S1) was 7.98 μmol mL^{-1}, and it was assigned that the number of S atom in peak S1 equals one. Similarly, the number of S atom in peak S2 and S4 was assigned to be one. Thus, peptide concentration was calculated to be 8.27 ± 0.85 μmol mL^{-1} from this chromatogram.

Taking account of tryptic digestion efficiency and dilution factor of 2, protein concentration was calculated as two times of peptide concentration after divided by tryptic digestion efficiency, and analytical results of protein concentration of β-casein were 258, 248, and 266 μmol L^{-1}, respectively. The average and the expanded uncertainty of measurement (combined standard uncertainty divided by positive square root *n* and multiplied by a coverage factor of 2) was 257±10 μmol L^{-1}, and theoretical value (265 μmol L^{-1}) existed within the range of expanded uncertainty of measurement.

Otherwise, it was reported that 15, 17, 18, 19, and 35 serine residues were modified to phosphoserine. We thus expected that we could obtain two peptides which contain P atom(s); one sequence is FQSEEQQQTEDELQDK (peptide a) with one phosphorylation site, and the other is ELEELNVPGEIVESLSSSEESITR (peptide b) with four phosphorylation sites (Fig. 8.8).

Two peaks for $^{31}P^{16}O$ signal were detected at retention time of 55.6 (peak P1) and 63~67 min (peak P2), and P concentrations of each peak were 6.39 and 33.9 μmol L^{-1}, respectively (Fig. 8.7). Peak P2 was eluted as a wider peak compared to peak P1. Profrock [37] reported phosphopeptide analysis for trypsin-digested β-casein using CE-ICPMS. In the Profrock's paper, two major P peaks were detected, and one of them, tetra-phosphorylated peptide (sequence

```
                    *  ***                       *
  1 RELEELNVPG EIVESLSSSE ESITRINKKI EKFQSEEQQQ
 41 TEDELQDKIH PFAQTQSLVY PFPGPIPNSL PQNIPPLTQT
 81 PVVVPPFLQP EVMGVSKVKE AMAPKHKEMP FPKYPVEPFT
121 ESQSLTLTDV ENLHLPLPLL QSWMHQPHQP LPPTVMFPPQ
161 SLVSLSQSKV LPVPQKAVPY PQRDMPIQAF LLYQEPVLGP
201 VRGPFPIIV
```

* Phosphorylation site of β-casein

⌐ Trypsin cleves sites of β-casein

Fig. 8.8 Amino acid sequence of *β*-casein from bovine milk (Uniprot ID; P02666, aa 16-224) and sequence positions where trypsin cleaves C-terminal to arginine and lysine residues. Boldface letters indicate S-containing amino acids

RELEELNVPGEIVESLSSSEESITR), was detected as a wide peak and like our chromatogram. When we calculate the phosphoric acid base per peptide, these values were 0.78 ± 0.08 and 4.13 ± 0.40, respectively. Profrock [37] reported that 4 serine residue of sequence ELEELNVPGEIVESLSSSEESITR was modified to phosphoserine (ELEELNVPGEIVEpSLpSpSpSEESITR) by a combination analysis of ICPMS and MALDI-MS. Navaza et al. [38] also reported the same results with a combination analysis of ICPMS and ESI-MS. These results were considered to support our quantitative results.

Finally, whether this method could be applied to the actual sample or not was examined by using milk as a test sample. Some bands containing P were detected by LA-ICPMS. Of these bands, the band 3 around MW:25 kDa was suggested to correspond to β-casein and was subjected to nanoHPLC-ICPMS analysis in the same manner as described above.

Chromatograms similar to Fig. 8.7 were obtained (Fig. 8.9). Peptide concentration was calculated to be 3.78 μmol L^{-1}, and a number of P atom in peak P1, P2, and P3 were calculated to be 0.55 ± 0.04, 0.98 ± 0.08, and 3.86 ± 0.30. Because P was not detected at retention time of 49.9 min in Fig. 8.7, P in peak P1 in Fig. 8.9 was considered to be contamination from other bands. We could detect mono- and tetra-phosphorylated peptide from crude sample.

β-casein concentration is expected to contain 88.2 mmol L^{-1} of the protein (6.50 mg mL^{-1}) in the milk. In this study, protein concentration of β-casein was calculated to be 86.0 mmol L^{-1} and showed a good agreement with expected value.

8.4 Conclusions

NanoHPLC-ICPMS enabled us to detect and quantify S in peptides. In particular, the introduction of O_2 gas into an ORC is useful for determining S in peptides. Even if a make-up solution was added, the peak width of the ICPMS chromatograms did not change. Moreover, using a make-up solution could eliminate any matrix effect due to the introduction of acetonitrile.

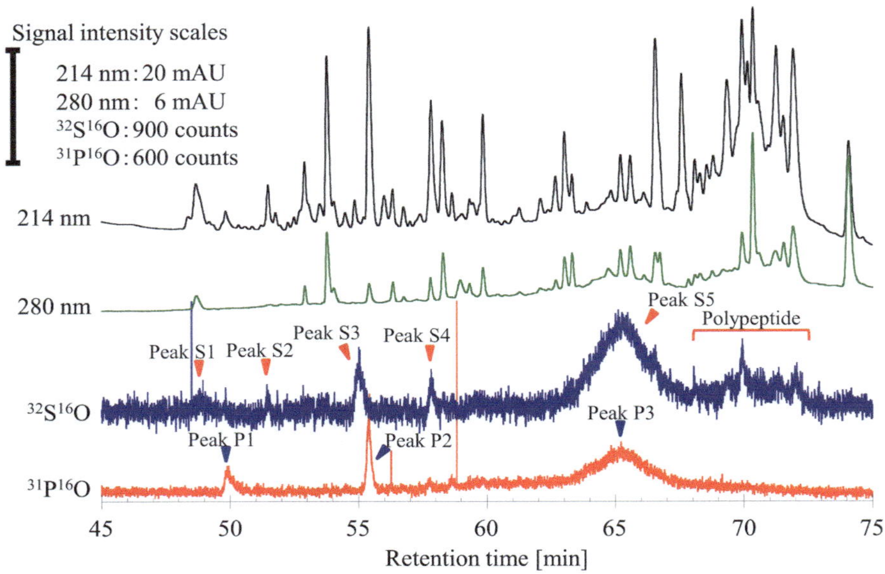

Fig. 8.9 NanoHPLC-UV-ICPMS chromatograms for tryptic digested phosphorylated protein corresponding to β-casein from milk under introduction of O_2 gas into octopole reaction cell

Protein quantification was achieved via peptide quantification, after taking into account the tryptic digestion efficiency, by using ICPMS to determine the S amount in the peptides. In other words, we showed that the number of S atoms in a sequence/chromatographic peak could be assigned by calculating verification scores after determining the S concentration by using nanoHPLC-ICPMS. The nanoHPLC-ICPMS technique thus provides information regarding the S amount of a peptide that can be used to speculate as to the S atom number in a peak/sequence. And simultaneous detection of P and S using nanoHPLC-ICPMS provides information about the number of P in peptide. This technique could, therefore, support LC-MS(/MS) use in proteomics.

Acknowledgments We are grateful to Mr. Nobuo Shibuya for providing the capillary column. This study was supported in part by a joint research project titled "Protein quantification and quantitative phosphorylation analysis by sulfur and phosphorus determinations with using ICP-MS" conducted at the Institute of Science and Engineering of Chuo University.

References

1. Bettmer J, Bayon MM, Encinar JR, Sanchez MLF, de la Campa MDF, Medel AS (2009) The emerging role of ICP-MS in proteomic analysis. J Proteomics 72:989–1005
2. Lowry OH, Rosebrough NJ, Farr AL, Randall RJ (1951) Protein measurement with the folin phenol reagent. J Biol Chem 193:265–275

3. Bradford MM (1976) A rapid and sensitive method for the quantitation of microgram quantities of protein utilizing the principle of protein-dye binding. Anal Biochem 72:248–254
4. Flores R (1978) A rapid and reproducible assay for quantitative estimation of proteins using bromophenol blue. Anal Biochem 88:605–611
5. Ma CQ, Li KA, Tong SY (1998) Spectrophotometric micromethod for protein determination with tetrachloro tetraiodo fluorescein. Anal Lett 31:1021–1036
6. Xu W, Wei YC, Xing D, Chen Q (2008) A novel chemiluminescence technique for quantitative measurement of low concentration human serum albumin. Anal Sci 24:115–119
7. Miller I, Crawford J, Gianazza E (2006) Protein stains for proteornic applications: which, when, why? Proteomics 6:5385–5408
8. Gygi SP, Rist B, Gerber SA, Turecek F, Gelb MH, Aebersold R (1999) Quantitative analysis of complex protein mixtures using isotope-coded affinity tags. Nat Biotechnol 17:994–999
9. Schmidt A, Kellermann J, Lottspeich F (2005) A novel strategy for quantitative proteornics using isotope-coded protein labels. Proteomics 5:4–15
10. Ross PL, Huang YLN, Marchese JN, Williamson B, Parker K, Hattan S et al (2004) Multiplexed protein quantitation in Saccharomyces cerevisiae using amine-reactive isobaric tagging reagents. Mol Cell Proteomics 3:1154–1169
11. Johnson LN (2009) The regulation of protein phosphorylation. Biochem Soc Trans 37:627–641
12. Kettenbach AN, Gerber SA (2011) Rapid and reproducible single-stage phosphopeptide enrichment of complex peptide mixtures: application to general and phosphotyrosine-specific phosphoproteomics experiments. Anal Chem 83:7635–7644
13. Prange A, Schaumlöffel D (2002) Hyphenated techniques for the characterization and quantification of metallothionein isoforms. Anal Bioanal Chem 373:441–453
14. Profrock D, Leonhard P, Prange A (2003) Determination of sulfur and selected trace elements in metallothionein-like proteins using capillary electrophoresis hyphenated to inductively coupled plasma mass spectrometry with an octopole reaction cell. Anal Bioanal Chem 377:132–139
15. Miyayama T, Ogra Y, Osima Y, Suzuki KT (2008) Narrow-bore HPLC–ICP–MS for speciation of copper in mutant mouse neonates bearing a defect in Cu metabolism. Anal Bioanal Chem 390:1799–1803
16. Shigeta K, Matsumura K, Suzuki Y, Shinohara A, Furuta N (2008) Distribution and dynamic pathway of selenium species in selenium-deficient mice injected with ^{82}Se-enriched selenite. Anal Sci 24:1117–1122
17. Suzuki Y, Hashiura Y, Matsumura K, Matsukawa T, Shinohara A, Furuta N (2010) Dynamic pathways of selenium metabolism and excretion in mice under different selenium nutritional statuses. Metallomics 2:126–132
18. Suzuki Y, Sakai T, Furuta N (2012) Isolation of Selenoprotein-P and determination of Se concentration incorporated in proteins in human and mouse plasma by tandem heparin affinity and size-exclusion column HPLC-ICPMS. Anal Sci 28:221–224
19. Rappel C, Schaumlöffel D (2008) The role of sulfur and sulfur isotope dilution analysis in quantitative protein analysis. Anal Bioanal Chem 390:605–615
20. Schaumloffel D, Giusti P, Preud'Homme H, Szpunar J, Lobinski R (2007) Precolumn isotope dilution analysis in nanoHPLC-ICPMS for absolute quantification of sulfur-containing peptides. Anal Chem 79:2859–2868
21. Wind M, Wegener A, Eisenmenger A, Kellner R, Lehmann WD (2003) Sulfur as the key element for quantitative protein analysis by capillary liquid chromatography coupled to element mass spectrometry. Angew Chem Int Ed 42:3425–3427
22. Zinn N, Krüger R, Leonhard P, Bettmer J (2008) μLC coupled to ICP–SFMS with post-column isotope dilution analysis of sulfur for absolute protein quantification. Anal Bioanal Chem 391:537–543
23. Tanner SD, Baranov VI, Bandura DR (2002) Reaction cells and collision cells for ICP-MS: a tutorial review. Spectrochim Acta B 57:1361–1452

24. Laemmli UK (1970) Cleavage of structural proteins during the assembly of the head of bacteriophage T_4. Nature 227:680–685
25. Giusti P, Lobinski R, Szpunar J, Schaumloffel D (2006) Development of a nebulizer for a sheathless interfacing of nanoHPLC and ICPMS. Anal Chem 78:965–971
26. Bandura DR, Baranov VI, Tanner SD (2002) Detection of ultratrace phosphorus and sulfur by quadrupole ICPMS with dynamic reaction cell. Anal Chem 74:1497–1502
27. Sturup S, Bendahl L, Gammelgaard B (2006) Optimization of LC-DRC-ICP-MS for the speciation of selenotrisulfides with simultaneous detection of sulfur and selenium as oxides combined with determination of elemental and isotope ratios. J Anal At Spectrom 21:201–203
28. Rampler E, Dalik T, Stingeder G, Hann S, Koellensperger G (2012) Sulfur containing amino acids – challenge of accurate quantification. J Anal At Spectrom 27:1018–1023
29. Anderson NL, Anderson NG (2002) The human plasma proteome: history, character, and diagnostic prospects. Mol Cell Proteomics 1:845–867
30. Giusti P, Schaumlöffel D, Preud'homme H, Szpunar J, Lobinski R (2006) Selenopeptide mapping in a selenium–yeast protein digest by parallel nanoHPLC-ICP-MS and nanoHPLC-electrospray-MS/MS after on-line preconcentration. J Anal At Spectrom 21:26–32
31. Tastet L, Schaumloffel D, Bouyssiere B, Lobinski R (2008) Identification of selenium-containing proteins in selenium-rich yeast aqueous extract by 2D gel electrophoresis, nanoHPLC-ICPMS and nanoHPLC-ESI MS/MS. Talanta 75:1140–1145
32. Ficarro SB, McCleland ML, Stukenberg PT, Burke DJ, Ross MM, Shabanowitz J, Hunt DF, White FM (2002) Phosphoproteome analysis by mass spectrometry and its application to Saccharomyces cerevisiae. Nat Biotechnol 20:301–305
33. Larsen MR, Thingholm TE, Jensen ON, Roepstorff P, Jorgensen TJD (2005) Highly selective enrichment of phosphorylated peptides from peptide mixtures using titanium dioxide microcolumns. Mol Cell Proteomics 4:873–886
34. Kinoshita E, Yamada A, Takeda H, Kinoshita-Kikuta E, Koike T (2005) Novel immobilized zinc(II) affinity chromatography for phosphopeptides and phosphorylated proteins. J Sep Sci 28:155–162
35. Koch J, Gunther D (2011) Review of the state-of-the-art of laser ablation inductively coupled plasma mass spectrometry. Appl Spectrosc 65:155a–162a
36. Muller-Renaud SP, Dupont D, Dulieu P (2004) Quantification of beta-casein in milk and cheese using an optical immunosensor. J Agric Food Chem 52:659–664
37. Profrock D (2010) Hyphenation of capillary-LC with ICP-MS and parallel on-line micro fraction collection for MALDI-TOF-TOF analysis – complementary tools for protein phosphorylation analysis. J Anal At Spectrom 25:334–344
38. Navaza AP, Encinar JR, Sanz-Medel A (2007) Absolute and accurate quantification of protein phosphorylation by using an elemental phosphorus standard and element mass spectrometry. Angew Chem Int Ed 46:569–571

Chapter 9
Analysis of Drug Active Pharmaceutical Ingredients and Biomolecules Using Triple Quadrupole ICP-MS

Naoki Sugiyama and Yasuyuki Shikamori

Abstract Triple quadrupole ICP-MS (ICP-QQQ) can measure heteroatoms such as sulfur (S), phosphorus (P), and chlorine (Cl) at much lower levels than previously possible, making it highly suitable for the measurement of molecular compounds in life science research, including drug development. The tandem MS (MS/MS) configuration of the instrument, with a quadrupole positioned either side of the reaction cell, allows for much better control of any ion-molecular reactions that take place in the cell compared to conventional, single quadrupole ICP-QMS. As a number of molecular compounds of interest in life science research contain S, P, or Cl heteroatoms, the superior control of interferences provided by ICP-QQQ for the measurement of these challenging elements at trace levels was evaluated. The samples analyzed included drug active pharmaceutical ingredients (APIs) and peptide/phosphor peptides.

Keywords API • Active pharmaceutical ingredient • Drug analysis • Protein quantification • Phosphor peptide quantification • ICP-QQQ • ICP-MS/MS • Reaction cell

9.1 Introduction

In this chapter, the relatively new technique of triple quadrupole ICP-MS (ICP-QQQ) is described. Details are also given for the measurement of drug active pharmaceutical ingredients (APIs) and biomolecules such as proteins and phospho-proteins using the heteroatoms contained in the compounds. The improved detection capability of ICP-QQQ promises to make it the most suitable ICP-MS for life science research.

N. Sugiyama (✉) • Y. Shikamori
Agilent Technologies International Japan, Ltd., 9-1 Takakura-cho, Hachioji, Tokyo 192-0033, Japan
e-mail: naoki_sugiyama@agilent.com

Y. Ogra, T. Hirata (eds.), *Metallomics*, DOI 10.1007/978-4-431-56463-8_9

With its high elemental and species-independent sensitivity, wide dynamic range, isotopic analysis capability, and high matrix tolerance, ICP-MS is already used by some life science researchers. Organic compounds that contain an ICP-MS detectable element can be quantified at trace levels on the basis of the compound's known stoichiometry. ICP-MS has been used in a number of studies regarding metalloproteins, which play a pivotal role in living organisms. Examples include metallothionein [1–3], superoxide dismutase (SOD) [4, 5], and ceruloplasmin [6] which contains copper (Cu) and/or zinc (Zn). The exceptional sensitivity of recent ICP-MS instrumentation has led to detection limits in the femtogram (fg)/g (ppq) range for elements with a mass number higher than 80. These elements tend to be relatively free from spectral interferences. The high sensitivity of ICP-MS has also been used to advantage to measure anticancer drug APIs and metabolites like cisplatin, which contains platinum (Pt) [7–9], as well as alternative drug candidates that contain ruthenium (Ru) [10, 11], which are expected to have less severe side effects.

For compounds that don't contain a metal/metalloid element, a "tagging" technique has been developed, where ICP-detectable elements such as iodine (I) [12], selenium (Se) [13], lanthanides [14, 15], or gold (Au) nanoparticles [16] are added into the target compound. The "tag" element is then detected by ICP-MS allowing the compound to be quantified. Tanner et al. [17] developed a novel technique to profile proteins in a single cell. A reagent was used to introduce different lanthanides to multiple target proteins using immunoassay. The profile of each protein in the cell was then obtained by measuring the lanthanides by ICP-MS. However, the complex process required to introduce the tag into target compound, possible different bioavailability of the tag-added compound from the original, or concern about tag introduction efficiency that must be high for accurate quantification are drawbacks of the technique.

A more straightforward approach to measure metal/metalloid-free compounds is to utilize a hetero element such as S, P, Cl, or Br as a natural tag. A number of proteins and peptides contain S, which is present in two proteinogenic amino acids: cysteine and methionine. Nucleic acids such as RNA and DNA contain P, and many drug APIs contain S, P, or halogens like Cl. It is a challenge for conventional ICP-QMS to measure these nonmetal elements at trace levels because of their low ionization rates (due to their high ionization potentials) and significant spectral interferences. However, ICP-QQQ, with its excellent spectral interference removal capability using MS/MS mode, allows the measurement of S, P, and Cl at much lower levels. Using this approach, it is possible to measure and quantify compounds via a heteroatom, without the need to introduce an artificial tag into the compound.

Quantitative analysis is one of the most valuable attributes of ICP-MS. For the technique to be more widely adopted by life science researchers, quantitative analysis of target compounds needs to be both accurate and simple. Due to the high temperature of the ICP ion source, compounds are readily atomized. Consequently, the sensitivity of ICP-MS is relatively matrix and compound independent compared to LC-MS, which is, in general, the first choice MS technique used in "omics" studies. While the high-temperature plasma ion source of ICP-MS doesn't provide species information, unlike LC-MS, it does provide species-independent elemental sensitivity, which may allow compound quantitation without the need for compound-specific calibration standards. Life science applications including drug development studies are an emerging and interesting field for ICP-MS, especially for ICP-QQQ that can measure more organic compounds directly than conventional ICP-QMS.

9.2 ICP-QQQ

9.2.1 ICP-QQQ

Most current single quadrupole ICP-QMS instruments are fitted with a collision/reaction cell (CRC) in order to resolve spectral interferences. Although ICP-QMS can be used without a cell gas (called no gas mode), introducing a suitable gas into the cell removes or, at least, alleviates spectral interferences and improves detection limits. Interferences are reduced either by "collision" or "reaction" with a gas within the cell, hence the term collision/reaction cell.

Operating the CRC in collision mode using an inert gas such as helium (He) ensures that the analyte ion is discriminated from the interfering ion based on the difference of their ionic cross section. An interfering polyatomic ion is generally larger than the interfered analyte ion. For example, $^{40}Ar^{35}Cl^{+}$ is larger than $^{75}As^{+}$ and undergoes a greater number of collisions in the cell. As each collision causes energy loss, the interfering polyatomic ions lose more energy than the analyte ions and are subsequently filtered from the mass spectrum. This is achieved by discriminating between the two different energies using a process known as kinetic energy discrimination (KED). A different voltage is applied to the quadrupole and cell, so lower energy polyatomic ions cannot pass the potential barrier [18]. Helium collision mode is commonly used in general applications due to its universal effectiveness in interference removal and ease of use. Also, in contrast to reactive gases, helium collision mode doesn't require application-specific optimization. However, the improvement in signal-to-noise (interference) ratio is typically limited to one or two orders of magnitude, and He mode is largely ineffective for interferences caused by atomic isobars and doubly charged ions.

An alternative approach is to use reaction chemistry in the CRC of an ICP-QMS. However, since all ions formed in the plasma enter the reaction cell of a single quadrupole system, the reaction chemistry is uncontrolled, often leading to erroneous analytical results. Using a quadrupole ion guide as a dynamic reaction cell to prevent low-mass ions from entering the cell has been successful in some applications [19, 20], but not in all cases [21, 22].

The full potential of reaction cell chemistry can only be realized by ICP-QQQ which uses an additional quadrupole MS (Q1) in front of the CRC to control which ions can enter the cell and react. The second quadrupole (Q2, positioned after the

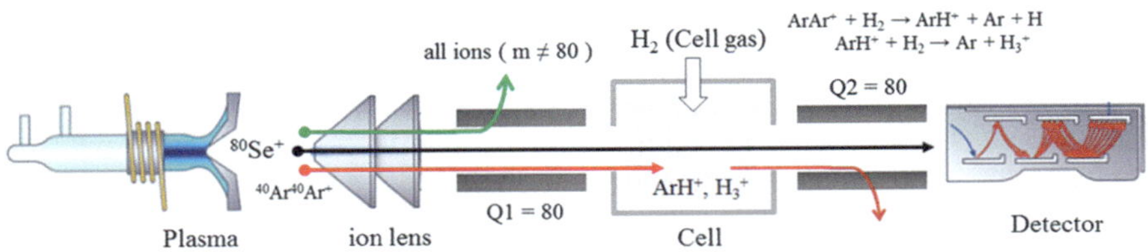

Fig. 9.1 ICP-QQQ operation in on-mass mode to remove $^{40}Ar^{40}Ar^+$ interference on $^{80}Se^+$

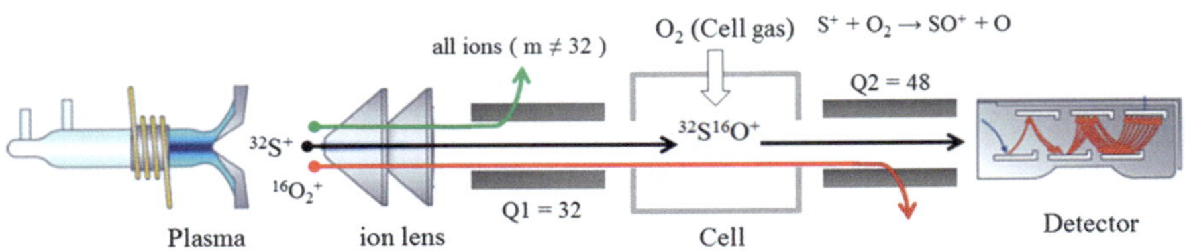

Fig. 9.2 ICP-QQQ operation in mass-shift mode to remove $^{16}O_2^+$ interference on $^{32}S^+$

cell) selects the analyte ions that are passed to the detector. This double mass selection (MS/MS) is the fundamental reason for the superior control of interferences provided by ICP-QQQ. The configuration corresponds to the IUPAC definition of a triple quadrupole MS, as shown in Figs. 9.1 and 9.2. That is:

> A tandem mass spectrometer comprising two transmission quadrupole mass spectrometers in series, with a (non-selecting) RF–only quadrupole (or other multipole) between them to act as a collision cell: [23]

The CRC consists of a series of lenses and a multipole RF ion guide which has multiple independent cell gas flow control systems. For example, the Agilent 8800 ICP-QQQ uses an octopole RF ion guide with a four-channel cell gas flow system used for the introduction of He (collision) gas and three reaction gases: hydrogen (H_2), oxygen (O_2), and ammonia (NH_3). Each gas can be used alone or as mixture (mixed in the cell). Q1 can be operated as a simple ion guide (strictly speaking, a mass filter with the wide mass window) or as a 1-amu window mass filter. The former setting is referred to as single quadrupole scan mode and is used for conventional operation in no gas or He collision mode, emulating ICP-QMS. The latter setting is known as MS/MS scan mode and is used with a reactive cell gas. Using Q1 to select which ions enter the cell based on their mass-to-charge ratio (m/z) ensures that the reactions within the cell are controlled and relatively sample independent. Consequently, ICP-QQQ with MS/MS has opened up new possibilities for analysts to address more applications using controlled reaction chemistry.

9.2.2 *How MS/MS Works with a Reaction Gas*

There are two types of reaction gas cell methods available with ICP-QQQ: *on-mass* where the setting of both quadrupoles is the same (Q1 = Q2) and *mass shift* (Q1 ≠ Q2). The choice of method depends on the difference in the chemical reactivity (reaction rate) between the interfering ion and the analyte ion with the cell gas.

When the interfering ion reacts readily with the cell gas, it is removed from its original *m*/*z* at a faster rate than the analyte ion. The analyte ion can then be measured “on-mass” at its true isotope mass, free from interference. An example of the on-mass method is given in Fig. 9.1, which illustrates how ^{80}Se is measured using H_2 cell gas to remove the interference from $^{40}Ar^{40}Ar^+$. $ArAr^+$ reacts quickly with H_2 to form $2Ar + H_3^+$, while Se^+ has a relatively low reaction rate with H_2. In this way, ^{80}Se can be measured using H_2 cell gas via mass pair of (Q1, Q2) = (80, 80). The approach can also be used with ICP-QMS for the analysis of simple matrix samples. However, if bromine (Br) is present in the sample, as is often the case with bio fluid samples, $^{79}BrH^+$ will form on reaction with H_2 in the cell and interfere with $^{80}Se^+$ at *m*/*z* 80. This is not a problem with ICP-QQQ since all ions with *m*/*z* ≠ 80, including $^{79}Br^+$, are rejected by Q1.

In contrast, mass shift is used when the analyte ion reacts readily with the reaction gas to form a new reaction product ion, while the interfering ion reacts slowly or not at all with the cell gas, so it does not contribute significantly to the signal at the new mass of the analyte product ion. An illustration is given in Fig. 9.2 using O_2 cell gas to remove the $^{16}O_2^+$ interference on ^{32}S. The analyte ion S^+ reacts with O_2 to form SO^+, while the O_2^+ interference reacts more slowly with O_2. Consequently, ^{32}S can be measured with a dramatically improved S/N ratio as $^{32}SO^+$ using O_2 cell gas via mass pair of (Q1, Q2) = (32, 48). If the background noise at the *m*/*z* of the product ion is low, even slow reaction rates between the analyte and cell gas to produce product ions will lead to a significant improvement in the S/N and detection limit (DL). Mass-shift mode is less suitable for ICP-QMS, which allows all ions formed in the plasma to enter the cell, including ions at the same *m*/*z* as any product ions. For example, if a sample contained calcium (Ca), titanium (Ti), or carbon (C) at high concentration, as in reverse phase LC-ICP-MS applications, $^{48}Ca^+$, $^{48}Ti^+$, and $^{36}Ar^{12}C^+$ would overlap the signal of the product ion $^{32}S^{16}O^+$, rendering the ICP-QMS method ineffective. With ICP-QQQ, in contrast, all of the potential interference ions ($^{48}Ca^+$, $^{48}Ti^+$ and $^{36}Ar^{12}C^+$) would be removed by Q1. The differences between ICP-QQQ and ICP-QMS mass-shift methods are illustrated in Fig. 9.3 using the effect of Ca on the determination of S as an example. A mass spectrum of 2 ppb S was acquired with and without 50 ppb Ca, using O_2 mass shift with MS/MS and single quad scan to emulate conventional ICP-QMS. It can be seen that $^{32}SO^+$ suffers from a large positive error by the overlap from Ca in single quad scan mode, while MS/MS provides a theoretically matched isotopic pattern for S without any influence from Ca.

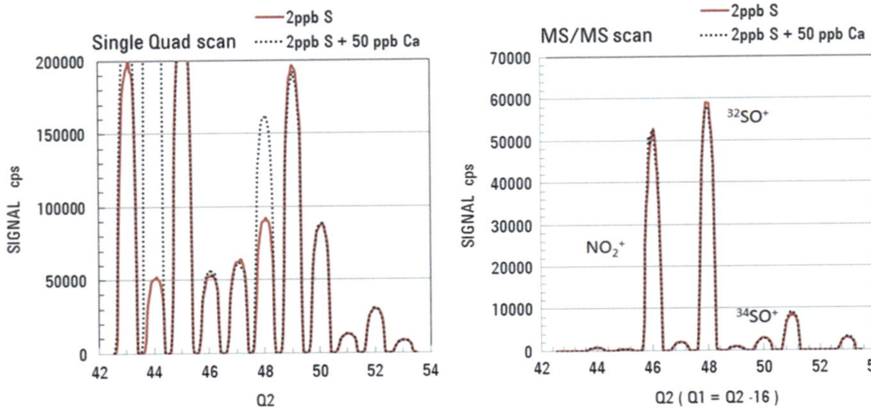

Fig. 9.3 Spectrum of 2 ppb sulfur acquired using O_2 mass-shift mode with and without the presence of 50 ppb Ca. *Left* single quad scan mode, *right* MS/MS scan mode

9.2.3 *Reaction Cell Gas Selection*

The most common ion-molecular reactions observed in the CRC are given below (not classified by reaction mechanism). B and $B_1 \cdot B_2$ represent cell gas molecules, and A^+ is an analyte ion or interfering ion which may be polyatomic, doubly charged or an atomic isobar.

- Charge transfer: $A^+ + B \rightarrow A + B^+$
- Atom transfer: $A^+ + B_1 \cdot B_2 \rightarrow A^+ \cdot B_1 + B_2$
- Association: $A^+ + B + C \rightarrow A^+ \cdot B + C$

Theoretically, the charge transfer reaction can proceed when IP (A) > IP (B), where IP stands for ionization potential. Atom transfer is observed in condensation or hydrogen atom transfer reactions, which will proceed if the affinity of A^+ toward B_1 is greater than affinity of B_2 toward B_1. Examples of atom transfer reactions using $B_1 = O$ [24], $B_1 = H$ [25], and $B_1 = F$ [26] can be found in the literature. Association reaction is known as clustering. For the product ion $A^+ \cdot B$ to be stable, it requires another collision with C to release excess energy before the product ion breaks up. C may be molecule of B or molecule of buffer gas, He. A number of atomic ions react with NH_3, CO_2, or CH_4 to form complex cluster ions in a sequence reaction: $A^+ \cdot B_n$, $n = 1,2,3 \ldots$

A number of reaction gases have been investigated including hydrogen, oxygen [24], ammonia [27], methane [28], ethylene [29], methyl fluoride [30], methyl chloride [31], nitrogen [32], nitrous oxide [33], carbon oxide [34], carbon dioxide [29], and others. However, the three most commonly used and effective gases for ICP-MS/ICP-QQQ are H_2, O_2, and NH_3.

Hydrogen (H_2) H_2 is an effective reaction gas used in on-mass methods to remove argide interferences such as $^{40}Ar^+$ on $^{40}Ca^+$, $^{38}ArH^+$ on $^{39}K^+$, $^{40}ArC^+$ on $^{52}C^+$, $^{40}ArO^+$

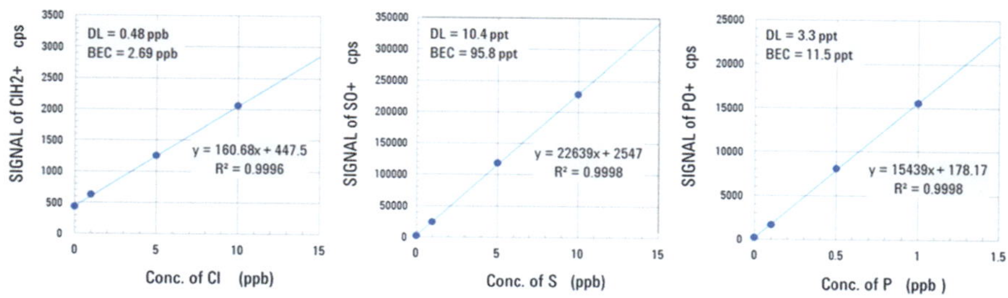

Fig. 9.4 From the *left*, (**a–c**): (**a**) calibration curve of ^{35}Cl in UPW acquired using ICP-QQQ in MS/MS mass-shift mode with H_2 cell gas, octopole bias = −18 V, H_2 flow rate = 5.0 ml/min, KED = 0 V. (**b**) Calibration curve of ^{32}S in UPW acquired using ICP-QQQ in MS/MS mass-shift mode with O_2 cell gas: octopole bias = −3 V, O_2 flow rate = 0.41 ml/min, KED = −8 V. (**c**) Calibration curve of ^{31}P in UPW acquired using ICP-QQQ in MS/MS mass-shift mode with O_2 cell gas: octopole bias = −3 V, O_2 flow rate = 0.41 ml/min, KED = −8 V

on $^{56}Fe^+$, $^{40}Ar^{18}OH^+$ on $^{59}Co^+$, $^{40}ArCl^+$ on $^{75}As^+$, and $^{40}Ar^{40}Ar^+$ on $^{80}Se^+$. Argide ions react with H_2 at a relatively fast rate, while nearly all elemental ions exhibit low reactivity with H_2. However, a few applications use H_2 effectively in mass-shift mode. An example is Cl analysis. Cl is a challenging element for ICP-QMS since both isotopes suffer intense interferences: $^{35}Cl^+$ suffers an overlap from $^{16}O^{18}OH^+$ and $^{37}Cl^+$ suffers overlaps from $^{36}ArH^+$, $^{18}O^{18}OH^+$, and $(H_3O)H_2O^+$. However, as Cl^+ reacts with H_2 to form ClH_2^+, ^{35}Cl can be measured using ICP-QQQ in MS/MS mass shift with H_2 via the mass pair of (Q1, Q2) = (35, 37). A calibration curve of ^{35}Cl in ultrapure water (UPW) is shown in Fig. 9.4a. A DL of 0.48 ppb and back equivalent concentration (BEC) of 2.69 ppb were achieved.

Oxygen (O_2) **O_2** is highly suited to mass-shift methodology. A number of atomic ions react with O_2 to form oxide ions by O-atom transfer. O-atom affinity of atomic ions is shown in Fig. 9.5, with the O-atom affinity of the O atom (5.2 eV) indicated by the dotted line. Theoretically, atomic ions that have a higher O-atom affinity than the O atom react with O_2 to form oxide ions. However, due to the contribution of collisional energy, atomic ions with a slightly lower O-atom affinity than the O atom actually form oxide ions in the reaction cell. Calibration curves of ^{32}S and ^{31}P in UPW obtained using O_2 mass-shift mode are shown in Fig. 9.4b, c. DLs of 10.4 ppt and 3.3 ppt and BECs of 95.8 ppt and 11.5 ppt were obtained for S and P, respectively.

Another cell gas, nitrous oxide (N_2O), is a more effective gas for O-atom transfer than O_2, since N_2 has a lower O-atom affinity (1.6 eV) than the O atom (5.2 eV). N_2O reacts with more atomic ions to form oxide ions.

O_2 is also used in on-mass methods to remove interferences based on the oxide ions of refractory elements. Some of the oxide ions are highly reactive with O_2 and are readily converted to dioxide and trioxide ions in the reaction cell. Unless the analyte ion reactivity toward O_2 is also high, the element can be measured on-mass with the oxide ion interference removed or reduced. The method is suitable to

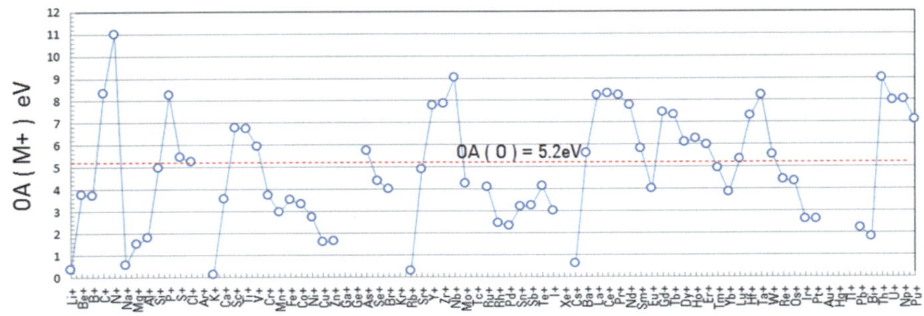

Fig. 9.5 O-atom affinity of atomic ions

resolve the molybdenum oxide ion (MoO^+) interference on cadmium (Cd) and zirconium oxide ion (ZrO^+) overlap on silver (Ag). Neither Cd^+ nor Ag^+ react readily with O_2, while neither MoO^+ [35] nor did ZrO^+ [36] react at a fast rate to be effectively converted to dioxide and trioxide ions.

Ammonia (NH_3) NH_3 is highly reactive due to its lone electron pair. With an IP of 10.2 eV, which is lower than other common cell gases such as H_2 (15.4 eV) and O_2 (12.1 eV), NH_3 is more likely to resolve an interference via a charge-exchange reaction. One example is the removal of $^{35}Cl^{16}O^+$ interference on $^{51}V^+$. In the reaction cell filled with NH_3, ClO^+ (IP of ClO = 10.9 eV) loses its charge via a charge-exchange reaction, while V^+ analyte ions (IP of V = 6.74 eV) remain mostly unaffected:

$$ClO^+ + V^+ + NH_3 \rightarrow ClO + V^+ + NH_3^+$$

NH_3 is also an effective reaction gas to be used in mass-shift methods. The approach is often very versatile, although method development is more complex than on-mass methodology. NH_3 reacts with many atomic ions, M^+, to form various product ions, described as $M^+(NH_n)(NH_3)_m$, where $n = 0, 1, 2$ and $m = 0, 1, 2, 3 \ldots$, via condensation and association reaction mechanisms. As long as there is a product ion that is free from interference, the product ion can be used to measure the analyte. Balcaen et al. [37] applied ICP-QQQ NH_3 mass-shift to the measurement of trace titanium (Ti) in blood and serum. All interferences on $^{48}Ti^+$, including the most challenging atomic isobaric overlap by ^{48}Ca, were resolved by measuring product ion Ti $(NH_3)_6{}^+$. NH_3 mass-shift methods can be very powerful when combined with the MS/MS capability of ICP-QQQ. Without the first quadrupole mass filter (Q1) available to remove any undesired precursor ions, coexisting elements in the sample would react with NH_3 to produce a number of cluster product ions, some of which would likely overlap with the analyte product ion to be measured.

9.3 Application of ICP-QQQ for Drug API Analysis and Life Science Research

ICP-QQQ is expected to complement LC-MS in "omics" studies to provide a means for simpler and more accurate quantitative analysis. With drastically improved sensitivity and ability to determine challenging elements, ICP-QQQ is opening up more application opportunities for analysts.

9.3.1 *Drug API Analysis by LC-ICP-QQQ*

Identification and quantification of drug APIs and metabolites are key tasks carried out during the many stages of drug development, including QA/QC. LC-MS is widely used for these applications. When the API is a metal-containing compound such as cisplatin, ICP-MS has also been used due to its unrivaled sensitivity to platinum (Pt). For the analysis of compounds containing more challenging elements such as S, P, and Cl at trace levels, ICP-QQQ can be used. Raeve et al. [38] applied ICP-QQQ to the absolute quantification of a molecular-targeted drug, trastuzumab. Trastuzumab is a monoclonal antibody (mAb) with a molecular mass of 148,000, comprising 44 sulfur atoms. The API sample was microwave-digested and was quantified using an isotope dilution (ID) method via ^{34}S-enriched H_2SO_4 spiked into the sample. A recovery of 97.8 % of the expected concentration was reported. Quantification of the API was achieved without the need for compound-specific calibration standards, unlike the general quantification method using LC-MS.

Drug development is an involved process which requires in vitro or in vivo studies including fundamental chemical characterization, toxicology tests, pharmacokinetic studies, bioavailability tests, etc. Each study would benefit from a faster, simpler, and more accurate technique for the quantification of the drug candidate API and/or the metabolites, all attributes of ICP-QQQ. A feasibility study of the analysis of drug APIs by LC-ICP-QQQ was carried out.

Three sulfonamide drug APIs, sulfamethizole, sulfamethazine, and sulfamethoxazole, were analyzed using LC-ICP-QQQ. The drugs are used for the treatment of bacterial infections as they inhibit the synthesis of bacterial folate that is essential for DNA synthesis in bacteria. Each sulfonamide is a small molecule, which contains one or two S atoms, and has a molecular mass of 250 – 280 Da. For the identification of the APIs, an HPLC (Agilent 1260 infinity bio-inert HPLC) fitted with a reverse phase (RP) column (Agilent ZORBAX plus C18 2.1×100mm) was coupled to an ICP-QQQ (Agilent 8800). The ICP-QQQ was equipped with a narrow injector torch (1.5 mm id injector). An optional gas (20 % oxygen in Ar) was added to the same injector gas flow used for the organic solvent LC mobile phase to prevent carbon buildup on the interface cones. An isocratic mobile phase consisting of 13 % acetonitrile with 0.1 % formic acid was used at a flow rate of 0.4 ml/min. The injection volume was 20 μl. An O_2 mass-shift method for ICP-QQQ was

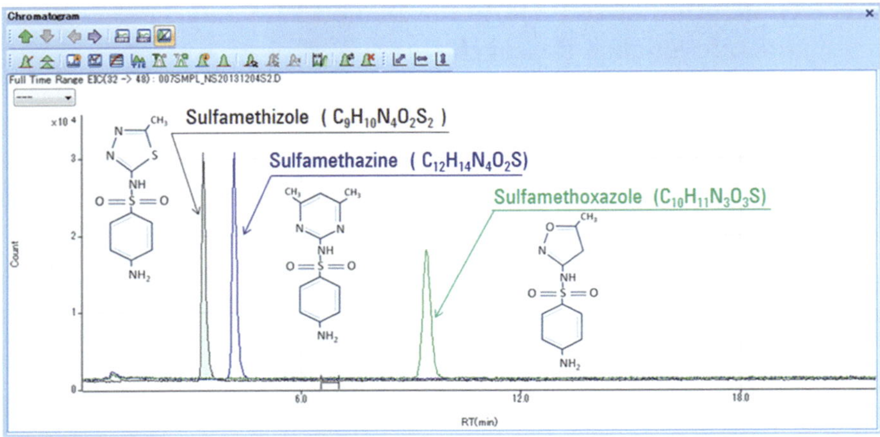

Fig. 9.6 Chromatogram of sulfonamide drugs using reverse phase LC-ICP-QQQ. Sample contains S drug APIs at a concentration of 100 ppb (as S)

optimized to detect S contained in the sulfonamides using the following parameters: O_2 cell gas flow rate = 0.3 ml/min, octopole bias = −4.0 V, and KED voltage = −8.0 V. From the chromatogram shown in Fig. 9.6, each API was identified via its retention time. The method detection limit for the compound sulfamethizole was calculated to be 23 nM (6.3 ppb as the compound and 1.5 ppb as the element S).

A similar approach (using the same configuration) was used to measure P- and Cl-containing drug APIs in two commercial drugs: ZOMETA® and Catapres®. ZOMETA® comprises a P-containing API and zoledronic acid monohydrate ($C_5H_{10}N_2O_7P_2/H_2O$). The drug is used for the treatment of hypercalcemia (high level of blood calcium). The compound was identified and quantified via its P content using anion ion-pair LC-ICP-QQQ. P, as well as S, and can be measured using an O_2 mass-shift method via a mass pair of (Q1, Q2) = (31, 47). Figure 9.7 shows the calibration curve and a chromatogram of the sample. The isocratic LC mobile phase consisted of a 70:30 mixture of A: 6 mM tetra-butyl-ammonium bromide and 5 mM acetic acid adjusted to PH 6.5 with NH_3(aq) and B: 95 % MeOH. The sample was prepared by diluting a 5 ml vial of commercially supplied ZOMETA® (which contained 4.264 mg of the API) by 2000-fold with the LC mobile phase. A five-point calibration curve from 100 μg/l (ppb) to 2 mg/l (ppm) was prepared using a zoledronic acid monohydrate standard purchased from Sigma-Aldrich (St. Louis, MO, US). From the calibration, the concentration of the API was determined to be 433 ng/ml, equating to a recovery of 102 %. The MDL for the drug compound was calculated to be 25 nM (144 pg, 7.2 ppb as the compound and 1.5 ppb as the element P).

Catapres® is a commercial drug used for the treatment of hypertension. Its API is a Cl-containing compound, clonidine hydrochloride ($C_9H_9Cl_2N_3 \cdot HCl$). The drug API was identified and quantified using RP-HPLC-ICP-QQQ to measure Cl. Figure 9.8 shows the calibration curve and chromatogram of the analysis. As

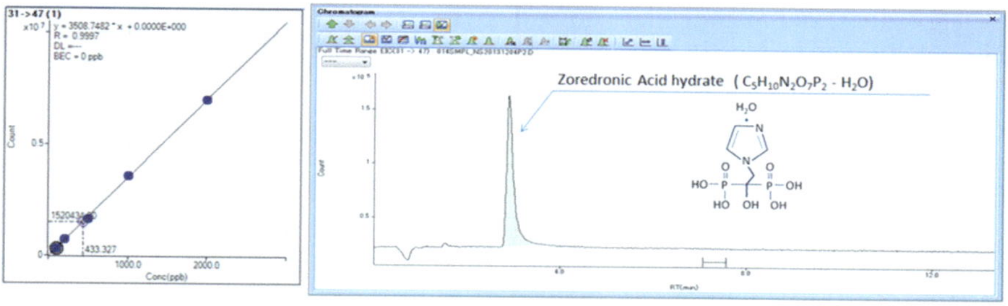

Fig. 9.7 Chromatogram of a P-containing drug API using LC-ICP-QQQ

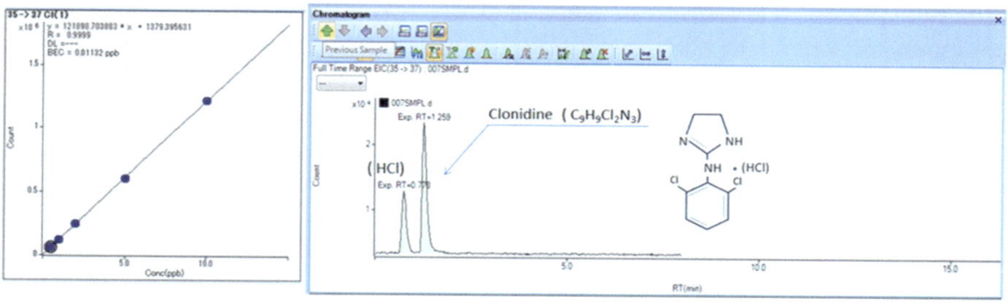

Fig. 9.8 Chromatogram of a Cl-containing drug API using LC-ICP-QQQ

mentioned previously, mass-shift mode using H_2 cell gas is recommended for Cl analysis via a mass pair of (Q1, Q2) = (35, 37). Cl^+ (*m/z* 35) is converted to ClH_2^+ (*m/z* 37) via reaction with H_2 cell gas at 3 ml/min flow rate. An isocratic mobile phase of 20 % acetonitrile with 0.1 % formic acid was used. A tablet of Catapres® (containing 75 μg of the API) was dissolved in 50 ml ultrapure water (Organo PURIC-ω), sonicated for 60 min, filtered with a 0.2 μm filter, and injected into the LC as a sample. A calibration curve was prepared by analyzing clonidine hydrochloride standards purchased from Sigma-Aldrich (St. Louis, MO, US). The concentration of the API was found to be 1444 ng/ml, which is a recovery of 96 %. The MDL for the drug compound was calculated to be 146 nM (780 pg, 39 ppb as compound and 15 ppb as the element Cl).

9.3.2 Peptide and Phosphor Peptide Analysis by Capillary-LC-ICP-QQQ

The objective of life science research is to find ever more suitable treatments for disorders. Several "omics" research fields have developed according to the kind of compound the research targets: genomics, transcriptomics, proteomics, metabolomics, and metallomics. A number of research themes are ongoing around the world to elucidate the cause or the pathway of various diseases and to finally find effective remedies. In omics studies, LC-MS is widely used for the identification, structural analysis, and quantification of target protein/peptides together with classical techniques like Western blot, ELISA, and modified techniques. However, the quantification of protein/peptides is still often challenging and ICP-MS is expected to fill the gap. The high-temperature plasma ionization source of ICP-MS provides an advantage over soft ionization sources such as ESI or MALDI of LC-MS, with its matrix-independent and species-independent sensitivity. The matrix-independent sensitivity of ICP-MS resolves, or at least alleviates, a well-known issue of ESI-MS, which is signal suppression by a co-eluting compound. As a result, ICP-MS provides a more accurate means of quantification of protein/peptides. The species-independent sensitivity of ICP-MS allows for compound-independent calibration (CIC), and CIC can be used to quantify even multiple protein/peptides using a single standard without the need for reference standards, which are often expensive or difficult to obtain.

A feasibility study has demonstrated the capability of ICP-QQQ for the analysis of protein/peptides and phosphor peptides [39]. Phosphorylation is one of the most important post transcript modification (PTM) components that plays a pivotal role in controlling the function of proteins in living organisms. Precise and accurate analysis of the PTM is of paramount importance in proteomics. As detailed previously, ICP-QQQ can measure protein/peptides and phosphor peptides via an S or P heteroatom. To do this, a micro-flow LC (Agilent 1200 series capillary LC) was coupled to an ICP-QQQ (Agilent 8800 with 1.5 mm id injector torch) via a total consumption nebulizer (Agilent G3680A), as shown in Fig. 9.9. The nebulizer was used with a specially designed single-pass spray chamber (not chilled), providing 100 % sample introduction efficiency at the low LC flow rate. The LC was fitted with a micro-bore RP column (Agilent ZORBAX SB C18, 5 μm, 0.3 × 150 mm).

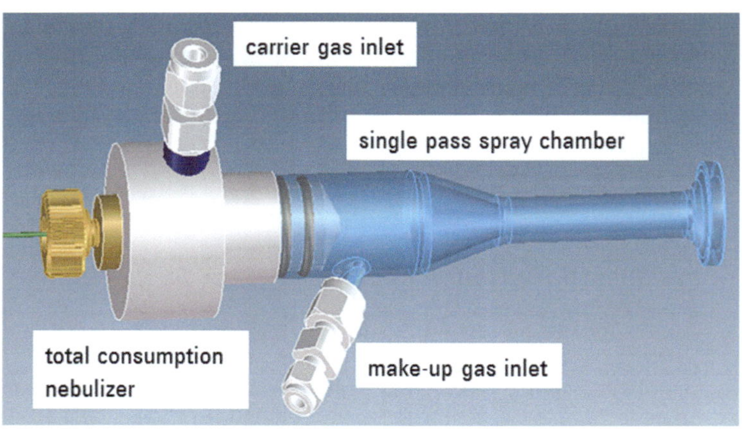

Fig. 9.9 Illustration of total consumption nebulizer and single-pass spray chamber

The end of the column was connected directly to the total consumption nebulizer via a capillary tube. The mobile phases A and B consisted of water and acetonitrile containing 0.1 % formic acid. A gradient flow was used at a flow rate of 5 μl: 0–3 min A:B = 99:1 isocratic and 3–35 min A:B = 40:60 linear. The injection volume was 1 μl. ICP-QQQ was used in O_2 mass-shift mode for the determination of S and P. The instrument was optimized by introducing 50 % acetonitrile containing an inorganic sulfur standard and maximizing the SO^+ signal. Figure 9.10 shows a chromatogram of a sample containing two peptides and two phosphor peptides at a concentration of 45 ng/ml (as S and P). The sample also contains methionine and bis(4-nitrophenyl) phosphate (BNPP) as internal standards at a concentration of 105 ng/ml (as S and P). The sample peptides have amino acid sequences of A**C**TPER**M**AE and VP**M**LK, which contain the S-containing amino acid cysteine and/or methionine. The amino acid sequences of the sample phosphor peptides are LRRA**p**SLG and KRS**p**YEEHIP, which contain phosphorylated serine and tyrosine, respectively. From the signal of the internal standard, the method DL was calculated to be 0.18 ng/ml as S and 0.10 ng/ml as P (11 fmol S and 6.6 fmol P). Future work will concentrate on the absolute quantification of peptide/phosphor peptides using CIC.

9.4 Conclusions

ICP-QQQ operating in MS/MS mode realizes the potential of using reactive cell gases to eliminate problematic spectral interferences. The first MS restricts which ions enter the cell, so that the reaction can be controlled, ensuring consistent and reliable reaction cell performance. Hence, elements considered challenging for conventional ICP-QMS, such as S, P, and Cl, can now be measured at lower levels using ICP-QQQ than previously possible. This provides more opportunities for the analysis of bio molecules in life science research and drug development processes,

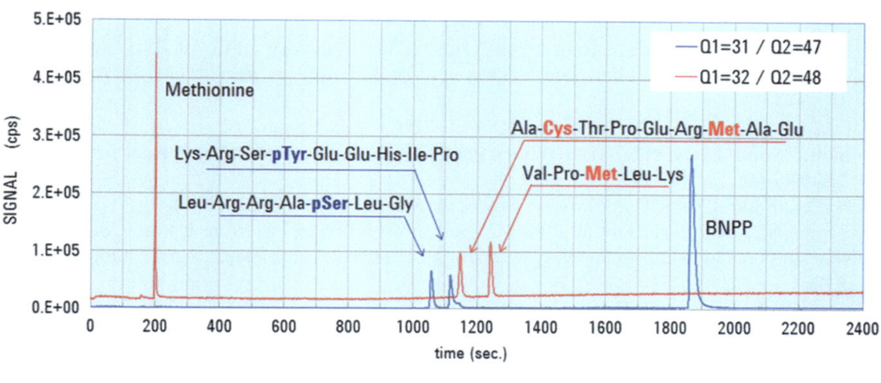

Fig. 9.10 Chromatogram of peptides and phosphor peptides using micro-flow LC-ICP-QQQ

since a number of the compounds of interest contain one of these heteroatoms. Due to the high-temperature ionization source, ICP-MS has an advantage in quantitative analysis over LC-MS as it provides simpler, faster, and more accurate quantitative analytical measurements. ICP-MS is expected to complement LC-MS for emerging life science research with ICP-QQQ becoming the technique of choice due to its superior detection capability of heteroatoms, while remaining applicable for conventional metals analysis, including the measurement of metal-tagged biomolecules.

To highlight the capabilities of ICP-QQQ for the analysis of drug APIs, S-, P-, and Cl-containing drug APIs were analyzed with excellent sensitivity using LC-ICP-QQQ. Drug development is a lengthy process comprising many stages. As ICP-QQQ provides improved quantitative analysis, the technique could contribute to more efficient and faster drug development. ICP-QQQ coupled with micro-flow LC was used to analyze peptides and phosphor peptides via the heteroatoms of S and P – again with good sensitivity. More widespread use of ICP-QQQ is expected in order to improve accuracy and sensitivity in quantitative analysis especially the degree of expressed protein or phosphorylation.

Future work will focus on the application of ICP-QQQ for actual research projects. To fully realize the advantage of ICP-MS in quantitative analysis, further study on compound-independent calibration (CIC) is needed in real applications. Quantifying biomolecules like proteins without the need for a standard compound would be of great benefit in life science research, including drug development. It remains a challenge currently to apply CIC generally, especially to gradient flow LC-ICP-MS, so this is an area of development.

References

1. Murray K, Boyd R, Eberlin M (2013) IUPAC recommendation 2013. Pure Appl Chem 85 (7):1515–1609
2. Tanner SD, Baranov VI, Bandura DR (2002) Reaction cells and collision cells for ICP-MS: a tutorial review. Spectrochim Acta B 57:1361–1452
3. Navaza A, Ruiz-Encinar J, Ballesteros A et al (2009) Capillary HPLC–ICPMS and tyrosine iodination for the absolute quantification of peptides using generic standards. Anal Chem 81:5390–5399
4. Ornatsky OI, Kinach R, Bandura DR et al (2008) Development of analytical methods for multiplex bio-assay with inductively coupled plasma mass spectrometry. J Anal At Spectrom 23:463–469
5. Agilent 8800 ICP-QQQ application handbook 5991-2802EN (2013):16–18
6. Tanner SD, Li C, Vais V et al (2004) Chemical resolution of Pu+ from U+ and Am+ using a band-pass reaction cell inductively coupled plasma mass spectrometer. Anal Chem 76:3042–3048
7. Quemet A, Vitorge P, Cimas A et al (2013) Reactivity of lanthanoid mono-cations with ammonia: a combined inductively coupled plasma mass spectrometry and computational investigation. Int J Mass Spectrom 334:27–37

8. Esteban-Fernández D, Bierkandt F, Linscheid M (2012) MeCAT labeling for absolute quantification of intact proteins using label-specific isotope dilution ICP-MS. J Anal At Spectrom 27:1701–1708
9. Vanhaecke F, Balcaen L, Deconinck I et al (2003) Mass discrimination in dynamic reaction cell (DRC)-ICP-mass spectrometry. J Anal At Spectrom 18:1060–1065
10. He Q, Zhu Z, Jin L et al (2014) Detection of HIV-1 p24 antigen using streptavidin–biotin and gold nanoparticles based immunoassay by inductively coupled plasma mass spectrometry. J Anal At Spectrom 29:1477–1482
11. Cheng P, Koyanagi GK, Bohme DK (2006) Gas-phase reactions of atomic lanthanide cations with CO2 and CS2: room-temperature kinetics and periodicities in reactivity. J Phys Chem A 110(47):12832–12838
12. Corte Rodriguez M, López Fernández L, Garcia Ferdandez A et al (2015) Elemental and molecular mass spectrometric strategies for probing interactions between DNA and new Ru(II) complexes containing phosphane ligands and either a tris(pyrazol-1-yl)borate or a pyridine bis (oxazoline) ligand. J Anal At Spectrom 30:172–179
13. Balcaen L, Bolea-Fernandez E, Resano M et al (2014) Accurate determination of ultra-trace levels of Ti in blood serum using ICP-MS/MS. Anal Chim Acta 809(27):1–8
14. Koyanagi GK, Baranov VI, Tanner SD et al (2000) An inductively coupled plasma/selected-ion flow tube mass spectrometric study of the chemical resolution of isobaric interferences. J Anal At Spectrom 15:1207–1210
15. Diez Fernández S, Sugiyama N, Ruiz-Encinar J et al (2012) Triple quad ICPMS (ICPQQQ) as a new tool for absolute quantitative proteomics and phosphoproteomics. Anal Chem 84:5851–5857
16. Sucharová J (2011) Optimisation of DRC ICP-MS for determining selenium in plants. J Anal At Spectrom 26:1756–1762
17. Moller LH, Gabel-Jensen C, Franzyk H et al (2014) Quantification of pharmaceutical peptides using selenium as an elemental detection label. Metallomics 6:1639–1647
18. Simpson LA, Thomsen M, Alloway BJ et al (2001) A dynamic reaction cell (DRC) solution to oxide-based interferences in inductively coupled plasma mass spectrometry (ICP-MS) analysis of the noble metals. J Anal At Spectrom 16:1375–1380
19. Bolea-Fernandez E, Balcaen L, Resano M et al (2015) Interference-free determination of ultra-trace concentrations of arsenic and selenium using methyl fluoride as a reaction gas in ICP–MS/MS. Anal Bioanal Chem 407:919–929
20. Chang C, Liu H, Jiang S (2003) Bandpass reaction cell inductively coupled plasma mass spectrometry for the determination of silver and cadmium in samples in the presence of excess Zr, Nb and Mo. Anal Chim Acta 493:213–218
21. Esteban-Fernández D, Verdaguer JM, Ramirez-Camacho R et al (2008) Accumulation, fractionation, and analysis of platinum in toxicologically affected tissues after cisplatin, oxaliplatin, and carboplatin administration. J Anal Toxicol 32(2):140–146
22. Lavrov VV, Blagojevic V, Koyanagi GK et al (2004) Gas-phase oxidation and nitration of first-, second-, and third-row atomic cations in reactions with nitrous oxide: periodicities in reactivity. J Phys Chem A 108:5610–5624
23. Bolea-Fernandez E, Balcaen L, Resano M et al (2014) Potential of methyl fluoride as a universal reaction gas to overcome spectral interference in the determination of ultratrace concentrations of metals in biofluids using inductively coupled plasma-tandem mass spectrometry. Anal Chem 86:7969–7977
24. Ferrarello C, Ruiz-Encinar J, Centineo G et al (2002) Comparison of three different ICP-MS instruments in the study of cadmium speciation in rabbit liver metallothionein-1 using reversed-phase HPLC and post-column isotope dilution analysis. J Anal At Spectrom 17:1024–1029
25. Miyayama T, Ogra Y, Suzuki K (2007) Separation of metallothionein isoforms extracted from isoform-specific knockdown cells on two-dimensional micro high-performance liquid

chromatography hyphenated with inductively coupled plasma-mass spectrometry. J Anal At Spectrom 22:179–182
26. Falta T, Heffeter P, Mohamed A et al (2011) Quantitative determination of intact free cisplatin in cell models by LC-ICP-MS. J Anal At Spectrom 26:109–115
27. Campenhout KV, Infante HG, Adams F et al (2004) Induction and binding of Cd, Cu, and Zn to metallothionein in carp (Cyprinus carpio) using HPLC-ICP-TOFMS. Toxicol Sci 80:276–287
28. Raeve P, Bianga J (2015) Fast and accurate absolute-quantification of proteins and antibodies using Isotope Dilution-Triple Quadrupole ICP-MS. Agilent Technologies application note 5991-6118EN
29. Lelie H, Liba A, Bourassa M et al (2011) Copper and zinc metallation status of copper-zinc superoxide dismutase from amyotrophic lateral sclerosis transgenic mice. J Biol Chem 286 (4):2795–2806
30. Hermann G, Heffeter P, Falta T et al (2013) In vitro studies on cisplatin focusing on kinetic aspects of intracellular chemistry by LC-ICP-MS. Metallomics 5(6):636–647
31. Nuevo Ordónez Y, Deitrich CL, Montes-Bayón M et al (2011) Species specific isotope dilution versus internal standardization strategies for the determination of Cu, Znsuperoxide dismutase in red blood cells. J Anal At Spectrom 26:150–155
32. Tayler VF, Evans RD, Cornett RJ (2007) Determination of 90Sr in contaminated environmental samples by tuneable bandpass dynamic reaction cell ICP–MS. Anal Bioanal Chem 387:343–350
33. Tanner SD, Bandura DR, Ornatsky O et al (2008) Flow cytometer with mass spectrometer detection for massively multiplexed single-cell biomarker assay. Pure Appl Chem 80 (12):2627–2641
34. Martinčič A, Milačič R, Vidmar J et al (2014) New method for the speciation of ruthenium-based chemotherapeutics in human serum by conjoint liquid chromatography on affinity and anion-exchange monolithic disks. J Chromatogr A 1371:168–176
35. Olesik JW, Jones DR (2006) Strategies to develop methods using ion-molecule reactions in a quadrupole reaction cell to overcome spectral overlaps in inductively coupled plasma mass spectrometry. J Anal At Spectrom 21:141–159
36. Koyanagi GK, Caraiman D, Blagojevic V et al (2002) Gas-phase reactions of transition-metal ions with molecular oxygen: room-temperature kinetics and periodicities in reactivity. J Phys Chem A 106:4581–4590
37. Olesik JW, Gray PJ (2014) Advantages of N2 and Ar as reaction gases for measurement of multiple Se isotopes using inductively coupled plasma-mass spectrometry with a collision/ reaction cell. Spectrochim Acta B 100:197–210
38. Lopez-Avila V, Sharpe O, Robinson WH (2006) Determination of ceruloplasmin in human serum by SEC-ICPMS. Anal Bioanal Chem 386:180–187
39. McCurdy E, Woods G (2004) The application of collision/reaction cell inductively coupled plasma mass spectrometry to multi-element analysis in variable sample matrices, using He as a non-reactive cell gas. J Anal At Spectrom 19:607–615

Chapter 10
Highly Sensitive Analysis of Proteins and Metabolites by Metal Tagging Using LC-ICP-MS

Daigo Iwahata

Abstract Sensitive and selective analytical methods are a fundamental requirement in science, including biochemistry. For example, when there are significant differences in the amounts of proteins and metabolites in a biological sample, all of them must be considered for proteomics and metabolomics, even if their contents are very low. Now, the analytical technique with metal tag using liquid chromatography–inductively coupled plasma mass spectrometry (LC-ICP-MS) attracts attention as a highly sensitive method. It is used in inductively coupled plasma mass spectrometry (ICP-MS), which is one of the most sensitive instruments for the detection and determination of organic compounds, proteins, and metabolites. The organic compounds are usually unable to be detected with ICP-MS. Then, they are artificially derivatized by metal tag reagent before detection. The derivatization with metal tag reagents is performed for the purpose of building the bond between the metal and the analytes. The metal of the metal tag is able to select for the easy detection and high sensitivity of ICP-MS. The derivatization is almost never used for inorganic analysis. But, it is a well-known preparation in chromatography. It is able to convert the nature of the analytes as necessary. Especially, addition of a metal tag has the advantage to use ICP-MS for sensitive and highly selective detection. Therefore, this method by metal tag with using LC-ICP-MS is expected as the newest ultra-trace analytical technique for proteins and metabolites.

Keywords Metal tag • Derivatization • LC-ICP-MS • Pre-column • Speciation • Protein • Metabolite • Determination

---

D. Iwahata (✉)
Institute for Innovation, Ajinomoto Co., Inc, 1-1, Suzuki-cho, Kawasaki-ku, 210-8681 Kawasaki-shi, Kanagawa, Japan
e-mail: daigo_iwahata@ajinomoto.com

Y. Ogra, T. Hirata (eds.), *Metallomics*, DOI 10.1007/978-4-431-56463-8_10

10.1 Introduction

Inductively coupled plasma mass spectrometry (ICP-MS) is one of the most sensitive analytical techniques. Recently, this method has been applied for the analysis of biomolecules owing to its high sensitivity and elemental capability. Detection of a natural metal complex with biomolecules is known as the heteroatom analytical technique [1–3]. This method is very useful for the determination and first screening of biomolecules in the sample. However, many of the beneficial biomolecules for science do not contain metal and, hence, are unable to be detected with ICP-MS. Therefore, a new analytical method for popular biomolecules has been developed that involves the tagging of non-heteroatom biomolecules with metal reagents before detection [4, 5]. This is different from the known inorganic analytical methods and is a new analytical method for biomolecules and other organic compounds.

These new analytical methods require the preparation of the derivatives of the analytes by using metal tag reagents. Because the derivatization reaction takes a certain amount of time, this analytical method will increase the time compared with the non-derivatizing method. However, after the reaction, the metals or the heteroatoms in the derivatives can be detected sensitively and selectively with ICP-MS. It is the significant merit for the quantitative analysis. Therefore, these artificial metal labeling techniques are useful for enabling the analysis of various biological samples.

The first report of the combination of an immunoreaction and ICP-MS was described in 2001 by Zhang et al. [6], for the determination of thyroid hormones in human serum. The authors used biotinylated antibodies and Eu^{3+}-labeled streptavidin. In 2002, Baranov et al. implemented a gel filtration-based ICP-MS immunoassay to separate and quantify various labeled antibodies [7], while Zhang et al. used antibodies conjugated with colloidal gold nanoparticles for a sandwich-type immunoreaction [8]. However, liquid chromatography–inductively coupled plasma mass spectrometry (LC-ICP-MS) was not used in these studies. Zhang and Baranov analyzed the metals by off-line ICP-MS after acidic digestion. They regarded the derivatized antibody as a common metal-containing compound. ICP-MS was only used to assist the determination of the trace samples. Their methods were extensions of the traditional inorganic analytical methods.

In most of the traditional inorganic analytical methods, the pretreatment of a digestion by mineral acids is common. The organic matrix is sometimes observed to have an impact on the analytical value of ICP-MS, which is called the matrix effect. Additionally, ICP-MS instruments are susceptible to instability [9]. Therefore, the removal of all matrix interferences is very important for the accurate determination of the analytes. Currently, ICP-MS is recognized as the most widely used and powerful elemental technique for trace amounts of liquid samples. ICP-MS has a more powerful ion source and is able to decompose and ionize most of the compounds in the samples [10, 11]. The combination of analytical methods with ICP-MS exhibits lower matrix effects compared with conventional bioanalytical techniques (e.g., bioassay, chemiluminescence, UV-VIS and fluorescence spectrometry, and liquid chromatography–mass spectrometry (LC-MS)). Such methods are expected to provide fast, sensitive, and robust screening and determination of biomolecules.

In 2004, Whetstone et al. introduced a cysteine-specific labeling strategy, referred to as an element-coded affinity tag for peptides and proteins [12]. The reagent consists of a derivative of the bifunctional chelating agent 1,4,7,10-tetraazacyclododecane-*N*,*N*′,*N*″,*N*‴-tetraacetic acid (DOTA) and contains a bromoacetamido reactive group to specifically target cysteine residues. In recent approaches, two strong metal chelators (DOTA and diethylenetriaminepentaacetic acid (DTPA)) were used as metal tags for LC-ICP-MS [13–16]. The derivatives of DOTA or DTPA are detected and determined directly without the off-line sample digestion. These metal tags were suitable for the analysis of large molecular weight compounds using size exclusion column chromatography.

However, Iwahata et al. established a highly sensitive amino acid analytical method using reversed-phase LC-ICP-MS. In 2007, their study was reported at the international metallomics conference (Metallomics 2007; held at Nagoya, Japan). They derivatized amino acids with the metal tag reagent, bis(2,2′-bipyridine)-4′-methyl-4-carboxybipyridine-ruthenium *N*-succinimidyl ester [17]. The derivatized amino acids were detected by ICP-MS. They also determined the branched-chain amino acids in the biological samples [18]. This study was the first analysis of metabolites with metal tags using LC-ICP-MS. The metal tag for amino acid analysis was different from that used for protein analysis. Therefore, the dedicated design of the metal tag reagent of amino acid analysis is required. The metal tag and the derivative with amino acid are shown in Figs. 10.1 and 10.2.

This chapter describes the new analytical method "metal tags analysis using LC-ICP-MS" that is based on the technique of biomolecule labeling with metal tags.

10.2 Analysis Using Metal Tags

10.2.1 Metal Tags and the Metal Tag Reagents

The metal tag is the functional group for ICP-MS detection. The tag contains a metal atom, a metal colloid, a nanoparticle, or other metal forms. The metal tag reagent is the compound used to join the metal tag to the analyte and is one of the new derivative reagents.

The metal tag reagent contains two essential components. First is the reactive site with a functional group, and second is the metal to enable detection with ICP-MS. The reactive site is selected for each functional group in the analytes (e.g., succinimidyl ester for amino group, bromoacetamido for cysteine residues, and bromophenacyl ester for carboxylic group). The amino group and thiol group

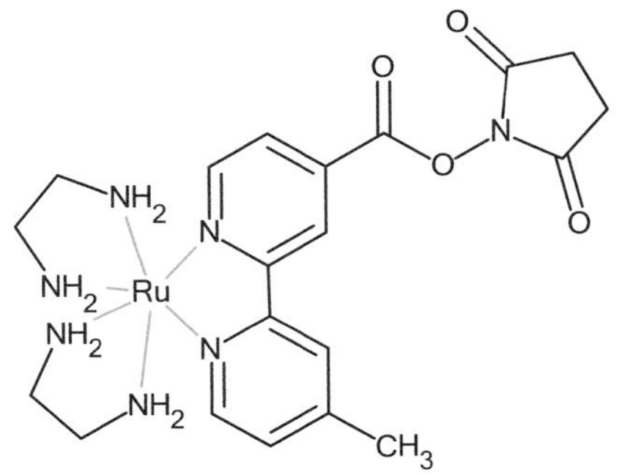

Bis(ethylenediamine)-4'-methyl-4-carboxybipyridine
-ruthenium *N*-succinimidyl ester (=ECRS)

Fig. 10.1 The metal tag for the amino acid analysis

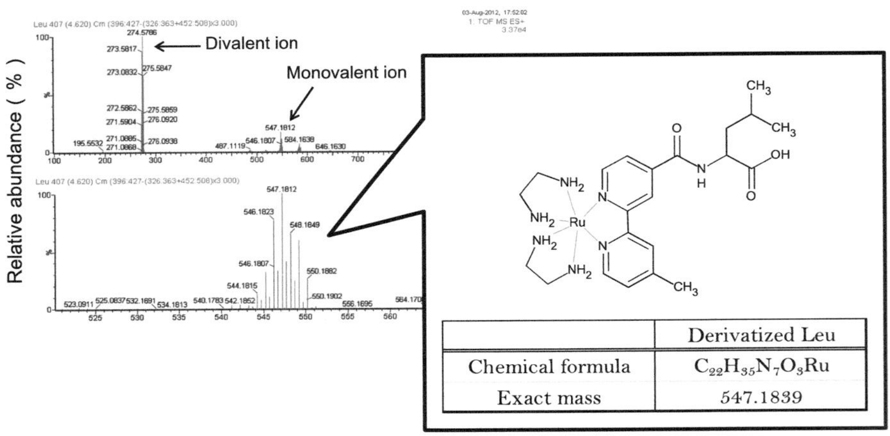

Fig. 10.2 The mass spectra of the derivatized leucine with the metal tag. Leucine was derivatized with ECRS. The mass spectrum was acquired using LC-Q-TOF-MS

exhibit extremely high reactivity to the derivative reagent, and they are selected as the first targets to insert the metal tags. Currently, there are many commercially available solutions for the derivatization of the analytes. The main commercial derivative reagents are shown in Table 10.1. These reagents have been widely applied for the biomolecules analysis using LC.

The bond between the metal moiety and the analyte needs to be maintained in an aqueous solution, such as the mobile phase, and enables separation of the derivatives by the column. It is essential that the derivatives with the metal tags are stable compounds, in terms of their chemical and physical properties. The length of the bond is important for the design of the metal tag reagents. This length will affect the separation ability by LC. When the structure of the metal tag is bulky and close to the analyte, the separation among the derivatives will become more difficult. In that

Table 10.1 The derivative reagents for HPLC analysis

Derivative reagents	Reactive groups	Detections
Nitorobenzenes	Primary and secondary amines, polyamine	UV-VIS
Ninhydrin	Primary amine	UV-VIS
Disuccinimido carbonate	Primary and secondary amines	UV-VIS
4-*N*, *N*-dimethylam inoazobenzene-4′-sulfonyl chloride	Primary and secondary amines, thiol, imidazole, phenol and aliphatic hydroxyl group	UV-VIS
Acetic anhydride	Tertiary aliphatic amine	UV-VIS
Benzoil chloride	Di- and polyamines	UV-VIS
Dimethylaminobenzaldehyde	Primary aromatic amine	UV-VIS
Phenylisocyanate	Primary and secondary aromatic amines (active hydrogen atom)	UV-VIS
o-phthalaldehyde	Primary amine	UV-VIS, fluorescence
Fluorescamine	Primary amine	UV-VIS, fluorescence
Sulfonyl chloride	Primary and secondary amines	UV-VIS, fluorescence
Carbonyl chroride and fluoride	Primary, secondary and tertiary amines, and alcohol	UV-VIS, fluorescence
1,2,-naphtoquinone-4-sulfonate	Primary and secondary amines	UV-VIS, fluorescence
Isocyanate and isothiocyanate	Primary, secondary and tertiary amines, urea, but also react with alcohol, water, Phenol and carboxylic acid	UV-VIS, fluorescence
Succinimidyl	amino acid	UV-VIS, fluorescence
Benzofurazan	Primary and secondary amines	UV-VIS, fluorescence
Oxazole	Amine, thiol and alcohol	UV-VIS, fluorescence

case, a long bond has an advantage for the separation. However, a short bond is sometimes observed which decreases the separation ability.

The metal of the metal tag is required for the easy detection and high sensitivity of ICP-MS. In addition, negligible quantity existing in biological samples and the environment improves the technique and is advantageous for the highly sensitive analysis. The metal is not confined to only one atom per molecule and participates with multiple atoms and structures. Nanocolloids and particles are also used. However, it is desirable that the metal in the metal tag is observed by ICP-MS to enable quantitative analysis.

Three recent approaches have been used to build the metal tags. These are tags based on metal chelates, covalently bound compounds, and nanoparticles. The details are described in Sect. 10.3. The structure of the metal tag reagent is shown in Fig. 10.3.

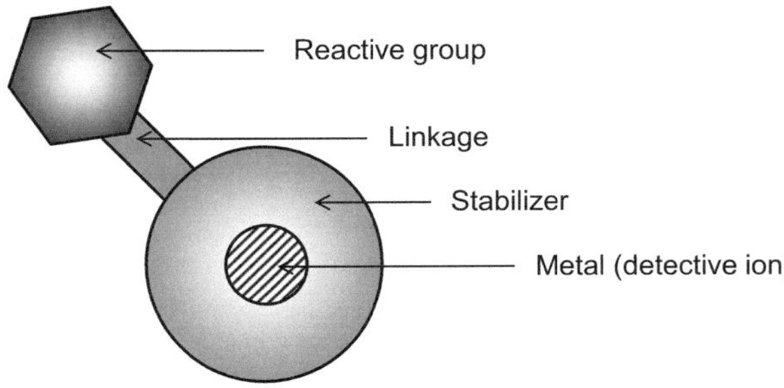

Fig. 10.3 The structure of the metal tag. The metal tags for the derivatization contain the metal and the reactive group inevitably

10.2.2 Derivatization by the Metal Tag Reagents

The derivatization with metal tag reagents is performed for the purpose of building the bond between the metal and the analytes. The metal tags for the derivatization inevitably contain the metal and the reactive group. The metal for the tag is desirable to be sensitive and easily detected by ICP-MS. ICP-MS is able to ionize and detect most elements in the periodic table. Therefore, there are many relevant choices for the metal tag according to the analytical purpose, although each metal exhibit their own benefits for various applications. For example, many transition metals and other metals have stable isotopes, and they are suitable for isotope ratio analysis. Metal tags with different stable isotopes are used for simultaneous determination. The derivatives with the isotope metal tags have equal or similar chemical and physical properties. This is very important for the protein and peptide analysis.

Platinum, gold, indium, and lanthanides are suitable for the analysis of biological samples, because they are scarce in living things. The derivatized analytes by the metal tag reagents with their metals are detectable at low concentration, without contamination from the experimental environment, by ICP-MS.

The reactive group builds the bond with the analytes by chemical, enzymatic, antibody, or another physical reaction. The most popular reactive groups can be used for the metal tag reagents. However, the reactive groups only need to preserve the strong bond with the metal and the analyte during analysis by LC-ICP-MS. The functional groups that are the targets for derivatization are the amino group, carboxyl group, hydroxyl group, thiol group, carbonyl group, and so on. The derivatization reaction of the analyte with the metal tag reagent is shown in Fig. 10.4.

The other parts of the metal tag are the stabilizer and the linkage. They are not absolutely necessary for the metal tag, but the stabilizer adds solubility and robustness to the metal in the solution, and the linkage contributes to the activity of the reactive group.

When the metal tag is the chelate compound, most stabilizers behave as ligands. This is probably easy to understand by way of an example. The ligands surround the center metals and contribute to maintaining the chelate compounds. The chelate

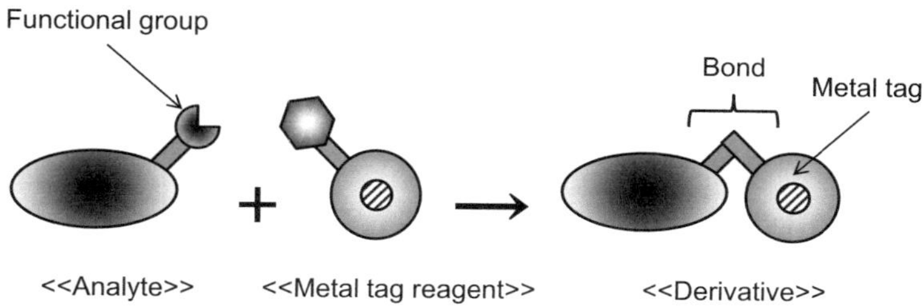

Fig. 10.4 The derivatization reaction of the analyte with the metal tag. The analyte has a functional group on the surface of the molecule. The derivatization forms a bond between the analyte and the metal tag

compounds with the stabilizers have a large influence because they occupy most of the metal tag.

The length of the linkage should be carefully monitored in the design of the metal tag reagents as the length will affect the reactivity of the derivatization. When the structure of the metal tag is bulky, a long linkage is preferential for the derivatization reaction. A short linkage has sometimes been used which decreases the reactivity owing to steric hindrance. Hence, a sufficient length of the linkage can be assured by using an alkyl chain or similar between the metal tag and the reactive group.

10.2.3 Separation of the Derivatives

In chromatography, derivatization is a well-known preparation even though it is almost never used for inorganic analysis. This technique requires efficient and effective reagents that can modify the behavior of complex compounds and allow for their detection. Currently, derivatization enables the analysis of many organic compounds by gas chromatography (GC), GC-MS, capillary electrophoresis (CE), CE-MS, LC, LC-MS, and LC-ICP-MS. The biggest reason why derivatization is chosen is the effective chemical modification of a compound by derivatizing its functional group (e.g., amino group, carboxyl group, hydroxyl group, thiol group, and carbonyl group) to aid in the use of chromatographic analysis.

The derivatization of biomolecules is a widely used approach in bioanalysis. It yields a modification intended to render the analyte of interest more suitable for characterization. Many biomolecules do not always exhibit a specific absorption or possess a fluorescent group, ionization group, or metal. Hence, derivatization is required for the sensitive and specific detection for analysis. Especially, addition of a metal tag has the advantage to use ICP-MS for ultra-sensitive and highly selective detection.

There are the two derivatization modes for the analysis using LC. These are pre-column and post-column derivatization. Although post-column derivatization is generally important for the fixed and low sensitivity analytical method (e.g., the ninhydrin method for amino acid), pre-column derivatization is effective for the highly sensitive analytical method.

In the post-column method, the mixing of an analyte solution with the reagent solution takes place just before the detector by the use of a T junction. The advantages of this technique are the good quantitative performance and reproducibility through an automated system. Furthermore, this technique is less influenced by sample matrix effects, because the analytes are separated before the derivatization reaction. This technique does not require pre-sample cleanup and facilitates the wide application of the samples.

However, a critical problem of the post-column derivatization is that this method employs an excess of reagent. In the analytical method with metal tag using ICP-MS, the excess metal is induced into the detector. This will be a very serious problem because the excess of metal causes an extremely high background signal for a long stretch of time which will hinder precise and accurate analysis.

In the pre-column method, analytes are derivatized before separation with the column, and the excess reagents also need to be removed or separated. The reagents for the derivatization reactions are mixed directly with the analytical samples. The reaction efficiency of the analytes is influenced directly by the sample matrix and the other reaction conditions. However, these factors do not have any influence on the post-column method, because this method uses the metal tag reagent only for the derivatization reaction. The amount of reagent consumed can be minimized by the use of a small reaction system or by the improvement of the sensitivity. In addition, unreacted derivatizing reagent is separated from the analytes in the various columns and so will not be an obstacle to the sensitive analysis. Therefore, pre-column derivatization is considered to be appropriate for the analysis of samples with high sensitivity.

The derivatization reactions of the pre-column mode are all simple and short. The majority of reactions are carried out at room temperature or weak heating conditions. These points are very important as most biomolecules are unstable under the high temperature. Hence, the derivatization reaction must be performed under mild conditions and in a short time.

In many cases using the pre-column method, reversed-phase chromatography is selected to separate the derivatives, as such techniques provide good separation ability. The pre-column derivatization is able to convert the nature of the analytes as necessary. For example, the pre-column derivatives can be modified with hydrophobic reagents to enable separation by reversed-phase chromatography. The reversed-phase LC-ICP-MS is recommended as the analytical method with metal tags owing to the high separation ability.

10.3 Design of Metal Tags

10.3.1 Tags Based on Metal Chelates

Metal chelates are most generally metal tags. Chelation is the most popular form of bound substances between a metal and an organic compound. Chelation involves the formation or presence of two or more separate coordinate bonds between a polydentate (multiple bonded) compound and one or multiple metal ions. The organic compounds are called the ligands or the chelators or the chelating agents and so on. These ligands are ions or molecules that bind to a central metal atom to form a coordination complex. The molecules of the ligands are important as they are the metal tag. They are attachable to the reactive group through some part of the structure. Generally, the bond between the metal and the ligand contains the formal donation of one or more of the electron pairs, although the nature of the bond with the metal and the ligand can range from covalent to ionic. These bonds are sufficiently tight to form the metal tag for the analysis using LC-ICP-MS.

The binding of the metal to the ligand can vary greatly (e.g., ethylenediamine, ethylenediaminetetraacetic acid, tetraazacyclododecane, bipyridine, ferrocene). The popular metal tags of metal chelates are diethylenetriamine pentaacetic acid (DTPA) [19–22] and DOTA. Their chelation complexes are sufficiently strong to remain coordinated during the LC separation. The stronger chelate complex is desirable for use in analysis. The chelators form a strong complex with trivalent metal ions (e.g., Fe^{3+}, In^{3+}, Tl^{3+}, and lanthanide ions), and the value of their equilibration constant is over 20. Many complexes with divalent metal ions (e.g., Ca^{2+}, Cu^{2+}, Fe^{2+}, Mn^{2+}, Zn^{2+}) are around ten. If the solute and pH of the buffer to be treated are carefully selected, they are sufficiently stable that the metal ions will not dissociate from the complex under the popular LC conditions (e.g., size exclusion chromatography, normal-phase chromatography, reversed-phase chromatography, hydrophilic interaction chromatography, and ion exchange chromatography).

Tags based on metal chelates are used widely from large molecules (e.g., protein, peptide, and nucleotide) to small molecules (e.g., amino acid, dipeptide, and organic acid). The nature of the metal chelates, which is hydrophilic or hydrophobic, is very important to separate and dissolve the derivatives. In the case of the protein, the hydrophilic metal chelate is suitable for derivatization, because many proteins are hydrophobic. If a hydrophobic metal tag is used for derivatization of proteins, the derivatives will deposit as the sediment. In contrast, many amino acids and organic acids are hydrophilic. The hydrophobic metal chelate is suitable for their derivatization. A reasonable hydrophobicity is needed to retain and separate the low molecular compounds in the reversed-phase column. The hydrophilicity of the metal chelate is able to be adjusted for the number and type of the functional group containing the ligands. The hydrophilicity and hydrophobicity of the chelation moiety is very important for the design of the metal tags [23].

10.3.2 Tags Based on Heteroatoms Binding with Covalent Bonds

The covalent bond involves the sharing of electron pairs between atoms and is usually very stable and strong. Such bonding is suitable for the metal tag using ICP-MS. However, covalent bonding is usually observed between nonmetallic elements, although there are also a few exceptions where such a bond is observed with metals.

There are a few metals and heteroatoms which are detectable by ICP-MS that also form covalent bond compounds, for example, mercury, tin, phosphorus, sulfur, and halogens. The covalent bond is sufficiently stable for analysis with LC-ICP-MS. When the biomolecule is derivatized by mercury, there is no need for a special metal tag reagent. The mercury ion can easily react with the thiol groups of the analytes.

In recent years, reactions between thiol compounds and metals have been used to build self-assembled monolayers (SAM) and other reactions to achieve cross-links. SAM of organic molecules are molecular assemblies formed on the surfaces of metallic plates and particles of mercury, gold, silver, copper, platinum, palladium, and various metals by adsorption. In many cases, the compounds form the monolayer that is strongly bound with the substrate. These reactions and the reagents for the SAM with thiol groups and metals are able to be applied in the design of the metal tag.

The mercury ion and the low molecular weight mercury compounds (methyl mercury and the other alkyl mercury compounds) are used to derivatize the proteins and peptides [24]. These compounds have very simple chemical formulas as the metal tag reagents and are easily separated by LC so that they are free from steric hindrance. However, mercury has high ionization energy (10.44 eV). Therefore, only 4 % is ionized so most of the mercury remains in the atomic state, in the argon plasma. Furthermore, it is known that mercury absorbs on the surface of the plastic products. Therefore, care must be taken for the trace analysis, because the sample container and the experimental apparatus can cause contamination and absorption.

Heteroatoms of phosphorus, sulfur, and halogens are also able to be used as the metal tag. Andrew et al. analyzed carboxylic acids by pre-column derivatization with the metal tag containing phosphorus, and the detection limit of derivatized maleic acid was 0.25 nmol per injection. However, phosphorus is not easily ionized inside the ICP, and the nitrogen oxide ion interferes with the detection of the phosphate ion at $m/z = 31$. To detect the phosphorus ion by ICP-MS, sector field (SF)-ICP-MS or the quadrupole ICP-MS attached to the collision cell was required, because of the higher sensitivity and better resolution than the normal quadrupole ICP-MS. The metal tags including the other heteroatoms are not very sensitive. Therefore, it is more beneficial to detect the heteroatoms in a natural tag (e.g., phosphorylated protein and peptide, phospholipid, sulfonated protein, arsenosugar, and selenoprotein) rather than the metal tag.

10.3.3 Tags Based on Nanoparticles

Nanoparticles are particles from 1 to 100 nm in size. In nanotechnology, a particle is defined as a small object that behaves as a whole unit with respect to its transport and properties. The nanoparticles are made from various materials, such as metal, silicon, clay, semiconductor, and so on. They have some unique physical, chemical, and biological properties. These properties are different from those of the corresponding bulk material, and nanoparticles are widely used in biological experiments. In particular, the reagents including metal nanoparticles are often used in immunoassays.

The specific interaction with the antigen has been widely used for the immunoassay. An immunoassay is a biochemical test that measures the presence or concentration of a protein in a solution. Immunoassays employ a variety of different labels to allow for detection of antibodies and antigens. These labels can be enzymes, radioactive isotopes, DNA reporters, fluorogenic reporters, electrochemiluminescent tags, and the metal tags (metal nanoparticles) for the detection using ICP-MS. The experimental methods for the metal tag immunoassay are no different to those for the conventional immunoassay, and the metal-labeled antibodies are added to the sample vials and mixed in the buffer solution. The mixed solutions are incubated for some hours at room temperature. The sample vials are then washed as necessary, and the reacted solution is subjected to LC-ICP-MS.

The analytical techniques for the nanoparticles using ICP-MS have improved. Initially, the nanoparticles were digested with various inorganic acids. However, the recent development of the analytical technique for the nanoparticles by ICP-MS has provided an analytical means for the direct quantification of the nanoparticles. The new quadrupole ICP-MS and time-of-flight ICP-MS are capable of very fast data acquisition. These methods are able to count the numbers of nanoparticles in the thin solution. For determination of this method, a suspended solution with the standard nanoparticles is needed.

The metal nanoparticles for the metal tags are mainly gold, platinum, and lanthanoids. The europium nanoparticle was initially used for the metal tag conjugated with an antibody. The antibody labeling with subsequent detection of the antigen and antibody interaction is used as the strategy for protein labeling with the metal tag. The metal tags of the antibody are connected on the chemical modification. Most of the tags are attached through the thiol group. These bonds are mainly used to attach the surface of the metal nanoparticle to the proteins. The reaction of the derivatization is similar to the covalent bond between the metal ion and the thiol group.

10.4 Biomolecule Analysis by Metal Tags

10.4.1 Proteins

The largest reason of the derivatization of biomolecules is to obtain a highly sensitive quantitative detection method [25–27]. For example, although there are differences in the amounts of proteins and metabolites in a biological sample, all of these compounds must be considered for proteomics and metabolomics, even if their contents are very low. Sensitive analysis is typically used for the time course analyses of the intrinsic components, including the proteins and the metabolites derivatized from the same individual. Sensitive and selective analytical techniques are a fundamental requirement in science.

The proteins are a popular analytical target for the metal tag. Several commercial metal tag reagents for the proteins have been developed and are commercially available. The popular metal tag reagents for proteins are 1,4,7,10-tetraazacyclododecane-*N*,*N*′,*N*″,*N*‴-tetraacetic acid succinimide ester (SCN-DOTA) [24, 28] and DTPA. They are both bifunctional chelating reagents and have been used to label the amino group of the proteins and chelate with lanthanoid ions and the other metal ions. The bifunctional reagents are chelated separately before derivatization with the amino group. Their metals are then able to undergo the analytical purpose. However, the excess metal should be removed from the derivatives, as this will cause an increase in the background signal.

It is necessary to denature the tertiary structure of the proteins from their native states before the derivatization. The reactive sites of the proteins are not always present on the surface. The majority of hydrophobic proteins contain hydrophilic reactive sites. The inner reactive sites are exposed by the use of denaturing agents: urea, guanidinium chloride, 2-mercaptoethanol, dithiothreitol, and so on. The denatured proteins build many bonds with the metal tags, and their analytical sensitivities are improved.

The process of the derivatization of the proteins is very simple and easy. The sample solution containing the proteins and the metal tag reagent are mixed in a suitable buffer solution. The buffer solution for the derivatization is always around neutral pH. These conditions are similar to the biological environment, and the protein and the derivatized protein are easy to denature. After mixing for some time, the reaction is completed. The mixture is stored in a closed container at low temperature and then subjected to LC-ICP-MS.

Size exclusion chromatography and affinity chromatography are used to analyze the protein derivatized with a metal tag for LC-ICP-MS. Both techniques play a key role in the purification of proteins and enzymes, polysaccharides, nucleic acids, and other biological macromolecules. Also known as gel filtration, size exclusion chromatography is the simplest and mildest of all chromatography techniques and separates molecules based on the differences in size. However, affinity chromatography separates the molecules based on the reversible interaction between the target protein and the specific ligand attached to a chromatography matrix. In affinity chromatography, the sample is investigated under conditions that favor specific binding to the ligand. Elution is performed specifically, using a competitive ligand, or non-specifically, by changing the pH, ionic strength or polarity. The eluent conditions can also sometimes obstruct the analysis by the metal tag with LC-ICP-MS.

Size exclusion chromatography is a nonbinding method, so the buffer composition does not directly affect separation. This means that separation can be performed in the presence of buffer additives and within a broad pH and ionic strength. Furthermore, size exclusion chromatography has an inactive solid phase. It is suitable to keep the chelate compound without dissolution. ICP-MS is more acceptable than electrospray ionization mass spectrometry (ESI-MS) in terms of the pH and ionic strength. For this reason, size exclusion chromatography is usually chosen as the front separator of LC-ICP-MS.

10.4.2 Amino Acids

Amino acids are the fundamental components of proteins and are essential for life. They account for 20 % of the body. The most common material is water, which accounts for 60–70 % of body weight, followed by amino acids (proteins). Thus, a person weighing 50 kg consists of about 10 kg of proteins, which make up important tissues of the body, such as muscles, gastrointestinal tract, internal organs, hemoglobin in the blood, and the collagen in hair and skin. The constituents of these proteins are amino acids. They are obtained from the diet and are used throughout the body.

In addition, amino acids work as hub compounds in metabolic pathways [29–31], cell signaling, and other cellular phenomena. Therefore, the concentration of amino acids in biofluids can provide information on the presence of various diseases and health conditions. For example, the amino acid metabolism of a cirrhosis patient is relatively decreased in the liver and increased in the skeletal muscle [32, 33]. As a result, the ratio of branched-chain amino acids (valine, leucine, and isoleucine) in the serum will be low, because they are mainly metabolized in the skeletal muscle. The change in the serum amino acid balance is explained by the Fisher formula: (Ile + Leu + Val)/(Phe + Tyr) [34]. The Fisher formula is used as an index of the hobnailed liver [35]. In the case of a liver fibrosis patient, the serum amino acid balance reflects the unique ratio.

Amino acid analysis is very important in biochemistry, food science, nutrition, and other fields. In 1958, Stein, Moore, and Spackmann developed an amino acid analyzer system for the colorimetric determination of amino acids by producing a purple color with the ninhydrin reagent [36], which was first reported by Ruhemann in 1911 [37], after stepwise separation of the amino acids while raising the pH of the citric acid buffer solution with a cation exchange resin. The analyzer is now capable of assaying not only protein hydrolysates but also amino acids in biological fluids, with the progressive improvement of HPLC pumping performance and the development of more effective column resins, although the sensitivity is not very high, and the detection limit is a few picomoles. Apart from the ninhydrin method, a variety of other techniques for converting amino acids to sensitive analyzable fluorescent derivatives have been developed since the late 1970s. The reagents used to produce the derivatives include ortho-phthalaldehyde (OPA), 9-aminoquinolyl-*N*-hydroxysuccinimidyl carbamate (AQC), and 4-fluoro-7-nitrobenzofurazan (NBD-F). The sensitivity of detection is within the range from sub-picomole to femtomole [38, 39]. Recently, mass spectrometry has also been used for detection in the analysis, by which the amino acids are detected either directly or after pre-column derivatization [40, 41].

The amino acids have the amino group as the functional group that is the same as the proteins. Hence, the process of the derivatization is nearly equal. The sample solution containing the amino acids and the metal tag reagent are mixed in the suitable buffer solution. After mixing for a certain time, the reaction is completed. The mixture was stored in a closed container at low temperature and was subjected to LC-ICP-MS. However, the peptide bonds of the amino acid derivatives are weak under basic conditions and are easily broken in the mixed solution. Then, after the derivatization, the solution is mixed with an acidic solution to result in a weakly acidic or neutral condition.

The derivative reaction for the amino acids will be completed quicker than that of the proteins. Generally, the small compounds are able to easily undergo the chemical reaction.

The reaction time of the derivatization is reduced by heating. The amino acids are derivatized for several minutes. The large compounds, such as proteins, require times of more than several hours for the derivatization. It is difficult to shorten the reaction time of proteins, as most proteins are unsuitably denatured and precipitated by heating.

The metal tag reagents for the amino acids have a different nature from those of the protein. The proteins are hydrophobic and the metal tags for the proteins are hydrophilic to dissolve in the sample solution. For example, DOTA and DTPA have the hydrophilic group, carboxylic acid. In contrast, most of the amino acids are highly hydrophilic compounds, and the metal tags for the amino acids are hydrophobic. The amino acid analysis is mainly performed by cation exchange and the reversed-phase chromatography. The automatic analyzer system for the determination of the amino acids with the cation exchange resin and the ninhydrin reagent has been used for a long time. However, the pre-column derivatization has been developed and performed for the sensitive analysis since the late 1970s. As the cation exchange chromatography may resolve the metal compound, reversed-phase chromatography is used to separate the amino acid derivatives with metal tag reagents. However, the amino acids are too hydrophilic to remain in the reversed-phase column. Then, the hydrophobic metal tag reagents are synthesized. The metal tags play both the roles of the detective target and the modifier of hydrophobicity. The metal tag reagents for the amino acids have been established as the most highly sensitive analytical method.

10.4.3 Organic Acids

The organic acids are organic compounds with acidic properties. They are various sizes from low to large molecular weight, and the most common compounds are the carboxylic acids, whose acidity is associated with their carboxyl group. The low molecular organic acids are hydrophilic and soluble in water. In biology, they are used as intermediate metabolites. They are very important compounds in the metabolic pathway. One of the most famous metabolic pathways is the citric acid cycle, also known as the tricarboxylic acid (TCA) cycle or the Krebs cycle. This pathway is ever present in all aerobes and plays an important role to produce the energy for life by oxidation. It is well known that adenosine triphosphate (ATP) and nicotinamide adenine dinucleotide (NADH) are the energy currency that is generated from the enzymatic reactions of the TCA cycle. The TCA cycle is a key metabolic pathway that connects the carbohydrate, the fat, and the protein at the center. It also provides precursors of certain amino acids. Hence, it is interesting and important to analyze the organic acids in the TCA cycle for the study of the metabolism.

However, there have not been many reports of the organic acid analysis with metal tags. Andrew et al. analyzed carboxylic acids by pre-column derivatization with the reagent tris(2,4,6-trimethoxyphenyl)phosphonium propylamine bromide on phosphorus [42]. The detection limit of the derivatized maleic acid was 0.25 nmol per injection. Phosphorus is not easily ionized inside the ICP, and the nitrogen oxide ion interferes with the detection of the phosphate ion at $m/z = 31$. To detect the phosphorus ion, sector field (SF) ICP-MS or quadrupole ICP-MS attached to the collision cell must be used. 4-(*N,N*-dimethylaminosulfonyl)-7-piperazino-2,1,3-benzoxadiazole was used as the tag reagent for the organic acids. Dicarboxylic acids, adipic acid, succinic acid, and oxaloacetic acid were derivatized and detected by a sulfur ion using LC-ICP-MS. The chromatograms of the organic acid derivatives and the tag are shown in Fig. 10.5.

The derivatization of the carboxylic group in the organic acids is more complicated than that of the amino group. The organic acids in the sample solution were added to the tag reagent solution in the presence of a condensation agent. These are mixed in the buffer, and the reaction accelerator is supplied as needed. The mixed solution is heated and is suspended to be derivatized by the reaction inhibitor. At last, the reacted solution was subjected to LC-ICP-MS.

In the separation process, the organic acids that have the alkyl bonds until C5 are strongly hydrophilic. They are desirable for analysis using reversed-phase LC after derivatizing with the hydrophobic reagents. The organic acids that have alkyl bonds from C5 to C10 are suitable for the hydrophilic metal tag reagent for the derivatization. The organic acids including the long bonds over C10 are too hydrophobic and so are difficult to separate using reversed-phase LC.

As another experiment, the organic phosphates have been specifically derivatized. In 2006, Koike et al. developed the phosphate-binding tag (Phos-tag)

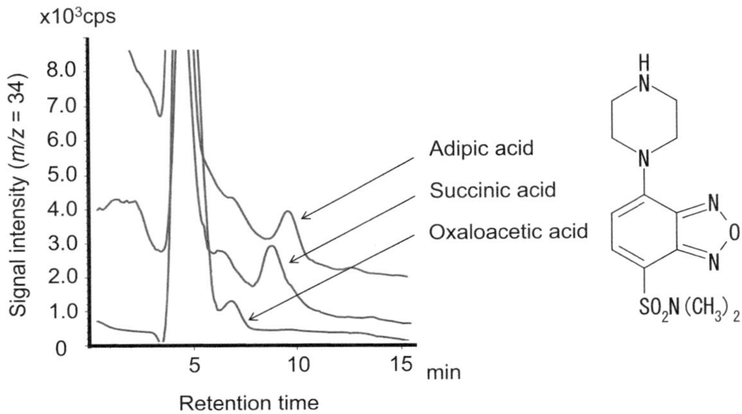

Fig. 10.5 The chromatograms of the organic acid derivatives using LC-ICP-MS. Derivative reagent: 4-(*N*,*N*-dimethylaminosulfonyl)-7-piperazinobenzofurazan. Detection element: sulfur ($m/z = 34$). Column: C18 (ODS). Mobile phase: (a) 20 mM acetate buffer (pH 5.8) and (b) 60 % acetonitrile/water

[43]. This tag was used to analyze phosphorylated proteins. The Phos-tag contained the alkoxide-bridged dinuclear metal (i.e., Zn^{2+} or Mn^{2+}) complexes [44].

Phos-tag is a novel labeling system for analysis of phosphorylation. The Phos-tag is a functional molecule that binds phosphorylated ions. Under physiological conditions of pH 5–8, the Phos-tag forms a stable complex with anionic substituents, particularly phosphoric esters with a single ester bond. The affinity for the phosphate dianion, PO_4^{2-}, is much higher than other negative ions making binding selective for phosphate groups.

This technique is very unique in that the compounds including only phosphate groups are detected. Phosphorylation is the addition of a phosphoryl group (PO_3^{2-}) to a molecule. The phosphorylation of the protein is a reversible reaction that is an important regulatory mechanism and is related to carcinogenesis and neuropathogenesis and other diseases when the mechanism is abnormal. There are thousands of different kinds of proteins in the organism, and it is estimated that 1/10 to 1/2 of proteins are phosphorylated. Hence, the methods for analyzing the phosphorylation of proteins are very important.

Moreover, the phosphorylated organic acids, particularly the small molecular weight compounds, are valuable for diverse biological and pathological processes. For example, it is well known that the sugar phosphate, one of the small phosphorylated compounds, is very popular in living things. The sugar phosphate is the initial compound in two major metabolic pathways, glycolysis and the pentose phosphate pathway. In the future, these compounds will be analyzed by the metal tag using LC-ICP-MS, and it is expected that these studies will make a substantial contribution to the understanding of the metabolic pathways.

References

1. Sanz-Medel A (2005) From metalloproteomics to heteroatom-tagged proteomics. Anal Bioanal Chem 381:1
2. Shah M, Caruso JA (2005) Inductively coupled plasma mass spectrometry in separation techniques: recent trends in phosphorus speciation. J Sep Sci 28:1969
3. Szpunar J (2005) Advances in analytical methodology for bioinorganic speciation analysis: metallomics, metalloproteomics and heteroatom-tagged proteomics and metabolomics. Analyst 130:442
4. Engelhard C (2011) Inductively coupled plasma mass spectrometry: recent trends and developments. Anal Bioanal Chem 399:213
5. Tholey A, Schaumlöffel D (2010) Metal labeling for quantitative protein and proteome analysis using inductively-coupled plasma mass spectrometry. TrAC Trends Anal Chem 29:399
6. Zhang C, Wu F, Zhang Y, Wang X, Zhang X (2001) A novel combination of immunoreaction and ICP-MS as a hyphenated technique for the determination of thyroid-stimulating hormone (TSH) in human serum. J Anal At Spectrom 16:1393
7. Quinn ZA, Baranov VI, Tanner SD, Wrana JL (2002) Simultaneous determination of proteins using an element-tagged immunoassay coupled with ICP-MS detection. J Anal At Spectrom 17:892
8. Zhang C, Wu F, Zhang X (2002) ICP-MS-based competitive immunoassay for the determination of total thyroxin in human serum. J Anal At Spectrom 17:1304
9. Date AR, Gray AL (1989) Inductively coupled plasma mass spectrometry. Blackie, New York
10. Jarvis KE, Gray AL, Houk RS (1992) Inductively coupled plasma mass spectrometry. Blackie, New York
11. Montaser A (1997) Inductively coupled plasma mass spectrometry. Wiley-VCH, New York
12. Whetstone PA, Butlin NG, Corneillie TM, Meares CF (2004) Element-coded affinity tags for peptides and proteins. Bioconjug Chem 15:3
13. Koellensperger G, Groeger M, Zinkl D, Petzelbauer P, Hann S (2009) Quantification of elemental labeled peptides in cellular uptake studies. J Anal At Spectrom 24:97
14. Kretschy D, Groger M, Zinkl D, Petzelbauer P, Koellensperger G, Hann S (2011) High-throughput flow injection analysis of labeled peptides in cellular samples – ICP-MS analysis versus fluorescence based detection. Int J Mass Spectrom 307:105
15. Ahrends R, Pieper S, Ku A, Weisshoff H, Hamester M, Lindemann T, Scheler C, Lehmann K, Taubner K, Linscheid MW (2007) A metal-coded affinity tag approach to quantitative proteomics. Mol Cell Proteomics 6:1907
16. Schwarz G, Mueller L, Beck S, Linscheid MW (2014) DOTA based metal labels for protein quantification: a review. J Anal At Spectrom 29:221
17. Iwahata D, Hirayama K, Miyano H (2008) A highly sensitive analytical method for metal-labelled amino acids by HPLC/ICP-MS. J Anal At Spectrom 23:1063
18. Iwahata D, Tsuda M, Aigaki T, Miyano H (2011) Quantitative analysis of branched amino acids in a single fruit fly by LC-ICP-MS after pre-column derivatization with a metal tag reagent. J Anal At Spectrom 26:2461
19. Crotti S, Granzotto C, Cairns WR, Cescon P, Barbante C (2011) Elemental labeling for the identification of proteinaceous-binding media in art works by ICP-MS. J Mass Spectrom 46:1297
20. Westerberg DA, Carney PL, Rogers PE, Kline SJ, Johnson DK (1989) Synthesis of novel bifunctional chelators and their use in preparing monoclonal antibody conjugates for tumor targeting. J Med Chem 32:236
21. Friedman M, Orlova A, Johansson E, Eriksson TL, Hoiden-Guthenberg I, Tolmachev V, Nilsson FY, Stahl S (2008) Directed evolution to low nanomolar affinity of a tumor-targeting epidermal growth factor receptor-binding affibody molecule. J Mol Biol 376:1388

22. Brechbiel MW, Gansow OA, Atcher RW, Schlom J, Esteban J, Simpson DE, Colcher D (1986) Synthesis of 1 – (p -1sothiocyanatobenzyI) derivatives of DTPA and EDTA. Antibody labeling and tumor-imaging studies. Inorg Chem 25:2112
23. Iwahata D, Nakamura K, Yamada R, Miyano H, Yamada N (2013) A new metal tag for highly selective and sensitive analyses of amino acids and dipeptides by HPLC/ICP-MS. J Anal Sci Methods Instrum 3:80
24. Bomke S, Sperling M, Karst U (2010) Organometallic derivatizing agents in bioanalysis. Anal Bioanal Chem 397:3483
25. Fernandez DE, Scheler C, Linscheid MW (2011) Absolute protein quantification by LC-ICP-MS using MeCAT peptide labeling. Anal Bioanal Chem 401:657
26. Pröfrock D, Prange A (2012) Inductively coupled plasma-mass spectrometry (ICP-MS) for quantitative analysis in environmental and life sciences: a review of challenges, solutions, and trends. Appl Spectrosc 66:843
27. Sanz-Medel A, Montes-Bayón M, Bettmer J, Encinar JR (2012) ICP-MS for absolute quantification of proteins for heteroatom-tagged, targeted proteomics. TrAC Trends Anal Chem 40:52
28. Jakubowski N, Waentig L, Hayen H, Venkatachalam A, Bohlen AV, Roos PH, Manz A (2008) Labelling of proteins with 2-(4-isothiocyanatobenzyl)-1,4,7,10-tetraazacyclododecane-1,4,7,10-tetraacetic acid and lanthanides and detection by ICP-MS. J Anal At Spectrom 23:1497
29. Wagner A, Fell D (2001) The small world inside large metabolic networks. Proc R Soc B 268:1803
30. Scaglia F (2010) New insights in nutritional management and amino acid supplementation in urea cycle disorders. Mol Genet Metab 100:S72
31. Shikata N, Maki Y, Nakatsui M, Mori M, Noguchi Y, Yoshida S, Takahashi M, Kondo N, Okamoto M (2010) Determining important regulatory relations of amino acids from dynamic network analysis of plasma amino acids. Amino Acids 38:179
32. Bolster DR, Jefferson LS, Kimball SR (2004) Regulation of protein synthesis associated with skeletal muscle hypertrophy by insulin-, amino acid- and exercise induced signalling. Proc Nutr Soc 63:351
33. Holecek M (2010) Three targets of branched-chain amino acid supplementation in the treatment of liver disease. Nutrition 26:482
34. Fischer JE, Baldessarini RJ (1971) False neurotransmitters and hepatic failure. Lancet 298:75
35. Soeters PB, Fischer JE (1976) Insulin, glucagon, amino acid imbalance, and hepatic encephalopathy. Lancet 308:880
36. Spackman DH, Stein WH, Moore S (1958) Automatic recording apparatus for use in the chromatography of amino acids. Anal Chem 30:1190
37. Siegfried R (1911) Triketohydrindene hydrate. Hydrindantin and its analogues. J Chem Soc 99:1306
38. Santa T, Takeda A, Uchiyama S, Fukushima T, Homma H, Suzuki S, Yokosu H, Lim CK, Imai K (1998) N-(4-nitro-2,1,3-benzoxadiazoyl-7-yl)-N-methyl-2-aminoacetohydrazide (NBD-CO-Hz) as a precolumn fluorescent derivatization reagent for carboxylic acids in high-performance liquid chromatography. J Pharm Biomed Anal 17:1065
39. Uchiyama S, Santa T, Okiyama N, Fukushima T, Imai K (2001) Fluorogenic and fluorescent labeling reagents with a benzofurazan skeleton. Biomed Chromatogr 15:295
40. Shimbo K, Yahashi A, Hirayama K, Nakazawa M, Miyano H (2009) Multifunctional and highly sensitive precolumn reagents for amino acids in liquid chromatography/tandem mass spectrometry. Anal Chem 81:5172
41. Shimbo K, Oonuki T, Yahashi A, Hirayama K, Miyano H (2009) Precolumn derivatization reagents for high-speed analysis of amines and amino acids in biological fluid using liquid chromatography/electrospray ionization tandem mass spectrometry. Rapid Commun Mass Spectrom 23:1483

42. Cartwright AJ, Jones P, Wolff JC, Evans EH (2005) Detection of phosphorus tagged carboxylic acids using HPLC-SF-ICP-MS. J Anal At Spectrom 20:75
43. Kinoshita E, Kinoshita EK, Takiyama K, Koike T (2006) Phosphate-binding tag, a new tool to visualize phosphorylated proteins. Mol Cell Proteomics 5:749
44. Kinoshita EK, Kinoshita E, Koike T (2009) Phos-tag beads as an immunoblotting enhancer for selective detection of phosphoproteins in cell lysates. Anal Biochem 389:83

Part III
Application of Metallomics in Molecular Biology, Medicine and Pharmaceutical Sciences

Chapter 11
Comprehensive Element Analysis of Prokaryotic and Eukaryotic Cells as well as Organelles by ICP-MS

Tomonari Umemura, Yuichiro Matsui, Shinnosuke Sakagawa, Taku Fukai, Eiji Fujimori, Hidetoshi Kumata, and Motohide Aoki

Abstract Elemental abundances of prokaryotic and eukaryotic cells as well as organelles were provided as the number of atoms per cell (or organelle). In order to examine the specific uptake of vital elements by cells, *E. coli* cells were grown in different media and harvested at several time points during the cell growth and proliferation. The major-to-ultratrace elements in the cells were determined by micro-flow injection (μFI)-ICP-MS after microwave-assisted acid digestion of less than 10 mg of dry weight sample. Cyanobacteria and chlorella were also cultivated to compare elemental abundance differences among species. In addition, to further examine the localization and utilization of vital elements in the cells, chloroplasts and mitochondria were isolated from the tissues of spinach leaves, potato tuber, and bovine liver by density-gradient centrifugation using Percoll. The concentration data obtained by ICP-MS measurements were normalized to a per cell (or organelle) basis, taking into account the cell size and water content. Among about 60 elements detectable by ICP-MS, more than 30 elements were quantitatively determined, and it was clearly found that essential elements participating in specific vital functions such as redox reactions were highly accumulated in the microorganisms and organelles.

T. Umemura (✉)
Department of Molecular Life Sciences, School of Life Sciences, Tokyo University of Pharmacy and Life Sciences, 1432-1 Horinouchi, Hachioji, Tokyo 192-0392, Japan

Department of Applied Chemistry, Graduate School of Engineering, Nagoya University, Furo-cho, Chikusa-ku, Nagoya 464-8603, Japan
e-mail: tumemura@toyaku.ac.jp

Y. Matsui • H. Kumata • M. Aoki
Department of Molecular Life Sciences, School of Life Sciences, Tokyo University of Pharmacy and Life Sciences, 1432-1 Horinouchi, Hachioji, Tokyo 192-0392, Japan

S. Sakagawa • T. Fukai • E. Fujimori
Department of Applied Chemistry, Graduate School of Engineering, Nagoya University, Furo-cho, Chikusa-ku, Nagoya 464-8603, Japan

Y. Ogra, T. Hirata (eds.), *Metallomics*, DOI 10.1007/978-4-431-56463-8_11

Keywords ICP-MS • Metallomics • Microbe • Organelle • Mitochondria • Chloroplast • *E. coli* • Cyanobacteria

11.1 Introduction

Essentiality and toxicity of trace elements in living cells and organisms have attracted sustained interest over the past several decades [1–4]. So far, such biometal and toxicological research has focused exclusively on several nutrient and toxic elements, but the combination of different elements exhibits a synergistic, antagonistic, or additive effect on cell viability. In addition, notorious elements may play a vital role in cellular function at a very low concentration level. Therefore, there is a great need for comprehensive element analysis of living organisms [5, 6]. However, it is still difficult to determine elemental abundance generally ranging over ten orders of magnitude, which is far greater than the measurement capability of current technologies.

As an example, elemental abundance of a fertilized salmon egg is presented in Fig. 11.1 [6]. Most of the elements were determined by inductively coupled plasma sector field mass spectrometry (ICP-SF-MS) after microwave-assisted acid digestion of the whole egg. Carbon and N were measured by a CHN corder without digestion, and H and O were estimated from the water content. It is noted that a

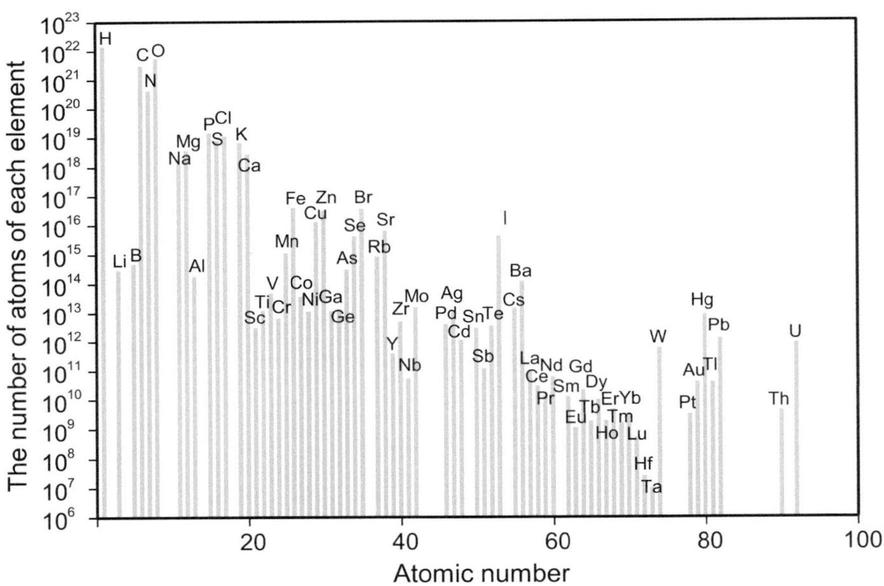

Fig. 11.1 Elemental abundance of a fertilized salmon egg

variety of elements are present in the small spherical egg with a diameter of only 5 mm. The total number of atoms of the determined 68 elements amounted to 2.3×10^{22}, which was comparable with the number of stars in the universe (approximately 3×10^{23} stars) [7]. The elements except for 11 elements (O, C, H, N, Ca, P, S, K, Na, Cl, and Mg) are often referred to as trace elements, but from the perspective of the organisms, this is a misnomer. In the salmon egg, even the rare earth elements are contained at a level of more than 10^6 atoms, as can be seen in Fig. 11.1. At a rough estimate, 10^6 atoms/egg of salmon corresponds to the concentration of about 1 ag/mL in the acid-digested analysis solution, which is below or close to the detection limits of ICP-SF-MS. Consequently, quite low abundant elements such as Be, F, Si, Rh, Rh, In, Re, Os, Ir, and Bi could not be detected, but the data in Fig. 11.1 strongly support that all the elements occurring in nature are present in the single egg, some of which may be incidentally incorporated as the impurities from the surrounding growth environment.

This elemental abundance data also provides a useful clue as to elemental bioavailability and/or potential risk of exposure to hazardous elements. Figure 11.2 shows the comparison of the elemental abundances of the salmon egg and seawater in mol%. It can be seen that most of the essential elements such as Fe, Zn, Mn, Cu, and Se are highly accumulated in the salmon egg by several orders of magnitude.

Salmon egg is, in some ways, a single cell that forms an individual organism. However, it has yolk as a nutrient source for the developing embryo, and then

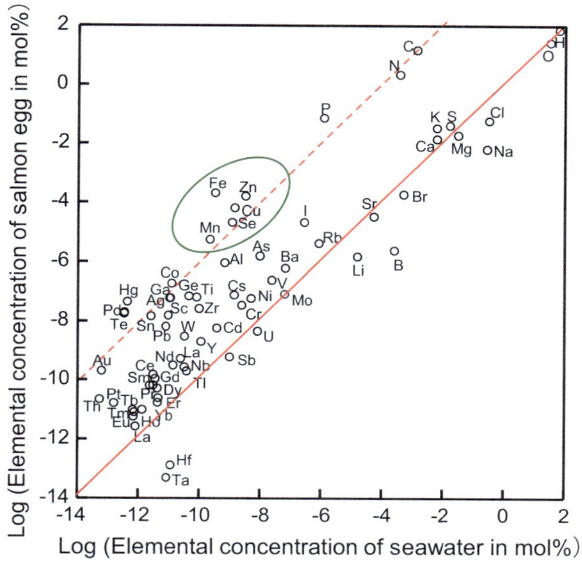

Fig. 11.2 Comparison of elemental abundances of a salmon egg and seawater

salmon egg is, in the real sense of the term, not a single cell. In order to investigate elemental abundance of a real single cell, in the present paper, some prokaryotic and eukaryotic cells were cultivated in different media and their multielement analysis was carried out by ICP-MS. Two types of ICP-MS methods were exploited for the purpose: one is a slurry nebulization time-resolved ICP-MS for a single-cell analysis [8, 9], and the other is a traditional solution-based ICP-MS after acid digestion of a certain amount (at least 1 mg in dry weight) of cells [10, 11]. As a result of our preliminary study, slurry nebulization ICP-MS technique enabled us to perform a real single-cell analysis of a few major elements such as P, Mg, and Fe, but the signals of almost all the other elements were buried in background noise signals. Therefore, in the present investigation, traditional solution-based ICP-MS method was adopted with the purpose of comprehensive element analysis of the cells. Furthermore, in order to examine vital elements localized in subcellular organelles, tissue cells from spinach leaves, bovine liver, and potato tuber were subdivided into organelles (mitochondria and chloroplasts), and their multielement analysis was carried out.

11.2 Experimental

11.2.1 Reagents

4-(2-Hydroxyethyl)-1-piperazineethanesulfonic acid (HEPES), disodium dihydrogen ethylenediaminetetraacetate dehydrate (EDTA-2Na), and polyvinylpyrrolidone (PVP-40) were purchased from Sigma-Aldrich (St. Louis, MO, USA). Tryptone, yeast extract, bovine serum albumin (BSA), sucrose, D-mannitol, and D-sorbitol were obtained from Merck (Darmstadt, Germany). Percoll was purchased from GE Healthcare (Pittsburgh, PA, USA). Nitric acid (60 %, density 1.38 g/mL) and hydrogen peroxide (30 %, 1.11 g/mL) used for acid digestion were of ultrapure grade from Kanto Chemicals (Tokyo, Japan). A series of multielement standard solutions were prepared by mixing single-element standard solutions (1000 μg/mL) for atomic absorption spectrometry from Wako Pure Chemicals (Osaka, Japan). Distilled deionized water (18 MΩ · cm resistivity) was prepared by a Milli-Q element purification system (Nihon Millipore Kogyo, Japan). Standard reference material (SRM 1577c bovine liver) from the National Institute of Standards and Technology (NIST, MD, USA) and certified reference material (CRM No. 3 chlorella) from the National Institute for Environmental Studies (NIES, Tsukuba, Japan) were used to validate the analytical method. All other reagents and solvents were of the highest purity commercially available from Sigma-Aldrich (St. Louis, MO, USA) or Tokyo Chemical Industry (Tokyo, Japan) and used without further purification.

11.2.2 Samples

11.2.2.1 Cell Strains and Culture Conditions

E. coli K-12 stain MG1655 was kindly donated by Ajinomoto Co., Inc. (Kawasaki, Japan). *E. coli* MG1655 cells were grown in Luria broth (LB)-rich medium and in M9 minimal medium at 37 °C. LB medium was composed of 10 g/L NaCl, 10 g/L tryptone, and 5 g/L yeast extract, while M9 medium was composed of 6 g/L Na_2HPO_4, 3.0 g/L KH_2PO_4, 0.5 g/L NaCl, 1.0 g/L NH_4Cl, 250 mg/L $MgSO_4.7H_2O$, 15 mg/L $CaCl_2$, 5.4 mg/L $FeCl_3$, 10 mg/L thiamine, and 2.0 g/L glucose.

A glucose-tolerant strain of cyanobacterium *Synechocystis* sp. PCC6803 was presented by Prof. M. Tsuzuki (Tokyo University of Pharmacy and Life Sciences). PCC6803 cells were cultured photoautotrophically at 30 °C in BG-11 medium (Allen, 1959) buffered with 30 mM HEPES-NaOH (pH 7.5) [12].

Chlorella kessleri NIES-2160 cells purchased from NIES were grown under aeration and light irradiation at 25 °C in fivefold diluted Gamborg's B5 medium [13] buffered with 20 mM HEPES-NaOH (pH 7.5) [13].

The cells grown in the medium were harvested by centrifugation and washed with a physiological saline solution three times by centrifuging at 2000 *g* for 5 min. Collected cells were weighed before and after freeze-drying (FDS-1200, Tokyo Rikakikai Co., Ltd., Japan) to determine water content.

11.2.2.2 Isolation of Subcellular Organelles

Spinach (*Spinacia oleracea* L.), potato (*Solanum tuberosum* L.), and bovine liver were purchased from local supermarkets to obtain chloroplasts and mitochondria. Isolation of the organelles is basically composed of three steps: (1) homogenization, (2) filtration and centrifugation, and (3) purification. All operations were carried out at 4 °C to prevent enzymatic damage. The detailed procedures for isolation of each organelle from the eukaryotic tissue cells are described below.

11.2.2.2.1 Isolation of Chloroplast from Spinach Leaves

The isolation of chloroplasts from spinach leaves was carried out with a slight modification of the method of Sato et al. (1997) [14]. First of all, 20 g of spinach leaves was torn into small pieces and then transferred into 100 mL of an isotonic buffer solution (330 mM D-sorbitol, 30 mM HEPES-KOH (pH 7.4), 2 mM EDTA-2Na, 0.1 % (w/v) BSA) to prevent osmotic damage. The leaf tissue was grounded with an electric blender three times for 5 s at cooling intervals of 10 s. The homogenate was filtered through four layers of cheesecloth, and the filtrate was centrifuged at 2000 *g* for 10 min. The supernatant was discarded, while the pellet was gently resuspended in 20 mL of the above buffer solution. Subsequently,

Percoll gradient centrifugation was carried out for further purification of the coarse chloroplasts. The suspension was placed on top of the Percoll layers, which consisted of 10 mL of Percoll solution A (40 %) and 5 mL of Percoll solution B (80 %) in a 50-mL tube, and centrifuged at 2500 *g* for 10 min. Intact chloroplasts were recovered as a band at the interface of the 40 % and the 80 % Percoll layers. They were collected using a pipette, and the suspension was diluted with a washing buffer solution (330 mM D-sorbitol, 30 mM HEPES-KOH (pH 7.4), 2 mM EDTA-2Na). By centrifuging at 2500 *g* for 10 min, purified chloroplasts were recovered as a pellet. The washing procedure was repeated once more to remove all Percoll.

11.2.2.2.2 Isolation of Mitochondria from Potato Tuber Tissues

Mitochondria from potato tuber were isolated following the procedure of Douce et al. [15]. Twenty grams of potato tuber was cut into 1-cm cubes and homogenized in 100 mL of a grinding buffer solution (300 mM D-mannitol, 50 mM HEPES-KOH (pH 7.2), 1 mM EDTA-2Na, 0.1 % (w/v) BSA, and 0.6 % (w/v) PVP-40). After filtration of the homogenate with four layers of cheesecloth, the filtrate was spun at 2000 *g* for 3 min to eliminate cellular debris. The pellet was discarded, while the supernatant was decanted into another polypropylene tube and spun at 10,000 *g* for 20 min. The obtained pellet containing mitochondria was gently resuspended in a small amount (2 mL) of another buffer solution (300 mM D-mannitol, 50 mM HEPES-KOH (pH 7.2), 1 mM EDTA-2Na). The suspension was placed on top of the Percoll layers, which were composed of 10 mL of Percoll solution A (5 %), 10 mL of Percoll solution B (10 %), and 5 mL of Percoll solution C (35 %) in a 50-mL tube, and spun at 10,000 *g* for 30 min. The purified mitochondria were recovered at the interface between the 10 % and 35 % Percoll layers. The mitochondria fraction was diluted with ten times its volume of a washing buffer solution (300 mM D-mannitol, 50 mM HEPES-KOH (pH 7.2), 1 mM EDTA-2Na) and then recovered as a pellet after centrifugation at 10,000 *g* for 10 min. The pellet was washed one more time.

11.2.2.2.3 Isolation of Mitochondria from Bovine Liver Tissues

Mitochondria from bovine liver were isolated according to the methods of Sims (1990) [16] and Garcia (2010) [17]. Twenty grams of bovine liver cut into 1-cm cubes was homogenized in 100 mL of a grinding buffer solution (250 mM sucrose, 20 mM HEPES-KOH (pH 7.4), 1 mM EDTA-2Na, 1 % (w/v) BSA), and the homogenate was filtered with four layers of cheesecloth. The filtrate was spun at 1000 *g* for 10 min, and the supernatant was centrifuged twice more at 1000 *g* for 10 min. After that, crude mitochondria were precipitated by centrifuging at 9000 *g* for 15 min. The pellet resuspended in 5 mL of the grinding buffer solution was placed on top of the Percoll layers consisting of 10 mL of Percoll solution A (10 %), 10 mL of Percoll solution B (23 %), and 5 mL of Percoll solution C (43 %)

and spun at 30,000 *g* for 30 min. The mitochondria were recovered as a band at the interface of the 23 % and 43 % Percoll layers and then washed twice with 20 mL of a washing buffer solution (250 mM sucrose, 20 mM HEPES-KOH (pH 7.4), 1 mM EDTA-2Na).

11.2.3 Apparatus

The size and shape of the cells and organelles were monitored on an Olympus BX50 microscope (Tokyo, Japan) and recorded with an Olympus DP11-P digital camera. For fluorescence observation of mitochondria, an Olympus IX71S1F-3 microscope equipped with a WIG filter set was used. Cell counting was performed using an improved Neubauer hemocytometer (Deep 1/10 mm, LABART, Germany).

A domestic microwave oven (National NE-EZ2, Japan) was mainly used for acid digestion of very small amount of sample (less than 10 mg in dry weight), while a dedicated microwave extraction system (ETHOS E, Milestone General, Italy) was also used [10].

Several ICP-MS instruments were used because of the limitations of available instruments; a model HP4500 (Agilent Technologies, Tokyo, Japan) and a model ELAN DRC II (Perkin Elmer, Toronto, Canada) were ICP-Q-MS with a quadrupole mass spectrometer, while a model Element2 (Thermo Fisher Scientific, MA, USA) was ICP-SF-MS with a double-focusing sector field mass spectrometer. In order to reduce sample consumption, a micro-flow injection (μFI) system was coupled to the ICP-MS instruments. The μFI system was composed of a metal-free double-plunger pump (model KP-11, Ogawa & Co., Ltd., Japan), an inert microinjection valve (model V1-12, Ogawa & Co., Ltd.) with a 100-μL PEEK sample loop, a demountable capillary-attached micronebulizer, and a temperature-controllable cyclonic spray chamber (IsoMist, Glass Expansion, Australia). The details of the system are described in our previous paper [11]. An ICP-AES instrument of model IRIS Intrepid II XSP (Thermo Fisher Scientific, MA, USA) was also adjunctively used for the determination of major and minor elements.

11.2.4 Downsized Microwave-Assisted Acid Digestion

The yield of purified organelles from biological tissues is about 0.1 %, and only 10 mg can be obtained from 10 g of tissue. Generally, insufficiency of sample amounts causes the deterioration of accuracy and precision in quantitative analysis. For the analysis of such precious biological samples, it is quite important to develop the procedure for reducing sample consumption as much as possible. In order to minimize the loss in the sample preparation, in the present experiments, a 1-mL PTFE vial with a screw cap (Sanplatec, Japan) was used as an alternative to the vessel for microwave-assisted acid digestion. The small-scale digestion conveniently reduces the risk of explosion. Then, a domestic microwave oven was used

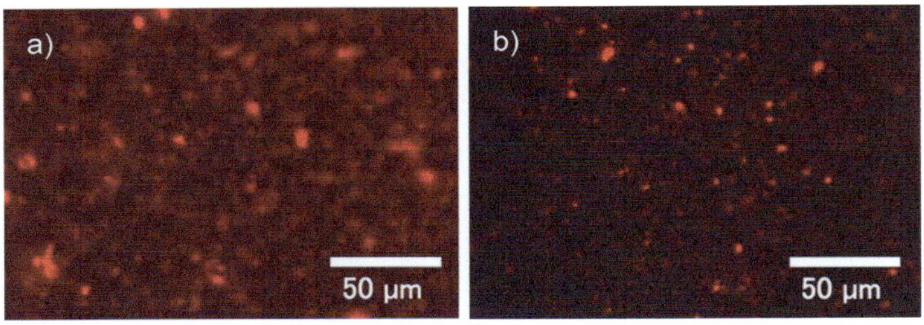

Fig. 11.3 Fluorescence micrographs of MitoRed-stained mitochondria before (**a**) and after (**b**) Percoll centrifugation

instead of the dedicated digestion apparatus. In the digestion procedure, several milligrams of dry weight (freeze-dried) sample and 100 times its volume of HNO_3-H_2O_2 (70:30, v/v) were added to the 1-mL PTFE vial, and microwave heating was carried out simply by applying 500 W for 5 min. Further details are described in our previous paper [10]. After the addition of internal standard and appropriate dilution, the digested sample solution was analyzed by μFI-ICP-MS.

11.3 Results and Discussion

11.3.1 Observation of Intact Cells and Organelles

In order to check if the harvested cells and isolated organelles were not disrupted, their size and shape were examined by a microscope. As an example, fluorescence micrographs of mitochondria before and after Percoll centrifugation are shown in Fig. 11.3. Mitochondria were not directly visible and then stained with MitoRed. It can be seen that the mitochondria after Percoll centrifugation are certainly purified and remain unbroken. The purified mitochondria have a spherical or bean-like body with a diameter of less than 1 μm. In the same manner, green chloroplasts with a diameter of about 1 μm were also observed by a microscope. The data obtained by microscope investigation are summarized in Table 11.1, together with those on water content (%). Water content was determined by the difference in sample weights before and after freeze-drying.

11.3.2 Determination of Major-to-Ultratrace Elements in Unicellular Microorganisms and Subcellular Organelles

The quantities of the harvested cells and purified organelles were around 10–100 mg in wet weight. The wet samples were freeze-dried and then digested with HNO_3-H_2O_2 (100 μL per 1-mg dry weight) under microwave heating. The

Table 11.1 The data on size and water content (%) of microbes and organelles

Microbes and organelles	Size (μm)	Weight (pg)	Water content (%)
Escherichia coli MG1655	Rod shape, 0.5 μm in diameter and 2 μm in length	1	70
Synechocystis sp. PCC6803 (cyanobacteria)	1–3 μm	8	70
Chlorella kessleri NIES-2160	2–10 μm	200	78
Chloroplast from spinach	Lens shape, 5–8 μm in diameter and 1–3 μm in thickness	40	89
Mitochondria from bovine liver	0.5–2 μm	0.5	87
Mitochondria from potato	0.5–2 μm	0.5	82

concentrations of major-to-ultratrace elements in the acid-digested samples were determined by ICP-MS. Due to the small sample amount and considerable background signals derived from the culture medium, about half of the elements in the periodic table could not be detected, but even so more than 30 elements were determined and the concentration data were converted to a per cell (or organelle) basis, taking into account the cell size and water content. Some representative data are summarized in Table 11.2. These values were the mean of three independent experiments, and the relative standard deviations were less than 20 % for almost all of the listed elements. Sodium (Na) was removed from the list, because physiological saline was used for washing.

11.3.3 Elemental Abundance of E. coli

In order to examine the specific uptake of vital elements by cells, *E. coli* cells were cultivated in two different media (LB and M9 media) and harvested in the late exponential growth phase. The third and fourth columns in Table 11.2 present the elemental abundances of *E. coli* cells grown in LB and M9 media, respectively.

It can be seen that the differences in the elemental abundances between *E. coli* cells grown in LB and M9 media are not large. Especially, the levels of essential minerals were in good agreement with each other. As can be seen in Table 11.2, such major elements were presented at quite high levels of more than 10^5 atoms per cell. The amounts of P, K, S, and Mg were on the order of 10^8 atoms per cell (e.g., 4.2×10^8, 2.6×10^8, 1.7×10^8, and 0.75×10^8/an *E coli* cell grown in LB, respectively), 10^6 atoms for Ca and Fe (6.2×10^6 and 3.3×10^6/cell), and 10^5 atoms for Zn and Mn (6.5×10^5 and 1.9×10^5/cell). These values were well consistent with those of Rouf [18] and O'halloran [19].

In order to examine temporal change in the elemental abundance, *E. coli* cells were also harvested every 1 h during the cell growth and proliferation. As a result of their multielement analysis, the variation between the minimum and maximum levels of the essential minerals remained within the range of two to three times. The results also agree with the findings of O'halloran et al. [19], who reported that the total concentration of Zn(II) was tightly controlled in bacteria cells. On the other hand, the levels of nonessential elements varied widely. For example, *E. coli* cells harvested earlier contained more than ten times higher levels of REEs compared to the cells harvested in the late exponential growth phase.

In order to check if the variations are merely dependent on the difference of elemental composition of the growth environment, multielement analysis of the culture media was also carried out. The concentrations of the elements in both the LB and M9 culture media are shown in Table 11.3. It can be seen that both the media contain a variety of elements as impurities at parts per trillion (ppt) to parts per billion (ppb) levels. The concentrations of Sc, V, Ni, Cu, and lanthanides were below or near the detection limits, but these elements were apparently detected in *E. coli* cells. These results mean that such heavy metals are accumulated by *E. coli*

Table 11.2 The number of atoms of each element in a single cell and in a single organelle

Atomic number	Symbol	*E. coli* MG1655 in LB	*E. coli* MG1655 in M9	*Synechocystis* sp. PCC6803 in BG-11	*Chlorella kessleri* NIES-2160 in Gamborg's B5	Chloroplast from spinach	Mitochondria from potato tuber	Mitochondria from bovine liver
12	Mg	75,000,000	110,000,000	120,000,000	1,500,000,000	320,000,000	310,000	130,000
15	P	420,000,000	560,000,000	470,000,000	23,000,000,000	990,000,000	37,000,000	16,000,000
16	S	170,000,000	150,000,000	400,000,000	11,000,000,000	430,000,000	4,400,000	11,000,000
19	K	260,000,000	420,000,000	410,000,000	18,000,000,000	760,000,000	15,000,000	9,300,000
20	Ca	6,200,000	3,800,000	33,000,000	3,600,000,000	14,000,000	80,000	42,000
21	Sc	420	260					
22	Ti	3900		83,000	9,700,000	470,000	7400	1400
23	V	3200				7000	220	
24	Cr	6000	4600	9700	1,500,000	15,000	260	58
25	Mn	190,000	44,000	5,000,000	10,000,000	3,100,000	3000	2500
26	Fe	3,300,000	1,200,000	76,000,000	970,000,000	23,000,000	170,000	57,000
27	Co	5500	2200	26,000	8300	1200	22	66
28	Ni	13,000	16,000	1900	18,000	8900	160	79
29	Cu	60,000	70,000	950,000	5,900,000	1,200,000	5500	450
30	Zn	650,000	500,000	1,600,000	94,000,000	1,300,000	4800	29,000
34	Se			3600	56,000			
37	Rb	31,000	2400		210,000	39,000	1200	
38	Sr	11,000	2400	24,000	1,100,000	24,000	86	
39	Y	850	280	330	59,000	920	17	2
40	Zr					120,000	920	1500
42	Mo	30,000	13,000	4700	180,000		180	1300
46	Pd	36				10,000	150	280
47	Ag	7		190	17,000		7	11
48	Cd	270	800			16,000	110	9

(continued)

Table 11.2 (continued)

Atomic number	Symbol	*E. coli* MG1655 in LB	*E. coli* MG1655 in M9	*Synechocystis* sp. PCC6803 in BG-11	*Chlorella kessleri* NIES-2160 in Gamborg's B5	Chloroplast from spinach	Mitochondria from potato tuber	Mitochondria from bovine liver
50	Sn	30		3200	89,000	1400	11	5
51	Sb	9		140	42,000	650	8	
55	Cs	90					17	
56	Ba	4400	5000	29,000	6,400,000	38,000	200	
57	La	600	320	340	5300	480	8	1.1
58	Ce	700	540	3700	33,000	810	20	3.9
59	Pr	100	60	24	3500	180	3	1.4
60	Nd	380		1600	44,000	660	15	0.9
62	Sm	90			9000	25	1.0	
63	Eu	25			2700		0.2	
64	Gd	80		110	14,000	67	1.5	
65	Tb	11		3	1800	25	0.4	0.1
66	Dy	65		14	8300	130	2.2	0.6
67	Ho	13		21	1400	100	1.5	1.2
68	Er	33		18	2900	110	1.7	0.8
69	Tm	4		13	370	57	0.8	0.7
70	Yb	28		30	1100	140	2.1	1.3
71	Lu	8						
74	W	330	1000			110		
79	Au	0.25						
81	Tl	45				2600	87	75
82	Pb	60	120	15,000	1,700,000	4200	46	3
90	Th	26						
92	U	12	12					

Table 11.3 Analytical results for the elements present in LB and M9 media

Atomic number	Symbol	Concentration in LB medium (μg/L)			Concentration in M9 medium (μg/L)		
12	Mg	6310	±	240	24,100	±	250
15	P	150,000	±	4000	1,730,000	±	20,000
16	S	103,000	±	4000	30,100	±	200
19	K	739,000	±	41,000	883,000	±	6100
20	Ca	8030	±	340	4050	±	80
21	Sc	0.085	±	0.014		n.d. (<0.02)	
22	Ti	13.8	±	5.3	2.74	±	0.17
23	V	79.7	±	4.5		n.d. (<0.04)	
24	Cr	4.58	±	0.11	2.11	±	0.16
25	Mn	18.4	±	1	2.65	±	0.09
26	Fe	340	±	30	1500	±	110
27	Co	11.9	±	0.5	0.08	±	0.01
28	Ni	3.9	±	0.3		n.d. (<0.8)	
29	Cu	15.3	±	0.2		n.d. (<0.2)	
30	Zn	1080	±	20	9.7	±	1.6
37	Rb	183	±	10	46.6	±	0.5
38	Sr	42.3	±	1.9	4.6	±	0.1
39	Y	0.138	±	0.014	0.029	±	0.0003
42	Mo	5.62	±	0.29	1.2	±	0.01
48	Cd	0.169	±	0.025	0.079	±	0.004
51	Sb	0.268	±	0.052	0.02	±	0.003
55	Cs	1.27	±	0.06	0.0035	±	0.0013
56	Ba	10.7	±	1.1	8.98	±	0.17
57	La	0.151	±	0.012		<0.002	
58	Ce	0.259	±	0.074	0.01	±	0.0034

(continued)

Table 11.3 (continued)

Atomic number	Symbol	Concentration in LB medium (μg/L)			Concentration in M9 medium (μg/L)		
59	Pr	0.0297	±	0.0034	0.001	±	0.0006
60	Nd	0.122	±	0.014		<0.003	
62	Sm	0.0254	±	0.0044		<0.006	
63	Eu	0.0032	±	0.0010		n.d. (<0.001)	
64	Gd	0.025	±	0.005		<0.003	
65	Tb	0.0037	±	0.0005		n.d. (<0.003)	
66	Dy	0.0219	±	0.0019		<0.003	
67	Ho	0.0111	±	0.0024		n.d. (<0.002)	
68	Er	0.0143	±	0.0016		<0.003	
69	Tm	0.0016	±	0.0004		n.d. (<0.0001)	
70	Yb	0.0111	±	0.0017		n.d. (<0.0008)	
71	Lu	0.0021	±	0.0004		n.d. (<0.002)	
74	W	0.765	±	0.057		<0.3	
81	Tl	0.218	±	0.010		<0.01	
82	Pb	0.00937	±	0.00629		<0.004	
83	Bi	0.000102	±	0.000042		n.d. (<0.0002)	
90	Th	0.021	±	0.001		<0.01	
92	U	0.0979	±	0.0071	0.0035	±	0.0011

cells from the undetectable level in the culture media to the detectable level in the cells. Generally, *E. coli* is known to accumulate a variety of heavy metal ions and also to have a high tolerance for such metal ions. Consequently, lanthanides may be incidentally incorporated into the cells together with essential elements and subsequently dispersed among the multiplying cells during the proliferation. Similar trends were also observed for so-called nonessential elements such as Cd and Ti. Essential elements restrict the proliferation and then their concentrations fall into a tight range, whereas nonessential elements might be accumulated up to a permissible level regardless of the bioavailability.

11.3.4 Elemental Abundance of Cyanobacteria

In order to examine the difference in element uptake between prokaryotic species, cyanobacteria (*Synechocystis*) was also grown, and elemental abundance data are shown in the fifth column in Table 11.2. As can be seen in Table 11.2, cyanobacteria also contained a variety of elements (33 elements). Phosphorous, S, K, Mg, Ca, Fe, Ca, Mn, Zn, and Cu were present at levels higher than about 10^6 atoms/cell, while Ti, Co, Sr, Ba, Pb, and Cr existed at levels of more than 10^4 atoms/cell. Selenium, Mo, and Ni were observed on the order of several hundreds, and heavy lanthanides were present at the level of several tens or less. Some of these elements may be incidentally taken from the growth environment, but others may be selectively incorporated into the cell for special vital function. Especially, the contents of Fe and Mn in cyanobacteria are more than ten times higher than those in *E. coli*. Cyanobacteria are a group of photosynthetic bacteria, and they multiply through photosynthesis. The significance of Fe and Mn in photosynthesis may account for their accumulation in cyanobacteria [20].

11.3.5 Elemental Abundance of Chlorella

In order to examine elemental abundance differences between prokaryotic and eukaryotic cells, eukaryotic chlorella was also cultivated and the multielement analysis was carried out. The results are shown in the sixth column in Table 11.2. The size of eukaryotic chlorella was about 20 times larger than that of prokaryotic cyanobacterium in volume, and the number of the atoms basically increased in proportion to the cell size. Chlorella is also a photosynthetic organism, and then, as expected, it contained distinct high levels of Fe, Zn, Cu, and Mn, which are known to play important roles in redox and antioxidant systems. Titanium (Ti), Ba, Cr, and Sr were also observed at relatively high levels in chlorella, although it was not clear that they are selectively incorporated into the cell.

11.3.6 Elemental Abundance of Chloroplast

The localization of trace elements at subcellular levels has received considerable attention in recent years [21–29]. As a way to examine elemental localization in a cell, in the present paper, isolation of some organelles from cellular tissue was carried out, as described in the experimental section. Elemental abundance of chloroplasts from spinach leaves is shown in the seventh column in Table 11.2. The concentrations were successfully determined within relative standard deviations less than 20 % for all the listed elements and converted to a per organelle basis, taking into account the organelle size and its water content.

As can be seen in Table 11.2, chloroplast also had high levels of metals including Fe, Mn, Cu, and Zn. Chloroplast is one of the most important organelles in plant cells, and it is generally considered to have originated from cyanobacteria through endosymbiosis [20, 30, 31]. As might have been expected, good correlation between the elemental abundances of cyanobacteria and chloroplast was observed in Fig. 11.4. The correlation coefficient (r) was 0.97. The levels of nonessential elements in chloroplast are relatively higher than those in cyanobacteria. This may be due to the accumulation by spinach leaf tissues.

Fig. 11.4 Comparison of elemental abundances of cyanobacteria and chloroplasts

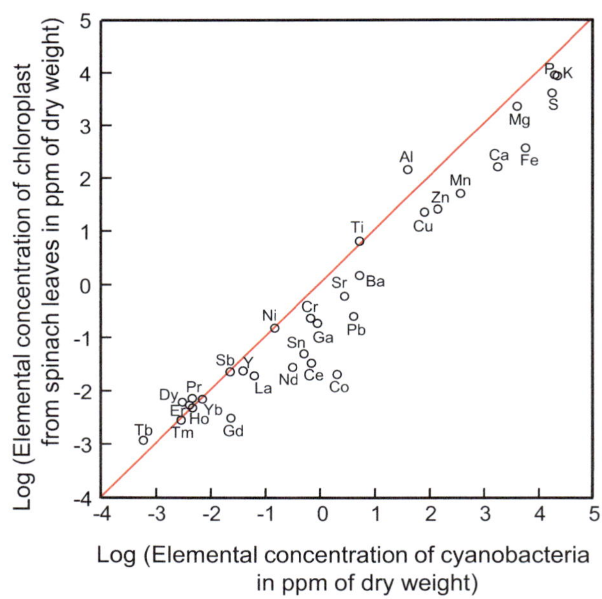

11.3.7 Elemental Abundance of Mitochondria

Mitochondria are the most important organelle in the cells of all eukaryotes. Thus, mitochondria were isolated from bovine liver and potato tuber. Their elemental abundances are shown in the eighth and ninth columns in Table 11.2. Elemental abundances of mitochondria are significantly influenced by the raw materials, but, as a whole, similar elemental abundance pattern was observed ($r_{=}0.94$). Mitochondria are the main source of reactive oxygen species, which may damage nuclear and mitochondrial DNA [32, 33]. Then, they contain a number of metalloenzymes that convert superoxide radicals and hydrogen peroxide to less toxic oxygen species. As a result, Fe, Mn, Cu, and Zn, which are involved in redox and antioxidant systems, may be accumulated in mitochondria.

In terms of biological taxonomy, mitochondria are considered to have originated from proteobacteria, which are a major group of gram-negative bacteria including *E. coli*. Good correlation between the elemental abundances of mitochondria and *E. coli* was certainly obtained ($r = 0.87$), while elemental abundance of the mitochondria was more similar to that of cyanobacteria ($r = 0.92$).

11.4 Conclusion

Comprehensive element analysis of prokaryotic and eukaryotic cells and subcellular organelles was performed by μFI-ICP-MS after microwave-assisted acid digestion of less than 10 mg of freeze-dried sample. Reliable data on elemental abundances of unicellular microorganisms and subcellular organelles were provided, although it's to be noted that the numbers are estimates and they will vary depending on cell source and growth environment. It is expected that these omics data will contribute to a variety of scientific and industrial innovations, including cellular regulation, environmental microbiology, and the field of food chemistry. The evolutionary pathway of organisms may also be traced back through the classification of element requirements.

Acknowledgments This work was partially supported by Grant-in-Aid for Scientific Research (B) (No. 21350041) and Grant-in-Aid for Scientific Research (A) (No. 25248035) from Japan Society for the Promotion of Science.

References

1. Williams RJP (2001) Chemical selection of elements by cells. Coord Chem Rev 216–217:583–595. doi:10.1016/S0010-8545(00)00398-2
2. Haraguchi H (2004) Metallomics as integrated biometal science. J Anal At Spectrom 19:5–14. doi:10.1039/b308213j

3. Salt DE (2004) Update on plant ionomics. Plant Physiol 136:2451–2456. doi:10.1104/pp.104.047753
4. Szpunar J (2005) Advances in analytical methodology for bioinorganic speciation analysis: metallomics, metalloproteomics and heteroatom-tagged proteomics and metabolomics. Analyst (Cambridge, UK) 130:442–465. doi:10.1039/b418265k
5. Rodushkin I, Odman F, Olofsson R, Axelsson MD (2000) Determination of 60 elements in whole blood by sector field inductively coupled plasma mass spectrometry. J Anal At Spectrom 15:937–944. doi:10.1039/b003561k
6. Haraguchi H, Ishii A, Hasegawa T, Matsuura H, Umemura T (2008) Metallomics study on all-elements analysis of salmon egg cells and fractionation analysis of metals in cell cytoplasm. Pure Appl Chem 80:2595–2608. doi:10.1351/pac200880122595
7. Marov MY (2014) The structure of the universe. The fundamentals of modern astrophysics. Springer, New York, pp 279–294
8. Miyazaki Y, Yasui T, Inagaki K, Okamoto Y, Kaji N, Umemura T, Tokeshi M, Baba Y (2014) Detection of metallic elements in a single cancer cell using microfluidic devices coupled with ICP-MS. In: International conference miniaturized systems chemistry and life sciences, 16th Chemical and Biological Microsystems Society, pp 767–769
9. Groombridge AS, Miyashita S, Fujii S, Nagasawa K, Okahashi T, Ohata M, Umemura T, Takatsu A, Inagaki K, Chiba K (2013) High sensitive elemental analysis of single yeast cells (Saccharomyces cerevisiae) by time-resolved inductively-coupled plasma mass spectrometry using a high efficiency cell introduction system. Anal Sci 29:597–603. doi:10.2116/analsci.29.597
10. Takasaki Y, Watanabe M, Yukawa H, Sabarudin A, Inagaki K, Kaji N, Okamoto Y, Tokeshi M, Miyamoto Y, Noguchi H, Umemura T, Hayashi S, Baba Y, Haraguchi H (2011) Estimation of the distribution of intravenously injected adipose tissue-derived stem cells labeled with quantum dots in mice organs through the determination of their metallic components by ICPMS. Anal Chem (Washington, DC) 83:8252–8258. doi:10.1021/ac202053y
11. Takasaki Y, Inagaki K, Sabarudin A, Fujii S, Iwahata D, Takatsu A, Chiba K, Umemura T (2011) Multielement analysis of micro-volume biological samples by ICP-MS with highly efficient sample introduction system. Talanta 87:24–29. doi:10.1016/j.talanta.2011.09.022
12. Allen MB (1959) Studies with Cyanidium caldarium, an anomalously pigmented chlorophyte. Arch Mikrobiol 32:270–277
13. Gamborg OL, Miller RA, Ojima K (1968) Nutrient requirements of suspension cultures of soybean root cells. Exp Cell Res 50:151–158. doi:10.1016/0014-4827(68)90403-5
14. Sato N, Misumi O, Shinada Y, Sasaki M, Yoine M (1997) Dynamics of localization and protein composition of plastid nucleoids in light-grown pea seedlings. Protoplasma 200:163–173. doi:10.1007/BF01283292
15. Douce R, Bourguignon J, Brouquisse R, Neuburger M (1987) Isolation of plant mitochondria: general principles and criteria of integrity. Methods Enzymol 148:403–415
16. Sims NR (1990) Rapid isolation of metabolically active mitochondria from rat brain and subregions using Percoll density gradient centrifugation. J Neurochem 55:698–707
17. Garcia J, Han D, Sancheti H, Yap L-P, Kaplowitz N, Cadenas E (2010) Regulation of mitochondrial glutathione redox status and protein glutathionylation by respiratory substrates. J Biol Chem 285:39646–39654. doi:10.1074/jbc.M110.164160
18. Rouf MA (1964) Spectrochemical analysis of inorganic elements in bacteria. J Bacteriol 88:1545–1549
19. Outten CE, O'Halloran TV (2001) Femtomolar sensitivity of metalloregulatory proteins controlling zinc homeostasis. Science 292:2488–2492
20. Shcolnick S, Keren N (2006) Metal homeostasis in cyanobacteria and chloroplasts. Balancing benefits and risks to the photosynthetic apparatus. Plant Physiol 141:805–810. doi:10.1104/pp.106.079251

21. Mueller L, Traub H, Jakubowski N, Drescher D, Baranov VI, Kneipp J (2014) Trends in single-cell analysis by use of ICP-MS. Anal Bioanal Chem 406:6963–6977. doi:10.1007/s00216-014-8143-7
22. Smart KE, Smith JAC, Kilburn MR, Martin BGH, Hawes C, Grovenor CRM (2010) High-resolution elemental localization in vacuolate plant cells by nanoscale secondary ion mass spectrometry. Plant J 63:870–879
23. Wang D, Bodovitz S (2010) Single cell analysis: the new frontier in "omics". Trends Biotechnol 28:281–290. doi:10.1016/j.tibtech.2010.03.002
24. Brehm-Stecher BF, Johnson EA (2004) Single-cell microbiology: tools, technologies, and applications. Microbiol Mol Biol Rev 68:538–559. doi:10.1128/MMBR.68.3.538-559.2004
25. Gao Y, Lin Y, Zhang B, Zou D, He M, Dong B, Hang W, Huang B (2013) Single-cell elemental analysis via high irradiance femtosecond laser ionization time-of-flight mass spectrometry. Anal Chem (Washington, DC) 85:4268–4272. doi:10.1021/ac400317q
26. Wang H, Wang B, Wang M, Zheng L, Chen H, Chai Z, Zhao Y, Feng W (2015) Time-resolved ICP-MS analysis of mineral element contents and distribution patterns in single cells. Analyst (Cambridge, UK) 140:523–531. doi:10.1039/C4AN01610F
27. Verboket PE, Borovinskaya O, Meyer N, Gunther D, Dittrich PS (2014) A new microfluidics-based droplet dispenser for ICPMS. Anal Chem (Washington, DC) 86:6012–6018. doi:10.1021/ac501149a
28. Ward SK, Heintz JA, Albrecht RM, Talaat AM (2012) Single-cell elemental analysis of bacteria: quantitative analysis of polyphosphates in Mycobacterium tuberculosis. Front Cell Infect Microbiol 3:63. doi:10.3389/fcimb.2012.00063
29. Shigeta K, Koellensperger G, Rampler E, Traub H, Rottmann L, Panne U, Okino A, Jakubowski N (2013) Sample introduction of single selenized yeast cells (*Saccharomyces cerevisiae*) by micro droplet generation into an ICP-sector field mass spectrometer for label-free detection of trace elements. J Anal At Spectrom 28:637–645. doi:10.1039/c3ja30370e
30. Blaby-Haas CE, Merchant SS (2013) Sparing and salvaging metals in chloroplasts. In: Met. Cells. Wiley, pp 51–63
31. Merchant SS (2007) Trace metal utilization in chloroplasts. Adv Photosynth Respir 23:199–218
32. Pierrel F, Cobine PA, Winge DR (2007) Metal ion availability in mitochondria. BioMetals 20:675–682. doi:10.1007/s10534-006-9052-9
33. Kalogeris T, Bao Y, Korthuis RJ (2014) Mitochondrial reactive oxygen species: a double edged sword in ischemia/reperfusion vs preconditioning. Redox Biol 2:702–714. doi:10.1016/j.redox.2014.05.006

Chapter 12
Iron Isotope Signature in Red Blood Cell Samples from Japanese Female Donors of Various Ages

Yu-ki Tanaka, Kohshiro Takata, Takahiro Kawasaki, Atsuko Shinohara, Kazuko Ishikawa-Takata, and Takafumi Hirata

Abstract We have measured $\delta^{56}Fe/^{54}Fe$ and $\delta^{57}Fe/^{54}Fe$ values in red blood cell (RBC) samples collected from male and female donors of various ages. To achieve this, effects of both Fe recovery through chemical separation procedures and Fe oxidation state in solutions for analysis on the measured Fe isotope ratios were rigorously evaluated. The resulting Fe isotope ratios ($^{56}Fe/^{54}Fe$ and $^{57}Fe/^{54}Fe$) in RBCs from female donors were significantly higher than those from male donors, suggesting that Fe isotopic composition in human RBCs can be different between male and female, as reflected by the inherent difference in Fe absorption efficiency from the diet. Regarding RBC samples from female donors, Jaouen and Balter (Am J Phys Anthropol 153:280–285, 2014) were first to report changes in Fe isotope signature in elderly donors of European population. This can be explained by possible changes in Fe absorption efficiency due to menopause. This is not true for the Japanese females. Our Fe isotope data showed that Fe isotope data for young females ($\delta^{56}Fe/^{54}Fe = -2.65 \pm 0.34$ ‰, $\delta^{57}Fe/^{54}Fe = -3.90 \pm 0.53$ ‰; 2SD,

Y.-k. Tanaka • K. Takata • T. Kawasaki
Laboratory for Planetary Sciences, Kyoto University, Kitashirakawa Oiwake-cho, Sakyo-ku, Kyoto 606-8502, Japan

A. Shinohara
Department of Epidemiology and Environmental Health, Faculty of Medicine, Juntendo University, 2-1-1 Hongo, Bunkyo-ku, Tokyo 113-8421, Japan

Research Institute for Cultural Studies, Seisen University, 3-16-21 Higashi Gotanda, Shinagawa-ku, Tokyo 141-8642, Japan

K. Ishikawa-Takata
Division of Health Promotion, National Institute of Health and Nutrition, 1-23-1 Tayama, Shinjuku-ku, Tokyo 162-8636, Japan

T. Hirata (✉)
Laboratory for Planetary Sciences, Kyoto University, Kitashirakawa Oiwake-cho, Sakyo-ku, Kyoto 606-8502, Japan

Geochemistry Research Center, The University of Tokyo, 1-1-3 Hongo, Tokyo 113-0033, Japan
e-mail: hrt1@eqchem.s.u-tokyo.ac.jp

Y. Ogra, T. Hirata (eds.), *Metallomics*, DOI 10.1007/978-4-431-56463-8_12

$n = 18$) did not vary significantly from elderly donors ($\delta^{56}Fe/^{54}Fe = -2.51 \pm 0.49$ ‰, $\delta^{57}Fe/^{54}Fe = -3.68 \pm 0.73$ ‰; 2SD, $n = 10$). Obvious discrepancy in the difference of the Fe isotope signature between Japanese and European elderly female donors can be explained by the differences in food culture or Fe nutritional status. In this study, the possible linkage between Fe isotope ratios and nutritional status of donors is discussed on the basis of the Fe isotope signature.

Keywords Isotope ratio • Iron • Red blood cell • ICP-MS • Aging • Menstruation • Postmenopause

12.1 Introduction

Iron (Fe) is one of the most important inorganic nutrients in most organisms. The biochemical function of Fe includes oxygen transport (cell respiration), photosynthesis, and/or nitrogen fixation [1]. The total amount of Fe in the human body is approximately 4 g for adult male and 2.5 g for adult female [2], and approximately 60–75% of total Fe is incorporated into hemoglobin [3, 4]. The remaining 10–30 % of Fe is stored mainly in the liver or spleen as ferritin, and other Fe measuring <10 % and <0.1 % are present as myoglobin in the muscle and transferrin in serum, respectively [3, 4]. Despite its biological importance, it is well recognized that excess Fe, particularly the ferric form, can be toxic to human and animals. Free Fe (III) can induce cell wall destruction through the Fenton reaction [5], and thus, both Fe overload and deficiency can be the major cause of such diseases as hereditary hemochromatosis (Fe overload in the body) or Fe deficiency anemia [6, 7]. To maintain the biochemical function of Fe and minimize Fe toxicity, Fe concentrations in tissues, cells, or fluids are rigorously controlled by a homeostatic function [8].

Iron concentration in blood or fluids can be mainly dependent on the Fe absorption efficiency at the intestine. Typical daily intake of Fe is 18.1 mg/day for male and 13.6 mg/day for female in the USA [9], and the average absorption efficiency of Fe is 6 % for male and 13 % for female [2]. This suggests that the total amount of Fe absorbed per day becomes approximately 1.1 mg for male and 1.8 mg for female, which would be almost equivalent to the amount of Fe loss per day [10, 11].

Nutritional Fe in food is absorbed by intestinal epithelial cells (Fig. 12.1). Two major Fe species in food are heme Fe (Fe bound to protein), which is present in animal-based food as Fe(II), and nonheme Fe, which is present in plant-based food as the inorganic complex of Fe(III). Shayeghi et al. (2005) reported that heme Fe is absorbed in the intact form through HCP1 (heme Fe transporter) [12]. In contrast, Qiu et al. (2006) indicated that HCP1 plays a major role in folate transport [13]. Nonheme Fe(III) is reduced to Fe(II) by Dcytb (ferrireductase) [14] on the epithelial cell surface, and this is followed by Fe absorption by epithelial cells using a divalent metal-ion transporter (DMT1) [15, 16]. In the epithelial cells, Fe atoms released by heme protein or inorganic complexes exist as ferritin (protein with

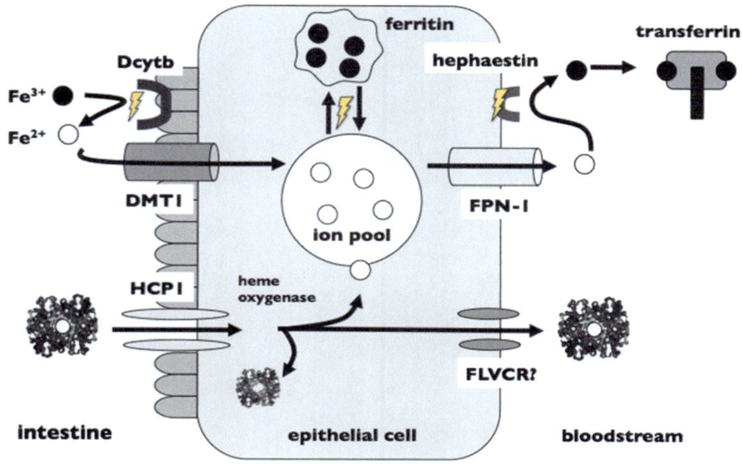

Fig. 12.1 Absorption processes of Fe in the intestinal epithelial cell. Nonheme Fe is reduced to ferrous and then transported into the cell through DMT1. Heme Fe is absorbed in intact form through HCP1. In the epithelial cell, a part of ferrous Fe, which is temporary stored in the cellular labile Fe pool, oxidized to ferric form to be incorporated into ferritin. Remaining ferrous Fe is removed from cells to bloodstream through ferroportin-1 (FPN-1) and incorporated into transferrin after oxidized to ferric form by copper-contained protein like "hephaestin"

Fe(III)) [17], and remaining Fe in the cellular labile Fe pool is transported from the cells to bloodstream with ferroportin-1 located on the basolateral surface of the cells [18–20]. In contrast, heme Fe is exported to the bloodstream by means of feline leukemia virus subgroup C receptor (FLVCR) [21, 22]. The released Fe from the epithelial cells is oxidized by hephaestin, a copper-containing protein, to synthesize transferrin [23].

In the human body, most Fe, in combination with transferrin, is mainly delivered to the liver, muscle, and red blood cells (RBCs) (Fig. 12.2). Transferrin is taken up by cells by endocytosis, and ferric Fe extracted from transferrin is reduced to ferrous form and stored in an ion pool. In liver cells, the majority of Fe is present in ferritin in the ferric form, and functional Fe proteins, such as hemoglobin or myoglobin, are synthesized by using ferrous Fe in the ion pool of RBCs or muscle cells.

Several studies have revealed that hepcidin, which is produced in liver cells, plays an important role for regulating Fe absorption in the human body [24–26]. Hepcidin is used for the decomposition of ferroportin-1, and the amount of hepcidin reflects the amount of Fe stored in the liver [24–28]. Knockout mice lacking hepcidin gene expression showed symptoms of Fe overload, likely with hemochromatosis, whereas transgenic mice expressing liver hepcidin exhibited Fe deficiency anemia [29, 30]. In human, it is widely recognized that hereditary hemochromatosis is strongly related to genetic anomaly in the synthesis of hepcidin [6]. The amounts of proteins involved in Fe absorption or utilization, such as DMT1, ferritin, ferroportin-1, or transferrin receptor, are controlled

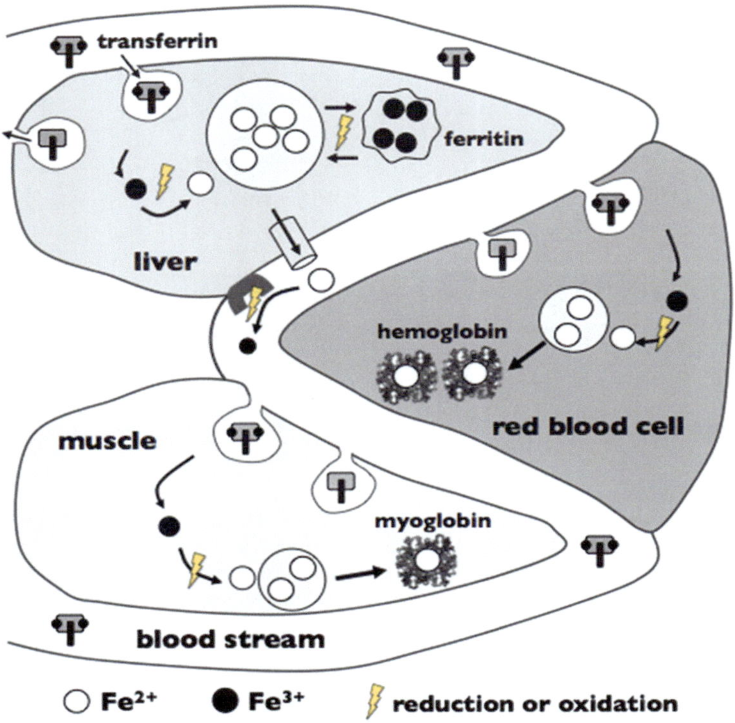

Fig. 12.2 Schematic diagram of Fe metabolism for the human body. Most of absorbed Fe at the epithelial cells transported to the liver, red blood cells, and muscle. In the liver cell, ferrous Fe in ion pool is partially oxidized to ferric form to be incorporated to ferritin. In the red blood cell and muscle cell, heme protein-binding Fe is synthesized using ferrous Fe in ion pool

posttranscriptionally at the mRNA level by the IRP/IRE system to regulate Fe concentration properly [20, 31, 32]. On the other hand, no excretory pathway is known to regulate Fe concentration in human tissues, although Fe is removed from the body through bile excretion or the flaking of epithelial cells [33].

To investigate Fe metabolism, isotope tracer experiments using both radioactive isotopes and enriched stable isotopes have been conducted (e.g., Mellon and Fairweather-Tait, 1997; Ulusoy and Whitley, 1999) [34, 35]. In the labeling using radioactive isotopes, benchmark information concerning elemental metabolism has been unearthed. However, this technique poses the potential risk of radioactive exposure not only to patients but also to investigators. Moreover, the labeling technique using radioactive isotopes is applicable to studies of elemental metabolism for only a short period. In contrast, studies using enriched stable isotopes can provide information of elemental metabolism for a long period. However, the total amount of isotopes loaded into the donors would be comparable to the total amount of elements in the body. This suggests that the loading of large amounts of stable isotopes can induce changes in absorption efficiency and/or elemental metabolism

[34, 35]. To overcome these analytical and empirical difficulties, changes in the isotopic composition of native elements due to the isotope effect have been used to monitor elemental metabolism or evaluate the nutritional status of elements.

In equilibrium reactions, the magnitude of isotope fractionation would be dependent on three important parameters: (1) temperature of the system, (2) relative mass difference between two isotopes ($\Delta m/m_1m_2$), and (3) difference in strength of chemical bonds between reactants and products (Δk_i) (Eq. 12.1) [36, 37]. For Fe, large variations in isotope ratios were found in chemical or biological reactions associated with redox reactions [38–40]. This is mainly due to the large difference in the strength of chemical bonds between the reactants and the products:

$$\ln K_{\text{ex}} = \frac{1}{24}\left(\frac{h}{2\pi kT}\right)^2 \frac{\Delta m_i}{m_i} \Delta k_i \tag{12.1}$$

Over the last decades, Fe isotopes in several organisms in the trophic level, including plants [41, 42], mammals [41, 43–45], and human [46–57], have been reported. Zhu et al. (2002) reported Fe isotope ratios in hemoglobin samples from herbivores, such as pig, horse, or bovine [43]. Walczyk and Blanckenburg (2002) reported Fe isotope ratios in terrestrial plant, animal, and human samples [41]. These pioneering studies revealed that the ^{56}Fe/^{54}Fe ratios of almost all biological samples were lowered by approximately 0.5 ‰/amu with the increase of trophic levels on the food chain. Light Fe isotopes were enriched in pig gastrointestinal tissue samples relative to feed, and heavy Fe isotopes were enriched in pig gastrointestinal tissue samples compared to blood samples [44]. The variation of isotopic compositions could be a result of the preferential absorption of light Fe isotopes and the storage of heavy Fe isotopes as ferritin in epithelial cells [41, 44, 50].

Even in the human body, the Fe isotope ratio was different among tissues or organs (Fig. 12.3) [41, 47, 48, 50, 51, 57]. This is also true for plants [42] and other animals [43–45]. Bone samples from herbivore and human showed higher ^{56}Fe/^{54}Fe values than other tissue samples [45, 51]. Liver samples have higher ^{56}Fe/^{54}Fe values than blood or muscle samples [44, 47, 50, 57]. Moreover, the Fe isotope ratio in plasma varied significantly from that in RBCs [50]. Overall variations of the ^{56}Fe/^{54}Fe ratio exceeded 1 %, reflecting differences in the Fe oxidation state among the organs. Ohno et al. (2004) reported that the ^{56}Fe/^{54}Fe ratio for individuals did not vary significantly over 1 year. The invariability of the ^{56}Fe/^{54}Fe ratio over 1 year could be due to the long turnover time of Fe in vivo [46]. Despite this, the measured ^{56}Fe/^{54}Fe ratio varied significantly among individuals. In particular, the ^{56}Fe/^{54}Fe ratios in RBCs of female donors were higher than those of males on average [41, 46–48, 50, 53, 55–57]. Gender differences in isotopic composition could be attributed to the difference in Fe absorption efficiency [2]. The above studies suggest that Fe isotope ratio in RBCs can be used as a marker for evaluating the long-term nutritional status of Fe or the risk of Fe metabolic anomaly.

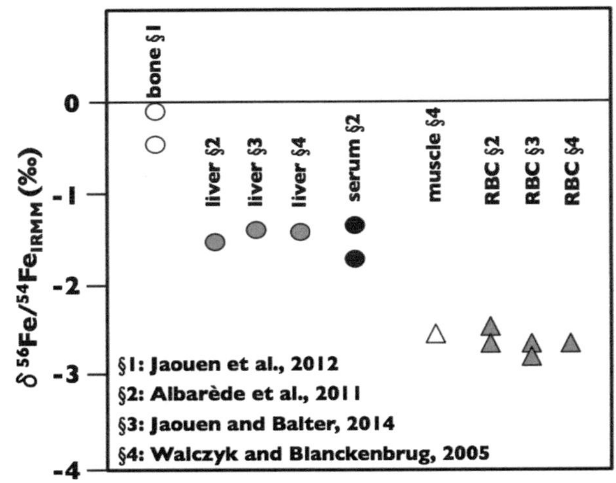

Fig. 12.3 Summary of $^{56}Fe/^{54}Fe_{IRMM}$ ratio for various organs in human body

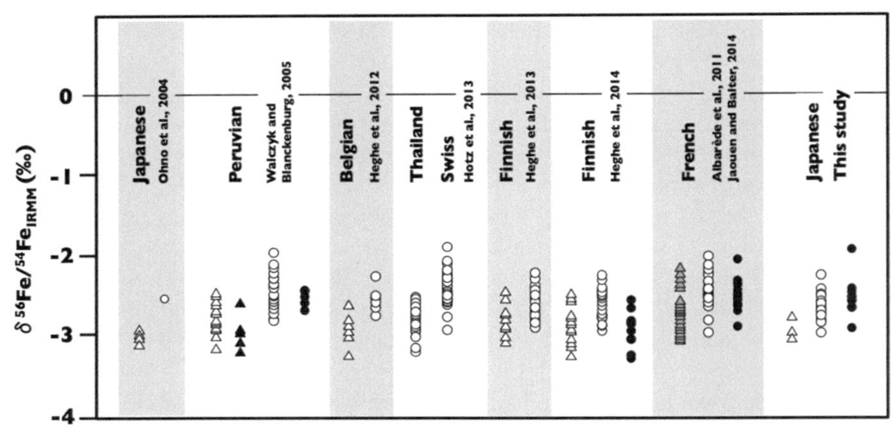

Fig. 12.4 Measured $^{56}Fe/^{54}Fe_{IRMM}$ isotope ratios for human red blood cell

The measured $^{56}Fe/^{54}Fe$ ratios were significantly higher in RBC samples from hemochromatosis patients than in those from healthy persons [47, 48, 56]. Hotz et al. (2012) reported that the $^{56}Fe/^{54}Fe$ ratio became higher up to 1 % with proceeding a bloodletting [52]. Stenberg et al. (2005) suggested that hemochromatosis patients and healthy adults could be distinguished on the basis of the combination of Fe and Zn isotope ratios in whole blood samples [49].

A comparison of Fe isotope ratios among racial groups can be made on the basis of previously reported data (Fig. 12.4). As regards RBC samples from young males in their 20s to 30s, Caucasian ($n = 17$; −2.75 %) [48] had higher $^{56}Fe/^{54}Fe$ ratio than Japanese ($n = 4$; −3.05 ‰) [46] or Belgian ($n = 6$; −2.93 ‰) [53]. In females, the $^{56}Fe/^{54}Fe$ ratio of Swiss ($n = 39$; −2.4 ‰) [54] was higher than that of Japanese ($n = 1$; −2.55 ‰) [46] or Belgian ($n = 6$; −2.56 ‰) [53]. In contrast, the measured $^{56}Fe/^{54}Fe$ ratio in Thai female ($n = 28$, −2.8 ‰) [54] was the lowest on average. Changes in Fe isotope ratios reflect differences in Fe isotope ratios in the diet and Fe absorption efficiency among countries or racial groups.

Regarding a possible link between Fe metabolism and age, particularly in postmenopausal females, changes in Fe absorption efficiency can induce variations in the isotope ratios. Several studies have revealed that the $^{56}Fe/^{54}Fe$ ratios in postmenopausal women were significantly lower than those in young women [55, 57]. As mentioned in the earlier section, the Fe isotope ratio can reflect both Fe absorption efficiency and nutritional status, and therefore, it is natural to consider that the $^{56}Fe/^{54}Fe$ ratio in female RBCs would change after menopause.

Despite the obvious difference in Fe isotope signature between young and elderly female donors, no clear data supporting changes in the nutritional status of Fe and other nutrients in postmenopausal women were given in previous studies. Because of this, the mechanism of Fe isotope signature changes in elderly female donors is still unknown. Moreover, it should be noted that the previous studies focused only on European postmenopausal women. To investigate the possible cause of changes in the $^{56}Fe/^{54}Fe$ ratio with age, a series of Fe isotopic analyses were carried out on RBC samples collected from Japanese female donors of various ages (18–82 years old). More importantly, biological data reflecting Fe nutritional status were measured in the blood samples in order to discuss the possible cause of changes in the Fe isotope ratios. Our results indicate that the Fe isotope ratio would serve as an alternative choice for the evaluation of the nutritional status of Fe in the human body.

12.2 Experimental

12.2.1 Instrumentation

Iron isotope analysis was carried out with a multiple collector-inductively coupled plasma mass spectrometer (MC-ICPMS). The MC-ICPMS system used in this study was a Nu Plasma 500 (Nu Instruments, Ltd., Wrexham, UK) installed at the Laboratory for Planetary Science, Kyoto University. Three Fe isotopes (^{54}Fe, ^{56}Fe, and ^{57}Fe) were simultaneously monitored using three Faraday cups equipped with a preamplifier system utilizing a conventional 10^{11} ohm feedback register. Several polyatomic interferences, such as $^{40}Ar^{14}N^+$, $^{40}Ar^{16}O^+$, or $^{40}Ar^{16}O^1H^+$, could cause systematic errors in the Fe isotope ratio measurements using the conventional ICP-MS instruments. Among the polyatomic interferences, the production of ArO^+ and $ArOH^+$ ions could be effectively reduced by minimizing water loading on the ICP. To achieve this, a desolvating nebulizer system (Aridus II, CETAC, Omaha,

USA) was used for solution introduction. The resulting signal intensities of ArO^+ and $ArOH^+$ were <1.5 mV and <0.1 mV, respectively, and were almost 1/50–1/100 of the signal intensities in the conventional solution nebulization technique. No significant difference was noted in the background intensity from $^{40}Ar^{14}N^+$ between the system with a desolvating nebulizer and that without it. This suggested that nitrogen would mainly originate in N_2 in air through the engulfment at the ICP. Residual $^{40}Ar^{14}N^+$, $^{40}Ar^{16}O^+$, or $^{40}Ar^{16}O^1H^+$ signals were corrected by the on-peak baseline subtraction technique. The signal intensities of these polyatomic interferences were carefully monitored before and after the analysis of sample or standard solutions, and the resulting signal intensity was used for blank subtraction.

In the Fe isotope analysis using the ICP-MS technique, the mass discrimination effect on Fe isotopes was 1.5–3 %/amu [58]. This would be mainly due to the large space charge effect within the plasma or interface regions. The mass discrimination effect was externally corrected by applying the standard-sample-standard bracketing method using an IRMM-014 isotope standard. The isotope ratio measurement of RBC samples consisted of five cycles of 50 isotope ratio measurements each at the integration time of 5 s. The resulting Fe isotope data were expressed in delta (δ) notation relative to the Fe isotope reference material IRMM-014, as defined in Eq. 12.2. Details of the instrument and the operation parameters are listed in Table 12.1.

Table 12.1 Instrumentation and operational settings

1	MC-ICP-MS instrument	Nu instruments Nu Plasma 500
2	ICP ion source	
	ICP frequency	27.12 MHz
	RF power	1.35 kW forward, <3 W ref.
	Cooling gas flow rate	15.0 L min^{-1}
	Auxilliary gas flow rate	0.70 L min^{-1}
	Nebulizer gas flow rate	0.85–1.00 L min^{-1}
3	Desolvating nebulizer system	Cetac ARIDUS II
	Spray chamber temperature	110 °C
	Membrane filter temperature	160 °C
	Sweep gas flow rate	5.0–7.0 L min^{-1}
	Sample uptake rate	100 μL min^{-1}
4	Mass spectrometer	
	Ion energy	4000 V
	Extraction	2700–2900 V
	Analysis mode	Static
	Ion detection	Analogue by three Faraday cups
	Typical transmission	1.2–1.5 V per μg ml^{-1}
5	Data acquisition	
	Blank mode	On-peak baseline subtraction
	Integration time	5 s
	Number of cycles	50–60 cycles per run

$$\delta^{56}Fe/^{54}Fe_{IRMM} = \left(\frac{(^{56}Fe/^{54}Fe)_{Sample}}{(^{56}Fe/^{54}Fe)_{IRMM-014}} - 1 \right) \times 1000 \qquad (12.2)$$

12.2.2 *Effects of Fe Valency in Solution on Measured Isotopic Composition*

Zhu et al. (2002) reported that the measured $^{56}Fe/^{54}Fe$ ratio could change in solutions having different proportions of Fe valency [43]. To investigate this, we measured the $^{56}Fe/^{54}Fe$ ratios in Fe solutions having various Fe^{2+}/Fe^{3+} ratios. Approximately 0.1 g of high-purity Fe powder (Puratronic powder 99.999 %, Lot number 21639; Johnson Matthey Chemicals, Royston, UK) was dissolved in 100 ml of 2 N HCl (TAMAPURE AA-100; TAMA Chemicals, Kawasaki, Japan). Four 300 μg/ml Fe solutions were prepared from this mother solution by the following treatment methods: (a) oxidation with H_2O_2 and evaporation to dryness and then redissolution in 0.1 wt% HCl, (b) evaporation to dryness and then redissolution in 0.1 wt% HCl, (c) simple dilution without drying or oxidation procedures, and (d) preparation of 1:1 mixture of the solutions in (b) and (c). The preparation and measurement of each of the four solutions were repeated three times to evaluate the repeatability of the treatment methods.

The molar fractions of Fe^{2+} relative to total Fe (i.e., $Fe^{2+}/(Fe^{3+} + Fe^{2+})$) for the 12 solutions for analysis were measured by colorimetry using 1,10-phenanthroline. The contents of Fe^{2+} were determined by measuring absorbance at 483 nm using a visible spectrophotometer (Digilab Hitachi U-2800 double beam UV/Visible spectrophotometer; Hitachi High-Technologies, Tokyo, Japan). For the colorimetric calibration using ferroin, approximately 1 mg of 1,10-phenanthroline was added to 30 μg of Fe. Then, buffer solution (pH=4) was added to stabilize complex formation. To determine total Fe ($Fe^{3+} + Fe^{2+}$), Fe was reduced to Fe^{2+} with hydroxylamine chloride, and total Fe was determined by measuring the absorbance of ferroin (at 483 nm). The absorbance of the mixture of 1,10-phenanthroline and 3 μg/ml (30 μg) of Fe^{3+}, prepared by dissolving Fe_2O_3 (Kanto Chemical, Japan), was measured as the blank value for Fe^{2+} for the colorimetry. A total of 12 solutions were subjected to the Fe isotope analyses using the MC-ICPMS system. Isotope ratios of 7 μg/ml of each sample were measured relative to the $^{56}Fe/^{54}Fe$ and $^{57}Fe/^{54}Fe$ ratios obtained from 7 μg/ml of IRMM-014 solution prepared by the identical procedure for solution (a).

To evaluate changes in the measured Fe isotope ratios, measurements of $^{56}Fe/^{54}Fe$ and $^{57}Fe/^{54}Fe$ ratios were carried out under two plasma conditions having different ICP incident powers and sweep gas flow rates (1150 W, 6.57 L/min and 1350 W, 5.45 L/min). Figure 12.5 shows that the measured $^{56}Fe/^{54}Fe$ ratio is correlated linearly with the molar fraction of Fe^{2+}. The resulting $\delta^{56}Fe$ values for the three solutions of (a) were 0.40 ± 0.04 ‰ (2SD, $n = 3$) and 0.38 ± 0.03 ‰ (2SD, $n = 3$) at

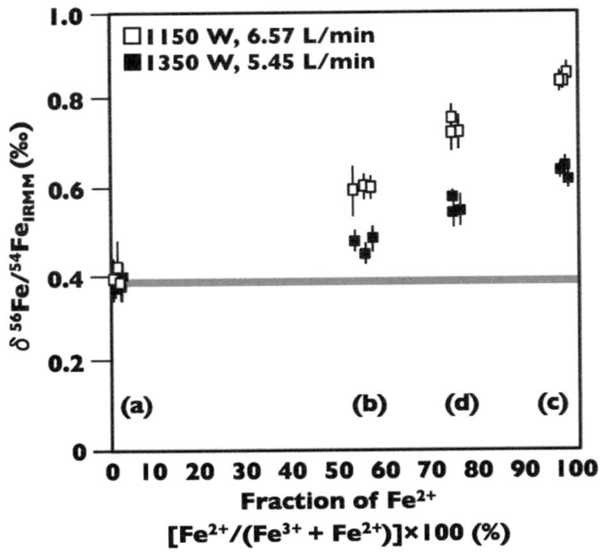

Fig. 12.5 Effect of valency of Fe onto the Fe_{IRMM} measured isotope ratio

each instrumental setting. For solution (c), which showed nearly 90 % difference in the molar fraction of Fe^{2+} from the oxidized Fe solution (IRMM-014 or solution (a)), the δ^{56}Fe values were significantly higher than those for solution (a).

Figure 12.5 shows the effect of oxidation status of Fe onto the measured $^{56}Fe/^{54}Fe$ ratio, suggestive of clear dependence in the measured $^{56}Fe/^{54}Fe$ ratio on the oxidation status of Fe in the analytical solution. While the effect of the oxidation status was strongly dependent upon ICP operational conditions, the measured $^{56}Fe/^{54}Fe$ ratio obtained from Fe^{2+}-bearing solution was about 50 % higher than that obtained from Fe^{3+}-bearing solution, when the ICP incident power of 1350 W and sweep gas flow rate of 5.45 L/min were adopted. This could be explained by the difference in the distance traveled from the ionization point in the ICP to the ion extraction point (i.e., orifice of sampling cone). Because the total dissociation energy of $Fe(III)Cl_3$ would be greater than that of $Fe(II)Cl_2$, mainly as a result of the larger number of Fe-Cl bonds, the overall energy for the dissociation-atomization-ionization process of Fe(III) could be higher than that of Fe(II). The high ionization energy of Fe(III) would result in the long traveling distance in the ICP, suggesting closeness to the ion extraction point (Fig. 12.6). Since some ions, produced in the ICP, can be lost through either by collision with Ar atoms or by space charge effect, a long traveling distance can enhance the mass discrimination at the ion source. The large isotope fractionation found in the Fe^{2+}-bearing solutions can be explained by the contribution of preferential transmission of Fe isotopes through scattering. However, the present mechanism for the mass discrimination effect in the ICP-MS might be oversimplified, and therefore, this must remain as a possibility. The data suggest clearly that great care must be given to the oxidation

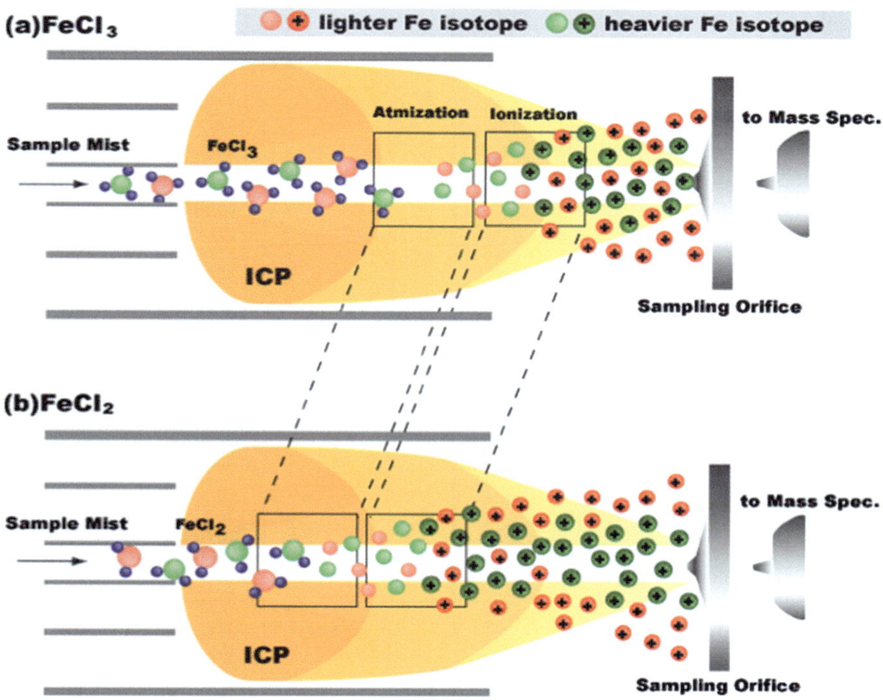

Fig. 12.6 Schematic diagram for mechanisms of changes in the measured Fe isotopes. Difference in overall energy for dissociation-atomization-ionization process can result in different traveling distances from ion source to the sampling position (See text for details)

state of Fe in the analysis solutions. In this study, Fe in the solution for analysis was completely oxidized to Fe(III) by the addition of hydrogen peroxide (H_2O_2). The resulting solution was then heated to dryness and redissolved in 2 M HCl and finally diluted with 0.1 wt% HCl prior to the isotope analysis.

12.2.3 Reproducibility of Analysis

The reproducibility of Fe isotope analysis was evaluated on the basis of repeated isotope ratio measurements of JMC (Johnson Matthey Chemicals) Fe solution. Relative deviations of measured $^{56}Fe/^{54}Fe$ and $^{57}Fe/^{54}Fe$ ratios from those of IRMM-014 were monitored for 17 months. As discussed in the previous section, Fe in both JMC and IRMM-014 solutions for analysis was oxidized by H_2O_2 and redissolved in 0.1 wt% HCl. The overall average of 42 analysis sessions was 0.38 ± 0.04‰ (2SD) for δ^{56}Fe and 0.57 ± 0.08‰ (2SD) for δ^{57}Fe. These values showed good agreement with the values reported by Ohno et al. (2004). The Fe isotope

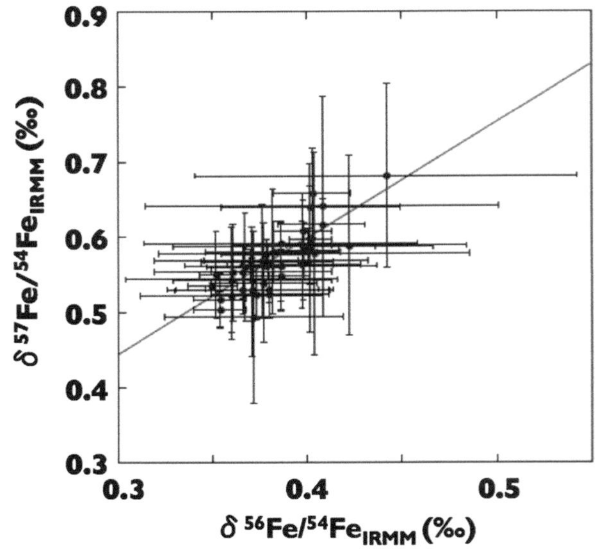

Fig. 12.7 Three isotope diagram ($^{57}Fe/^{54}Fe_{IRMM}$ and $^{56}Fe/^{54}Fe_{IRMM}$) for JMC Fe solution

data obtained here were plotted on the three-isotope diagram (plot of $\delta^{56}Fe/^{54}Fe$ ratio against $\delta^{57}Fe/^{54}Fe$: Fig. 12.7). The data points fall close to a straight line, defined as mass-dependent fractionation line, suggesting that the contribution of mass spectrometric interferences by ArN^+, ArO^+, and $ArOH^+$ signals could be negligibly small after the on-peak baseline subtraction approach employed in this study.

12.2.4 Sample

Blood samples were collected from healthy Japanese female volunteers, classified into three groups: young females (21–22 years old, $n = 8$), young rhythmic gymnasts (18–21 years old, $n = 10$), and elderly postmenopausal females (68–82 years old, $n = 10$). For comparison, three blood samples from healthy young males (22–27 years old) were also collected. Blood sampling was conducted according to the regulations of the ethical code of the National Institute of Health and Nutrition for female donors and of Juntendo University for male donors. Venous blood was drawn after an overnight fast. Blood was centrifuged with heparin, and RBC sample was washed with cold 5 % glucose solution. The collected RBC sample was stored at −4 °C and then transported to Kyoto University. Simultaneously, only for female donors, hemoglobin, serum Fe, total iron-binding capacity (TIBC), unsaturated iron-binding capacity (UIBC), and ferritin were analyzed by LSI Medience Corporation (Tokyo, Japan).

12.2.5 Sample Preparation

Approximately 1 ml of RBC sample was heated to dryness on a hot plate at 100 °C for 3 h. The resulting sample cake was decomposed by adding a mixture of 6 ml of concentrated EL-grade HNO_3 (Electronic Grade L, Kanto Chemical, Tokyo, Japan) and 1 ml of H_2O_2 (AA-10, Tama Chemicals, Kawasaki, Japan) under heating at 220 °C using a microwave digestion system (ETHOS-TC, Milestone Srl. Sorisole, Italy). This was followed by chemical separation of Fe using ion chromatography. The basic procedure for the chemical separation/purification of Fe was based on the procedures reported by Ohno et al. (2004) [46]. The dissolved sample was heated to dryness and then redissolved in 7 M EL-grade HCl (Electronic Grade L, Kanto Chemical, Tokyo, Japan). The resulting solutions were loaded onto a polypropylene column (i.d. 9 mm × 42 mm) filled with 1.8 ml of Bio-Rad AG-MP1 resin (100–200 mesh, Bio-Rad Lab., Hercules, USA). Major components, such as Na, Mg, K, or Cu, were eluted by the addition of 30 ml of 7 M HCl, and Fe was eluted with 10 ml of 2 M HCl. The average recovery of Fe was 99.6 ± 1.4 % ($n = 2$).

All of the chemical decomposition and separation procedures were carried out in a class 1000 clean room. Teflon vessels or polypropylene bottles used in this study were pre-cleaned successively with 3 M HNO_3 and deionized water. The chemical blank of Fe for the present chemical separation procedure was 230 ± 140 ng (2SD, $n = 3$). Because the typical amount of Fe collected from 1 ml of RBC sample was approximately 1000 μg, the contribution of blank Fe could be neglected (<0.03 %).

To evaluate the contribution of isotope fractionation during ion chromatography [59], we measured changes in the $^{56}Fe/^{54}Fe$ ratio for each Fe in eluent. In Fig. 12.8, the accumulated $^{56}Fe/^{54}Fe$ ratios were plotted against the volume of eluent, and the results suggested clear isotope fractionation during the ion chromatography. The isotope fractionation of Fe would be smaller than the analytical uncertainties determined in this study (<0.05 ‰) if the recovery of Fe was higher than 97 %. The recovery of Fe through the ion chromatography (99.6 %) was significantly

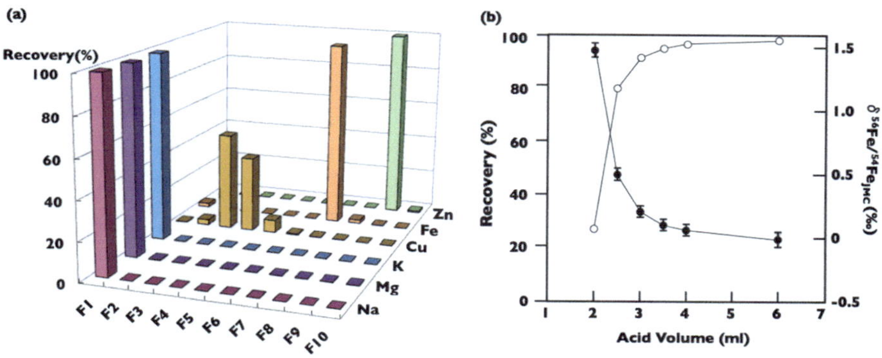

Fig. 12.8 Chromatographic separation for Na, Mg, K, Cu, Fe_{IRMM} and Zn

higher than the critical value (97 %), suggesting that the isotope fractionation during the Fe separation was negligible.

12.2.6 Dietary Survey

The average dietary intake of elderly postmenopausal and young female subjects was assessed by asking the subjects to fill out a self-administered diet history questionnaire (DHQ). The DHQ is a validated 16-page structured questionnaire that assesses dietary habits in the preceding 1-month period [60, 61]. The average dietary intake of the female rhythmic gymnasts was assessed on the basis of 1-week dietary records. After the subjects filled out the dietary records and took photographs of the food they consumed, a registered dietitian checked the records and calculated nutrient intake using the Standard Tables of Food Composition in Japan, 2010.

12.3 Results and Discussion

12.3.1 Gender Difference in Fe Isotope Ratio

The Fe isotope ratios in RBC samples from male and female donors are summarized in Table 12.2 and plotted in Fig. 12.4. Despite the very limited number of male samples, a significant difference in Fe isotope ratios was noted between male ($\delta^{56}Fe/^{54}Fe = -2.92 \pm 0.28$ ‰, $\delta^{57}Fe/^{54}Fe = -4.32 \pm 0.46$ ‰; 2SD, $n = 3$) and female ($\delta^{56}Fe/^{54}Fe = -2.60 \pm 0.41$ ‰, $\delta^{57}Fe/^{54}Fe = -3.82 \pm 0.63$ ‰; 2SD, $n = 28$) with the t-test value of $p < 0.05$. The difference in the averaged $\delta^{56}Fe/^{54}Fe$ values of male and female was approximately 0.3 ‰ for Japanese, consistent with the data for Japanese [46], Peruvian [47], and Belgian [53] populations.

We discuss the possible cause of the systematic difference in $\delta^{56}Fe/^{54}Fe$ values between male and female. Recently, Hotz et al. (2013) measured Fe absorption efficiency in the intestine using ^{57}Fe or ^{58}Fe isotope tracers, and they claimed that the changes of the Fe absorption efficiency could be the major source of the variations in the Fe isotope ratio [54]. Most of Fe in the dietary food can present as either heme Fe or nonheme Fe. Since the Fe(II), present as heme Fe, can be directly absorbed as is (heme Fe form), the magnitude of the isotope fractionation during absorption would be negligibly small. This is mainly because of very small relative mass difference of the Fe isotopologue of heme Fe (Mw >60,000). In contrast, nonheme Fe in the diet would not be absorbed as is. Nonheme Fe would be reduced to Fe(II) by Dcytb prior to absorption.

Bigeleisen and Mayer (1947) predicted that the isotope fractionation can be enhanced under the oxidation-reduction reaction, in which Δk_i value (Eq. 12.1) can

Table 12.2 Summary of the resulting $\delta^{56}Fe$ and $\delta^{57}Fe$ for female RBC, together with biological data for individual donors

	Isotope signature						Biological parameter					Nutritional survey data								
	$\delta^{56}Fe/^{54}Fe$ (‰)		2σ (‰)	$\delta^{57}Fe/^{54}Fe$ (‰)		2σ (‰)	Hb (g/dL)	Serum Fe (μg/dL)	TIBC (μg/dL)	TSAT (%)	Serum Ferritin (μg/L)	Energy (kcal)	Protein (g)	Lipid (g)	Hydrocarbon (g)	Fe (mg)	Folate (μg)	Vitamin B12 (μg)	Vitamin C (mg)	Fiber (g)
Young athlete (member of Rhythmic Gymnastics Sports Club) women																				
A-1	−2.69	±	0.03	−4.01	±	0.05	12.9	308	352	87.5	23.0	2457	71.2	90.3	335.9	8.6	356.1	4.4	349.4	14.4
A-2	−2.59	±	0.03	−3.80	±	0.05	12.7	186	336	55.4	53.8	1501	45.6	58.0	198.9	7.8	252.6	2.8	144.5	9.5
A-3	−2.72	±	0.03	−4.02	±	0.05	12.4	156	294	53.1	66.6	1836	62.0	56.9	266.0	6.0	280.3	3.8	110.9	11.8
A-4	−2.42	±	0.03	−3.61	±	0.05	12.7	159	339	46.9	14.7	1840	47.3	72.1	249.9	6.4	270.3	2.2	146.3	12.1
A-5	−2.72	±	0.03	−4.01	±	0.05	12.7	225	313	71.9	19.7	2163	63.4	58.9	340.0	8.4	313.5	3.6	179.7	12.8
A-6	−2.82	±	0.07	−4.06	±	0.56	13.7	139	300	46.3	21.5	1805	56.0	66.9	242.6	7.1	301.2	3.3	139.8	13.5
A-7	−2.48	±	0.02	−3.60	±	0.03	11.6	134	462	29.0	8.5	1389	49.5	53.9	175.0	5.7	229.0	3.0	110.6	10.2
A-8	−2.65	±	0.02	−3.95	±	0.04	13.6	111	312	35.6	18.8	1541	53.6	55.9	204.0	6.3	243.8	3.5	109.2	12.1
A-9	−2.49	±	0.02	−3.63	±	0.04	12.3	162	343	47.2	13.1	1614	56.8	48.1	236.4	7.0	223.2	3.7	88.5	13.2
A-10	−2.56	±	0.04	−3.73	±	0.06	13.0	191	346	55.2	16.0	2187	63.9	72.1	311.9	8.3	215.1	2.9	183.0	13.5
Mean	−2.61			−3.84			12.8	177	340	52.8	25.6	1833	56.9	63.3	256.1	7.2	268.5	3.3	156.2	12.3
Young women																				
Y-1	−2.25	±	0.03	−3.22	±	0.05	12.8	25	368	6.8	2.9	1975	74.8	59.6	277.3	7.5	338.1	7.0	101.7	12.2
Y-2	−2.80	±	0.03	−4.15	±	0.05	12.6	130	289	45.0	103.0	1754	68.4	54.0	243.8	6.1	182.9	5.0	70.6	9.4
Y-3	−2.67	±	0.03	−3.89	±	0.06	12.5	123	286	43.0	26.8	1540	52.8	50.8	205.5	4.5	188.8	3.7	42.4	10.7
Y-4	−2.60	±	0.03	−4.02	±	0.04	14.7	68	360	18.9	13.9	1596	53.1	70.5	186.3	6.6	322.4	4.1	85.4	10.5
Y-5	−2.69	±	0.02	−3.98	±	0.04	12.3	65	320	20.3	34.1	2061	71.4	78.9	270.9	8.9	498.0	5.5	245.4	21.1
Y-6	−2.83	±	0.03	−4.15	±	0.06	12.6	102	344	29.7	31.8	4109	162.4	141.5	536.0	16.8	671.6	27.2	311.2	28.9
Y-7	−2.96	±	0.04	−4.34	±	0.06	13.4	75	336	22.3	52.7	2807	91.6	111.6	354.3	11.0	522.4	11.1	173.0	19.5
Y-8	−2.74	±	0.02	−4.08	±	0.04	12.1	109	334	32.6	62.6	1700	64.3	60.0	218.6	6.5	271.6	6.2	66.8	12.3
Mean	−2.69			−3.98			12.9	87	330	27.3	41.0	2193	79.9	78.4	286.6	8.5	374.5	8.7	137.1	15.6
Postmenopausal women																				
E-1	−2.89	±	0.05	−4.26	±	0.10	12.1	87	296	29.4	120.0	1732	82.6	67.4	197.9	13.8	680.1	9.5	215.8	20.6
E-2	−1.93	±	0.03	−2.82	±	0.04	12.1	45	364	12.4	8.0	1248	50.1	42.1	161.8	6.2	330.0	8.9	130.2	10.8
E-3	−2.66	±	0.03	−3.97	±	0.05	13.5	145	321	45.2	30.8	1637	71.3	61.1	194.3	8.6	401.3	9.9	126.3	12.0
E-4	−2.56	±	0.03	−3.70	±	0.05	12.8	120	344	34.9	42.4	2052	104.7	66.0	259.9	11.7	626.7	13.1	224.3	19.3
E-5	−2.52	±	0.05	−3.66	±	0.10	12.5	67	382	17.5	7.4	2200	94.4	71.7	293.2	11.7	633.7	14.4	267.6	22.9
E-6	−2.56	±	0.02	−3.75	±	0.04	12.7	87	291	29.9	22.6	1847	85.7	58.6	236.2	11.5	633.8	11.1	227.6	17.1
E-7	−2.63	±	0.02	−3.83	±	0.04	11.8	59	342	17.3	8.5	2615	136.7	96.8	296.1	19.2	969.2	32.7	316.9	25.5

(continued)

Table 12.2 (continued)

	Isotope signature						Biological parameter					Nutritional survey data									
	$\delta^{56}Fe/^{54}Fe$ (‰)		2σ (‰)	$\delta^{57}Fe/^{54}Fe$ (‰)		2σ (‰)	Hb (g/dL)	Serum Fe (μg/dL)	TIBC (μg/dL)	TSAT (%)	Serum Ferritin (μg/L)	Energy (kcal)	Protein (g)	Lipid (g)	Hydrocarbon (g)	Fe (mg)	Folate (μg)	Vitamin B12 (μg)	Vitamin C (mg)	Fiber (g)
E-8	−2.44	±	0.03	−3.54	±	0.05	15.0	106	314	33.8	34.3	1593	53.0	51.5	227.9	6.6	399.3	6.0	186.2	12.3
E-9	−2.40	±	0.02	−3.60	±	0.04	13.7	131	271	48.3	88.3	1458	58.5	40.4	213.0	6.2	318.9	8.7	164.7	11.0
E-10	−2.50	±	0.02	−3.69	±	0.04	13.6	101	358	28.2	87.0	1894	89.2	80.2	205.8	12.6	666.7	9.5	246.4	22.7
Mean	−2.51			−3.68			13.0	95	328	29.7	44.9	1828	82.6	63.6	228.6	10.8	566.0	12.4	210.6	17.4

become higher. In fact, oxidation-reduction experiments demonstrated that the precipitate produced by partial reduction of ferric Fe had lower $^{56}Fe/^{54}Fe$ ratios than the initial solution (ferric Fe) [38, 39]. Moreover, ferrous Fe produced by Fe-reducing bacteria was isotopically lighter than the initial ferrihydrite substrate (ferric Fe) [40]. These data suggests that the $^{56}Fe/^{54}Fe$ ratio can change through the reduction in the absorption process at the intestine. In addition, Hotz et al. (2011) and Albarède et al. (2011) suggested that preferential accumulation of heavy Fe isotopes as ferritin through the oxidation of ferrous Fe in the epithelial cell would also contribute to the wide variation in the $^{56}Fe/^{54}Fe$ ratio for RBCs [44, 50].

The magnitude of the isotope fractionation would be dependent upon reaction efficiency. The measured isotope fractionation can be further enhanced, when the reactions partly proceeded. These features lead us to consider that the difference in Fe absorption efficiency between male (6 %) and female (13 %) would be the main cause of the gender difference in Fe isotope ratio.

After absorption by epithelial cells, a part of Fe is bound to transferrin for transport to the liver. In liver cells, transferrin taken up by endocytosis releases ferric Fe, and then ferric Fe is reduced to the ferrous form for transport by DMT1 and stored in the Fe pool temporarily [62]. Ferrous Fe, which is not used for the synthesis of functional proteins or enzymes (excess Fe), would be oxidized to the ferric form and stored as ferritin. Several studies have indicated that Fe present in storage organs, such as the liver, spleen, or bone marrow, shows significant enrichment of heavy Fe isotope compared to Fe in other tissues in the human and animal body [41, 44]. Hotz et al. (2012) suggested that the Fe isotope ratio in blood could vary with the proportion of Fe distribution between the liver and RBCs [52]. As the fraction of oxidized Fe in liver tissue increases, the Fe isotope ratio in the liver would become similar to that of circulating Fe in blood. If this were the case, the remaining ferrous Fe in the Fe pool would be depleted in the heavy Fe isotope, reflecting the low Fe isotope ratio in RBCs. Thus, the higher proportion of Fe stored in the liver can result in the low Fe isotope ratio in RBCs in male compared to that in female.

12.3.2 Effects of Age

The $\delta^{56}Fe/^{54}Fe$ and $\delta^{57}Fe/^{54}Fe$ values for females of various ages are summarized in Table 12.2 and plotted in Fig. 12.4. The data demonstrated no significant variation in $\delta^{56}Fe/^{54}Fe$ between young women aged 18 to 22 years old ($\delta^{56}Fe/^{54}Fe = -2.65 \pm 0.34$ ‰, $\delta^{57}Fe/^{54}Fe = -3.90 \pm 0.53$ ‰; 2SD, $n = 18$) and elderly postmenopausal women aged 68 to 82 years old ($\delta^{56}Fe/^{54}Fe = -2.51 \pm 0.49$ ‰, $\delta^{57}Fe/^{54}Fe = -3.68 \pm 0.73$ ‰; 2SD, $n = 10$). The data presented here are consistent with the data reported by Krayenbuehl et al. (2005) [48]. In contrast, Jaouen and Balter (2014) and Van Heghe et al. (2014) claimed that there is a clear difference in the measured $\delta^{56}Fe/^{54}Fe$ and $\delta^{57}Fe/^{54}Fe$ values between young and elderly females [55, 57].

As discussed before, because an excess of Fe in the human body causes various diseases, it is natural to consider that Fe absorption efficiency would be lowered after menopause, which would occur around age 51 [63]. The lowering of the Fe absorption efficiency would result in a larger magnitude of the isotope fractionation, and therefore, the $\delta^{56}Fe/^{54}Fe$ and $\delta^{57}Fe/^{54}Fe$ values of elderly postmenopausal females would be reduced to the same level as those of males. In fact, serum ferritin levels, which are an indicator of Fe storage, would be the same for female donors over age 50 and young males (20–30 years old) [64]. This suggests that Fe absorption efficiency would no longer be high even for the elderly postmenopausal females. However, this is not the case for Japanese female. The resulting Fe isotope data of the elderly postmenopausal female donors did not vary significantly from those of the young female donors. The age difference of donors would be considered as the reason for the obvious discrepancy in the Fe isotope signature. The ages of the postmenopausal female donors in this study ranged from 68 to 82 years old, which were clearly higher than those in previous studies: 49–68 years old [55] and 55–66 years old [57]. The age difference could affect the change of Fe bioavailability in the gastrointestinal environment, as follows.

Neither intestinal motility nor the morphological features of intestinal cells would reflect the donor age [65–68]. Elderly persons tend to suffer from atrophic gastritis, which leads to achlorhydria or hypochlorhydria [69, 70], as a result of *Helicobacter pylori* infection [71] or use of proton pump inhibitors [72, 73]. Atrophic gastritis results in reduced gastric acid secretion, which can induce malabsorption of nutrients due to bacterial overgrowth [74]. In addition, the incomplete digestion of food or the insolubility of ferric Fe due to the lack or decrease of gastric acid would lead to reduced Fe bioavailability in the gastrointestinal environment [73, 75]. It is recognized that *H. pylori* infection can induce Fe-deficient anemia, mainly due to impaired gastric acid secretion or the removal of Fe from lactoferrin by bacterial function [70, 76]. The low bioavailability of Fe results in higher Fe intake efficiency from the diet. A stable isotope study using enriched ^{58}Fe isotope has revealed that the absorption efficiencies in children of developing countries, who present with decreased gastric acid output due to *H. pylori* infection, slightly increased relative to those of healthy controls [76].

Iron bioavailability can be evaluated on the basis of nutritional information. Iron isotope studies indicated no significant difference in nutritional information, such as the amount of major nutrient (e.g., protein, hydrocarbon, lipid, fiber, and minerals including Fe), among donors of different ages. Moreover, the amounts of Fe taken by the elderly female donors in this study were larger than the average amount taken by the same-age population. These data suggested that the amount of Fe in the diet did not vary measurably before and after menopause. Nevertheless, Fe availability in the elderly female donors could have been decreased by *H. pylori* infection and/or medication. To compensate for the loss of approximately 1 mg of

Fe per day, the elderly female donors would absorb and utilize nonheme Fe more efficiently. The enhancement of the overall absorption efficiency results in the smaller Fe isotope fractionation from the diet. The lack in variation of the $^{56}Fe/^{54}Fe$ ratios for the elderly female donors relative to the young female donors could also be explained by the release of heavy Fe isotopes from the liver. These explanations were supported by the small difference in serum ferritin level between the elderly and young female donors, implying similar amounts of Fe stored for these groups.

12.3.3 Differences Among Race

We discuss the racial difference in Fe isotope ratios in female RBC samples. The $\delta^{56}Fe/^{54}Fe$ values for Swiss (−2.4‰, $n = 39$) [54] and Belgian (−2.56‰, $n = 6$) young females [53] were significantly higher than those for Japanese (this study, −2.65‰, n = 18) and Thai (−2.8‰, $n = 28$) [54] young females. Serum ferritin level in Thai young females (62 ± 44 μg/L) [54] was slightly higher than those in Swiss (26 ± 22 μg/L) [54] and Japanese (32 ± 26 μg/L) young females. Serum ferritin levels in postmenopausal African-American females were higher than those in postmenopausal Caucasian and Hispanic females [64]. The obvious racial differences in serum ferritin levels could originate in hereditary or environmental (e.g., food, nutritional status) factors. For Swiss, Thai, and Japanese females, $\delta^{56}Fe/^{54}Fe$ values and serum ferritin levels showed a negative correlation on average (Fig. 12.9). The possible linkage between $\delta^{56}Fe/^{54}Fe$ values and serum ferritin levels among the different racial groups implies that the Fe isotope signature found in this study can reflect the absorption efficiency of Fe at the intestine and/or proportion of Fe stored in the liver. Nevertheless, as serum ferritin levels could change due to acute or chronic inflammation or an infection, the correlation between $\delta^{56}Fe/^{54}Fe$ values and serum ferritin levels must be examined further.

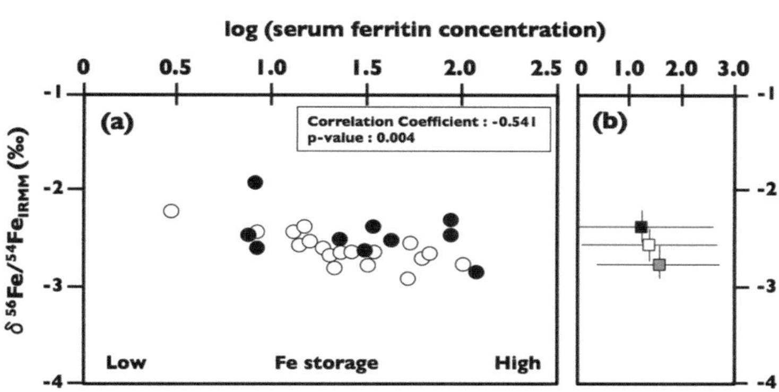

Fig. 12.9 Correlationship between $\delta^{56}Fe/^{54}Fe_{IRMM}$ and concentration of serum ferritin

12.3.4 Effect of Biological and Nutritional Data on Fe Isotopes

Several biological parameters, such as serum ferritin or transferrin saturation (TSAT), were measured to investigate Fe metabolic status. Iron deficiency is characterized by low serum ferritin level (<12 μg/L) or low TSAT (<15 %) [77]. Among the samples analyzed in this study, both serum ferritin levels and TSAT were low in two donors (Y-1 and E-2) (Table 12.2). Based on biological data, these donors would be chronically classified into the Fe-deficient state. In addition, the resulting $\delta^{56}Fe/^{54}Fe$ values of Y-1 (−2.25 ‰) and E-2 (−1.93 ‰) donors were slightly higher than the overall average for Japanese females (−2.65 ‰, $n = 18$), suggesting higher Fe absorption efficiency from the diet. However, it should be noted that there is no significant difference in the concentrations of hemoglobin. Except for the two female donors having very low TSAT values (Y-1 and E-2), the TSAT values may not reflect serum ferritin levels that are widely used to evaluate Fe storage. This implies that TSAT values can reflect short-term changes in Fe nutritional status, whereas Fe isotope signature can reflect long-term Fe metabolism.

Vitamin C (L-ascorbic acid) or animal protein in the diet can enhance nonheme Fe absorption by increasing solubility through the reduction of ferric Fe to ferrous Fe [78]. However, the biological data obtained in this study indicated no significant relationship between the concentration of vitamin C or protein and $\delta^{56}Fe/^{54}Fe$ values, suggesting the very minimal effect of vitamin C and protein on Fe absorption efficiency. Other important nutrients are vitamin B_{12} and folate. It is widely recognized that both vitamin B_{12} and folate are strongly related to the synthesis of DNA encoding the differentiation of erythroblasts into RBCs [79]. For these parameters, there was no correlation between vitamin B_{12} or folate and hemoglobin concentrations and $\delta^{56}Fe/^{54}Fe$ values. The lack of correlation between Fe isotope signature and biological data for blood samples, including TSAT, suggests that the biological data reflect the daily variation of Fe metabolism, whereas the $\delta^{56}Fe/^{54}Fe$ values reflect long-term Fe metabolism. Moreover, the difference in daily nutritional intake between donors did not influence $\delta^{56}Fe/^{54}Fe$ values in blood, similar to the result showing identical Fe isotope ratios between omnivores and vegetarians [53].

12.3.5 Conclusive Remarks

We have developed an analytical technique for measuring $^{56}Fe/^{54}Fe$ and $^{57}Fe/^{54}Fe$ ratios from biological samples using a multiple collector-ICP mass spectrometer. Iron isotope ratio measurements on Fe fraction in the ion chromatography revealed that higher than 97 % Fe recovery would be required to reduce the magnitude of the Fe isotope fractionation to a value smaller than the analytical uncertainties determined in this study (<0.05 % for $^{56}Fe/^{54}Fe$). More importantly, the measured $^{56}Fe/^{54}Fe$ ratio was found to vary by approximately 0.4 % among Fe solutions for analysis having various Fe oxidation states.

The resulting $^{56}Fe/^{54}Fe$ ratios in RBCs from Japanese male donors were significantly lower than those from Japanese female donors. Gender differences in the measured $^{56}Fe/^{54}Fe$ ratios were consistent with previous data of Walczyk and Blanckenburg (2005) and Van Heghe et al. (2012). In contrast, no significant difference in $\delta^{56}Fe/^{54}Fe$ values was found between young female donors and elderly postmenopausal female donors. This is inconsistent with the report of Jaouen and Balter (2014) or Van Heghe et al. (2014). The clear discrepancy in the Fe isotope signature of elderly female donors could be attributed to the changes in Fe metabolism in two stages. The first stage is the change in Fe isotope ratio at early postmenopause, which would be associated with the decrease of the Fe isotope ratio originating from the decrease in Fe absorption efficiency and the increase in Fe storage efficiency. The second stage is the change in Fe absorption and storage efficiencies, which are, respectively, increased and decreased in the course of decreasing Fe availability due to impairment of gastrointestinal function with age. Iron isotope signature shows great potential for use in the evaluation of Fe nutritional status in both young and elderly females in the Fe-deficient state.

Except serum ferritin level, biological data and nutritional parameters did not reflect the Fe isotope signature. These results suggest that $\delta\,^{56}Fe/^{54}Fe$ values would have the potential to become the signature for evaluating long-term Fe metabolism as well as serum ferritin levels, whereas the other parameters would be valid for short-term Fe status.

Acknowledgments We are grateful to Drs. Takeshi Ohno (Gakushuin University, Japan), Momoko Chiba (Juntendo University, Japan), Thomas Walczyk (National University of Singapore, Singapore) and Frank Vanhaecke (Ghent University, Belgium) for many scientific discussions and advices on the biological samples and nutritional studies. This work was financially supported by a Grant-in-Aid for Scientific Research to TH (A26247094) from the Ministry of Education, Culture, Sports, Science and Technology, Japan, and the Ministry of Agriculture, Forestry and Fisheries of Japan.

Compliance with Ethical Standards

Conflict of interest. The authors declare that they have no conflict of interest.
Ethical approval. All procedures performed in studies involving animals were in accordance with the ethical standards of the National Institute of Health and Nutrition and Juntendo University.
Informed consent. Informed consent was obtained from all individual participants included in the study.

References

1. Raven JA (1990) Prediction of Fe and Mn use efficiencies of phototrophic growth as a function of light availability for growth and C assimilation pathway. New Phytol 116:1–17
2. Yip R (2002) Chapter 30 Iron. In: Bowman BA, Russell RM (eds) Present knowledge in nutrition. ILSI Press, Washington, DC, pp 311–328
3. Cook JD (1979) In: Bothwell TH, Charlton RW, COOK JD, Finch CA (eds) Iron metabolism in man. Blackwell Scientific, Oxford, pp 427–429
4. Bothwell TH, Charlton RW (1981) Iron deficiency in women, vol 3. The Nutrition Foundation, Washington, DC, pp 68
5. Halliwell B, Gutteridge JMC (1999) Role of free radicals and catalytic metal ions in human disease: an overview. Methods Enzymol 186:1–85
6. Pietrangelo A (2007) Hemochromatosis: an endocrine liver disease. Hepatology 46:1291–1301
7. Yip R, Dailman PR (1988) The roles of inflammation and iron deficiency as causes of anemia. Am J Clin Nutr 48:1295–1300
8. Hentze MW, Muckenthaler MU, Andrews NC (2004) Balancing acts: molecular control of mammalian Iron metabolism. Rev Cell 117:285–297
9. U.S. Department of Agriculture, Agricultural Research Service. What We Eat in America 2011–2012
10. Green R, Charlton R, Seftel H, Bothwell TH, Mayet F, Adams B, Finch CA, Layrisse M (1968) Body iron excretion in man: a collaborative study. Am J Med 45:336–353
11. Hunt JR, Zito CA, Johnson LK (2009) Body iron excretion by healthy men and women. Am J Clin Nutr 89:1792–1798
12. Shayeghi M, Latunde-Dada GO, Oakhill JS, Laftah AH, Takeuchi K, Halliday N, Khan Y, Warley A, McCann FE, Hider RC, Frazer DM, Anderson GJ, Vulpe CD, Simpson RJ, McKie AT (2005) Identification of an intestinal heme transporter. Cell 122:789–801
13. Qiu A, Jansen M, Sakaris A, Hee Min S, Chattopadhyay S, Tsai E, Sandoval C, Zhao R, Akabasa MH, Goldman ID (2006) Identification of an intestinal folate transporter and the molecular basis for hereditary folate malabsorption. Cell 127:917–928
14. McKie AT, Barrow D, Latunde-Dada GO, Rolfs A, Sager G, Mudaly E, Mudaly M, Richardson C, Barlow D, Bomford A, Peters TJ, Raja KB, Shirali S, Hediger MA, Farzaneh F, Simpson RJ (2001) An iron-regulated ferric reductase associated with the absorption of dietary iron. Science 291:1755–1759
15. Gunshin H, Mackenzie B, Berger UV, Gunshin Y, Romero MF, Boron WF, Nussberger S, Gollan JL, Hediger MA (1997) Cloning and characterization of a mammalian proton-coupled metal-ion transporter. Nature 388:482–488
16. Fleming MD, Trenor CC III, Su MA, Foernzler D, Beier DR, Dietrich WF, Andrews NC (1997) Microcytic anaemia mice have a mutation in Nramp2, a candidate iron transporter gene. Nat Genet 16:383–386
17. Harrison PM (1977) Ferritin: an iron-storage molecule. Semin Hematol 14:55–70
18. Donovan A, Brownlie A, Zhou Y, Shepard J, Pratt SJ, Moynihan J, Paw BH, Drejer A, Barut B, Zapata A, Lawk TC, Brugnarak C, Lux SE, Pinkus GS, Pinkus JL, Kingsley PD, Palis J, Fleming MD, Andrews NC, Zon LI (2000) Positional cloning of zebrafish ferroportin1 identifies a conserved vertebrate iron exporter. Nature 403:776–781
19. McKie AT, Marciani P, Rolfs A, Brennan K, Wehr K, Barrow D, Miret S, Bomford A, Peters TJ, Farzaneh F, Hediger MA, Hentze MW, Simpson RJ (2000) A novel duodenal iron-regulated transporter, IREG1, implicated in the basolateral transfer of Iron to the circulation. Mol Cell 5:299–309
20. Abboud S, Haile DJ (2000) A novel mammalian iron-regulated protein involved in intracellular iron metabolism. J Biol Chem 275(26):19906–19912

21. Quigley JG, Yang Z, Worthington MT, Phillips JD, Sabo KM, Sabath DE, Berg CL, Sassa S, Wood BL, Abkowitz JL (2004) Identification of a human heme exporter that is essential for erythropoiesis. Cell 118:757–766
22. Keel SB, Doty RT, Yang Z, Quigley JG, Chen J, Knoblaugh S, Kingsley PD, De Domenico I, Vaughn MB, Kaplan J, Palis J, Abkowitz JL (2008) A heme export protein is required for red blood cell differentiation and iron homeostasis. Science 319:285–288
23. Vulpe CD, Kuo Y-M, Murphy TL, Cowley L, Askwith C, Libina N, Gitschier J, Anderson GJ (1999) Hephaestin, a ceruloplasmin homologue implicated in intestinal iron transport, is defective in the sla mouse. Nat Genet 21:195–199
24. Krause A, Neitz S, Mägert HJ, Schulz A, Forssmann WG, Schulz-Knappe P, Adermann K (2000) LEAP-1, a novel highly disulfide-bonded human peptide, exhibits antimicrobial activity. FEBS Lett 480:147–150
25. Park CH, Valore EV, Waring AJ, Ganz T (2001) Hepcidin, a urinary antimicrobial peptide synthesized in the liver. J Biol Chem 276:7806–7810
26. Pigeon C, Ilyin G, Courselaud B, Leroyer P, Turlin B, Brissot P, Lorèal O (2001) A new mouse liver-specific gene, encoding a protein homologous to human antimicrobial peptide hepcidin, is overexpressed during iron overload. J Biol Chem 276:7811–7819
27. Ganz T (2003) Hepcidin. A key regulator of iron metabolism and mediator of anemia of inflammation. Blood 102:783–788
28. Nemeth E, Tuttle MS, Powelson J, Vaughn MB, Donovan A, Ward DM, Ganz T, Kaplan J (2004) Hepcidin regulates cellular iron efflux by binding to ferroportin and inducing its internalization. Science 306:2090–2093
29. Nicolas G, Bennoun M, Devaux I, Beaumont C, Grandchamp B, Kahn A, Vaulont S (2001) Lack of hepcidin gene expression and severe tissue iron overload in upstream stimulatory factor 2 (USF2) knockout mice. PNAS 98:8780–8785
30. Nicolas G, Bennoun M, Porteu A, Mativet S, Beaumont C, Grandchamp B, Sirito M, Sawadogo M, Kahn A, Vaulont S (2002) Severe iron deficiency anemia in transgenic mice expressing liver hepcidin. PNAS 99:4596–4601
31. Frazer DM, Wilkins SJ, Becker EM, Murphy TL, Vulpe CD, McKie AT, Anderson GJ (2003) A rapid decrease in the expression of DMT1 and Dcytb but not Ireg1 or hephaestin explains the mucosal block phenomenon of iron absorption. Gut 52:340–346
32. Liebold EA, Guo B (1992) Iron-dependent regulation of ferritin and transferrin receptor expression by the iron- responsive element binding protein. Annu Rev Nutr 12:345–368
33. McCance RA, Widdowson EM (1937) Absorption and excretion of iron. Lancet 2:680–684
34. Mellon FA, Fairweather-Tait SJ (1997) Stable isotope methods for studying nutrient mineral metabolism in humans. Endeavour 21(1):12–18
35. Ulusoy U, Whitley JE (1999) Determination of intestinal uptake of iron and zinc using stable isotopic tracers and rare earth markers. Nutr Res 19(5):675–688
36. Bigeleisen J, Mayer MG (1947) Calculation of equilibrium constants for isotopic exchange reactions. J Chem Phys 15:261–267
37. Urey HC (1947) The thermodynamic properties of isotopic substances. J Chem Soc 562–581
38. Johnson CM, Skulan JL, Beard BL, Sun H, Nealson KH, Braterman PS (2002) Isotope fractionation between Fe(III) and Fe(II) in aqueous solutions. EPSL 195:141–153
39. Welch SA, Beard BL, Johnson CM, Braterman PS (2003) Kinetic and equilibrium Fe fractionation between aqueous Fe(II) and Fe(III). Geochim Cosmochim Acta 67(22):4231–4250
40. Beard BL, Johnson CM, Cox L, Sun H, Nealson KH, Aguilar C (1999) Iron isotope biosignatures. Science 285(17):1889–1892
41. Walczyk T, von Blanckenburg F (2002) Natural iron isotope variations in human blood. Science 295(15):2065–2066
42. Guelke M, von Blanckenburg F (2007) Fractionation of stable iron isotopes in higher plants. Environ Sci Technol 41(6):1896–1901

43. Zhu XK, Guo Y, Williams RJP, O'Nions RK, Matthews A, Belshaw NS, Canters GW, de Waal EC, Weser U, Burgess BK, Salvato B (2002) Mass fractionation processes of transition metal isotopes. EPSL 200:47–62
44. Hotz K, Augsburgerb H, Walczyk T (2011) Isotopic signatures of iron in body tissues as a potential biomarker for iron metabolism. J Anal At Spectrom 26:1347–1353
45. Jaouen K, Pons M-L, Balter V (2013) Iron, copper and zinc isotopic fractionation up mammal trophic chains. EPSL 374:164–172
46. Ohno T, Shinohara A, Kohge I, Chiba M, Hirata T (2004) Isotopic analysis of Fe in human red blood cells by multiple collector-ICP-mass spectrometry. Anal Sci 20:617–621
47. Walczyk T, von Blanckenburg F (2005) Deciphering the iron isotope message of the human body. Int J Mass Spectrom 242:117–134
48. Krayenbuehl P-A, Walczyk T, Schoenberg R, von Blanckenburg F, Schulthess G (2005) Hereditary hemochromatosis is reflected in the iron isotope composition of blood. Blood 105(10):3812–3816
49. Stenberg A, Malinovsky D, Ohlander B, Andren H, Forsling W, Engstrom LM, Wahlin A, Engstrom E, Rodushkin I, Baxter DC (2005) Measurement of iron and zinc isotopes in human whole blood: preliminary application to the study of HFE genotypes. J Trace Elem Med Bio 19:55–60
50. Albarède F, Telouk P, Lamboux A, Jaouen K, Balter V (2011) Isotopic evidence of unaccounted for Fe and Cu erythropoietic pathways. Metallomics 3:926–933
51. Jaouen K, Balter V, Herrscher E, Lamboux A, Telouk P, Albarède F (2012) Fe and Cu stable isotopes in archeological human bones and their relationship to sex. Am J Phys Anthropol 148:334–340
52. Hotz K, Krayenbuehl P-A, Walczyk T (2012) Mobilization of storage iron is reflected in the iron isotopic composition of blood in humans. J Biol Inorg Chem 17:301–309
53. Van Heghe LV, Engström E, Rodushkin I, Cloquet C, Vanhaecke F (2012) Isotopic analysis of the metabolically relevant transition metals Cu, Fe and Zn in human blood from vegetarians and omnivores using multi-collector ICP-mass spectrometry. J Anal At Spectrom 27:1327–1334
54. Hotz K, Walczyk T (2013) Natural iron isotopic composition of blood is an indicator of dietary iron absorption efficiency in humans. J Biol Inorg Chem 18:1–7
55. Van Heghe LV, Deltombe O, Delanghe J, Depyperec H, Vanhaecke F (2014) The influence of menstrual blood loss and age on the isotopic composition of Cu, Fe and Zn in human whole blood. J Anal At Spectrom 29:478–482
56. Van Heghe LV, Delanghe J, Vlierberghec HV, Vanhaecke F (2013) The relationship between the iron isotopic composition of human whole blood and iron status parameters. Metallomics 5:1503–1509
57. Jaouen K, Balter V (2014) Menopause effect on blood Fe and Cu isotope compositions. Am J Phys Anthropol 153:280–285
58. Hirata T (1996) Lead isotopic analyses of NIST standard reference materials using multiple collector inductively coupled plasma mass spectrometry coupled with a modified external correction method for mass discrimination effect. Analyst 121:1407–1411
59. Anbar AD, Roe JE, Barling J, Nealson KH (2000) Nonbiological fractionation of iron isotopes. Science 288:126–128
60. Sasaki S, Yanagibori R, Amano K (1998) Self-administered diet history questionnaire developed for health education: a relative validation of the test-version by comparison with 3-day diet record in women. J Epidemiol 8:203–215
61. Sasaki S, Ushio F, Amano K, Morihara M, Todoriki T, Uehara Y, Toyooka T (2000) Serum biomarker-based validation of a self-administered diet history questionnaire for Japanese subjects. J Nutr Sci Vitaminol 46:285–296
62. Gkouvatsos K, Papanikolaou G, Pantopoulos K (2012) Regulation of iron transport and the role of transferrin. Biochim Biophys Acta 1820(3):188–202

63. Kato I, Toniolo P, Akhmedkhanov A, Koenig KL, Shore R, Zeleniuch-Jacquotte A (1998) Prospective study of factors influencing the onset of natural menopause. J Clin Epidemiol 51:1271–1276
64. Zacharski LR, Ornstein DL, Woloshin S, Schwartz LM (2000) Association of age, sex, and race with body iron stores in adults: analysis of NHANES III data. Am Heart J 140(1):98–104
65. Fich A, Camilleri M, Phillips SF (1989) Effect of age on human gastric and small bowel motility. J Clin Gastroenterol 11:416–420
66. Madsen JL (1992) Effects of gender, age, and body mass index on gastrointestinal transit times. Dig Dis Sci 37:1548–1553
67. Lipski PS, Bennett MK, Kelly PJ, James OF (1992) Ageing and duodenal morphometry. J Clin Pathol 45:450–452
68. Corazza GR, Frazzoni M, Gatto MR, Gasbarrini G (1986) Ageing and small-bowel mucosa: a morphometric study. Gerontology 32:60–65
69. Krasinski SD, Russell RM, Samloff IM, Jacob RA, Dallal GE, McGandy RB, Hartz SC (1986) Fundic atrophic gastritis in an elderly population: effect on hemoglobin and several nutritional indicators. J Am Geriatr Soc 34(11):800–806
70. Cardenas VM, Mulla ZD, Ortiz M, Graham DY (2006) Iron deficiency and helicobacter pylori infection in the United States. Am J Epidemiol 163:127–134
71. Haruma K, Kamada T, Kawaguchi H, Okamoto S, Yoshihara M, Sumii K, Inoue M, Kishimoto S, Kajiyama G, Miyoshi A (2000) Effect of age and Helicobacter pylori infection on gastric acid secretion. J Gastroenterol Hepatol 15:277–283
72. Pereira SP, Gainsborough N, Dowling RH (1998) Drug-induced hypochlorhydria causes high duodenal bacterial counts in the elderly. Aliment Pharmacol Ther 12:99–104
73. Skikne BS, Lynch SR, Cook JD (1981) Role of gastric acid in food iron absorption. Gastroenterology 81:1068–1071
74. Hodgson HJF, Epstein O (2007) Malabsorption. Medicine 35(4):220–225
75. Jacobs P, Bothwell TH, Charlton RW (1964) Role of hydrochloric acid in iron absorption. J Appl Physiol 19:187–188
76. Sarker SA, Davidsson L, Mahmud H, Walczyk T, Hurrell RF, Gyr N, Fuchs GJ (2004) Helicobacter pylori infection, iron absorption, and gastric acid secretion in Bangladeshi children. Am J Clin Nutr 80:149–153
77. Beard JL, Hendricks MK, Perez EM, Murray-Kolb LE, Berg A, Vernon-Feagans L, Irlam J, Isaacs W, Sive A, Tomlinson M (2005) Maternal iron deficiency anemia affects postpartum emotions and cognition. J Nutr 135(2):267–272
78. Monsen ER, Hallberg L, Layrisse M, Hegsted DM, Cook JD, Mertz W, Finch CA (1978) Estimation of available dietary iron. Am J Clin Nutr 31(1):134–141
79. Koury MJ, Ponka P (2004) New insights into erythropoiesis: the roles of folate, vitamin B_{12}, and iron. Annu Rev Nutr 24:105–131

Chapter 13
Roles of Zinc Transporters in Cellular Transport of Cadmium and Manganese

Seiichiro Himeno and Hitomi Fujishiro

Abstract As cadmium (Cd) is a nonessential metal, there is no specific transport system for cellular entry of Cd in the organisms. The establishment of Cd-resistant cells from metallothionein-null mouse cells, application of multi-tracer technique, and microarray analyses have revealed that Cd^{2+} shares the pathway for cellular incorporation with Mn^{2+}, and the responsible transporters for this pathway were found to be ZIP8 (Zrt- and Irt-related protein 8) and ZIP14. Although other transport systems for iron or calcium are also utilized for cellular incorporation of Cd^{2+} and Mn^{2+}, characterization of ZIP8 and ZIP14 has demonstrated important physiological and pathological roles of these transporters in metal transport. We show here the significant roles of ZIP8 in segment-specific transport of Cd in proximal tubule of the kidney and the roles of ZIP14 and ZnT10 in Mn transport in neuronal cells in the presence of cytokine. Recently, critical roles of Mn transport systems have been highlighted by the findings of human diseases related to the mutation in ZIP8 and ZnT10. This chapter summarized historical background and recent advances in the studies on the roles of ZIP8 and ZIP14 in the transport of Cd^{2+} and Mn^{2+}.

Keywords Cadmium • Manganese • Zinc transporter • ZIP8 • ZIP14 • ZnT10 • Kidney • Brain

13.1 Introduction

Cadmium (Cd) is a heavy metal known as a causative agent for *Itai-itai* disease in Japan. This disease is characterized by painful (*Itai* in Japanese) bone damages including osteomalacia [1]. The primary cause of osteomalacia in this disease is the renal dysfunction called Fanconi syndrome which causes the loss of calcium (Ca) and phosphorus from the body due to the disturbance in reabsorption of these elements at the proximal tubule of the kidney. Cd accumulates in human

S. Himeno (✉) • H. Fujishiro
Laboratory of Molecular Nutrition and Toxicology, Faculty of Pharmaceutical Sciences, Tokushima Bunri University, Tokushima 770-8514, Japan
e-mail: himenos@ph.bunri-u.ac.jp

Y. Ogra, T. Hirata (eds.), *Metallomics*, DOI 10.1007/978-4-431-56463-8_13

kidney after a long-term consumption of Cd-containing diets, especially rice in Japan. The biological half-life of Cd in human kidney is assumed to be more than 20 years [2]. Thus, the elucidation of mechanisms underlying the uptake, retention, distribution, and excretion of Cd in the kidney is critical for understanding pathophysiology of Cd-induced renal damage. Until now, however, molecular mechanism of Cd transport at cellular level has not been fully elucidated.

Metallothionein (MT) is a low-molecular-weight sulfhydryl-rich protein and plays important roles in the protection against heavy metal toxicity through its high affinities for a variety of metals including Cd [3, 4]. Under normal conditions, cellular level of MT is kept at minimal level, but upon the intrusion of metals into cells, the transcription of MT gene and consequently the synthesis of MT protein are quickly accelerated. The detoxification of metals by MT is achieved through the tight binding of the metals within cells, but not through the excretion of the metals. Thus, after the induction of MT, cellular Cd concentrations are kept high, but cytotoxicity of Cd is prevented because free Cd ions cannot be readily released from MT. This mechanism is indeed beneficial to cells but also contributes to the long-lasting accumulation of Cd in cells and tissues. Unfortunately, the existence of this efficient Cd-binding molecule within cells has made it difficult for Cd researchers to elucidate the influx and efflux mechanisms of Cd.

To solve this problem, we attempted to find non-MT factors involved in Cd toxicity and transport and found that ZIP8 and ZIP14, the zinc transporter ZIP (Zrt- and Irt-related protein) family members, play important roles in Cd transport. Further characterization of these metal transporters by our group and others has demonstrated wider roles of these transporters than initially expected in the handling of Cd as well as manganese (Mn) in a variety of cells and tissues.

13.2 Cadmium Transport from a Viewpoint of Metallomics

13.2.1 Multiple Candidate Transporters for Cadmium Transport

To date, several candidates for Cd transport system have been proposed. Because Cd is not an essential element, it has been presumed that the transport systems for other essential elements are used for cellular incorporation of Cd. Transporters for Ca, iron (Fe), zinc (Zn), Mn, and magnesium (Mg) have so far been proposed to be involved in Cd uptake in mammalian cells.

As Ca has an ionic radius similar to Cd, earlier studies have focused on the roles of Ca channels in the transport of Cd. In the field of pharmacology regarding Ca channels, Cd has been used as an inhibitor of Ca channels, though the concentrations of Cd used as an inhibitor were more than 100 μM, which is a lethal dose to cells. Furthermore, Ca channels are expressed highly in neuronal cells, but these cells are not the major target of Cd toxicity. On the other hand, intestine is the tissue where Ca is efficiently absorbed from the diet. The experimental animals fed a Ca-deficient diet accumulated higher amounts of Cd from the diet, suggesting that the transport system for Ca is utilized for Cd absorption at least in the intestine [5]. Min et al. [6] showed that CAT1, calcium transporter 1, which is a member of Ca channel but behaves like a transporter, is responsible for intestinal Ca absorption.

Several studies have shown that animals fed an Fe-deficient diet accumulated higher amounts of Cd [7–9]. Intestinal Fe is known to be absorbed via DMT1, divalent metal transporter 1 also known as Nramp2, which was identified as a divalent Fe (Fe^{2+}) transporter but was found to have broad affinities for other divalent metals including Mn and Cd [10]. When an Fe-deficient diet was given to experimental animals, the expression of DMT1 was highly enhanced, suggesting the role of DMT1 in intestinal Cd absorption especially under anemic conditions [8, 11]. However, the roles of DMT1 in other tissues than the intestine remain unclear because subcellular localization of DMT1 was detected mainly in late endosomes in some cells [12].

Kidney accumulates high concentration of Cd after a long-term exposure. It has been assumed that Cd bound to MT in serum is filtered through the glomerulus of the kidney and reabsorbed in the proximal tubule cells via endocytosis [13]. The injection of Cd-MT complex directly to the blood circulation resulted in an efficient accumulation of Cd in the kidney [14, 15]. Thus, it has been considered that the primary route of renal Cd accumulation is the reabsorption of Cd-MT in the lumen into epithelial cells of proximal tubule, where Cd is retained for a long time. However, we have recently demonstrated that the Cd handling in the kidney may be more dynamic and complicated than previously expected. This will be discussed in more detail in Sect. 13.3.3.

Although fragmental information on the mechanism of Cd transport has been accumulated, molecular mechanism for cellular Cd incorporation remained unsolved. As Cd is thought to be transported via the systems for other essential metals, the identification of a metal that share the same pathway for cellular transport with Cd is required.

13.2.2 Establishment of Cadmium-Resistant Cells from Metallothionein-Null Cells

To identify endogenous factors for the protection against xenobiotic chemicals or drugs, the development of a drug-resistant cell line has been frequently utilized, and a number of transporters for drugs have been identified through this strategy. Also for Cd, a variety of resistant cells have been established and characterized. Most of the Cd-resistant cells so far established showed enhanced expression of MT, confirming an important role of MT for detoxification of Cd [16–18]. However, no information on Cd transport system has been provided by these resistant cells.

In general, the development of drug-resistant cells provides at least three types of cellular mechanisms involved in the protection against drug toxicity, reflecting the three strategic steps for acquiring resistance against the drug. The primary protective strategy for cells to survive when the drug enters cells is to reduce cellular concentrations of the drugs by expelling them out or by inhibiting their entry. A number of drug transporters have been identified by this way. If the cells could not reduce the cellular concentration of the drug, then the next strategy is to inactivate the drug within cells. Metabolic inactivation of drugs is a typical example of this strategy. Also, the trapping of metals by MT can be classified into this strategy. The third strategy is to repair the cellular damages quickly and efficiently. DNA repair enzymes could be categorized into this strategy.

According to this categorization of resistant factors, we attempted to develop a Cd-resistant cell line from the cells lacking the expression of MT, expecting the establishment of a cell line in which the alteration in Cd transport plays a primary role in acquiring Cd resistance. For this purpose, Yanagiya et al. [19] utilized SV40-immortalized MT-null cells derived from embryonic fibroblasts of MT-I and MT-II knockout mice [20]. Using these MT-null cells, Yanagiya et al. [19] established two lines of Cd-resistant MT-null cells and found that the resistance of these cells against Cd was conferred mainly by a marked decrease in Cd accumulation. These cells were the first Cd-resistant cells in which lowered Cd accumulation, but not enhanced MT expression, is the primary cause for acquisition of Cd resistance. The influx and efflux experiments revealed that a marked decrease in the initial uptake rate of Cd was the predominant reason for the decrease in Cd accumulation [19].

13.2.3 Metallomics Approach to Identify Cadmium Transporter

Since the uptake rate of Cd into cells was suppressed in Cd-resistant MT-null cells, Cd resistance of these cell lines might have been conferred by the "loss of function" of a certain metal incorporation system, which should be shared with other essential elements. To identify the element that is not efficiently incorporated into the Cd-resistant MT-null cells, we utilized the multi-tracer technique developed at RIKEN (the Institute for Physical and Chemical Research) in Japan [21, 22]. A solution of the multi-tracer is produced by the irradiation of target metal using high-energy heavy ions generated in the ring cyclotron in RIKEN (Fig. 13.1). The multi-tracer solution containing more than 20 radioisotopes was added to the media of Cd-resistant and parental cells for 120 min, and then the contents of all nuclides were determined simultaneously by a germanium detector.

By using this novel technique, we found that the incorporation of Mn^{2+} was extremely suppressed in MT-null Cd-resistant cells [23]. Although the term "metallomics" was not yet in use at that time, the multi-tracer technique is definitely

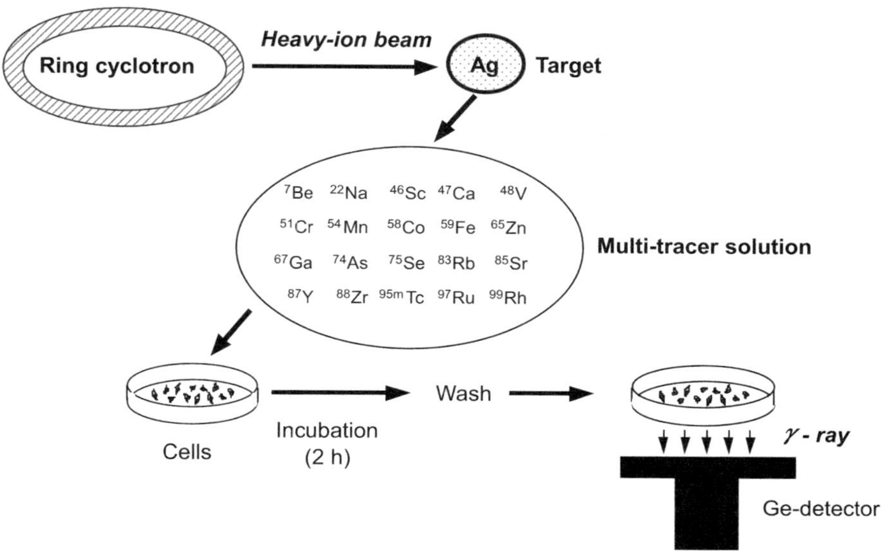

Fig. 13.1 Schematic presentation of multi-tracer technique. The solution of multi-tracer was produced by irradiation of target metal (Ag) using heavy-ion beam generated in the ring cyclotron in RIKEN. The multi-tracer solution was added to the media of cells for 120 min, and then the γ-ray emission from all nuclides can be determined simultaneously by a germanium detector

a metallomics approach and enabled us to identify the element, Mn, as the candidate metal involved in cellular Cd transport.

The decrease in Mn^{2+} uptake rate in Cd-resistant MT-null cells was also confirmed by using a single tracer of ^{54}Mn-labeled $MnCl_2$. In addition, we found that a decreased uptake of Mn^{2+} was observed only at doses less than 10 μM in Cd-resistant MT-null cells, suggesting that a high-affinity transport system for Mn^{2+} is involved in Cd^{2+} transport. In a competition assay in parental cells, the uptakes of Cd^{2+} and Mn^{2+} were mutually inhibited by each other, but not by other divalent metals such as Co^{2+}, Ni^{2+}, Fe^{2+}, or Cu^{2+}. Interestingly, the uptakes of Cd^{2+} and Mn^{2+} were significantly inhibited by Zn^{2+}. No mutual inhibition of Cd^{2+}, Mn^{2+}, and Zn^{2+} uptakes was observed in Cd-resistant MT-null cells, suggesting that the transport system having affinities for Mn^{2+}, Cd^{2+}, and probably Zn^{2+} is suppressed in Cd-resistant cells. Since mutual inhibition of Cd^{2+} and Mn^{2+} was observed also in several lines of cultured cells (HeLa, PC12, and Caco-2 cells), this transport system may exist commonly among mammalian cells.

The next question was, therefore, to identify the transporter which has affinities for Cd^{2+}, Mn^{2+}, and Zn^{2+} and is suppressed in Cd-resistant MT-null cells.

13.3 The Roles of ZIP8 and ZIP14 in Cadmium Transport

13.3.1 Identification of ZIP8 and ZIP14 as Cadmium Transporters

To identify the genes responsible for Cd resistance in Cd-resistant MT-null cells, we performed a microarray and subsequent real-time RT-PCR analyses and found that the expression of *Slc39a8* encoding ZIP8 and *Slc39a14* encoding ZIP14 was downregulated in MT-null Cd-resistant cells [24, 25]. Furthermore, the introduction of shRNA of ZIP8 into parental cells resulted in an approximately 60 % reduction in cellular Cd accumulation, whereas shRNA of DMT1, which also have affinities for Cd^{2+}, Mn^{2+}, and Zn^{2+}, did not alter Cd accumulation [26]. Among ZIP family members, ZIP8 and ZIP14 have a similarity in amino acid sequence, suggesting that both transporters have similar functions and affinities for metals [27]. These results suggest that ZIP8 and ZIP14 may be involved in the transport of Cd^{2+}, and the suppressed expression of these transporters resulted in lowered Cd accumulation in Cd-resistant MT-null cells.

During the time when we were attempting to identify the transporter involved in Cd transport, another group led by Prof. Nebert in Cincinnati University in the USA was also searching for the gene responsible for the strain difference in Cd-induced testicular hemorrhage in mice. It has been well known that injection of Cd into mice acutely causes testicular hemorrhage, but some strains of mice such as C57Bl/6 and A/J are resistant to Cd-induced hemorrhage even when very high dose of Cd was injected [28]. Dalton et al. [29] identified *Slc39a8* as the responsible gene determining the strain difference in sensitivity to Cd-induced testicular hemorrhage. Thus, two independent approaches have identified the same genes, *Slc39a8*, as a determinant of Cd resistance and transport [30].

Nebert and coworkers have shown that ectopic expression of ZIP8 in cultured cells resulted in an efficient uptake of Cd^{2+} as well as Mn^{2+} with Km values less than 10 μM for both Cd^{2+} and Mn^{2+} [31]. Similarly, ZIP14A and ZIP14B, the products of alternative splicing of *Slc39a14* gene, were expressed in mouse fetal fibroblast cells, and the uptakes of both Cd^{2+} and Mn^{2+} were found to be highly enhanced with low Km values [32]. These results demonstrated that ZIP8 and ZIP14 are the transporters that have high affinities for Cd^{2+} and Mn^{2+} in addition to Zn^{2+}.

13.3.2 Characterization of ZIP8 and ZIP14 as Cadmium Transporters

Wang et al. [33] produced transgenic mice in which BAC clone containing *Slc39a8* was expressed and determined whether ZIP8 is actually involved in Cd uptake into the target tissues of Cd in vivo. The results indicated that ZIP8 is highly expressed in the vasculature of the testis and apical membrane of proximal tubules in the kidney in ZIP8-transgenic mice. Administration of $CdCl_2$ caused severe testicular hemorrhage in ZIP8-transgenic B6 mice, but not in wild-type B6 (C57Bl/6J) mice, a strain resistant to Cd-induced testicular hemorrhage. Thus, the overexpression of ZIP8 in the kidney and testis actually contributed to the enhancement of Cd toxicity in vivo.

The roles of ZIP8 and ZIP14, especially ZIP8, in the transport of Cd^{2+} and Mn^{2+}, have been confirmed further by additional establishment of Cd-resistant cells. We developed Cd-resistant cells using immortalized mouse embryonic fibroblast cells derived from MT+/+ mice [34]. Although the expression of MT was highly induced in Cd-resistant MT+/+ cells, the accumulation of Cd was found to be less than 50 % of parental cells. Furthermore, Cd-resistant MT+/+ cells showed a cross-resistance to Mn^{2+} accompanied by the reduced accumulation of Mn in cells. The expression of ZIP8 was suppressed in Cd-resistant cells, confirming the important role of ZIP8 for the transport of both Cd^{2+} and Mn^{2+}. In addition to ZIP8, the expression of DMT1 and L-type Ca channel subunit ($Ca_V1.2$) was also suppressed in Cd-resistant MT+/+ cells. Since IC_{50} value (about 200 μM) of Cd in Cd-r MT+/+ cells was about 10 times higher than that of Cd-resistant MT-null cells, it is presumable that the suppression of multiple transporters in addition to ZIP8 was required for the survival of cells during the selection process of resistant cells with high concentration of Cd in the medium.

Further evidence for the roles of ZIP8 in Cd^{2+} and Mn^{2+} transport has been obtained by the study on RBL-2H3 (rat basophilic leukemia) cells [35]. In the screening of Cd-sensitive or resistant cell lines, we found that the IC_{50} value of $CdCl_2$ for RBL-2H3 cells was as low as 9 μM, close to that of MT-null mouse embryonic fibroblast cells (4 μM). In addition, RBL-2H3 cells showed about tenfold higher sensitivity to $MnCl_2$ compared with other cell lines. We found that RBL-2H3 cells showed the highest expression of ZIP8, the highest accumulation of Cd and Mn at 24 h, and the highest uptake efficiency of Cd^{2+} and Mn^{2+} among the rat cell lines examined. No difference was found in the excretion efficiency. Thus, RBL-2H3 cells may serve as a good model for studying the roles of ZIP8 in cellular transport of Cd^{2+} and Mn^{2+}.

By using RBL-2H3 cells, we examined the roles of HCO_3^- in transport of Cd^{2+} and Mn^{2+} via ZIP8. Liu et al. [36] reported that the addition of HCO_3^- to the medium enhanced the uptake of Cd^{2+}, and the addition of DIDS, which is an inhibitor of HCO_3^- uptake, resulted in a decrease in Cd^{2+} uptake, suggesting that Cd^{2+} is transported with HCO_3^- via ZIP8. To test whether ZIP8 is actually involved in HCO_3^--induced enhancement of Cd^{2+} and Mn^{2+} uptakes, Fujishiro et al. [35] examined the effects of siRNA knockdown of ZIP8 in RBL-2H3 cells, in which the addition of HCO_3^- in the medium also enhanced the uptakes of Cd^{2+} and Mn^{2+}. The downregulation of ZIP8 resulted in the decrease in the uptakes of Cd^{2+} and Mn^{2+} only in the presence of HCO_3^-, confirming the involvement of HCO_3^- in the transport of these metals by ZIP8. The involvement of HCO_3^- on ZIP14-mediated metal transport was also observed in *Xenopus* oocytes in which ZIP14 gene was expressed [37]. However, the downregulation of ZIP14 in RBL-2H3 cells did not alter the uptakes of Cd^{2+} and Mn^{2+} probably due to the compensatory effects of ZIP8 which is highly expressed in this cell line [35].

To further examine the roles of ZIP8, Fujishiro et al. [38] developed Cd-resistant RBL-2H3 (RBL-Cdr) cells and characterized their sensitivity to and transport of Cd^{2+} and Mn^{2+}. As expected, the expression of MT was upregulated. In addition to MT upregulation, the mRNA and protein levels of ZIP8 were markedly lowered in RBL-Cdr cells. Reflecting the low ZIP8 expression, RBL-2H3 cells showed suppressed accumulation of Cd in parallel with the decrease in Cd^{2+} uptake efficiency. Interestingly, RBL-Cdr cells showed a cross-resistance to $MnCl_2$ due to the reduced Mn accumulation in parallel with reduced efficiency of Mn^{2+} uptake efficiency. No change was observed in the expression of ZIP14 or DMT1, suggesting an important role of ZIP8 in the transport of both Cd^{2+} and Mn^{2+} in RBL-2H3 cells.

13.3.3 The Roles of ZIP8 and ZIP14 in Cadmium Transport in the Kidney

It has long been known that the target organ of Cd toxicity is the kidney, especially proximal tubule epithelial cells (PTECs). The next question, therefore, was whether ZIP8 and ZIP14 play a role in accumulation and toxicity of Cd in renal PTECs. Accumulating evidence has already shown that Cd-MT is deeply involved in renal Cd accumulation [3, 14, 15, 39]. As the molecular size of Cd-MT is about 7000, it has been shown that Cd-MT in blood circulation is readily filtered through the glomerulus, and the Cd-MT in the lumen of proximal tubule is reabsorbed by PTECs by megalin-dependent endocytosis in S1 and S2 segments of proximal tubule [13], where Cd accumulation and toxicity is clearly recognized. Indeed, this is the primary route of Cd accumulation in the kidney. However, the mechanisms of renal Cd accumulation other than Cd-MT endocytosis have been poorly understood.

To test the roles of ZIP8 in renal Cd accumulation, in vivo studies have been conducted in the mice in which ZIP8 is overexpressed. Wang et al. [33] produced ZIP8-transgenic mice using a BAC clone containing *Slc39a8* gene and showed that ZIP8 is highly expressed in the apical membrane of proximal tubules in the kidney. Administration of $CdCl_2$ caused enhanced renal damage in ZIP8-transgenic mice than in control mice. Schneider et al. [40] also produced ZIP8-transgenic mice with different copy numbers of *Slc39a8* genes and found that renal Cd accumulation was higher in transgenic mice after oral administration of Cd. These in vivo studies, however, did not elucidate molecular mechanisms of higher Cd accumulation by overexpression of ZIP8.

In an in vitro study, Fujishiro et al. [41] examined the roles of metal transporters by using immortalized mouse PTECs, named PT cells. In this study, a trans-well culture system was utilized to examine the efficiency of apical and basolateral transport of metals separately. Apical and basolateral uptakes of metals into PT cells in a trans-well culture system mimic the metal incorporation from the lumen to PTECs and that from blood to PTECs in the kidney. The results showed that the uptake of Cd^{2+} from apical side was higher than that from basolateral side. The apical uptake of Cd^{2+} was dose-dependently inhibited by Mn^{2+} and that of Mn^{2+} was similarly inhibited by Cd^{2+}, suggesting that the transporter(s) having affinity for both Cd^{2+} and Mn^{2+} is involved in the uptake of ionic forms of Cd and Mn at the apical surface of PTECs. To test which transporter is responsible for the uptake of Cd^{2+} and Mn^{2+}, siRNAs for ZIP8, ZIP14, and DMT1 were introduced into PT cells, and then the metal uptake was examined after the cells were transferred to and cultured in the trans-well culture system. The transfection of ZIP8 and ZIP14 siRNAs reduced the apical uptake of Cd^{2+} by about 50 %. The transfection of DMT1 siRNA also reduced both Cd^{2+} and Mn^{2+} uptakes, but to a less extent. These data suggest that ZIP8 and/or ZIP14 may play a significant role in the uptake of ionic forms of Cd and Mn in the lumen to PTECs in the kidney.

The application of the trans-well system also enabled the determination of Cd excretion from PT cells to the medium of apical or basolateral side [41]. The results showed that the efficiency of Cd excretion into the apical side was much higher that into the basolateral side. These data suggest a possibility that at least a part of Cd could be excreted into the lumen from PTECs in the kidney. As the excretion efficiency was examined in PT cells 1 h after the exposure to Cd, it seems unlikely that Cd was excreted in the form of Cd-MT, which could not be synthesized within 1 h.

To further examine the roles of ZIP8 and ZIP14 in the transport of Cd in the kidney, an in situ hybridization of ZIP8, ZIP14, and DMT1 was carried out in mouse kidney [41]. The results showed that the expression of ZIP8 was observed in PTECs especially in the boundary of cortex and outer medulla, where S3 segment of proximal tubule is located. On the other hand, expression of ZIP14 and DMT1 was observed diffusively in the whole kidney.

Based on the abovementioned in vitro and in vivo evidence, we proposed a hypothesis of dynamic handling of Cd in the kidney. As shown in Fig. 13.2, the Cd complexed with MT passes through the glomerulus into the lumen and is absorbed from the apical side of PTECs in S1 and S2 segments where megalin-dependent endocytosis occurs. Previously, the fate of the accumulated Cd in PTECs has not been examined based on the notion that Cd bound to MT remains in PTECs for a long time. By using a trans-well culture system, however, a novel possibility was proposed that a part of Cd accumulated in PTECs may be released into the lumen in the form of ionic Cd, which may then be absorbed into PTECs in S3 segment where ZIP8 is highly expressed.

Pharmacodynamics studies of urate have already shown that the urate filtered through the glomerulus into the lumen will be absorbed into PTECs in S1 and S2 segments, then released into the lumen, absorbed again in PTECs in S3 segment, and released again into the lumen [42]. In the case of Cd, this kind of multi-compartment model of renal handling has not been anticipated to happen because Cd was believed to be stably deposited in PTECs after the endocytosis of Cd-MT.

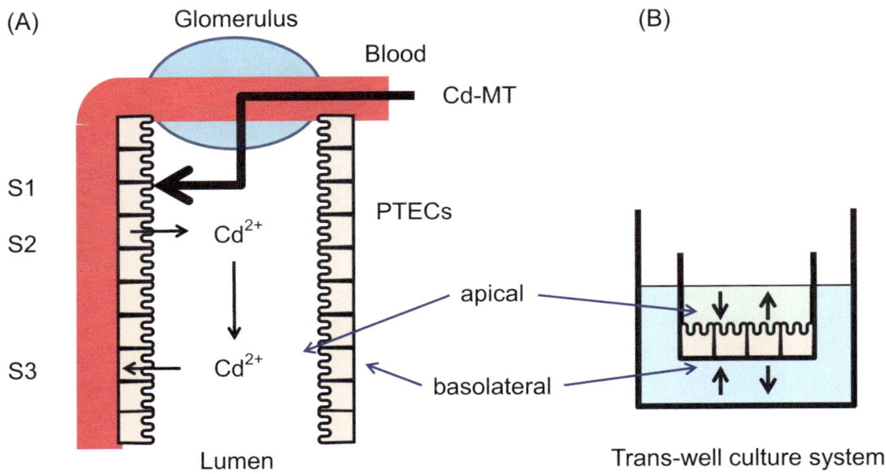

Fig. 13.2 Schematic presentation of hypothesis on Cd transport in the proximal tubule of the kidney. (**a**) Cd-MT in serum is filtered through the glomerulus and reabsorbed by endocytosis in PTECs in S1 and S2 segments. A part of Cd, possibly in the form of Cd^{2+}, is released into the lumen and then absorbed again in PTECs in S3 segment where the expression of ZIP8 is high. (**b**) To mimic the apical and basolateral transport of Cd in PTECs, a trans-well culture system was used. The cells seeded on the cup will gradually form tight junctions and apical and basolateral sides will be polarized. The influx and efflux of Cd in both sides can be determined by this culture system

However, a study on Cu and Cu-MT metabolism in the kidney has already suggested a possibility similar to our notion. Okabe et al. [43] showed that after the intravenous injection of Cu-MT, CuMT protein incorporated by endocytosis was detected in PTECs in S1 and S2 segments, but the mRNA synthesis of MT gene was detected later in PTECs in S3 segment, and suggested that Cu ions were released from the S1 and S2 segments into the lumen possibly via ATP7A or ATP7B, then reabsorbed into S3 segment, and induced MT mRNA transcription there. To test whether similar mechanisms are involved in Cd handling in the kidney or not, we are currently investigating the transport of Cd in S1, S2, and S3-derived PTECs.

13.4 The Roles of ZIP8 and ZIP14 in Manganese Transport

13.4.1 Interactions of Cd and Mn Transport in Nonmammalian Species

Before the discovery of ZIP8 and ZIP14 as a Cd/Mn co-transporter, no researchers in the field of Cd toxicity have considered that Cd is transported via the pathway for

Mn in mammals. However, in nonmammalian species including bacteria, yeast, and plants, accumulating evidence has shown the existence of the transporters having affinities for both Cd and Mn.

In bacteria, the Mn^{2+} transporter MntH in *Bacillus subtilis* [44] and MntA in *Lactobacillus plantarum* [45] were shown to have affinity for Cd^{2+}. The nucleotide sequence of MntH is homologous to Nramp family transporter and that of MntA is homologous to P-type cation-translocating ATPase family.

In yeast, SMF1 is a well-known metal transporter having affinities for various metals including Mn^{2+}, Cu^{2+}, Co^{2+}, and Cd^{2+}. The disruption of *smf1* gene resulted in a decrease in Mn^{2+} uptake, while the overexpression of *smf1* gene resulted in an increase in Mn^{2+} uptake [46]. The amino acid sequence of SMF1 showed a homology to mammalian Nramp family proteins.

Since Mn plays a crucial role in the photosynthetic reaction, this element is essential for the survival of plants and photosynthetic organisms. To identify the gene for Mn^{2+} transport in the plants, *smf1* null strain of *S. cerevisiae* was utilized. Expression of IRT1, the Fe^{2+} transporter gene of *Arabidopsis thaliana* in *smf1* null yeast restored the uptake of Mn^{2+} [47]. The enhanced uptake of Mn^{2+} by IRT1 in *smf1* null mutant yeast was inhibited by the addition of Fe^{2+}, Cu^{2+}, Zn^{2+}, and Cd^{2+}, suggesting that plant Fe^{2+} transporter IRT1 has broad affinities for metals including Mn^{2+} and Cd^{2+}. IRT1 has no homology to P-type ATPase or Nramp family proteins. It should be noted here that mammalian ZIP transporter was named Zrt- and Irt-related protein (ZIP) according to the similarity of the functions and structure to Irt.

Recently, a new piece of evidence demonstrated an important role of Mn^{2+} transport system for Cd^{2+} uptake in rice. Two groups in Japan independently found that Nramp5 in rice is responsible for the uptakes of Mn^{2+} and Cd^{2+} from the root. Sasaki et al. [48] showed that rice Nramp5 is localized at the distal side of both exodermis and endodermis cells of the root and knockout of Nramp5 resulted in a marked decrease in Mn accumulation in the grains and shoot of rice. Cd uptake in Nramp5−/− rice was almost completely suppressed. Ishikawa et al. [49] used a technique to induce mutations in rice grain by using carbon ion beam irradiation and selected two lines of mutated rice, in which Cd accumulation was extremely low even when cultured in Cd-contaminated soil. The mutated gene responsible for the lowered Cd uptake was identified as Nramp5 gene. Thus, two independent studies revealed that the uptake system for Mn by Nramp5 is used for Cd accumulation in rice. The disruption of Nramp5-mediated Cd transport in rice may pave a novel way to reduce the body burden of Cd in humans, especially among rice-eating Asians who accumulate considerable amounts of Cd in their kidney.

13.4.2 RBL-2H3 Cell Line as a Model of Cellular Manganese Transport

As mentioned in 3.2., RBL-2H3 cell line showed a high sensitivity to both Cd^{2+} and Mn^{2+} due to high accumulation of both metals via highly expressed ZIP8. To further elucidate the mechanisms underlying the sensitivity of RBL-2H3 cells to Mn cytotoxicity, Fujishiro et al. [38] developed an Mn-resistant cell line from parental RBL-2H3 cells by a stepwise increase in $MnCl_2$ concentration in the medium. The established Mn-resistant cell line (RBL-Mnr) showed a decrease in ZIP8 expression accompanied by a decrease in Mn accumulation. RBL-Mnr also showed a resistance to $CdCl_2$ cytotoxicity due to reduced accumulation of Cd. Thus, RBL-Mnr showed a cross-resistance to Cd, and RBL-Cdr showed a cross-resistance to Mn as mentioned in Sect. 13.3.2, both due to the decrease in metal accumulation by the suppression of ZIP8 expression.

In addition to the suppression of ZIP8 expression, there should be other changes in gene expression in RBL-Mnr. A microarray assay identified several genes downregulated or upregulated in RBL-Mnr cells compared with parental RBL-2H3 cells. Fujishiro et al. [50] found that the expression of S100A9 was markedly reduced in RBL-Mnr cells. S100A9 is the member of S100 protein and is highly expressed in neutrophils. Recent advances in the studies on the roles of S100 proteins in the protection against bacterial infection suggest that S100A9 together with S100A8 has an ability to bind Zn^{2+} and Mn^{2+} and causes bacteriocidal effects by sequestering these metals [51, 52]. Although the roles of the suppression of S100A9 expression in the acquisition of Mn tolerance remains unknown, further elucidation of interactions of these Mn-binding proteins and Mn cytotoxicity may shed light on novel roles of S100A9 proteins in cellular Mn handling.

13.4.3 Manganese Transport in Neuronal Cells

Mn is essential in many organisms, but excess amount of Mn is toxic to the nervous system. Occupational exposure to Mn dusts and fumes has long been known to cause manganism, a Parkinsonism-like syndrome [53]. However, the precise mechanism of Mn accumulation in the brain remains unsolved. Previously it was considered that the predominant chemical form of Mn in the blood circulation is Mn $^{3+}$, which is bound to transferrin similarly to Fe^{3+}, and is transported into neuronal cells via the endocytosis by transferrin receptor. However, speciation analyses of Mn in the serum and cerebrospinal fluid (CSF) showed that considerable amounts of Mn exist in the form of Mn-citrate especially in CSF [54]. Citrate can form a complex with Mn^{2+} [55]. As to the responsible transporter for neuronal Mn^{2+} uptake, many studies have considered DMT1 as a primary transporter [56, 57]. Indeed, DMT1 is known to have a high affinity for Mn^{2+} [10], and the enhanced expression of DMT1 in haemochromatosis showed increased accumulation of Mn [58]. To date, however, little information is available for the roles of ZIP8 or ZIP14 in the transport of Mn in the nervous system.

Fujishiro et al. [59] examined the roles of DMT1, ZIP8, and ZIP14 in the uptake of Mn^{2+} by using SH-SY5Y cells, a model cell line for dopaminergic cells. As expected, the transfection of DMT1 siRNA into SH-SY5Y cells resulted in the reduction of Mn^{2+} uptake. On the other hand, no effects of ZIP8 siRNA was observed probably due to a very low expression level of ZIP8 at basal level in this cell line. Interestingly, the transfection of ZIP14 siRNA resulted in a decrease in Mn^{2+} uptake by about 50 % even in the presence of DMT1, suggesting that ZIP14 may be at least partly involved in Mn^{2+} uptake in SH-SY5Y cells.

An earlier study showed that the expression of ZIP14 in the liver is enhanced by interleukin-6 (IL-6), and consequently the serum Zn is taken up by the liver, resulting in higher hepatic Zn and lower serum Zn [60]. These results explained the observation of low serum Zn levels at the time of inflammation. If the expression of ZIP14 is enhanced by IL-6 also in the nervous system, the incorporation of Mn^{2+} into neuronal cells by ZIP14 may be enhanced. To test this hypothesis, Fujishiro et al. [59] examined the effect of IL-6 on the expression of ZIP14 and Mn accumulation in SH-SY5Y cells. As expected, both mRNA and protein levels of ZIP14 were elevated by IL-6 treatment and accumulation of Mn at 24 h was increased to almost threefold in SH-SY5Y cells. On the other hand, expression of DMT1 was not changed by IL-6. Thus, although DMT1 may contribute significantly to the uptake of Mn^{2+} under normal conditions, the contribution of ZIP14 to Mn accumulation into neuronal cells may be enhanced in the presence of IL-6 at the time of inflammation.

These results provided new insights into the interactions of cytokines and metal accumulation in the brain of neurodegenerative diseases. It has been well documented that brain or CSF levels of cytokines including IL-6 are enhanced in a variety of neurodegenerative diseases such as Parkinsonism and Alzheimer's disease [61]. The enhanced production of cytokines including IL-6 in these patients may induce ZIP14 expression also in the brain and affect the metal deposition in the brain. Since recent studies showed that ZIP14 may also play a role in Fe transport [62], the upregulation of ZIP14 may affect the deposition of not only Zn and Mn but also Fe in neuronal cells.

On the other hand, exposure to Mn itself induces IL-6 expression in the liver [63] and in microglia [64]. Thus, it seems likely that during the process of brain damage development by occupational exposure to Mn, the neuronal Mn accumulation is further accelerated by IL-6-induced ZIP14. More attention should be paid to the interaction of cytokine-induced ZIP14 upregulation with aberrant metal deposition in the brain of neurodegenerative disease patients as well as workers exposed to Mn.

13.5 Human Diseases Related to Disturbances in Manganese Transport

13.5.1 ZnT10 Mutation and Hyperaccumulation of Manganese

In 2012, a new piece of evidence was provided on the roles of metal transporter in the development of brain diseases. Two independent studies have demonstrated that ZnT10 mutation is the cause for the hyperaccumulation of Mn in the liver and brain, resulting in manganism syndromes characterized by neurological disturbances similar to Parkinsonism [65, 66]. ZnT (encoded by *SLC30a* genes) is a family of Zn transporters involved in cellular transport of Zn especially from cytosol to the organelle or extracellular spaces [27]. Although the precise mechanism of Mn transport via ZnT10 remains to be clarified, the human cases of ZnT10 mutation and aberrant accumulation of Mn suggested that ZnT10 is a transporter responsible for Mn excretion from cells to extracellular spaces.

Currently, little information is available on the regulation mechanism of ZnT10 expression including the effects of cytokines. As mentioned in 4.3., treatment of SH-SY5Y cells with IL-6 enhanced cellular Mn accumulation to about threefold at 24 h [59]. However, the initial Mn uptake rate was increased to only to about 30 %. Therefore, there is a possibility that not only the uptake but also the excretion of Mn may be altered by IL-6 treatment. To test this possibility, the effects of IL-6 treatment on mRNA and protein levels of ZnT10 in SH-SY5Y cells were examined [59]. The results showed that IL-6 reduced the expression of ZnT10 by about 50 %, suggesting that the increase in 24-h Mn accumulation by IL-6 treatment was achieved not only by the increase in initial Mn uptake rate caused by ZIP14 upregulation but also by the decrease in the excretion of Mn caused by downregulation of ZnT10. It is noteworthy that the suppression of ZnT10 expression was found in the frontal cortex of Alzheimer's disease patient [67]. However, further studies are required to elucidate the regulatory mechanism of ZnT10 and its involvement in aberrant Mn accumulation in the brain.

13.5.2 ZIP8 Mutation and Disorders of Glycosylation

Recently, a clinical case of congenital disorders of glycosylation was found to be related to the mutation in ZIP8 [68]. An infant showing multiple metabolic disorders was diagnosed with congenital disorders of glycosylation as evidenced by abnormal glycosylation pattern of serum transferrin with the loss of galactose residues. Supplementation of a large dose of galactose was effective in curing the symptoms. The known important transporter for glycosylation is UDP-galactose transporter, encoded by *SLC35A2*, responsible for the Golgi import of UDP-galactose. However, no mutation was detected in *SLC35A2* gene. Whole exome sequencing was performed, and finally the mutation in *SLC39A8* encoding ZIP8 was identified as the likely candidate for causing defective glycosylation.

The transfer of galactose to acceptor proteins is carried out by a group of enzymes called galactosyltransferases (GalT) located in the Golgi apparatus. GalT activation is completely dependent on metal ions, and Mn^{2+} was shown to be the most potent activator. Since the metal-binding site of GalT has a high affinity for Mn^{2+} (Km value of 0.4 μM), the loss of Mn^{2+} leads to a marked decrease in enzyme activity. The serum and urinary concentrations of Mn in this patient were non-detectable levels, although the concentrations of Zn and other metals were within normal ranges. No biochemical changes, however, were observed in other Mn-dependent enzymes such as arginase or glutamine synthetase. The mutation of ZIP8 in this patient was identified as ^{340}Ile to Asn, which is located in the transmembrane domain V of this protein.

Similar cases of glycosylation disorder were found in Egypt with no detectable Mn in serum [69]. Currently, however, no explanation is available why ZIP8 mutation resulted in undetectable serum Mn level. Further studies are required for the elucidation of the link between ZIP8 mutation and severe deficiency of Mn leading to a specific disturbance in GalT activity.

13.6 Conclusion

More than a decade ago, we started a study to solve a simple question "how Cd enters cells." This simple question led us to the finding of Mn as a candidate metal that shares the pathway for cellular incorporation with Cd. This was achieved by using a multi-tracer technique, a powerful tool for metallomics studies. Genetic studies have enabled us to identify ZIP8 and ZIP14 as the transporters responsible for the cellular uptake of Cd^{2+} and Mn^{2+}. Recent studies showed that ZnT10 is a transporter for the excretion of Mn from cells. A number of studies have demonstrated that ZIP and ZnT family transporters play important roles as the regulators of Zn ions across the plasma membranes of cells as well as between the organelle and cytosol [70]. However, the important roles of ZIP8 and ZIP14 as well as ZnT10 were highlighted by their affinities for multiple metals. Recently, the human diseases related to the disturbances in Mn metabolism have been found to be caused by mutations in ZnT10 and ZIP8. These findings will further accelerate the studies on molecular mechanisms of regulation, subcellular localization, and pathophysiological roles of these multi-metal transporters.

Novel findings in the important roles of ZIP8 and ZIP14 in the transport of Cd and Mn have also accelerated the mechanistic studies in the field of heavy metal toxicology. As the toxicity manifestation of metals largely depends on the amounts of metals deposited in the tissues, cells, and organelles, further studies on the roles of metal transporters in dynamic handling of metals not only between cells and extracellular fluids but also between the organelles and cytosol are required in a future study.

References

1. Yanaga M, Enomoto S, Hirunuma R, Furuta R, Endo K, Tanaka A, Ambe S, Tozawa M, Ambe F (1996) Multitracer study on uptake and excretion of trace elements in rats. Appl Radiat Isot 47(2):235–240
2. Yanagiya T, Imura N, Enomoto S, Kondo Y, Himeno S (2000) Suppression of a high-affinity transport system for manganese in cadmium-resistant metallothionein-null cells. J Pharmacol Exp Ther 292(3):1080–1086
3. Schneider SN, Liu Z, Wang B, Miller ML, Afton SE, Soleimani M, Nebert DW (2014) Oral cadmium in mice carrying 5 versus 2 copies of the *Slc39a8* gene: comparison of uptake, distribution, metal content, and toxicity. Int J Toxicol 33(1):14–20
4. Fujishiro H, Doi M, Enomoto S, Himeno S (2011) High sensitivity of RBL-2H3 cells to cadmium and manganese: an implication of the role of ZIP8. Metallomics 3(7):710–718
5. Que Q, Helmann JD (2000) Manganese homeostasis in *Bacillus subtilis* is regulated by MntR, a bifunctional regulator related to the diphtheria toxin repressor family of proteins. Mol Microbiol 35(6):1454–1468
6. Hayden JA, Brophy MB, Cunden LS, Nolan EM (2013) High-affinity manganese coordination by human calprotectin is calcium-dependent and requires the histidine-rich site formed at the dimer interface. J Am Chem Soc 135(2):775–787
7. He L, Girijashanker K, Dalton TP, Reed J, Li H, Soleimani M, Nebert DW (2006) ZIP8, member of the solute-carrier-39 (SLC39) metal-transporter family: characterization of transporter properties. Mol Pharmacol 70(1):171–180
8. Matzapetakis M, Karligiano N, Bino A, Dakanali M, Raptopoulou CP, Tangoulis V, Terzis A, Giapintzakis J, Salifoglou A (2000) Manganese citrate chemistry: syntheses, spectroscopic studies, and structural characterizations of novel mononuclear, water-soluble manganese citrate complexes. Inorg Chem 39(18):4044–4051
9. Boycott KM, Beaulieu CL, Kernohan KD, Gebril OH, Mhanni A, Chudley AE, Redl D, Qin W, Hampson S, Küry S, Tetreault M, Puffenberger EG, Scott JN, Bezieau S, Reis A, Uebe S, Schumacher J, Hegele RA, McLeod DR, Gálvez-Peralta M, Majewski J, Ramaekers VT, Care4Rare Canada Consortium, Nebert DW, Innes AM, Parboosingh JS, Abou Jamra R (2015) Autosomal-recessive intellectual disability with cerebellar atrophy syndrome caused by mutation of the manganese and zinc transporter gene SLC39A8. Am J Hum Genet 97 (6):886–893
10. Fukada T, Kambe T (2011) Molecular and genetic features of zinc transporters in physiology and pathogenesis. Metallomics 3(7):662–674
11. Wang B, Schneider SN, Dragin N, Girijashanker K, Dalton TP, He L, Miller ML, Stringer KF, Soleimani M, Richardson DD, Nebert DW (2007) Enhanced cadmium-induced testicular necrosis and renal proximal tubule damage caused by gene-dose increase in a *Slc39a8*-transgenic mouse line. Am J Physiol Cell Physiol 292(4):C1523–C1535
12. Grady DL, Moyzis RK, Hildebrand CE (1987) Molecular and cellular mechanisms of cadmium resistance in cultured cells. Exp Suppl 52:447–456
13. Gunshin H, Mackenzie B, Berger UV, Gunshin Y, Romero MF, Boron WF, Nussberger S, Gollan JL, Hediger MA (1997) Cloning and characterization of a mammalian proton-coupled metal-ion transporter. Nature 388(6641):482–488
14. Guilarte TR (2010) Manganese and Parkinson's disease: a critical review and new findings. Environ Health Perspect 118(8):1071–1080
15. Kim J, Buckett PD, Wessling-Resnick M (2013) Absorption of manganese and iron in a mouse model of hemochromatosis. PLoS One 8(5):e64944
16. Klaassen CD, Liu J, Diwan BA (2009) Metallothionein protection of cadmium toxicity. Toxicol Appl Pharmacol 238(3):215–220
17. Liuzzi JP, Lichten LA, Rivera S, Blanchard RK, Aydemir TB, Knutson MD, Ganz T, Cousins RJ (2005) Interleukin-6 regulates the zinc transporter Zip14 in liver and contributes to the hypozincemia of the acute-phase response. Proc Natl Acad Sci U S A 102(19):6843–6848

18. Korshunova YO, Eide D, Clark WG, Guerinot ML, Pakrasi HB (1999) The IRT1 protein from Arabidopsis thaliana is a metal transporter with a broad substrate range. Plant Mol Biol 40 (1):37–44
19. Quadri M, Federico A, Zhao T, Breedveld GJ, Battisti C, Delnooz C, Severijnen LA, Di Toro ML, Mignarri A, Monti L, Sanna A, Lu P, Punzo F, Cossu G, Willemsen R, Rasi F, Oostra BA, van de Warrenburg BP, Bonifati V (2012) Mutations in SLC30A10 cause parkinsonism and dystonia with hypermanganesemia, polycythemia, and chronic liver disease. Am J Hum Genet 90(3):467–477
20. Suzuki T, Momoi K, Hosoyamada M, Kimura M, Shibasaki T (2007) Normal cadmium uptake in microcytic anemia *mk/mk* mice suggests that DMT1 is not the only cadmium transporter *in vivo*. Toxicol Appl Pharmacol 227(3):462–467
21. Hao Z, Chen S, Wilson DB (1999) Cloning, expression, and characterization of cadmium and manganese uptake genes from *Lactobacillus plantarum*. Appl Environ Microbiol 65 (11):4746–4752
22. Yu CW, Chen HC, Lin LY (1998) Transcription of metallothionein isoform promoters is differentially regulated in cadmium-sensitive and -resistant CHO cells. J Cell Biochem 68 (2):174–185
23. Ishikawa S, Ishimaru Y, Igura M, Kuramata M, Abe T, Senoura T, Hase Y, Arao T, Nishizawa NK, Nakanishi H (2012) Ion-beam irradiation, gene identification, and marker-assisted breeding in the development of low-cadmium rice. Proc Natl Acad Sci U S A 109(47):19166–19171
24. Fujishiro H, Yano Y, Takada Y, Tanihara M, Himeno S (2012) Roles of ZIP8, ZIP14, and DMT1 in transport of cadmium and manganese in mouse kidney proximal tubule cells. Metallomics 4(7):700–708
25. Sabolic I, Ljubojevic M, Herak-Kramberger CM, Brown D (2002) Cd-MT causes endocytosis of brush-border transporters in rat renal proximal tubules. Am J Physiol Renal Physiol 283(6): F1389–F1402
26. Enomoto S (2005) Development of multitracer technology and application studies on biotrace element research. Biomed Res Trace Elements 16(3):233–240
27. Nebert DW, Gálvez-Peralta M, Hay EB, Li H, Johansson E, Yin C, Wang B, He L, Soleimani M (2012) ZIP14 and ZIP8 zinc/bicarbonate symporters in *Xenopus* oocytes: characterization of metal uptake and inhibition. Metallomics 4(11):1218–1225
28. Roth JA, Garrick MD (2003) Iron interactions and other biological reactions mediating the physiological and toxic actions of manganese. Biochem Pharmacol 66(1):1–13
29. Horiguchi H, Aoshima K, Oguma E, Sasaki S, Miyamoto K, Hosoi Y, Katoh T, Kayama F (2010) Latest status of cadmium accumulation and its effects on kidneys, bone, and erythropoiesis in inhabitants of the formerly cadmium-polluted Jinzu River Basin in Toyama, Japan, after restoration of rice paddies. Int Arch Occup Environ Health 83(8):953–970
30. Himeno S, Yanagiya T, Fujishiro H (2009) The role of zinc transporters in cadmium and manganese transport in mammalian cells. Biochimie 91(10):1218–1222
31. Bosomworth HJ, Adlard PA, Ford D, Valentine RA (2013) Altered expression of ZnT10 in Alzheimer's disease brain. PLoS One 8(5):e65475
32. Park JH, Hogrebe M, Grüneberg M, DuChesne I, von der Heiden AL, Reunert J, Schlingmann KP, Boycott KM, Beaulieu CL, Mhanni AA, Innes AM, Hörtnagel K, Biskup S, Gleixner EM, Kurlemann G, Fiedler B, Omran H, Rutsch F, Wada Y, Tsiakas K, Santer R, Nebert DW, Rust S, Marquardt T (2015) SLC39A8 deficiency: A disorder of manganese transport and glycosylation. Am J Hum Genet 97(6):894–903
33. Valberg LS, Sorbie J, Hamilton DL (1976) Gastrointestinal metabolism of cadmium in experimental iron deficiency. Am J Physiol 231(2):462–467
34. Squibb KS, Pritchard JB, Fowler BA (1984) Cadmium-Metallothionein nephropathy: relationships between ultrastructural/biochemical alterations and intracellular cadmium binding. J Pharmacol Exp Ther 229(1):311–321
35. Chen P, Chakraborty S, Mukhopadhyay S, Lee E, Paoliello MM, Bowman AB, Aschner M (2015) Manganese homeostasis in the nervous system. J Neurochem 134(4):601–610

36. Vallee BL (1995) The function of metallothionein. Neurochem Int 27:23–33
37. Fujishiro H, Yoshida M, Nakano Y, Himeno S (2014) Interleukin-6 enhances manganese accumulation in SH-SY5Y cells: implications of the up-regulation of ZIP14 and the down-regulation of ZnT10. Metallomics 6(4):944–949
38. Okabe M, Nakayama K, Kurasaki M, Yamasaki F, Aoyagi K, Yamanoshita O, Sato S, Okui T, Ohyama T, Kasai N (1996) Direct visualization of copper-metallothionein in LEC rat kidneys: application of autofluorescence signal of copper-thiolate cluster. J Histochem Cytochem 44 (8):865–873
39. Kobayashi K, Kuroda J, Shibata N, Hasegawa T, Seko Y, Satoh M, Tohyama C, Takano H, Imura N, Sakabe K, Fujishiro H, Himeno S (2007) Induction of metallothionein by manganese is completely dependent on interleukin-6 production. J Pharmacol Exp Ther 320(2):721–727
40. Nordberg GF, Goyer R, Nordberg M (1975) Comparative toxicity of cadmium-metallothionein and cadmium chloride on mouse kidney. Arch Pathol 99(4):192–197
41. Tabuchi M, Yoshimori T, Yamaguchi K, Yoshida T, Kishi F (2000) Human NRAMP2/DMT1, which mediates iron transport across endosomal membranes, is localized to late endosomes and lysosomes in HEp-2 cells. J Biol Chem 275(29):22220–22228
42. Schäfer SG, Schwegler U, Schümann K (1990) Retention of cadmium in cadmium-naive normal and iron-deficient rats as well as in cadmium-induced iron-deficient animals. Ecotoxicol Environ Saf 20(1):71–81
43. Sasaki A, Yamaji N, Yokosho K, Ma JF (2012) Nramp5 is a major transporter responsible for manganese and cadmium uptake in rice. Plant Cell 24(5):2155–2167
44. Dalton TP, Miller ML, Wu X, Menon A, Cianciolo E, McKinnon RA, Smith PW, Robinson LJ, Nebert DW (2000) Refining the mouse chromosomal location of *Cdm*, the major gene associated with susceptibility to cadmium-induced testicular necrosis. Pharmacogenetics 10 (2):141–151
45. Kondo Y, Yanagiya T, Himeno S, Yamabe Y, Schwartz D, Akimoto M, Lazo JS, Imura N (1999) Simian virus 40-transformed metallothionein null cells showed increased sensitivity to cadmium but not to zinc, copper, mercury or nickel. Life Sci 64(11):PL145–PL150
46. Fujishiro H, Okugaki S, Yasumitsu S, Enomoto S, Himeno S (2009) Involvement of DNA hypermethylation in down-regulation of the zinc transporter ZIP8 in cadmium-resistant metallothionein-null cells. Toxicol Appl Pharmacol 241(2):195–201
47. Michalke B, Lucio M, Berthele A, Kanawati B (2013) Manganese speciation in paired serum and CSF samples using SEC-DRC-ICP-MS and CE-ICP-DRC-MS. Anal Bioanal Chem 405 (7):2301–2309
48. Filipov NM, Seegal RF, Lawrence DA (2005) Manganese potentiates in vitro production of proinflammatory cytokines and nitric oxide by microglia through a nuclear factor κB-dependent mechanism. Toxicol Sci 84(1):139–148
49. Yanagiya T, Imura N, Kondo Y, Himeno S (1999) Reduced uptake and enhanced release of cadmium in cadmium-resistant metallothionein null fibroblasts. Life Sci 65(14):PL177–PL182
50. Washko P, Cousins RJ (1977) Role of dietary calcium and calcium binding protein in cadmium toxicity in rats. J Nutr 107(5):920–928
51. Min KS, Ueda H, Tanaka K (2008) Involvement of intestinal calcium transporter 1 and metallothionein in cadmium accumulation in the liver and kidney of mice fed a low-calcium diet. Toxicol Lett 176(1):85–92
52. Smith JA, Das A, Ray SK, Banik NL (2012) Role of pro-inflammatory cytokines released from microglia in neurodegenerative diseases. Brain Res Bull 87(1):10–20
53. Fujishiro H, Okugaki S, Kubota K, Fujiyama T, Miyataka H, Himeno S (2009) The role of ZIP8 down-regulation in cadmium-resistant metallothionein-null cells. J Appl Toxicol 29 (5):367–373
54. Park JD, Cherrington NJ, Klaassen CD (2002) Intestinal absorption of cadmium is associated with divalent metal transporter 1 in rats. Toxicol Sci 68(2):288–294

55. Dalton TP, He L, Wang B, Miller ML, Jin L, Stringer KF, Chang X, Baxter CS, Nebert DW (2005) Identification of mouse SLC39A8 as the transporter responsible for cadmium-induced toxicity in the testis. Proc Natl Acad Sci U S A 102(9):3401–3406
56. Liu XF, Supek F, Nelson N, Culotta VC (1997) Negative control of heavy metal uptake by the *Saccharomyces cerevisiae* BSD2 gene. J Biol Chem 272(18):11763–11769
57. Tuschl K, Clayton PT, Gospe SM Jr, Gulab S, Ibrahim S, Singhi P, Aulakh R, Ribeiro RT, Barsottini OG, Zaki MS, Del Rosario ML, Dyack S, Price V, Rideout A, Gordon K, Wevers RA, Chong WK, Mills PB (2012) Syndrome of hepatic cirrhosis, dystonia, polycythemia, and hypermanganesemia caused by mutations in SLC30A10, a manganese transporter in man. Am J Hum Genet 90(3):457–466
58. Fujishiro H, Kubota K, Inoue D, Inoue A, Yanagiya T, Enomoto S, Himeno S (2011) Cross-resistance of cadmium-resistant cells to manganese is associated with reduced accumulation of both cadmium and manganese. Toxicology 280(3):118–125
59. Dantzler WH (2005) Challenges and intriguing problems in comparative renal physiology. J Exp Biol 208(Pt 4):587–594
60. Fujishiro H, Okugaki S, Nagao S, Satoh M, Himeno S (2006) Characterization of gene expression profiles of metallothionein-null cadmium-resistant cells. J Health Sci 52 (3):292–299
61. Wolff NA, Abouhamed M, Verroust PJ, Thévenod F (2006) Megalin-dependent internalization of cadmium-metallothionein and cytotoxicity in cultured renal proximal tubule cells. J Pharmacol Exp Ther 318(2):782–791
62. Elinder CG, Lind B, Kjellström T, Linnman L, Friberg L (1976) Cadmium in kidney cortex, liver, and pancreas from Swedish autopsies. Estimation of biological half time in kidney cortex, considering calorie intake and smoking habits. Arch Environ Health 31(6):292–302
63. Cigliano S, Remondelli P, Minichiello L, Mellone MC, Martire G, Bonatti S, Leone A (1996) Analysis of metal-regulated metallothionein and heat shock gene expression in HeLa-derived cadmium-resistant cells. Exp Cell Res 228(2):173–180
64. Jenkitkasemwong S, Wang CY, Coffey R, Zhang W, Chan A, Biel T, Kim JS, Hojyo S, Fukada T, Knutson MD (2015) SLC39A14 is required for the development of hepatocellular iron overload in murine models of hereditary hemochromatosis. Cell Metab 22(1):138–150
65. Clark HL, Jhingran A, Sun Y, Vareechon C, de Jesus CS, Skaar EP, Chazin WJ, Calera JA, Hohl TM, Pearlman E (2016) Zinc and manganese chelation by neutrophil S100A8/A9 (calprotectin) limits extracellular *Aspergillus fumigatus* hyphal growth and corneal infection. J Immunol 196(1):336–344
66. Liu Z, Li H, Soleimani M, Girijashanker K, Reed JM, He L, Dalton TP, Nebert DW (2008) Cd^{2+} versus Zn^{2+} uptake by the ZIP8 HCO_3^--dependent symporter: kinetics, electrogenicity and trafficking. Biochem Biophys Res Commun 365(4):814–820
67. Fukada T, Kambe T (eds) (2014) Zinc signals in cellular functions and disorders. Springer, Tokyo
68. Fujishiro H, Ohashi T, Takuma M, Himeno S (2013) Down-regulation of S100A9 and S100A10 in manganese-resistant RBL-2H3 cells. J Toxicol Sci 38(5):753–757
69. Girijashanker K, He L, Soleimani M, Reed JM, Li H, Liu Z, Wang B, Dalton TP, Nebert DW (2008) *Slc39a14* gene encodes ZIP14, a metal/bicarbonate symporter: similarities to the ZIP8 transporter. Mol Pharmacol 73(5):1413–1423
70. Fujishiro H, Ohashi T, Takuma M, Himeno S (2013) Suppression of ZIP8 expression is a common feature of cadmium-resistant and manganese-resistant RBL-2H3 cells. Metallomics 5 (5):437–444

Chapter 14
Link Between Metal Homeostasis and Neurodegenerative Diseases: Crosstalk of Metals and Amyloidogenic Proteins at the Synapse

Dai Mizuno and Masahiro Kawahara

Abstract Increasing evidence suggests that disruption of metal homeostasis contributes to the pathogenesis of various neurodegenerative diseases, including Alzheimer's disease, Lewy body diseases, vascular dementia, and prion diseases. Conformational changes of disease-related proteins such as ß-amyloid protein, α-synuclein, and prion proteins are well-established contributors to the synaptotoxicity, neurotoxicity, and pathogenesis of these diseases. Recent studies have revealed that these proteins are metalloproteins that coexist in synapses and play significant roles in the maintenance of metal homeostasis in synapses. Trace elements such as zinc (Zn), iron (Fe), copper (Cu), and aluminum (Al) bind to these proteins, thereby influencing their conformations and functions. Additionally, these metals have common binding sites; binding of metals to proteins is nonspecific. Therefore, metal-metal interactions at synapses contribute to the neurodegenerative processes. We present a current review of the role of trace elements in the functions and toxicity of disease-related proteins, as well as in the pathogenesis of neurodegenerative diseases. Possible therapeutic approaches related to metal homeostasis are discussed.

Keywords Calcium homeostasis • Iron homeostasis • Neurotoxicity • Alzheimer's disease • Prion disease

D. Mizuno
Department of Forensic Medicine, Faculty of Medicine, Yamagata University,
2-2-2 Iida-Nishi, Yamagata-shi, Yamagata 990-9585, Japan

M. Kawahara, Ph.D. (✉)
Department of Bio-Analytical Chemistry, Faculty of Pharmacy, Musashino University,
Nishi-Tokyo shi, Tokyo 202-8585, Japan
e-mail: makawa@musashino-u.ac.jp

Y. Ogra, T. Hirata (eds.), *Metallomics*, DOI 10.1007/978-4-431-56463-8_14

14.1 Introduction

The brain is a unique organ, representing only 2 % of the typical human body weight, but consuming 20 % of the total oxygen used by the body. The disproportionate oxygen metabolism of the brain contributes to increased susceptibility to oxidative stress. The brain is primary composed of two cell types: neurons and glial cells. Neurons generally do not regenerate and are maintained for the individual's entire life span. Therefore, neurons are susceptible to accumulation of toxic substances such as heavy metals. The blood-brain barrier (BBB), which is composed of tight junctions between endothelial cells, is one mechanism to prevent brain exposure to toxic substances. However, this barrier system is not perfect, and when the brain experiences long durations of toxic substance exposure, the substances may accumulate and result in adverse effects such as Minamata disease.

Essential trace elements including iron (Fe), zinc (Zn), and copper (Cu) exist in the brain and have crucial roles in normal brain functions, such as myelination, neurotransmitter synthesis, and neural information processing. Deficiency of these metals produces severe adverse effects on central nervous system functions, especially learning and memory. Moreover, growing evidence suggests that dyshomeostasis of metals, either excess or deficiency, is implicated in the pathogenesis of various neurodegenerative diseases including Alzheimer's disease (AD), prion diseases, Lewy body diseases (Parkinson's disease, dementia with Lewy bodies (PD/DLB), etc.), and vascular dementia [1, 2]. All of these diseases, except for vascular dementia, include abnormal deposition of disease-related proteins in the brain. The disease-associated proteins, termed amyloidogenic proteins, are ß-amyloid protein (AßP) in AD, prion protein in prion diseases, and α-synuclein in Lewy body diseases. Although the primary sequences are identical to typical proteins, the disease-associated proteins form fibril-like structures of oligomers with the ß-pleated sheet structures (amyloid fibrils) and are associated with neurotoxicity. Since the conformational changes to these amyloidogenic proteins are central to the pathogenesis, a new category, termed "conformational diseases" (protein-misfolding diseases), has proposed [3]. Metals are of particular interest as factors that influence conformational changes to the amyloidogenic proteins [4]. Recent studies have demonstrated that these amyloidogenic proteins or their precursor proteins (such as amyloid precursor protein; APP) possess the binding ability to metals and thereby participate in regulating metal homeostasis. Furthermore, metals and most of these amyloidogenic proteins are co-localized at the synapses which are the crucial node of neural networks. Table 14.1 summarizes the structure and properties of theses amyloidogenic proteins.

In the current neurometallomics study, we investigated the molecular mechanisms of neurodegenerative diseases, focusing on metal-protein interactions and metal-metal interactions at the synapse. Here, we describe the characteristics of three essential elements (Fe, Zn, Cu) and one toxic element (Al) in the brain, and thereafter, we review the functions and neurotoxicity of metals as well as the

Table 14.1 Characteristics of amyloidogenic proteins and the related peptides

Disease	The primary sequence of amyloidogenic protein or its fragment peptide	Binding metal	ß-sheet formation	Cytotoxicity	Channel formation	Functions
Alzheimer's disease	*AβP(1-42)*	Al	+	+	+	(Functions of APP)
	DAEFRHDSGYEVHHQKLVFFAEDVGSNKGAIIGLMVGGVVIA	Zn				Neuronal proliferation and development
		Cu				Neurite outgrowth
		Fe				Fe homeostasis
Prion disease	Prion protein: *PrP106-126*	Zn	+	+	+	SOD activity
	MANLGCWMLVLFVATWSDLGLCKKRPKPGG WNTGGSRYPGQGSPGGNRYPPQGGGGWGQPHGGGWGQPH GGGWGQPHGGGWGQPHGGGWGQGGGTHSQWNKPSKP *KTNMK HMAGAAAAGAVVGGLG*GYVL GSAMSRPIIH FGSDYEDRYYRENMHRYPNQVYYRPMDEYSNQNNFVHDC VNITIKQHTVT TTTKGENFTE TDVKMMERVVEQMCITQY ERESQAYYQRGS SMVLFSSPPVILLISFLIFL IVG	Cu				Cu homeostasis
		Fe				Zn homeostasis
		Mn				Ferrireductase activity
Lewy body disease	α-synuclein; *NAC* (a *fragment of α-synuclein)*	Cu	+	+	+	Dopamine release
	MDVFMKGLSKAKEGVVAAAEKTKQGVAEAA GKTKEGVLYVGSKTKEGVVHGVTTVAEKTK *EQVSNVGGAVVTGVTAVAHKTVEGAGNFAA* *ATGLV*KKDQKNESGFGPEGTMENSENMPVNPNNETY EMPPEEEYQDYDPEA	Fe				Ferrireductase activity
		Al				

The sequence of fragment peptide of each amyloidogenic protein (PrP106-126, NAC) is indicated in italic form
In "functions" of Alzheimer's disease, possible functions of APP are noted

implications for neurodegenerative disease pathogenesis, based on recent studies from our lab and those by other authors.

14.2 Properties of Metals in the Brain

14.2.1 Iron (Fe) in the Brain

Considerable amounts of trace elements such as Fe, Zn, and Cu exist in the brain, in addition to ubiquitous elements such as sodium (Na), potassium (K), calcium (Ca), and magnesium (Mg). The distribution of each metal differs across brain regions [5]. Among these trace elements, Fe is the most abundant in the brain and in the whole body. Fe is essential for numerous biological functions as an enzyme cofactor for metabolic processes such as the oxygen transport, oxidative phosphorylation, and energy transfer. Fe has critical roles in specialized brain functions such as the synthesis of dopaminergic neurotransmitters and the myelination. Therefore, Fe deficiency impairs learning abilities, especially in children or infants, and it produces impaired working ability or learning ability also in adults [6]. However, excess Fe can generate reactive oxygen species (ROS) that damage DNA, proteins, and lipids and can therefore be toxic to neurons [7]. Fe exists in two different forms: ferrous iron (Fe^{2+}) and ferric iron (Fe^{3+}). In general, oxidized Fe^{3+} is insoluble and exists extracellularly, whereas reduced Fe^{2+} is soluble and intracellularly located. Orally administrated Fe is primary absorbed from the gastrointestinal pathway by divalent metal transporter-1 (DMT-1) as Fe^{2+} [8]. Once it enters the circulation, Fe^{2+} ions are oxidized to Fe^{3+} by ferroxidases such as ferritin or ceruloplasmin. It binds with transferrin, an iron binding protein that binds two Fe^{3+}ions. Transferrin-bound iron (Fe^{3+}) crosses BBB *via* transferrin receptors and enters into the cells. Finally, Fe^{3+} is reduced to Fe^{2+} by ferrireductase and function as a cofactor for neuronal enzymes such as tyrosine hydroxylase, which is necessary for the dopamine synthesis. Thus, Fe levels as well as the ratio between Fe^{2+} and Fe^{3+} are strictly regulated in normal brains. Two amyloidogenic proteins, prion protein and α-synuclein, reportedly possess ferrireductase activity [9, 10]. Ferroportin, a transmembrane protein, controls Fe export from intracellular to extracellular compartments. Recent findings suggest that APP, a precursor protein of Alzheimer's AßP, binds to ferroportin and regulates Fe efflux [11].

The iron-responsive element/iron-regulatory protein (IRE/IRP) network also regulates Fe levels [12]. The mRNA encoding Fe-binding proteins such as ferritin or transferrin possess IRE domains in their mRNA. This system regulates production of Fe-binding proteins and prevents the formation of free Fe^{2+} and toxic free radicals. In Fe-deficient conditions, IRP binds to the IREs and inhibits their expression. As the concentration of free Fe^{2+} increases, binding of Fe to IRP and the expression of transferrin is downregulated, whereas that of ferritin is upregulated, thereby decreasing the amount of free Fe^{2+}. The expression of two amyloid-related proteins, APP and α-synuclein, is also controlled by Fe [13] (*see* Sects. 14.3, 14.4, and 14.5).

14.2.2 Zinc (Zn) in the Brain

Zn is the second most abundant element in the brain. Zn is essential for most organisms and has important roles in various physiological functions such as mitotic cell division, immune system functioning, synthesis of proteins and DNA, and as a cofactor to more than 300 enzymes or metalloproteins. Recent studies have revealed that Zn signaling plays crucial roles as a second messenger in various human biological systems [14]. Zn deficiency in children results in dwarfism, delayed mental and physical development, immune dysfunction, and learning disabilities [15]. Zn deficiency also produces learning disorders, taste disorders, and odor disorders also in adults [16].

The human body contains approximately 2g of Zn, primary located in the testes, muscle, liver, and brain tissues. In the brain, Zn occurs at the highest concentrations in the hippocampus, amygdala, cerebral cortex, thalamus, and olfactory cortex [17]. The total Zn content of the hippocampus is estimated to be 70–90 mg/kg (dry weight). Although some Zn in the brain binds firmly to metalloproteins or enzymes, a substantial fraction (approximately 10 % or more) either forms free Zn ions (Zn^{2+}) or is loosely bound and is histochemically detectable by staining using chelating reagents. This chelatable Zn is stored in the presynaptic vesicles of excitatory glutamatergic neurons and is secreted from these vesicles into synaptic clefts, together with glutamate, during neuronal excitation. The concentration of this secreted Zn is estimated to be 1–100 μM [18], although this value is still controversial. Secreted Zn modulates overall brain excitability by binding with *N*-methyl-D-aspartate (NMDA)-type glutamate receptors, GABA receptors, and glycine receptors. Recent studies have suggested that the secreted Zn^{2+} is critical to information processing, synaptic plasticity, learning, and memory [19, 20]. Indeed, Zn^{2+} in the hippocampus is essential for the induction of long-term potentiation (LTP), a form of synaptic information storage that has become a well-known paradigm for the mechanisms underlying memory formation [21].

There are two factors in the maintenance of Zn homeostasis: metallothioneins and Zn transporters. Metallothioneins are ubiquitous metal-binding proteins with 68 amino acids, which bind seven metal atoms (such as Zn, Cu, Cd, etc.) via its 20 cysteine residues. There are three types of metallothioneins: MT-1, MT-2, and MT-3. MT-1 and MT-2 are ubiquitously expressed throughout the whole body, whereas MT-3 is primary localized in the central nervous system [22]. Uchida et al. found that neuronal growth inhibitory factor (GIF) which inhibits neurite extensions and prevents neuronal death was decreased in the brains of AD patients and determined that GIF is equivalent to MT-3 [23]. Therefore, MT-3 (GIF) is implicated in AD-associated neuronal death [24].

Zn transporters also control Zn homeostasis by facilitating Zn influx when it is deficient and efflux when it is present in excess [25]. There are two types of mammalian Zn transporters: ZnT transporters and Zrt-, Irt-like protein (ZIP) transporters. ZnT transporters are involved with the gene family for solute carrier (*SLC30*) and decrease intracellular Zn via facilitation of Zn efflux from cells. There are 14 types of ZnT transporters in mammals, including ZnT-1 and ZnT-3, which are co-localized with chelatable Zn in the brain. ZnT-1 is a membrane protein with six transmembrane domains and is widely distributed in mammalian cells. ZnT-1 has a pivotal role in Zn efflux and in protection from excess Zn. The expression of ZnT-1 is induced after transient global ischemia and decreases following dietary Zn deficiency [26]. ZnT-3 is localized to the membranes of presynaptic vesicles, transports Zn into synaptic vesicles, and maintains high Zn concentrations in the vesicles. Although the physiological roles of ZnT-3 have not fully elucidated, ZnT-3 knockout mice have depleted synaptic Zn and impaired memory formation [27].

ZIP transporters are another type of Zn transporter encoded by *SLC39* genes. ZIP transporters increase cytosolic Zn by promoting transport from extracellular to intracellular compartments. Fourteen ZIP genes have been identified in mammal, and the ZIP transporters are localized to cell membranes or in the membranes of the Golgi apparatus or endoplasmic reticulum (ER). These transporters control Zn influx into subcellular organs, and Zn transporter mutations produce severe diseases such as Ehlers-Danlos syndrome [28].

14.2.3 Cu (Copper) in the Brain

Cu is essential for brain functions and is a cofactor for numerous enzymes, such as cytochrome C, superoxide dismutase, lysyl oxidase, and tyrosinase. Cu is involved in Fe homeostasis as a component of ceruloplasmin and has neuroprotective activity as a component of Cu/Zn superoxide dismutase, an endogenous antioxidant. Cu deficiency results in adverse effects on myelination. Cu is a redox-active metal and exists both as oxidized Cu^{2+} and reduced Cu^{+}. Thus, excess free Cu is toxic because it produces ROS and binds with the thiol groups of functional proteins. The Cu transporters, ATP7A and ATP7B, transport Cu in ATP dependently. Another Cu transporter, Ctr1, is also involved in neuronal Cu uptake. Mutations of these transporters are linked with neurodegenerative diseases such as Wilson disease and Menkes disease [29]. Recent studies suggest that Cu has modulatory effects on neuronal information processes. Intracellular Cu accumulates in synaptic vesicles and is then released into the synaptic clefts during neuronal excitation, similarly to Zn [30]. Its concentration is estimated to be approximately 15 μM. Although the physiological roles of the released Cu are still controversial, Cu reportedly blocks glutamate receptors and modulates neuronal spontaneous activity [31].

14.2.4 Aluminum (Al) in the Brain

Given that Al is implicated in many neurodegenerative diseases due to its peculiar chemical characteristics, it can be considered a neurotoxic metal. Al is the third most abundant element in the earth's crust. Despite its widespread distribution throughout the environment, Al is not essential to living organisms. In contrast, Al reportedly inhibits more than 300 essential enzymatic functions. Al is suspected to contribute to various neurodegenerative diseases including Alzheimer's disease (AD) [1, 32], amyotrophic lateral sclerosis (ALS) and Parkinsonism dementia (PD) in the Kii Peninsula and Guam [33], dialysis encephalopathy [34], and Gulf War syndrome [35].

Al exhibits only one oxidation state, Al^{3+}. In acidic solutions with $pH < 4$, Al^{3+} exists as a soluble octahedral hexahydrate $Al(H_2O)_6{}^{3+}$ (Al^{3+}). As the pH increases, its solubility decreases and $Al(OH)_3$ precipitates at neutral pH. Al^{3+} has affinity for negatively charged, oxygen-donor ligands. Inorganic and organic phosphates, carboxylate, and deprotonated hydroxyl groups form strong bonds with Al^{3+}. Owing to these chemical characteristics, Al^{3+} binds to the phosphate groups of DNA and RNA, affecting DNA topology and influencing the expression of various genes essential for brain functions. Al^{3+} also binds to the phosphate groups of nucleoside di- and triphosphates, such as ATP, and can thus influence energy metabolism. Al also inhibits the functions of various protein kinases and phosphatases. Furthermore, Al^{3+} has a very low ligand-exchange rate compared to other metals. For example, the ligand-exchange rate of Mg^{2+} is 10^5 times faster than that of Al^{3+}, and therefore, Al^{3+} inhibits enzymes with Mg^{2+} cofactors. Al^{3+} also inhibits biological processes involving rapid Ca^{2+} exchange: the exchange rate for Al^{3+} is 10^8 times slower than that of Ca^{2+}. Therefore, Al cannot participate in Ca^{2+}- or Mg^{2+}-related enzymatic reactions and has an extended half-life in the body.

The strong positive charges and a relatively small ionic radius of Al^{3+} in comparison to other metal ions such as Ca^{2+}, Zn^{2+}, and Na^+ facilitate Al^{3+} binding firmly to metal-binding amino acids (histidine, His; tyrosine, Tyr; arginine, Arg; etc.) or phosphorylated amino acids and thereby influences their conformations. Al is a well-known protein cross-linker and is therefore used as a leather-tanning agent.

Additionally, Al^{3+} has characteristics similar to Fe and binds to Fe-binding proteins such as transferrin and ferritin. Al^{3+} also stimulates Fe-induced membrane lipid peroxidation and causes oxidative damage in vitro and in vivo, although Al^{3+} does not directly affect peroxidation [36].

Overall, Al influences various processes induced by Ca, Mg, Fe, and other elements. Metal-metal interaction by Al will be discussed later in this chapter.

14.3 Alzheimer' s Disease (AD) and Metals

14.3.1 Amyloid Hypothesis

AD is a severe senile type of dementia that affects a large proportion of the elderly population worldwide. In Japan, there were more than 4,000,000 patients with senile dementia in 2013, and the number is increasing annually. The number of patients with mild cognitive incidence (MCI), which is the precursor stage of senile dementia, is also estimated to be more than 4,000,000. AD accounts for approximately half of patients with senile dementia.

AD is characterized by profound memory loss and the inability to form new memories. The pathological hallmarks of AD are the presence of numerous extracellular deposits, including senile plaques and neurofibrillary tangles (NFTs), and the selective loss of synapses and neurons in the hippocampal and cerebral regions [37]. Indeed, there is a strong correlation between the decrease in the number of synapses and the severity of memory impairment [38]. The major components of NFTs and senile plaques are phosphorylated tau protein and AßP, respectively.

Numerous biochemical, cell biological, and genetic studies support the idea termed "amyloid cascade hypothesis," which proposes that AßP accumulation and the consequent neurodegeneration are central to AD, although this hypothesis requires further investigation [39, 40]. Recent studies of the identified AßP species have indicated that the oligomerization of AßP and its conformational changes are critical to the neurodegeneration process.

AßP is a small peptide of 39–43 amino acid residues, which results from cleavage of a large precursor protein (APP; amyloid precursor protein) at the N-terminus by the ß-APP cleaving enzyme (BACE) and the intramembrane cleavage of its C-terminus by γ-secretase (Fig. 14.1). Genetic studies of early-onset cases of familial AD have indicated that APP mutations and AßP metabolism are associated with AD [41]. Presenilins are considered to be one of γ -secretases, and their mutations also account for the majority of cases of early-onset familial AD [42].

Yankner et al. reported that the first 40 amino acid residues of AßP (AßP (1–40)) causes the death of cultured rat hippocampal neurons or neurodegeneration in the brains of experimental animals [43]. AßP is a hydrophobic peptide with an intrinsic tendency to self-assemble and form SDS-stable oligomers in aqueous solutions. The monomeric form of AßP has a randomly coiled structure and is also less toxic. Oligomeric AßPs have ß-pleated sheet structures and form insoluble aggregates, termed amyloid fibrils. The neurotoxicity of AßP (1–40) peptides is enhanced by the process of "aging" (aggregation under incubation at 37 °C for several days), compared to freshly prepared AßP(1–40), and correlates with its ß-sheet contents [44]. Jarrett and Lansbury demonstrated that AßP formed oligomers via a nucleation-dependent process and that AßP (1-42) forms seeds for the aggregation of AßP (1-40) [45]. These results suggest that the ratio of AßP (1-40) and AßP (1-42) is critical to the pathogenesis. The point mutations of APP are located near

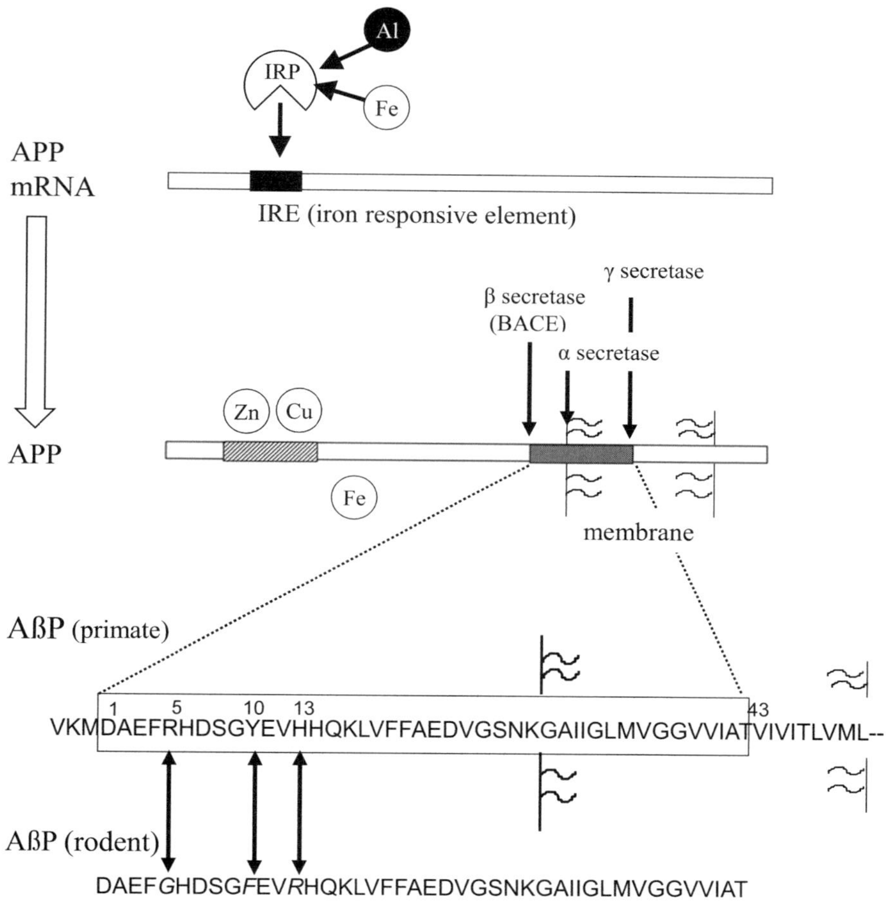

Fig. 14.1 Structure of AßP and its secretion from APP
AßP is secreted by the cleavage of the APP N-terminus by ß-secretase (BACE), followed by intramembrane cleavage of the C-terminus by γ-secretase. APP also binds to Cu or Zn. Human AßP and rodent AßP differ by three amino acids (Arg^5, Tyr^{10}, and His^{13}). The expression of APP is regulated by Fe and Al

the γ-secretase cleavage site of AßP (1-42) and influence the ratio of AßP (1-40) and AßP (1-42). Other mutations in presenilin genes influence the ratio of AßP (1-40) and AßP (1-42) by increasing production of AßP (1-42) in transfected cell lines.

14.3.2 Interaction Between Metals and AßP

AßP is secreted in the cerebrospinal fluid (CSF) of young individuals as well as in that of elderly individuals and non-dementia individuals [46]. Therefore, factors that accelerate or inhibit oligomerization may essentially contribute to the pathogenesis of AD. Several factors, such as the concentration of peptides, pH, composition of solvents, and temperature, influence the oligomerization processes. Oxidations, mutations, and racemization of AßP enhance its oligomerization. Additionally, substances including cholesterol or its oxidation products, apolipoprotein E, transthyretin, rifampicin, curcumin, aspirin, and ß-sheet breaker peptide, inhibit AßP oligomerization in vitro (Fig. 14.2) [47].

It is well established that rodent AßP exhibits a reduced tendency for oligomerization compared to human AßP in vitro, and AßP accumulation is rarely observed in the brains of rodents (rats or mice) compared to primates (humans or monkeys) [48]. As shown in Fig. 14.1, the amino acid sequences of human and rodent AßP are similar, but rodent AßP differs at three amino acids (Arg^5, Tyr^{10}, and His^{13}) compared to primate AßP. Considering that all of these amino acids are able to bind metals and metals are important determinants for the conformation of proteins, it is possible that trace elements such as Al, Zn, Cu, and Fe may play central roles in the accumulation of AßP in the human brain. Al is of particular interest since epidemiological Al is considered to be a risk factor of AD [49]. Al^{3+} has strong positive charges and a relatively small ionic radius in comparison with the other metal ions. Thus, Al^{3+} firmly binds to metal-binding amino acids and induces conformational changes to proteins. Exley et al. first demonstrated that Al induces a conformational change in AßP (1-40), using circular dichroism

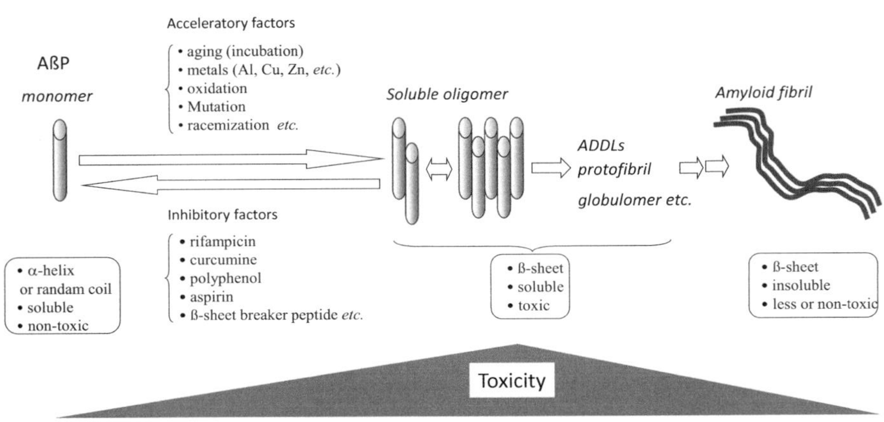

Fig. 14.2 Oligomerization of AßP. AßP monomers exhibit random or α-helix structures. However, under aging conditions or in the presence of acceleratory factors, AßP self-aggregates and forms several types of oligomers (SDS-soluble oligomers, ADDLS, globulomers, protofibrils, etc.) before finally forming insoluble aggregates (amyloid fibrils). Oligomeric soluble Aßs are toxic, although the monomeric and fibril aggregates are relatively nontoxic

(CD) spectroscopy [50]. Al also promotes aggregation of ^{125}I-labeled AßP (1-40), with similar findings as for Fe and Zn [51]. We have developed a system for investigating AßP polymerization that involves immunoblotting and precipitation. Using this system, we have demonstrated that Al enhances oligomerization of AßP (1-40) and forms SDS-stable oligomers in vitro [52, 53]. The aggregated AßP (1-40) is redissolved by adding an Al chelator, deferoxamine (DFO). The oligomerization induced by Al is more marked than that induced by other metals, including Zn, Fe, Cu, and Cd. Furthermore, we have demonstrated that Al-aggregated AßPs bind tightly to the surface of cultured neurons and form fibrillar deposits, meanwhile, Zn-aggregated AßPs are rarely observed on the surface of cultured neurons. Furthermore, Al inhibits AßP degradation resulting from conformational changes and enhances AßP accumulation.

Other trace metals such as Zn and Cu enhance AßP oligomerization. Bush et al. have demonstrated that Zn^{2+}, at a concentration similar to that in the CSF, produced AßP aggregation [54]. They reported that Cu^{2+} also binds to AßP, induces its aggregation, and increases ROS levels [55].

Despite all oligomers possess ß-pleated sheet structure, not all oligomers are equally neurotoxic. Recent studies using size-exclusion chromatography, gel electrophoresis, and atomic force microscopy revealed that several stable types of oligomers have been reported: naturally occurring soluble oligomers (dimmers or trimers), AßP-derived diffusible ligands (ADDLs), AßP globulomers, and protofibrils [4]. Hartley et al. separated aggregated AßP(1–40) into low-molecular-weight (mainly monomers), protofibrillar, and fibril fractions using size-exclusion chromatography and found that the protofibrillar fraction produced robust changes in the electrical activity of cultured neurons and was neurotoxic, but fibrils did not [56]. Walsh et al. found that SDS-stable oligomers exist in conditioned medium with cultured cells exogenously expressing the human APP gene [57]. The natural AßP oligomers obtained from the CSF of AD patients induced the loss of dendritic spines and synapses and blocked LTP [58].

The characteristics (size or shape) of AßP oligomers formed in the presence of Al, Zn Cu, and Fe are revealed to be identical by morphological analysis using atomic force microscopy [59]. Additionally, Sharma et al. revealed that Zn-aggregated AßPs are less toxic than Cu-aggregated AßP [60]. Meanwhile, Al-aggregated AßPs reportedly caused cytoskeletal changes, mitochondrial dysfunction, and increased production of ROS [61]. Bolognin et al. investigated aggregation and toxicity of AßP induced by Al, Cu, Fe, and Zn, finding that Al-aggregated AßPs induce overproduction of APP and tau, but AßP oligomers formed in the presence of other metals did not [62]. Everett et al. reported that AßP has the catalytic activity to reduce Fe^{3+} to redox-active Fe^{2+} and found that Al enhanced the reductive activity of AßP [63].

Metals can also participate in AßP-induced neurodegeneration pathways. Our previous studies, as well as numerous other studies, have demonstrated that AßPs are directly incorporated onto the surface of the cellular membrane and create unregulated pore-like channels that have cytotoxic effects [64–66]. These "amyloid channels" are giant multilevel pores that can facilitate the transport of large amounts of Ca^{2+} [67]. Zn binds to His residues, which are exposed to the internal site of the amyloid pore and inhibit Ca^{2+} influx [68, 69]. We found that AßP induced abnormal increases in $[Ca^{2+}]_i$ using a high-resolution multi-site video imaging system in conjunction with a Ca^{2+}-sensitive fluorescent dye (fura-2) [70]. Numerous studies, including those from our lab, have revealed that prion protein fragment peptide or α-synuclein also form amyloid channels on membranes [65, 66]. Zn can regulate Ca influx via these amyloid channels.

14.3.3 Interactions Between Metals and APP or Presenilin

Despite the wide distribution of APP in the brain, its physiological roles have not been elucidated. APP has distinct binding domains for Cu, Zn, and Fe [71]. Wong et al. found that APP does not possess ferroxidase activity, but binds to the Fe transporter ferroportin and enhances Fe efflux [11]. Furthermore, APP mRNA contains an IRE domain as well as ferritin, and its expression is regulated by Fe, as described in Fig. 14.1 [72]. Therefore, APP regulates Fe homeostasis, and Fe controls APP expression. There are other important findings implicating Fe homeostasis in the AD pathogenesis. Fe-related genes, such as transferrin C2 or the hemochromatosis gene, are risk factors for AD [73]. Imagawa et al. reported that Fe supplementation is effective for recovery of cognitive functions in patients with AD [74].

APP has two Cu-binding domains in its N-terminal and possesses the ability to reduce Cu^{2+} to Cu^{+} [75]. Cu induces the dimerization of APP, AßP production, and trafficking of APP from the ER to neurites [76]. Cu also influences APP processing and the expression of APP. These results suggest that APP has crucial roles in the maintenance of metal homeostasis, particularly Fe homeostasis. When this homeostasis is disrupted, increased Fe causes ROS production or increased APP causes AßP production, both initiate the degenerative processes. Indeed, APP knockout mice exhibited increased Cu levels in the brain [77]. Cicotosto et al. reported that knockout of APP or its analogue APLP2 in mice resulted in changes in the distribution of Cu, Zn, Fe, and Ca [78]. Decreased levels of ferroportin and accumulation of Fe in the brains of AD patients have been reported [79].

It is very possible that Al influences Fe homeostasis. As noted previously, Al has characteristics similar to Fe and binds to Fe-binding proteins, including ferritin and transferrin, or to Fe chelators such as DFO. Al^{3+} also binds to IRP [80, 81] and thus affects the expression of various Fe-regulated genes containing IREs, thereby causing elevation in the Fe concentration. Additionally, Al reportedly enhances the expression of APP [82–85]. Furthermore, Al influences the uptake of Fe into cultured neurons or glial cells [86].

Presenilins, one of the γ-secretases, and their mutations account for most cases of early-onset AD. Presenilins mainly exist in the ER and they are implicated in Ca homeostasis [87, 88]. Recent studies demonstrate that presenilins are also metal-binding proteins and participate in the neuronal uptake of Zn and Cu [89]. Presenilins also promote the uptake of dietary Cu [90].

14.4 Prion Disease and Metals

14.4.1 Overview

Prion diseases are fatal neurodegenerative diseases that include scrapie in sheep, bovine spongiform encephalopathy (BSE) in cattle, as well as Creutzfeldt-Jakob disease (CJD), Gerstmann-Sträussler-Scheinker syndrome (GSS), and kuru disease in humans [91]. The common pathological hallmarks are the spongiform degeneration of neurons and glial cells, and the accumulation of amyloidogenic prion protein (PrP) in the brain of the infected animals or person. Prion diseases are also called transmissible spongiform encephalopathies, since their characteristic infections can be initiated by the administration of pathogenic tissue. Although the molecular pathogenesis and transmission mechanisms of prion diseases are still controversial, it is widely accepted that the conformational conversion of normal cellular prion protein (PrP^C) to an abnormal scrapie-type isoform (PrP^{Sc}) is related to the transmissible characteristics of prion diseases. PrP^C is a 30–35 kDa cell surface glycoprotein anchored at the plasma membrane by a glycosylphosphatidylinositol (GPI) domain and is ubiquitously expressed throughout the entire body, particularly in the brain. Both PrP^C and PrP^{Sc} have the same characteristics in terms of chemical modifications as well as the same primary sequence. However, PrP^C differs from PrP^{Sc} in that PrP^{Sc} is resistant to protease digestion, has high ß-sheet secondary structure, and has a propensity to form insoluble amyloid fibrils. When the misfolded PrP^{Sc} enters the body, for example, by the administration of contaminated food, the protease-resistant PrP^{Sc} can then aggregate. Next, fibril formation induces other PrP^C molecules in the brain to misfold and aggregate.

14.4.2 Interactions Between Metals and PrP

The physiological roles of PrP^C are still controversial; however, increasing evidence suggests that PrP^C is a metalloprotein involved in metal homeostasis [92]. PrP^C contains 208 amino acid residues and possesses a highly conserved octa-repeat domain composed of multiple tandem copies of the eight-residue sequence PHGGGWGQ in its N-terminal (Fig. 14.3). In 1997, Brown et al. reported decreased levels of Cu in the brains of PrP-knockout mice compared to wild-type mice. The activity of Cu-dependent enzymes was also reduced in PrP-null mice [93]. Jackson et al. reported that PrP^C binds four Cu atoms at its octa-repeat domain and binds two additional Cu atoms to two other histidine (His) residues, His^{96} and His^{111} [94]. They also demonstrated that other metal ions including Zn^{2+}, Mn^{2+}, and Ni^{2+} can bind to these sites, although with lower affinity than Cu^{2+}. PrP^C reportedly transports Cu from the extracellular space to the intracellular space via endocytosis and regulates the intracellular concentration of Cu. PrP^C possesses or modulates Cu/Zn superoxide dismutase (Cu/Zn SOD) activity in the brain and has protective

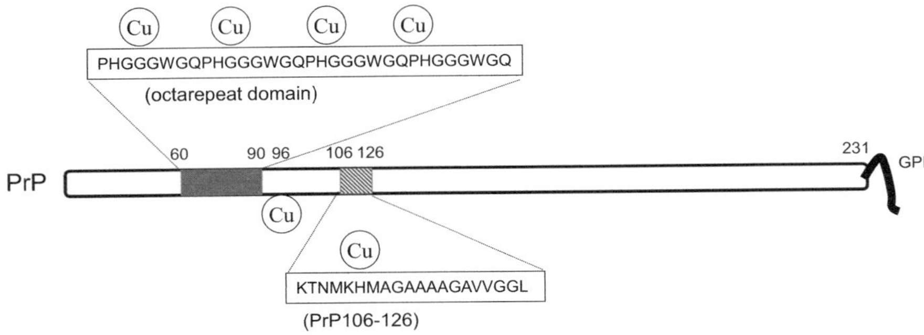

Fig. 14.3 The structure of the prion protein and its metal -binding properties

roles against oxidative stress [95]. Recent studies have suggested that PrP^C regulates the function of the NMDA-type glutamate receptor in a Cu-dependent manner [96]. In contrast, Cu^{2+} influences the gene expression and cellular trafficking of PrP [97]. It is also possible that the depletion of PrP^C and a subsequent increase in Cu could cause oxidative damage to neurons [98]. Furthermore, PrP-deficient neurons exhibit lower glutathione reductase activity and increased susceptibility to damage by hydrogen peroxide [99].

Zn^{2+} has the next highest affinity for binding with PrP, after Cu^{2+}. As discussed previously, Zn accumulates in synaptic vesicles, and a considerable amount of Zn (~100 μM) is released with glutamate during neuronal excitation. Since the concentration of Zn in the brain is much higher than the concentration of Cu (~15 μM), it is possible that Zn^{2+} influences PrP^C binding to Cu [100]. Moreover, recent bioinformatics analysis has revealed the evolutionary similarities between prion genes and genes encoding the ZIP transporters [101]. Of 14 ZIP transporters, sequence similarities have been reported between PrP^C and N-terminal ectodomains of ZIP5, ZIP6, and ZIP10. Furthermore, PrP^C co-localizes with ZIP5 in neuroblastoma cells [102]. Watt et al. found that PrP^C facilitates the uptake of Zn into neuronal cells and that the effect is mediated by α-amino-3-hydroxy-5-methyl-4-isoxazolepropionate (AMPA) receptors [103]. These authors hypothesized that PrP^C functions as a Zn sensor, detecting extracellular levels up to a particular threshold. Indeed, PrP^C reportedly attenuates Zn-induced neurotoxicity.

In addition to Cu and Zn, other metals such as Fe and Mn are associated with PrP. Singh et al. suggested that PrP^C functions as a ferrireductase that is responsible for reducing Fe^{3+} to bioavailable Fe^{2+} and modulating the cellular uptake of Fe [104]. PrP-knockout mice exhibit altered Fe metabolism [105]. Decreased levels of ferroxidase activity and transferrin have been observed in the cerebrospinal fluid of patients with CJD [106].

Furthermore, several studies have suggested that Mn may facilitate prion diseases. Johnson et al. investigated the levels of trace elements in prion-infected hamster brains using X-ray photoelectron emission microscopy with synchrotron radiation and found both reduced Cu and increased Mn in plaques composed of prion proteins [107]. Thackray et al. reported that PrP^C loses its SOD-like activity when Cu is replaced with Mn [108]. Furthermore, Mn enhances the stability of PrP in soils and increases its infectivity [109]. The risk of contracting prion disease in elk, termed "chronic wasting disease," has been associated with Mg deficiency and increased levels of Mn [110]. A recent epidemiological survey also suggested a relationship between the pathogenesis of CJD and Mn imbalance [111].

14.4.3 Toxicity of PrP Fragments and Metals

The neurodegenerative mechanisms of prion diseases include three possibilities: the "loss of normal protective functions of PrP^{C}," the "gain of toxic functions of PrP^{Sc}," or a combination of both. PrP^{C} is thought to regulate Cu homeostasis and to have antioxidant and cytoprotective effects against neurotoxicity induced by Cu^{2+} or free radicals. We have demonstrated that PrP fragment with octa-repeat domain attenuated Cu-induced neurotoxicity [112]. Thus, PrP^{C} depletion may initiate various neurodegenerative processes.

Meanwhile, PrP^{Sc} produces synaptic impairment and apoptosis in neurons and astrocytes in vitro and in vivo. To investigate the mechanisms of PrP^{Sc} neurotoxicity, researchers (including our lab) have used synthetic fragment peptides of PrP (PrP106–126), because methodological difficulties occur when attempting to use a whole prion protein due to its strongly infectious characteristics. PrP106–126 has been used as a model peptide of PrP^{Sc}, as it coincides with the proposed sequence for ß-sheet structures, forms aggregates of ß-sheet structures to produce amyloid fibrils that share several characteristics with PrP^{Sc}, and causes the apoptotic death of cultured neurons or glial cells [113]. Additionally, PrP106–126 has the ability to bind to metals including Cu^{2+} and Zn^{2+} [114].

We observed that PrP106–126 forms ß-sheet structures during the "aging" process (incubation at 37 °C for several days) using a thioflavin T (ThT) fluorescence assay, far-UV circular dichroism (CD) spectroscopy, and atomic force microscopy (AFM) imaging [112]. Furthermore, aged PrP106–126 produced significant neurotoxicity in primary cultured rat hippocampal neurons. These characteristics are quite similar to that of AßP. We have added various trace elements to solutions of PrP106–126 during the aging processes and evaluated the resulting conformational changes and neurotoxicity. The presence of either Zn^{2+} or Cu^{2+} during the aging process significantly attenuated PrP106–126 neurotoxicity, whereas the presence of Al^{3+}, Fe^{2+}, and Fe^{3+} did not produce significant changes. Additionally, we investigated the effects of these metal ions on the ß-sheet formation of PrP106–126. We have utilized the changes in fluorescence of ThT, which binds with pleated ß-sheet structures, to observe the oligomerization of PrP106–126. The ThT fluorescence for solutions of aged PrP106–126 increased compared with freshly dissolved PrP106–126 solutions. The ThT fluorescence of PrP106–126 incubated with Zn^{2+}, Fe^{2+}, or Fe^{3+} was significantly decreased compared with aged PrP106–126 alone. In particular, the addition of Cu^{2+} dramatically decreased ThT fluorescence levels similar to fresh PrP106–126. Aged PrP106–126 forms amyloid fibrils with a distinctively straight and long morphology on mica plates as observed by AFM, although no fiber-like structures could be observed in freshly prepared PrP106–126. However, PrP106–126 aged with Cu or Zn exhibited different morphological features compared to aged PrP106–126 alone. Although the CD spectra of aged PrP106–126 exhibited ß-sheet structures, the CD spectra of PrP106–126 aged with Cu^{2+} exhibited random coil-like structures. Our results suggest that Cu and Zn influenced the ß-sheet formation of PrP106–126 and thereafter attenuated its neurotoxicity. These findings are consistent with other studies that Cu inhibits ß-sheet formation by PrP111–126 [115]. Thakur et al. investigated the conformational changes of full-length PrP by NMR and reported that Cu did not produce oligomerization of PrP at physiological temperatures and that Cu may act as an attenuator in prion diseases [116]. Although it is widely accepted that Cu enhances the aggregation of other amyloidogenic proteins such as AßP and α-synuclein, the effects of Cu on the conformational changes to amyloidogenic proteins may be complex and may depend on peptide structures [117].

14.5 Lewy Body Diseases and Metals

Lewy body disease is a category that includes Parkinson's disease (PD), dementia with Lewy body (DLB), and multiple system atrophy (MSA). These diseases commonly exhibit abnormal cellular inclusions, termed Lewy bodies, which is the accumulation of α-synuclein, and therefore are termed "synucleopathy" [118]. DLB includes approximately 25 % senile dementia and has common pathological changes with AD, such as the deposition of senile plaques and tau protein. Moreover, the fragment peptide of α-synuclein, non-amyloid component (NAC), is co-accumulated with AßP in Alzheimer's senile plaques. The oligomerization and fibrillation of α-synuclein have been implicated in the formation of Lewy bodies and the etiology of Lewy body diseases. It is demonstrated that α-synuclein is a 140-amino acid protein, abundantly present in the brain and particularly located in the nucleus and in the presynaptic terminals.

Involvement of metals in Lewy body diseases and α-synuclein has been extensively researched, since Mn toxicity exhibits Parkinson-like symptoms [119] and Fe-rich regions such as the substantia nigra are particularly vulnerable in Parkinson's diseases [104]. Accumulation of Fe and Al has been reported in Lewy bodies of Parkinson patients. Furthermore, α-synuclein aggregation is accelerated by Cu and Al in vitro. Uversky et al. found the metals, such as Al and Mn, enhance the α-synuclein oligomerization [120]. Cu binds to α-synuclein in its N-terminal domain and C-terminal domain [121].

Increasing evidence suggests that α-synuclein primary exists in the presynaptic terminals of dopaminergic neurons and controls dopamine release, synaptic functions, and synaptic plasticity. Additionally, α-synuclein reportedly binds Fe and functions as a ferrireductase, which reduces Fe^{3+} to bioavailable Fe^{2+}

[9]. Furthermore, α-synuclein mRNA has an IRE domain as well as APP or ferritin, and Fe regulates its expression [122]. In the postmortem brains of patients with Parkinson patient, Fe levels as well as the ratio of Fe^{2+} to Fe^{3+} were changed [123].

14.6 Vascular Dementia and Metals

Vascular dementia (VD) is a degenerative cerebrovascular disease that accounts for approximately one third of senile dementia cases. Risk factors include aging, sex (male), diabetes, and high blood pressure. The most common type of VD is caused by a series of small strokes or ischemia [124]. Following transient global ischemia or stroke, the interruption of blood flow and the resulting oxygen-glucose deprivation induce long-lasting membrane depolarization and result in excessive glutamate release into synaptic clefts. The excess glutamate then overstimulates receptors including NMDA-type receptors, AMPA-type receptors, and kainate-type receptors. Finally, Ca^{2+} dyshomeostasis (the entry of large quantities of Ca^{2+} to glutamate-responsive neurons) triggers delayed death of vulnerable populations of neurons, such as pyramidal neurons in the hippocampus, which is associated with learning and memory. Development of an infarct and the subsequent cognitive dysfunction produce the pathogenesis of VD. Approximately 30 % of patients with stroke experience show symptoms of dementia within 3 months of the initial stroke.

Increasing evidence suggests that Zn is central to ischemia-induced neuronal death and the pathogenesis of VD [125]. In ischemic conditions, as much as 300 μM Zn is co-released with glutamate into the synaptic clefts following membrane depolarization [126] and exhibits apoptotic cell death. Furthermore, the chelatable Zn moved from presynaptic terminals into cell bodies of degenerated neurons [127]. The increase in intracellular Zn^{2+} levels ($[Zn^{2+}]_i$), namely, Zn translocation, occurs in vulnerable neurons in the CA1 or CA3 regions of the hippocampus after transient global ischemia. Administration of calcium EDTA (Ca EDTA), a membrane-impermeable Zn chelator, blocked Zn translocation, protected hippocampal neurons after transient global ischemia, and reduced the infarct volume [128]. At least three major routes of Zn^{2+} entry were identified: voltage-gated Ca^{2+} channels (VGLC), NMDA-type glutamate receptors, and AMPA/kainate-type glutamate receptors [129]. Although NMDA-type glutamate receptors are present in most neurons, the permeability of Zn^{2+} and Ca^{2+} through AMPA/kainate channels is greater for NMDA receptor.

In the normal physiological condition, hippocampal neurons typically express AMPA receptors with GluR2 subunits, which are poorly permeable to divalent cations including Ca^{2+} and Zn^{2+}. However, following ischemia, there is an acute reduction in GluR2 subunit expression, and neurons possess specific types of AMPA receptors with channels that are directly Ca^{2+}-permeable (Ca-AMPA/kainate channels; Ca-A/K-R). The appearance of Ca-AMPA/kainate channels result in increased permeability of Ca^{2+}, thereby enhancing toxicity. Therefore, the expression of Zn^{2+}-permeable Ca-AMPA/kainate channels and the entry of Ca^{2+} and/or Zn^{2+} through the channels are mediators of the delayed neuronal death that follows ischemia. Considering that Ca EDTA, a Zn chelator, attenuates ischemia-induced downregulation of the GluR2 gene [128], Zn is also implicated in the transcriptional regulation of Ca-AMPA/kainate channels.

We investigated the molecular mechanism of Zn-induced neuronal death using GT1-7 cells (immortalized hypothalamic neurons), which are much more sensitive to Zn than other neuronal cells [130, 131]. GT1-7 cells possess neuronal characteristics such as neurite extensions, secretion of gonadotropin-releasing hormone (GnRH), and expression of neuron-specific proteins or receptors including microtubule-associated protein 2 (MAP2), tau protein, neurofilament, synaptophysin, GABAA receptors, dopamine receptors, and L-type Ca^{2+} channels [132]. We demonstrated that the ER stress pathway, the mitochondrial energy pathway, and the disruption of Ca homeostasis contribute to Zn-induced apoptosis. Screening for substances that are protective against Zn neurotoxicity yielded findings that carnosine (ß-alanyl histidine) attenuates Zn-induced neuronal death [133].

14.7 Hypothesis: Crosstalk of Metals and Amyloidogenic Proteins at the Synapse

The evidence indicates that the amyloidogenic proteins (or their fragment peptides), including APP (AβP), prion protein (PrP106-126), and α-synuclein (NAC), share similarities such as oligomerization with β-pleated sheet structures, neurotoxicity, and the ability to binding metals, as shown in Table 14.1. Furthermore, recent findings demonstrate that these proteins are co-localized at the synapse. APP occurs primarily in the presynaptic membrane, PrP^C occurs in postsynaptic membranes, and α-synuclein occurs primarily in the presynaptic cytosol and to a lesser extent in membranes. Additionally, presenilins are predominantly located in the ER and also occur in the presynaptic and postsynaptic membranes.

The synapse is a local site of communications between neurons. It is small, but is the crucial node of brain neural networks. Neurotransmitters and metals (Zn or Cu) are co-released from synaptic vesicles in the presynaptic terminals to the synaptic clefts and bind to receptors in the postsynaptic domains (PSDs). Synapses are vulnerable regions for these neurodegenerative diseases, since synaptic plasticity is essential to memory formation. The synaptic cleft is conceptualized as a cylinder with 20 nm height and 200 nm radius [134]. Thus, the distance between presynaptic terminals and the postsynaptic membranes (~20 nm) may be small enough for proteins at the presynaptic terminal to interact with proteins in the postsynaptic terminal. Considering the small volume of synaptic clefts, it is plausible that neurotransmitters or metals are concentrated at synaptic clefts and their levels may be much higher than the level in the extracellular fluid. The concentration of glutamate in the synaptic cleft is estimated to reach a few millimolar range after 1 ms of neuronal depolarization, and the Zn concentration after depolarization is estimated to be 1–100 μM. Thus, amyloidogenic proteins must certainly interact with other proteins in a considerable amount of Zn and Cu.

Considering the combined evidence, we hypothesize that metal imbalance at the synapse contributes to the pathogenesis of neurodegenerative diseases. Under typical physiological conditions (Fig. 14.4), Zn can be released with glutamate and can bind with NMDA-type glutamate receptors or other receptors, inhibiting overall brain excitability. Secreted Zn can diffuse across the synaptic cleft, spill over to neighboring synapses, and influence the activity of neighboring synapses

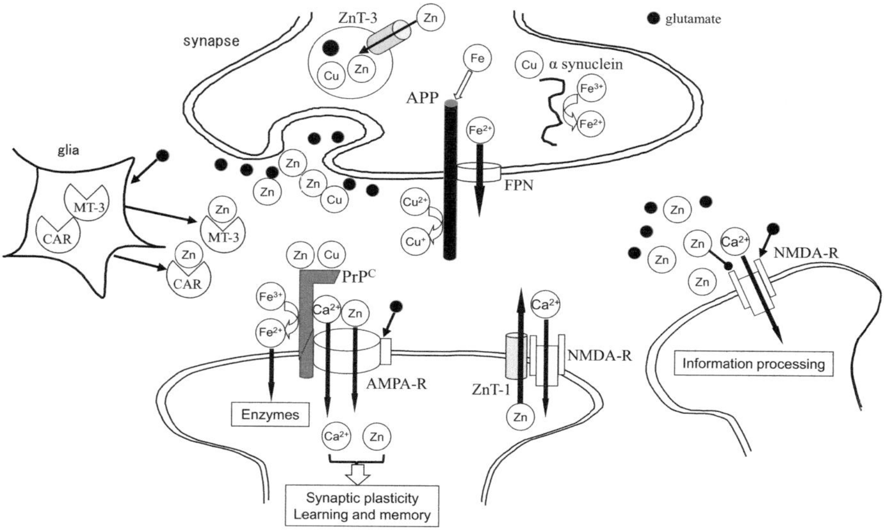

Fig. 14.4 Crosstalk between trace elements and amyloidogenic proteins at the synapse under normal condition. Zn and glutamate accumulate in synaptic vesicles and are released into synaptic clefts during neuronal excitation. Zn^{2+} regulates Ca^{2+} influx through glutamate receptors (NMDA-R, Ca-A/K), modulates neuronal information, and is implicated in the maintenance of synaptic plasticity and memory formation, similarly to Ca^{2+}. Zn has important roles in information processing in the targeted neuron and also in neighboring neurons. APP and α-synuclein exist in the presynaptic domain, and PrP^C is localized to the postsynaptic domain. These proteins are closely associated with the synaptic cleft and Zn and have cytoprotective roles via the regulation of metal homeostasis and protection from free radicals. PrP^C binds to AMPA-type glutamate receptors and regulates Zn^{2+} levels similarly to ZIP Zn transporters. Additionally, the ZnT-1 Zn transporter is localized to postsynaptic membranes that express NMDA-type glutamate receptors and regulates Zn homeostasis. Moreover, PrP^C has SOD activity and also regulates Cu^{2+} levels, which influence APP sequencing. APP converts Cu^{2+} to Cu^{+} and regulates Cu at the synapse. PrP^C and α-synuclein have ferrireductase activity, which converts Fe^{3+} to Fe^{2+}. APP binds to ferroportin and thereby regulates Fe^{2+} efflux. MT-3 and carnosine are released from glial cells, into synaptic clefts, and are also implicated in regulation of excess Zn. *ZnT-1* zinc transporter 1, *AMPA-R* AMPA-type glutamate receptor, *NMDA-R* NMDA-type glutamate receptor, *MT-3* metallothionein 3

dose-dependently. Since glutamate produces excitation and Zn produced inhibition, differing concentrations of glutamate and Zn in the adjacent synapses create precise modulation of neuronal activity. Therefore, it is possible that Zn may have neuromodulator roles, transmitting the spatiotemporal information of neuronal activity. This may enable "lateral inhibition," based on signaling contrast, and may be based on synaptic plasticity [135]. Synaptic Zn enters the postsynaptic neurons through Ca^{2+} channels and NMDA channels and regulates functions of various channels and receptors. Recent evidence suggests that the ZnT-1 transporter, which enhances Zn efflux to the extracellular compartment, is localized in postsynaptic membranes [136]. The ZnT-1 transporter binds with NMDA-type glutamate receptors and regulates the activity. In comparison, PrP^C, an analogue of ZIP zinc transporters, localizes in postsynaptic membranes binding with AMPA-type glutamate receptor, which facilitate Zn influx. Thus, it is likely that synaptic Zn levels are controlled by both ZnT-1 and PrP^C. MT-3 (GIF) secreted from neurons or glia may also regulate Zn homeostasis at synapses. Another contributor to Zn homeostasis is carnosine (CAR), an endogenous antioxidant and anti-cross-linking peptide, which is synthesized in glial cells [137].

Cu is also secreted at synaptic clefts following neuronal excitation. PrP^C binds to Cu at its N-terminal domain and regulates synaptic Cu levels. It is also possible that PrP^C provides Cu to APP or to NMDA-type glutamate receptor, thereby influencing the production of AβP or the neuronal excitability. APP also regulates Cu levels by reducing Cu^{2+} to Cu^{+}, and both APP and PrP^C reportedly attenuate Cu-induced toxicity.

Furthermore, APP controls Fe homeostasis by binding with ferroportin and promotes Fe efflux. In contrast, both PrP^C and α-synuclein possess ferrireductase activity, regulate the Fe^{2+}/Fe^{3+} ratio in synapses, and thereby control neurotransmitter synthesis.

We have demonstrated the complex and subtle interactions between metals and amyloidogenic proteins at the synapse. When the homeostasis of metals is disrupted, synapses and neuron degradations occur, which contribute to the pathogenesis of neurodegenerative diseases (Fig. 14.5).

Interactions between one metal and other metals may disrupt metal homeostasis. Varying types of metal ions usually share binding sites, although their binding constants differ. In particular, Cu^{2+} and Zn^{2+}share similar chemical characteristics and interact with each other. Similarly, Al^{3+} and Fe^{3+} share similar characteristics and interact with each other. Thus, the disruption of metal homeostasis may occur in the presence of toxic metals such as Al or the excess of essential metals such as Zn, Cu, or Fe.

In case of the pathogenesis of AD, Al enters into the brain. Al binds to Fe-binding proteins, inhibits IRP/IRE pathway as described previously, and influences the expression of Fe-binding proteins in the brain, including APP, α-synuclein, and ferritin, thereby increasing Fe levels. Overexpression of APP induced by Al results in accumulation of AβP. Al and other metals also accelerate AβP oligomerization and enhance the neurotoxicity. AβP oligomers may form amyloid channels on synaptic membranes, produce Ca dyshomeostasis, initiating

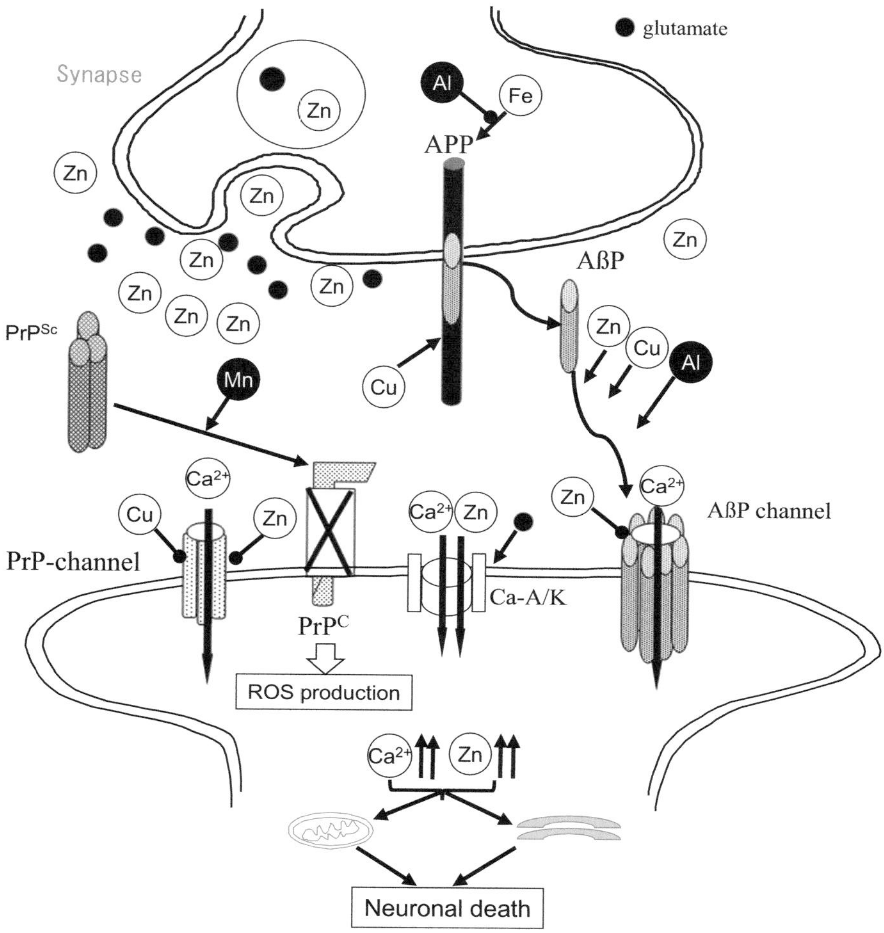

Fig. 14.5 Disruption of metal homeostasis and the pathogenesis of neurodegenerative diseases. When toxic metals such as Al cross the blood-brain barrier, Al binds various Fe-binding proteins and influences Fe homeostasis, which initiates formation of free radicals and upregulation of APP and results in overproduction of AßP. Al also causes the oligomerization and accumulation of AßP. These adverse effects may enhance the synaptotoxicity and neurotoxicity of AβP and produce AD pathogenesis. Oligomerized AβPs form pore-like structures (AβP channels) in membranes, cause Ca dyshomeostasis, and initiate apoptotic neuronal death. Conversely, Zn inhibits amyloid channels and has neuroprotective properties. However, excess Zn resulting from overexcitation during ischemia also disrupts metal homeostasis at the synapse, initiates neurodegeneration, and contributes to the pathogenesis of vascular dementia. When exogenous PrP^{Sc} enters into the brain, it stimulates conversion of endogenous PrP^{C} and results in PrP^{C} depletion and PrP^{Sc} accumulation. The loss of neuroprotective functions by PrP^{C} produces oxidative stress, enhances AßP neurotoxicity, and results in neuronal death. Accumulated PrP^{Sc} forms Ca^{2+}-permeable channels (PrP channel) in membranes and disrupts Ca homeostasis

synaptotoxicity and neurotoxicity, and produce the pathogenesis of AD. Zn inhibits AβP amyloid channel formations and neurotoxicity [69]. Zn also infounces Al-induced neurotoxicity [138]. In this context, Zn may have two faces in the pathogenesis of AD.

In case of the pathogenesis of prion diseases, pathogenetic PrP^{Sc} enters the brain primary by food contamination, triggers conformational changes to PrP^{C}, and causes conversion of PrP^{C} to PrP^{Sc}. PrP^{C} is neuroprotective, since it has antioxidant activity and attenuates AßP neurotoxicity. Loss of neuroprotective PrP^{C} may disrupt metal homeostasis. Increased Cu at the synapse produces the oxidative stress and influences the APP expression and processing. Thus, the depletion of normal PrP^{C} results in oxidative stress, enhances the neurotoxicity of AßP, and results in apoptotic death of neurons and glia. Additionally, accumulated PrP^{Sc} may form channels (PrP channel) in the membrane and thereby disrupt Ca homeostasis as well as AßP. Cu^{2+} and Zn^{2+} inhibit the conformational conversion from PrP^{C} to PrP^{Sc} and also attenuate PrP^{Sc}-induced Ca dyshomeostasis and neurotoxicity. However, Mn may enhance PrP^{Sc} toxicity by influencing the conversion of PrP^{C}.

In the case of total ischemia, excess Zn is released with glutamate, which produces disruption of Ca homeostasis, triggers mitochondrial energy depletion and ER stress, and contributes to neurodegeneration. Zn also enhances the expressions of Ca-permeable AMPA-type glutamate receptors and enhances Ca dyshomeostasis. Considering that vascular degeneration may be linked with the early-stage AD, Zn neurotoxicity may also be associated with AD.

14.8 Metal-Related Treatments for Neurodegenerative Diseases

Several studies have investigated metal chelators as potential treatment for neurodegenerative diseases. Clioquinol (5-chloro-7-iodo-8-quinolinol) and its derivatives have been assessed in therapeutic trials for AD. Clioquinol (quinoform), a chelator of Cu^{2+} or Zn^{2+}, inhibits oligomerization of AßP and attenuates the accumulation of amyloid in the brains of experimental animals [139]. Clinical trials using its analogue, PBT2, are under investigation [140]. Deferoxamine (DFO), a chelator of Al and Fe, attenuates the decline of daily living skills in AD patients [141]. Intranasal application of DFO reportedly attenuates memory loss and amyloid deposition [142]. Silicates, which couple with Al and reduce its toxicity, are also candidates for chelation therapy in AD [143].

Clioquinol treatment also had beneficial effects on scrapie-induced memory impairment [144]. It was also revealed that D-(-)-penicillamine, a Cu^{2+}-specific chelator, attenuated the pathogenesis of prion diseases in vivo [145].

Our screen for substances that protect against PrP106–126-induced neurotoxicity revealed that carnosine may be a candidate treatment modality for prion diseases [112]. Carnosine is a water-soluble dipeptide contained in mammalian muscles and in the brain, particularly in the olfactory bulbs [137]. Carnosine has antioxidant and anti-glycation properties as well as the ability to bind to metals. Additionally, carnosine has anti-cross-linking properties, inhibits the oligomerization of AßP, and attenuates neurodegeneration in a mouse model of AD [146]. We previously reported that carnosine inhibited Zn^{2+}-induced neuronal death after ischemia [2, 133]. Considering these beneficial characteristics, carnosine may have neuroprotective functions in the brain. Based on these findings, we have published a patent for carnosine as a possible basis for drugs designed to treat vascular dementia [147].

14.9 Conclusions

We have demonstrated the complex and subtle interactions between metals and amyloidogenic proteins at the synapse. This crosstalk is essential for normal brain functions, and therefore, the disruption of metal homeostasis may contribute to the conformational changes of amyloidogenic proteins and to the pathogenesis of neurodegenerative diseases. Our findings suggest that metal dyshomeostasis may be one of the common mechanisms for this pathogenesis and may elucidate the enigmatic roles of trace elements in neurodegenerative diseases and preventive drug development. In conclusion, our working hypothesis may contribute to enhanced understanding of the role of metals in the neurodegenerative diseases. Further research regarding "neurometallomics" is necessary, particularly in relation to the molecular mechanisms of synaptic degeneration and the quantitative analysis of neurometals.

References

1. Kawahara M, Kato-Negishi M (2011) Link between aluminum and the pathogenesis of Alzheimer's disease: the integration of the aluminum and amyloid cascade hypotheses. Int J Alzheimer Dis 2011:276393
2. Kawahara M, Mizuno D, Koyama H, Konoha K, Ohkawara S, Sadakane Y (2013) Disruption of zinc homeostasis and the pathogenesis of senile dementia. Metallomics 6:209–219
3. Carrell RW, Lomas DA (1997) Conformational disease. Lancet 350:134–138
4. Kawahara M (2010) Role of calcium dyshomeostasis via amyloid channels in the pathogenesis of Alzheimer's disease. Curr Pharm Des 16:2779–2789
5. Becker JSS, Matusch A, Palm C, Salber D, Morton K, Becker S (2010) Bioimaging of metals in brain tissue by laser ablation inductively coupled plasma mass spectrometry (LA-ICP-MS) and metallomics. Metallomics 2:104–111
6. Youdim MB (2008) Brain iron deficiency and excess; cognitive impairment and neurodegeneration with involvement of striatum and hippocampus. Neurotox Res 14:45–56
7. Crichton RR, Wilmet S, Legssyer R, Ward RJ (2002) Molecular and cellular mechanisms of iron homeostasis and toxicity in mammalian cells. J Inorg Biochem 91:9–18

8. Conrad ME, Umbreit JN (2000) Iron absorption and transport-an update. Am J Hematol 64:287–298
9. Davies P, Moualla D, Brown DR (2011) Alpha-synuclein is a cellular ferrireductase. PLoS One 6, e15814. doi:10.1371/journal.pone.0015814
10. Singh A, Haldar S, Horback K, Tom C, Zhou L, Meyerson H, Singh N (2013) Prion protein regulates iron transport by functioning as a ferrireductase. J Alzheimers Dis 35:541–552
11. Wong BX, Tsatsanis A, Lim LQ, Adlard PA, Bush AI, Duce JA (2014) β-Amyloid precursor protein does not possess ferroxidase activity but does stabilize the cell surface ferrous iron exporter ferroportin. PLoS One 9:e114174. doi: 10.1371/journal.pone.0114174. eCollection 2014
12. Muckenthaler MU, Galy B, Hentze MW (2008) Systemic iron homeostasis and the iron-responsive element/iron-regulatory protein (IRE/IRP) regulatory network. Annu Rev Nutr 28:197–213
13. Rogers JT, Bush AI, Cho HH, Smith DH, Thomson AM, Friedlich AL, Lahiri DK, Leedman PJ, Huang X, Cahill CM (2008) Iron and the translation of the amyloid precursor protein (APP) and ferritin mRNAs: riboregulation against neural oxidative damage in Alzheimer's disease. Biochem Soc Trans 36:1282–1287
14. Fukada T, Yamasaki S, Nishida K, Murakami M, Hirano T (2011) Zinc homeostasis and signaling in health and diseases: zinc signaling. J Biol Inorg Chem 16:1123–1134
15. Sandstead HH (2012) Subclinical zinc deficiency impairs human brain function. J Trace Elem Med Biol 26:70–73
16. Takeda A, Tamano H (2009) Insight into zinc signaling from dietary zinc deficiency. Brain Res Rev 62:33–44
17. Frederickson CJ, Suh SW, Silva D, Frederickson CJ, Thompson RB (2000) Importance of zinc in the central nervous system: the zinc-containing neuron. J Nutr 130:1471S–1483S
18. Vogt K, Mellor J, Tong G, Nicoll R (2000) The actions of synaptically released zinc at hippocampal mossy fiber synapses. Neuron 26:187–196
19. Takeda A, Fujii H, Minamino T, Tamano H (2014) Intracellular Zn(2+) signaling in cognition. J Neurosci Res 92:819–824
20. Ueno S, Tsukamoto M, Hirano T, Kikuchi K, Yamada MK, Nishiyama N, Nagano T, Matsuki N, Ikegaya Y (2002) Mossy fiber Zn2+ spillover modulates heterosynaptic N-methyl-D-aspartate receptor activity in hippocampal CA3 circuits. J Cell Biol 158:215–220
21. Pan E, Zhang XA, Huang Z, Krezel A, Zhao M, Tinberg CE, Lippard SJ, McNamara JO (2011) Vesicular zinc promotes presynaptic and inhibits postsynaptic long-term potentiation of mossy fiber-CA3 synapse. Neuron 71:1116–1126
22. Bolognin S, Cozzi B, Zambenedetti P, Zatta P (2014) Metallothioneins and the central nervous system: from a deregulation in neurodegenerative diseases to the development of new therapeutic approaches. J Alzheimers Dis 41:29–42
23. Uchida Y, Takio K, Titani K, Ihara Y, Tomonaga M (1991) The growth inhibitory factor that is deficient in the Alzheimer's disease brain is a 68 amino acid metallothionein-like protein. Neuron 7:337–347
24. Uchida Y (1994) Growth-inhibitory factor, metallothionein-like protein, and neurodegenerative diseases. Biol Signals 3:211–215
25. Fukada T, Kambe T (2011) Molecular and genetic features of zinc transporters in physiology and pathogenesis. Metallomics 3:662–674
26. Tsuda M, Imaizumi K, Katayama T, Kitagawa K, Wanaka A, Tohyama M, Takagi T (1997) Expression of zinc transporter gene, ZnT-1, is induced after transient forebrain ischemia in the gerbil. J Neurosci 17:6678–6684
27. Sindreu C, Palmiter RD, Storm DR (2011) Zinc transporter ZnT-3 regulates presynaptic Erk1/2 signaling and hippocampus-dependent memory. Proc Natl Acad Sci U S A 108:3366–3370

28. Bin BH, Fukada T, Hosaka T, Yamasaki S, Ohashi W, Hojyo S, Miyai T, Nishida K, Yokoyama S, Hirano T (2011) Biochemical characterization of human ZIP13 protein: a homo-dimerized zinc transporter involved in the spondylocheiro dysplastic Ehlers-Danlos syndrome. J Biol Chem 286:40255–40265
29. Kodama H, Fujisawa C, Bhadhprasit W (2010) Pathology, clinical features and treatments of congenital copper metabolic disorders–focus on neurologic aspects. Brain Dev 33:243–251
30. D'Ambrosi N, Rossi L (2015) Copper at synapse: release, binding and modulation of neurotransmission. Neurochem Int 90:36–45
31. Dodani SC, Firl A, Chan J, Nam CI, Aron AT, Onak CS, Ramos-Torres KM, Paek J, Webster CM, Feller MB, Chang CJ (2014) Copper is an endogenous modulator of neural circuit spontaneous activity. Proc Natl Acad Sci U S A 111:16280–16285
32. Exley C (2005) The aluminium-amyloid cascade hypothesis and Alzheimer's disease. Subcell Biochem 38:225–234
33. Yase Y, Yoshida S, Kihira T, Wakayama I, Komoto J (2001) Kii ALS dementia. Neuropathology 21:105–109
34. Alfrey AC, LeGendre GR, Kaehny WD (1976) The dialysis encephalopathy syndrome: possible aluminium intoxication. New Engl J Med 294:184–188
35. Petrik MS, Wong MC, Tabata RC, Garry RF, Shaw CA (2007) Aluminum adjuvant linked to Gulf War illness induces motor neuron death in mice. Neruomol Med 9:83–92
36. Oteiza PI (1994) A mechanism for the stimulatory effect of aluminum on iron-induced lipid peroxidation. Arch Biochem Biophys 308:374–379
37. Selkoe DJ (1991) The molecular pathology of Alzheimer's disease. Neuron 6:487–498
38. Terry RD, Masliah E, Salmon DP, Butters N, DeTeresa R, Hill R, Hansen LA, Katzman R (1991) Physical basis of cognitive alterations in Alzheimer's disease: synapse loss is the major correlate of cognitive impairment. Ann Neurol 30:572–580
39. Hardy JA, Higgins GA (1992) Alzheimer's disease: the amyloid cascade hypothesis. Science 256:184–185
40. Wirths O, Multhaup G, Bayer TA (2004) A modified ß-amyloid hypothesis: intraneuronal accumulation of the beta-amyloid peptide–the first step of a fatal cascade. J Neurochem 91:513–520
41. Goate A, Chartier-Harlin MC, Mullan M, Brown J, Crawford F, Fidani L, Giuffra L, Haynes A, Irving N, James L et al (1991) Segregation of a missense mutation in the amyloid precursor protein gene with familial Alzheimer's disease. Nature 349:704–706
42. Selkoe DJ, Wolfe MS (2007) Presenilin: running with scissors in the membrane. Cell 131:215–221
43. Yankner BA, Duffy LK, Kirschner DA (1990) Neurotropic and neurotoxic effects of amyloid ß protein: reversal by tachykinin neuropeptides. Nature 250:279–282
44. Pike CJ, Burdick D, Walencewicz AJ, Glabe CG, Cotman CW (1993) Neurodegeneration induced by beta-amyloid peptides in vitro: the role of peptide assembly state. J Neurosci 13:1676–1687
45. Jarrett JT, Lansbury PT Jr (1993) Seeding "one-dimensional crystallization" of amyloid: a pathogenic mechanism in Alzheimer's disease and scrapie? Cell 73:1055–1058
46. Fukuyama R, Mizuno T, Mori S, Nakajima K, Fushiki S, Yanagisawa K (2000) Age-dependent change in the levels of Aβ40 and Aβ42 in cerebrospinal fluid from control subjects, and a decrease in the ratio of Aβ42 to Aβ40 level in cerebrospinal fluid from Alzheimer's disease patients. Eur Neurol 43:155–160
47. Kawahara M, Negishi-Kato M, Sadakane Y (2009) Calcium dyshomeostasis and neurotoxicity of Alzheimer's beta-amyloid protein. Expert Rev Neurother 9:681–693
48. Dyrks T, Dyrks E, Masters CL, Beyreuther K (1993) Amyloidogenicity of rodent and human ß A4 sequences. FEBS Lett 324:231–236
49. Flaten TP (2001) Aluminium as a risk factor in Alzheimer's disease, with emphasis on drinking water. Brain Res Bull 55:187–196

50. Exley C, Price NC, Kelly SM, Birchall JD (1993) An interaction of β-amyloid with aluminium in vitro. FEBS Lett 324:293–295
51. Mantyh PW, Ghilardi JR, Rogers S, DeMaster E, Allen CJ, Stimson ER, Maggio JE (1993) Aluminum, iron, and zinc ions promote aggregation of physiological concentrations of beta-amyloid peptide. J Neurochem 61:1171–1174
52. Kawahara M, Muramoto K, Kobayashi K, Mori H, Kuroda Y (1994) Aluminum promotes the aggregation of Alzheimer's β-amyloid protein in vitro. Biochem Biophys Res Commun 198:531–535
53. Kawahara M, Kato M, Kuroda Y (2001) Effects of aluminum on the neurotoxicity of primary cultured neurons and on the aggregation of ß-amyloid protein. Brain Res Bull 55:211–217
54. Bush AI, Pettingell WH, Multhaup G, d Paradis M, Vonsattel JP, Gusella JF, Beyreuther K, Masters CL, Tanzi RE (1994) Rapid induction of Alzheimer Aß amyloid formation by zinc. Science 265:1464–1467
55. Atwood CS, Moir RD, Huang X, Scarpa RC, Bacarra NM, Romano DM, Hartshorn MA, Tanzi RE, Bush AI (1998) Dramatic aggregation of Alzheimer abeta by Cu(II) is induced by conditions representing physiological acidosis. J Biol Chem 273:12817–12826
56. Hartley DM, Walsh DM, Ye CP, Diehl T, Vasquez S, Vassilev PM, Teplow DB, Selkoe DJ (1999) Protofibrillar intermediates of amyloid ß-protein induce acute electrophysiological changes and progressive neurotoxicity in cortical neurons. J Neurosci 19:8876–8884
57. Walsh DM, Tseng BP, Rydel RE, Podlisny MB, Selkoe DJ (2000) The oligomerization of amyloid ß-protein begins intracellularly in cells derived from human brain. Biochemistry 39:10831–10839
58. Selkoe DJ (2008) Soluble oligomers of the amyloid beta-protein impair synaptic plasticity and behavior. Behav Brain Res 192:106–113
59. Chen WT, Liao YH, Yu HM, Cheng IH, Chen YR (2011) Distinct effects of Zn^{2+}, Cu^{2+}, Fe^{3+}, and Al^{3+} on amyloid-beta stability, oligomerization, and aggregation: amyloid-beta destabilization promotes annular protofibril formation. J Biol Chem 286:9646–9656
60. Sharma AK, Pavlova ST, Kim J, Kim J, Mirica LM (2013) The effect of Cu(2+) and Zn(2+) on the Aβ42 peptide aggregation and cellular toxicity. Metallomics 5:1529–1536
61. Drago D, Cavaliere A, Mascetra N, Ciavardelli D, di Ilio C, Zatta P, Sensi SL (2008) Aluminum modulates effects of beta amyloid(1-42) on neuronal calcium homeostasis and mitochondria functioning and is altered in a triple transgenic mouse model of Alzheimer's disease. Rejuvenation Res 11:861–871
62. Bolognin S, Zatta P, Lorenzetto E, Valenti MT, Buffelli M (2013) β-Amyloid-aluminum complex alters cytoskeletal stability and increases ROS production in cortical neurons. Neurochem Int 62:566–574
63. Everett J, Céspedes E, Shelford LR, Exley C, Collingwood JF, Dobson J, van der Laan G, Jenkins CA, Arenholz E, Telling ND (2014) Ferrous iron formation following the co-aggregation of ferric iron and the Alzheimer's disease peptide β-amyloid (1-42). J R Soc Interface 11:20140165. doi:10.1098/rsif.2014.0165
64. Arispe N, Diaz JC, Simakova O (2007) Abeta ion channels. Prospects for treating Alzheimer's disease with Abeta channel blockers. Biochim Biophys Acta 1768:1952–1965
65. Kawahara M, Ohtsuka I, Yokoyama S, Kato-Negishi M, Sadakane Y (2011) Membrane incorporation, channel formation, and disruption of calcium homeostasis by Alzheimer's ß-amyloid protein. Int J Alzheimer Dis:304583
66. Lal R, Lin H, Quist AP (2007) Amyloid beta ion channel: 3D structure and relevance to amyloid channel paradigm. Biochim Biophys Acta 1768:1966–1975
67. Arispe N, Rojas E, Pollard HB (1993) Alzheimer disease amyloid beta protein forms calcium channels in bilayer membranes: blockade by tromethamine and aluminum. Proc Natl Acad Sci U S A 90:567–571
68. Arispe N, Pollard HB, Rojas E (1996) Zn^{2+} interaction with Alzheimer amyloid beta protein calcium channels. Proc Natl Acad Sci U S A 93:1710–1715

69. Kawahara M, Arispe N, Kuroda Y, Rojas E (1997) Alzheimer's disease amyloid ß-protein forms Zn^{2+}-sensitive, cation-selective channels across excised membrane patches from hypothalamic neurons. Biophys J 73:67–75
70. Kawahara M, Arispe N, Kuroda Y, Rojas E (2000) Alzheimer's ß-amyloid, human islet amylin and prion protein fragment evoke intracellular free-calcium elevations by a common mechanism in a hypothalamic GnRH neuronal cell-line. J Biol Chem 275:14077–14083
71. Ayton S, Lei P, Bush AI (2013) Metallostasis in Alzheimer's disease. Free Radic Biol Med 62:76–89
72. Rogers JT, Randall JD, Cahill CM, Eder PS, Huang X, Gunshin H, Leiter L, McPhee J, Sarang SS, Utsuki T, Greig NH, Lahiri DK, Tanzi RE, Bush AI, Giordano T, Gullans SR (2002) An iron-responsive element type II in the 5′-untranslated region of the Alzheimer's amyloid precursor protein transcript. J Biol Chem 277:45518–45528
73. Namekata K, Imagawa M, Terashi A, Ohta S, Oyama F, Ihara Y (1997) Association of transferrin C2 allele with late-onset Alzheimer's disease. Hum Genet 101:126–129
74. Imagawa M, Naruse S, Tsuji S, Fujioka A, Yamaguchi H (1992) Coenzyme Q10, iron, and vitamine B6 in genetically-confirmed Alzheimer's disease. Lancet 340:671
75. White AR, Multhaup G, Maher F, Bellingham S, Camakaris J, Zheng H, Bush AI, Beyreuther K, Masters CL, Cappai R (1999) The Alzheimer's disease amyloid precursor protein modulates copper-induced toxicity and oxidative stress in primary neuronal cultures. J Neurosci 19:9170–9179
76. Baumkötter F, Schmidt N, Vargas C, Schilling S, Weber R, Wagner K, Fiedler S, Klug W, Radzimanowski J, Nickolaus S, Keller S, Eggert S, Wild K, Kins S (2014) Amyloid precursor protein dimerization and synaptogenic function depend on copper binding to the growth factor-like domain. J Neurosci 34:11159–11172
77. White AR, Reyes R, Mercer JF, Camakaris J, Zheng H, Bush AI, Multhaup G, Beyreuther K, Masters CL, Cappai R (1999) Copper levels are increased in the cerebral cortex and liver of APP and APLP2 knockout mice. Brain Res 842:439–444
78. Ciccotosto GD, James SA, Altissimo M, Paterson D, Vogt S, Lai B, de Jonge MD, Howard DL, Bush AI, Cappai R (2014) Quantitation and localization of intracellular redox active metals by X-ray fluorescence microscopy in cortical neurons derived from APP and APLP2 knockout tissue. Metallomics 6:1894–1904
79. Raha AA, Vaishnav RA, Friedland RP, Bomford A, Raha-Chowdhury R (2013) The systemic iron-regulatory proteins hepcidin and ferroportin are reduced in the brain in Alzheimer's disease. Acta Neuropathol Commun 1:55. doi:10.1186/2051-5960-1-55
80. Oshiro S, Kawahara M, Shirao M, Muramoto K, Kobayashi K, Ishige R, Nozawa K, Hori M, Yung C, Kitajima S, Kuroda Y (1998) Aluminum taken up by transferrin independent iron uptake affects the iron metabolism in rat cortical cells. J Biochem 123:42–46
81. Yamanaka K, Minato N, Iwai K (1999) Stabilization of iron regulatory protein 2, IRP2, by aluminum. FEBS Lett 462:216–220
82. Li XB, Zhang ZY, Yin LH, Schluesener HJ (2012) The profile of β-amyloid precursor protein expression of rats induced by aluminum. Environ Toxicol Pharmacol 33:135–140
83. Liang RF, Li WQ, Wang H, Wang JX, Niu Q (2013) Impact of sub-chronic aluminium-maltolate exposure on catabolism of amyloid precursor protein in rats. Biomed Environ Sci 26:445–452
84. Walton JR, Wang MX (2009) APP expression, distribution and accumulation are altered by aluminum in a rodent model for Alzheimer's disease. J Inorg Biochem 103:1548–1554
85. Wang L, Hu J, Zhao Y, Lu X, Zhang Q, Niu Q (2014) Effects of aluminium on β-amyloid (1-42) and secretases (APP-cleaving enzymes) in rat brain. Neurochem Res 39:1338–1345
86. Oshiro S, Kawahara M, Kuroda Y, Zhang C, Cai Y, Kitajima S, Shirao M (2000) Glial cells contribute more to iron and aluminum accumulation but are more resistant to oxidative stress than neuronal cells. Biochim Biophys Acta 1502:405–414
87. Green KN, LaFerla FM (2008) Linking calcium to Aß and Alzheimer's disease. Neuron 59:190–42008
88. Mattson MP (2010) ER calcium and Alzheimer's disease: in a state of flux. Sci Signal 3:pe10. doi:10.1126/scisignal.3114pe10

89. Greenough MA, Volitakis I, Li QX, Laughton K, Evin G, Ho M, Dalziel AH, Camakaris J, Bush AI (2011) Presenilins promote the cellular uptake of copper and zinc and maintain copper chaperone of SOD1-dependent copper/zinc superoxide dismutase activity. J Biol Chem 286:9776–9786
90. Southon A, Greenough MA, Ganio G, Bush AI, Burke R, Camakaris J (2013) Presenilin promotes dietary copper uptake. PLoS One 8, e62811
91. Prusiner SB (1997) Prion diseases and the BSE crisis. Science 278:245–251
92. Mizuno D, Koyama H, Ohkawara S, Sadakane Y, Kawahara M (2014) Involvement of trace elements in the pathogenesis of prion diseases. Curr Pharam Biotech 15:1049–1057
93. Brown DR, Qin K, Herms JW, Madlung A, Manson J, Strome R, Fraser PE, Kruck T, von Bohlen A, Schulz-Schaeffer W, Giese A, Westaway D, Kretzschmar H (1997) The cellular prion protein binds copper in vivo. Nature 390:684–687
94. Jackson GS, Murray I, Hosszu LL, Gibbs N, Waltho JP, Clarke AR, Collinge J (2001) Location and properties of metal-binding sites on the human prion protein. Proc Natl Acad Sci U S A 98:8531–8535
95. Brown DR (2009) Brain proteins that mind metals: a neurodegenerative perspective. Dalton Trans 21:4069–4076
96. Gasperini L, Meneghetti E, Pastore B, Benetti F, Legname G (2015) Prion protein and copper cooperatively protect neurons by modulating NMDA receptor through S-nitrosylation. Antioxid Redox Signal 22:772–784
97. Alfaidy N, Chauvet S, Donadio-Andrei S, Salomon A, Saoudi Y, Richaud P, Aude-Garcia C, Hoffmann P, Andrieux A, Moulis JM, Feige JJ, Benharouga M (2013) Prion protein expression and functional importance in developmental angiogenesis: role in oxidative stress and copper homeostasis. Antioxid Redox Signal 18:400–411
98. Rotilio G1, Carrì MT, Rossi L, Ciriolo MR (2000) Copper-dependent oxidative stress and neurodegeneration. IUBMB Life 50:309–314
99. White AR, Collins SJ, Maher F, Jobling MF, Stewart LR, Thyer JM, Beyreuther K, Masters CL, Cappai R (1999) Prion protein-deficient neurons reveal lower glutathione reductase activity and increased susceptibility to hydrogen peroxide toxicity. Am J Pathol 155:1723–1730
100. Stellato F, Spevacek A, Proux O, Minicozzi V, Millhauser G, Morante S (2011) Zinc modulates copper coordination mode in prion protein octa-repeat subdomains. Eur Biophys J 40:1259–1270
101. Schmitt-Ulms G, Ehsani S, Watts JC, Westaway D, Wille H (2009) Evolutionary descent of prion genes from the ZIP family of metal ion transporters. PLoS One 4, e7208
102. Pocanschi CL, Ehsani S, Mehrabian M, Wille H, Reginold W, Trimble WS, Wang H, Yee A, Arrowsmith CH, Bozóky Z, Kay LE, Forman-Kay JD, Rini JM, Schmitt-Ulms G (2013) The ZIP5 ectodomain co-localizes with PrP and may acquire a PrP-like fold that assembles into a dimer. PLoS One 8, e72446
103. Watt NT, Griffiths HH, Hooper NM (2013) Neuronal zinc regulation and the prion protein. Prion 7:203–208
104. Singh N, Haldar S, Tripathi AK, McElwee MK, Horback K, Beserra A (2014) Iron in neurodegenerative disorders of protein misfolding: a case of prion disorders and Parkinson's disease. Antioxid Redox Signal 21:471–484
105. Singh A, Kong Q, Luo X, Petersen RB, Meyerson H, Singh N (2009) Prion protein (PrP) knock-out mice show altered iron metabolism: a functional role for PrP in iron uptake and transport. PLoS One 4, e6115
106. Haldar S, Beveridge J, Wong J, Singh A, Galimberti D, Borroni B, Zhu X, Blevins J, Greenlee J, Perry G, Mukhopadhyay CK, Schmotzer C, Singh N (2013) A low-molecular-weight ferroxidase is increased in the CSF of sCJD cases: CSF ferroxidase and transferrin as diagnostic biomarkers for sCJD. Antioxid Redox Signal 19:1662–1675

107. Johnson CJ, Gilbert PU, Abrecht M, Baldwin KL, Russell RE, Pedersen JA, Aiken JM, McKenzie D (2013) Low copper and high manganese levels in prion protein plaques. Viruses 5:654–662
108. Thackray AM, Madec JY, Wong E, Morgan-Warren R, Brown DR, Baron T, Bujdoso R (2003) Detection of bovine spongiform encephalopathy, ovine scrapie prion-related protein (PrP^{Sc}) and normal PrP^c by monoclonal antibodies raised to copper-refolded prion protein. Biochem J 370:81–90
109. Davies P, Brown DR (2009) Manganese enhances prion protein survival in model soils and increases prion infectivity to cells. PLoS One 4:e7518
110. White SN, O'Rourke KI, Gidlewski T, VerCauteren KC, Mousel MR, Phillips GE, Spraker TR (2010) Increased risk of chronic wasting disease in Rocky Mountain elk associated with decreased magnesium and increased manganese in brain tissue. Can J Vet Res 74:50–53
111. Slivarichová D, Mitrová E, Ursínyová M, Uhnáková I, Koscová S, Wsólová L (2011) Geographic accumulation of Creutzfeldt-Jakob disease in Slovakia–environmental metal imbalance as a possible cofactor. Cent Eur J Public Health 19:158–164
112. Kawahara M, Koyama H, Nagata T, Sadakane Y (2011) Zinc, copper, and carnosine attenuate neurotoxicity of prion fragment PrP106-126. Metallomics 3:726–734
113. Bonetto V, Massignan T, Chiesa R, Morbin M, Mazzoleni G, Diomede L, Angeretti N, Colombo L, Forloni G, Tagliavini F, Salmona M (2002) Synthetic miniprion PrP106. J Biol Chem 277:31327–31334
114. Grasso D, Milardi D, La Rosa C, Rizzarelli E (2004) The different role of Cu^{++} and Zn^{++} ions in affecting the interaction of prion peptide PrP106-126 with model membranes. Chem Commun (Camb) 21:246–247
115. Inayathullah M, Satheeshkumar KS, Malkovskiy AV, Carre AL, Sivanesan S, Hardesty JO, Rajadas J (2013) Solvent microenvironments and copper binding alters the conformation and toxicity of a prion fragment. PLoS One 8, e85160
116. Thakur AK, Srivastava AK, Srinivas V, Chary KV, Rao CM (2011) Copper alters aggregation behavior of prion protein and induces novel interactions between its N- and C-terminal regions. J Biol Chem 286:38533–38545
117. Ward B, Walker K, Exley C (2008) Copper(II) inhibits the formation of amylin amyloid in vitro. J Inorg Biochem 102:371–375
118. Dettmer U, Selkoe D, Bartels T (2015) New insights into cellular α-synuclein homeostasis in health and disease. Curr Opin Neurobiol 36:15–22
119. Chen P, Chakraborty S, Mukhopadhyay S, Lee E, Paoliello MM, Bowman AB, Aschner M (2015) Manganese homeostasis in the nervous system. J Neurochem 134:601–610
120. Uversky VN, Li J, Fink AL (2001) Metal-triggered structural transformations, aggregation, and fibrillation of human alpha-synuclein. A possible molecular NK between Parkinson's disease and heavy metal exposure. J Biol Chem 276:44284–44296
121. Tavassoly O, Nokhrin S, Dmitriev OY, Lee JS (2014) Cu(II) and dopamine bind to α-synuclein and cause large conformational changes. FEBS J 281:2738–2753
122. Cahill CM, Lahiri DK, Huang X, Rogers JT (2009) Amyloid precursor protein and alpha synuclein translation, implications for iron and inflammation in neurodegenerative diseases. Biochim Biophys Acta 1790:615–628
123. Kienzl E, Jellinger K, Stachelberger H, Linert W (1999) Iron as catalyst for oxidative stress in the pathogenesis of Parkinson's disease? Life Sci 65:1973–1976
124. Lee J, Grabb MC, Zipfel GJ, Choi DW (2000) Brain tissue responses to ischemia. J Clin Invest 106:723–731
125. Weiss JH, Sensi SL, Koh JY (2000) Zn(2+): a novel ionic mediator of neural injury in brain disease. Trends Pharmacol Sci 21:395–401
126. Frederickson CJ, Klitenick MA, Manton WI, Kirkpatrick JB (1983) Cytoarchitectonic distribution of zinc in the hippocampus of man and the rat. Brain Res 27:335–339
127. Koh JY, Suh SW, Gwag BJ, He YY, Hsu CY, Choi DW (1996) The role of zinc in selective neuronal death after transient global cerebral ischemia. Science 272:1013–1016

128. Calderone A1, Jover T, Mashiko T, Noh KM, Tanaka H, Bennett MV, Zukin RS (2004) Late calcium EDTA rescues hippocampal CA1 neurons from global ischemia-induced death. J Neurosci 24:9903–9913
129. Weiss JH, Sensi SL (2000) Ca2+-Zn2+ permeable AMPA or kainate receptors: possible key factors in selective neurodegeneration. Trends Neurosci 23:365–371
130. Kawahara M, Kato-Negishi M, Kuroda Y (2002) Pyruvate blocks zinc-induced neurotoxicity in immortalized hypothalamic neurons. Cell Mol Neurobiol 22:87–93
131. Koyama H, Konoha K, Sadakane Y, Ohkawara S, Kawahara M (2012) Zinc neurotoxicity and the pathogenesis of vascular-type dementia: involvement of calcium dyshomeostasis and carnosine. J Clin Toxicol S3-002. doi:10.4172/2161-0495.S3-002
132. Mellon PL, Windle JJ, Goldsmith PC, Padula CA, Roberts JL, Weiner RI (1990) Immortalization of hypothalamic GnRH neurons by genetically targeted tumorigenesis. Neuron 5:1–10
133. Mizuno D, Kawahara M (2014) Carnosine: a possible drug for vascular dementia. J Vasc Med Surg 2:3. doi.org/10.4172/2329-6925.1000146
134. Schikorski T, Stevens CF (1997) Quantitative ultrastructural analysis of hippocampal excitatory synapses. J Neurosci 17:5858–5867
135. Wang X, Sommer FT, Hirsch JA (2011) Inhibitory circuits for visual processing in thalamus. Curr Opin Neurobiol 21:726–733
136. Mellone M, Pelucchi S, Alberti L, Genazzani AA, Di Luca M, Gardoni F (2015) Zinc transporter-1: a novel NMDA receptor-binding protein at the postsynaptic density. J Neurochem 132:159–168
137. Hipkiss AR (2009) Carnosine and its possible roles in nutrition and health. Adv Food Nutr Res 57:87–154
138. Singla N, Dhawan DK (2014) Zinc modulates aluminium-induced oxidative stress and cellular injury in rat brain. Metallomics 6:1941–1950
139. Zhang YH, Raymick J, Sarkar S, Lahiri DK, Ray B, Holtzman D, Dumas M, Schmued LC (2013) Efficacy and toxicity of clioquinol treatment and A-beta42 inoculation in the APP/PSI mouse model of Alzheimer's disease. Curr Alzheimer Res 10:494–506
140. Faux NG, Ritchie CW, Gunn A, Rembach A, Tsatsanis A, Bedo J, Harrison J, Lannfelt L, Blennow K, Zetterberg H, Ingelsson M, Masters CL, Tanzi RE, Cummings JL, Herd CM, Bush AI (2010) PBT2 rapidly improves cognition in Alzheimer's disease: additional phase II analyses. J Alzheimers Dis 20:509–516
141. Crapper McLachlan DR, Dalton AJ, Kruck TP, Bell MY, Smith WL, Kalow W, Andrews DF (1991) Intramuscular desferrioxamine in patients with Alzheimer's disease. Lancet 337:1304–1308
142. Fine JM, Renner DB, Forsberg AC, Cameron RA, Galick BT, Le C, Conway PM, Stroebel BM, Frey WH, Hanson LR (2015) Intranasal deferoxamine engages multiple pathways to decrease memory loss in the APP/PS1 model of amyloid accumulation. Neurosci Lett 584:362–367. doi:10.1016/j.neulet.2014.11.013
143. Exley C (2007) Organosilicon therapy in Alzheimer's disease? J Alzheimers Dis 11:301–302
144. Bareggi SR, Braida D, Pollera C, Bondiolotti G, Formentin E, Puricelli M, Poli G, Ponti W, Sala M (2009) Effects of clioquinol on memory impairment and the neurochemical modifications induced by scrapie infection in golden hamsters. Brain Res 1280:195–200
145. Sigurdsson EM, Brown DR, Alim MA, Scholtzova H, Carp R, Meeker HC, Prelli F, Frangione B, Wisniewski T (2003) Copper chelation delays the onset of prion disease. J Biol Chem 278:46199–46202
146. Corona C, Frazzini V, Silvestri E, Lattanzio R, La Sorda R, Piantelli M, Canzoniero LM, Ciavardelli D, Rizzarelli E, Sensi SL (2011) Effects of dietary supplementation of carnosine on mitochondrial dysfunction, amyloid pathology, and cognitive deficits in 3xTg-AD mice. PLoS One 6:e17971. doi:10.1371/journal.pone.0017971
147. Kawahara M, Konoha K, Nagata T, Sadakane Y (2007) Protective substances against zinc-induced neuronal death after ischemia: carnosine a target for drug of vascular type of dementia. Recent Patents CNS Drug Discov 2:145–149

Chapter 15
New Insight into Metallomics in Cognition

Atsushi Takeda and Hanuna Tamano

Abstract Brain zinc homeostasis is strictly controlled under healthy condition, indicating the importance of zinc for physiological functions in the brain. A portion of zinc exists in the synaptic vesicles and is released from glutamatergic (zincergic) neuron terminals. The zinc serves as a signal factor (as Zn^{2+}) in the intracellular (cytosol) compartment, in addition to the extracellular compartment. The dynamic crosstalk of synaptic Zn^{2+} signaling to intracellular Ca^{2+} signaling via calcium channels is involved in synaptic plasticity such as long-term potentiation (LTP) and cognitive activity. Intracellular Zn^{2+} signaling is critical for cognitive activity as well as intracellular Ca^{2+} signaling. On the other hand, excess intracellular Zn^{2+} signaling, which is induced by excess glutamatergic neuron activity, is involved in neuronal death in neurological disorders as well as excess intracellular Ca^{2+} signaling. The hypothalamic-pituitary-adrenal (HPA) axis activity, i.e., glucocorticoid secretion, can potentiate glutamatergic neuron activity and modify synaptic Zn^{2+} signaling. This activity is enhanced by stress, aging, and zinc-deficient diet. Synaptic Zn^{2+} signaling is critical in both functional and pathological aspects and involved in β-amyloid-mediated cognitive decline. This chapter summarizes the physiological significance of intracellular Zn^{2+} signaling in cognition, in association with other divalent metals.

Keywords Zinc • Calcium • Copper • Glutamate • Hippocampus • Memory • Hypothalamic-pituitary-adrenal axis

15.1 Introduction

The importance of zinc is known in human health [1–4]. Over 1000 proteins require zinc for their functions in microorganisms, plants, and animals. Zinc powerfully influences cell division and differentiation [5]. Zinc deficiency in children is a nutritional and health problem in both developing and developed countries

A. Takeda (✉) • H. Tamano
Department of Neurophysiology, School of Pharmaceutical Sciences, University of Shizuoka, 52-1 Yada, Suruga-ku, Shizuoka 422-8526, Japan
e-mail: takedaa@u-shizuoka-ken.ac.jp

Y. Ogra, T. Hirata (eds.), *Metallomics*, DOI 10.1007/978-4-431-56463-8_15

[6, 7]. An estimated 17.3 % of the world's population is at risk of inadequate zinc intake [8]. The marginal deficiency of zinc is not always correlated with the decrease in plasma zinc in humans. On the other hand, dietary zinc deficiency easily decreases plasma zinc levels in experimental animals [9]. Plasma zinc is thought to be a major pool of zinc in the living body to transfer zinc to the tissues and organs. Zinc homeostasis in the living body is tightly controlled by both intestinal absorption and intestinal and renal excretions. Zrt- and Irt-like proteins (ZIP), which are responsible for the movement of Zn^{2+} into the cytoplasm, and a family of zinc transporters (ZnT), which are responsible for the transport of Zn^{2+} out of the cytoplasm, are involved in the absorption and excretions of zinc to maintain zinc homeostasis in the living body [10–13].

Zinc concentration in the plasma is approximately 15 μM. Zinc transports from the plasma to the brain extracellular fluid and the cerebrospinal fluid is strictly regulated by the brain barrier system, i.e., the blood–brain and blood–CSF barrier (Fig. 15.1). It is reported that zinc concentration in the cerebrospinal fluid (CSF) is approximately 0.15–0.38 μM [14–16]. Extracellular zinc concentration in the brain is estimated to be less than 1 μM [17]. Zinc transport into the brain is critical for brain function. Zinc concentration in the brain is increased with brain development and reaches around 200 μM in the adults [18]. The brain barrier system serves for zinc homeostasis to maintain healthy brain [19, 20]. Zinc homeostasis in the brain is closely associated with neurological disorders [21, 22] and is spatiotemporally altered in the process of neurological disorders [23]. Zinc dyshomeostasis is closely linked to pathophysiology of neurological disorders.

Approximately 80 % of the total brain zinc exists as zinc metalloproteins, and the remaining part is histochemically reactive as revealed by Timm's sulfide-silver staining method (Fig. 15.2d) [24, 25], based on the data that the removal of ZnT-3

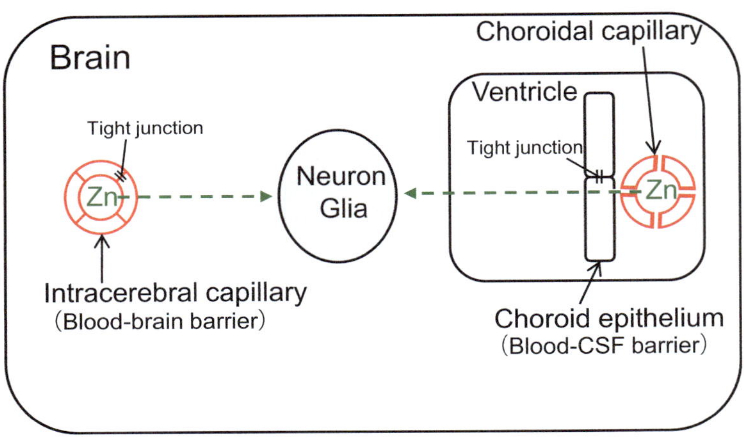

Fig. 15.1 The brain barrier system and zinc transport. The blood–brain and the blood–CSF barriers consist in the tight junction between the brain capillary endothelial cells and the tight junction between the choroidal epithelial cells, respectively, and are involved in zinc transport

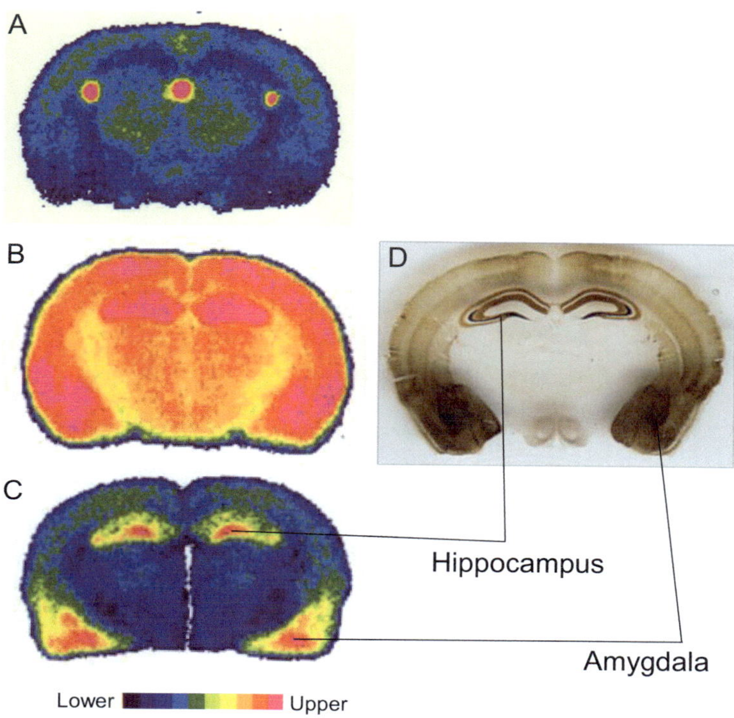

Fig. 15.2 ^{65}Zn concentration in the brain and vesicular zinc staining by Timm's method. ^{65}Zn concentration in coronal sections 1 h (**a**), 12 d (**b**), and 30 d (**c**) after intravenous injection of ^{65} $ZnCl_2$ into rats. Timm's stain in mouse coronal section (**d**)

protein, which is responsible for the movement of Zn^{2+} from the cytoplasm into synaptic vesicles [26], results in a 20 % reduction of the total amount of zinc in the brain [27]. Zinc is relatively concentrated in the hippocampus and amygdala (Fig. 15.2b, c) [28], which are enriched with histochemically reactive zinc (Fig. 15.2d). Histochemically reactive zinc predominantly exists in the presynaptic vesicles of glutamatergic (zincergic) neurons and serves as a signal factor (as Zn^{2+}) in the cytosolic compartment as well as the extracellular compartment. Zn^{2+} released from neuron terminals serves as an endogenous neuromodulator of several important receptors including α-amino-3-hydroxy-5-methyl-4-isoxazolepropionic acid (AMPA)/kainate receptor, N-methyl-D-aspartate (NMDA) receptors and γ-amino butyric acid (GABA) receptors [29, 30]. The Zn^{2+} is also taken up into presynaptic and postsynaptic neurons during synaptic excitation and serves for intracellular signaling [31, 32].

The hippocampus and amygdala are involved in cognitive and emotional behavior. Synaptic plasticity such as long-term-potentiation (LTP) is believed to be a cellular mechanism in learning and memory and has been widely studied at glutamatergic synapses. For the processing of the information on memory,

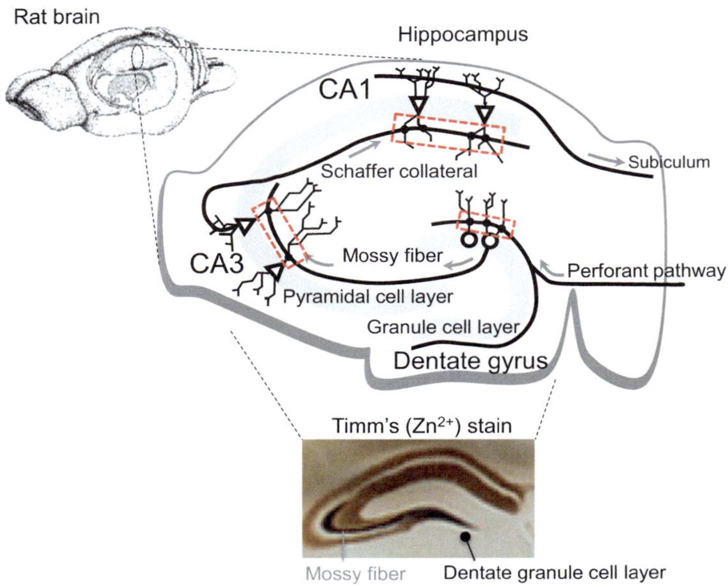

Fig. 15.3 Glutamatergic trisynaptic circuits in the hippocampus. Trisynaptic circuits, i.e., the perforant pathway, mossy fiber, and Schaffer collateral syanpses, which are surrounded by *dotted lines* are glutamatergic. A subclass of the three synapses contains zinc in the synaptic vesicles stained by Timm's method

glutamatergic neurons compose the neural circuit in the hippocampus and amygdala. The role of the neural circuit in processing memory has been extensively studied in the hippocampus (Fig. 15.3). Memory formation is linked to strengthening of synaptic connections between neurons, i.e., synaptic plasticity, within the dentate gyrus–CA3–CA1 trisynaptic circuits of the hippocampus. All of the mossy fibers and approximately 50 % of the Schaffer collateral, which were stained by Timm's method, contain zinc in the presynaptic vesicles (Fig. 15.3). The perforant pathway is minimally stained by Timm's method and the medial perforant pathway is non-zincergic.

It has been reported that hippocampal CA1 and CA3 are susceptible to neuronal death under neurological disorders and that Zn^{2+} released from zincergic neuron terminals is involved in neurodegeneration [33]. Neuronal death via excess Zn^{2+} influx into postsynaptic neurons as well as excess Ca^{2+} influx is observed in neurological disorders such as stroke/ischemia, temporal lobe epilepsy, Alzheimer's disease, and amyotrophic lateral sclerosis [34]. Therefore, homeostasis of intracellular Zn^{2+}, which is linked to Zn^{2+} release from zincergic neuron terminals, is critical for both functional and pathological aspects [35–38]. This chapter summarizes the physiological and pathological significance of intracellular Zn^{2+} signaling in cognition, in association with other divalent metals.

15.2 Zinc Transport Through the Brain Barrier System and Its Significance

A very stable environment via the brain barrier system is necessary for brain functions including cognition. The blood–brain barrier, which consists in the tight junction between the brain capillary endothelial cells (Fig. 15.1) [39], is completed approximately 2 weeks and 4 months after birth in rats and humans, respectively. The blood–CSF barrier, which consists in the tight junction between the choroidal epithelial cells (Fig. 15.1), is functional in early development [40]. Zinc is slowly transported into the brain through the blood–brain and the blood–CSF barriers [41, 42]. The cells forming the brain barrier system express ZIP and ZnT and play key roles to transport Zn^{2+} into the brain [43]. These transporters have specialized functions for cellular Zn^{2+} transport. However, their precise roles in Zn^{2+} transport across the brain barrier system are unknown [44, 30].

Zn^{2+} transport into the brain is linked to zinc concentration in the presynaptic vesicle in brain development. Timm's stain is hardly observed in the rat brain just after birth and the staining intensity increases along with brain maturation [45, 46], indicating that synaptic Zn^{2+} signaling is involved in not only the development and maturation of neural circuits but also brain functions. Zn^{2+} transport into the brain is necessary for brain functions after brain development. Dietary zinc deficiency readily reduces plasma zinc [9], while facilitating Zn^{2+} transport into the brain via changes in zinc transporter expression such as ZnT-1 and LIV-1 (SLC39A6 or ZIP6) is linked to the maintenance of zinc concentration in the brain [47]. However, chronic zinc deficiency decreases zinc concentration in the hippocampus of rats, like the case of feeding a zinc-deficient diet for 12 weeks [48]. In spite of the tight regulation of Zn^{2+} transport via the brain barrier system, hippocampal zinc is susceptible to chronic zinc deficiency. The decrease in plasma zinc may lead to the decrease in CSF zinc, which is important for Zn^{2+} transport into the hippocampus [42].

Zinc concentration in the brain is hardly affected by exposure to zinc. In a blood–brain barrier model, the expression of zinc transporters such as ZnT-1 and ZnT-2, also metallothioneins, and zinc-binding proteins is changed to respond to a moderately excessive zinc environment [49]. The brain barrier system protects brain parenchyma cells from toxic heavy metals such as cadmium, which is not usually transported into the brain. However, the brain capillary endothelial cells play an active part in the process of many neurological diseases. The blood–brain barrier is modified and/or disrupted in the process of neurological disorders such as neoplasia, stroke/ischemia, epilepsy, and dementia [50]. The increase in the blood–brain barrier permeability elicits the dyshomeostasis of essential metals such as zinc and iron, which has been implicated in many neurodegenerative diseases [51, 52].

15.3 Extracellular and Intracellular Zn^{2+} Homeostasis

At zincergic synapses such as mossy fiber synapses, the optimal range (level) of Zn^{2+} in the cytosolic compartment after synaptic excitation remains to be clarified, as well as in the extracellular compartment [37]. In both the extracellular and cytosolic compartments, basal (static) Zn^{2+} levels appear to be correlated with zinc concentration in the synaptic vesicles. Fluorescence intensity of extracellular and intracellular ZnAF-2, which estimates extracellular and intracellular Zn^{2+} levels, respectively [53], varied among areas in the hippocampus (Fig. 15.4) [54, 55]; the stratum lucidum and the hilus where mossy fibers exist are strongly stained with both fluorescence indicators, as well as Timm's staining for vesicular zinc (Fig. 15.3).

With regard to plasma zinc, Zn^{2+} concentration is estimated to be as low as $0.1 - 1$ nM [56] and potentially most susceptible to zinc deficiency, although plasma zinc is readily decreased in experimental animals by zinc deficiency. Zn^{2+} concentration in the brain extracellular fluid is estimated to be 10 nM [57] and is decreased by zinc deficiency; extracellular Zn^{2+} levels detected with ZnAF-2 are lower in the hippocampus of zinc-deficient rats than that of the control rats [55]. In the brain extracellular fluid, the high rate of Zn^{2+} concentration to the total zinc concentration seems to be associated with synaptic Zn^{2+} signaling. Extracellular Zn^{2+} concentration is dynamically changed at zincergic synapses and linked to intracellular Zn^{2+} signaling.

The basal Zn^{2+} concentrations are extremely low in the intracellular (cytosol) compartment ($<10^{-9}$ M) [58, 59]. ZnT proteins such as ZnT-1, ZnT-3, and ZnT-10 and ZIP such as ZIP4 and ZIP6 are involved in the control of Zn^{2+} levels in the cytosolic compartment, especially in the basal circumstance [60]. Some of these transporters transport cytosolic Zn^{2+} into different subcellular organelles, e.g., mitochondria, lysosomes, endosomes, and Golgi apparatus, probably to maintain the basal Zn^{2+} levels in the cytosolic compartment [61–63]. On the other hand, Zn^{2+} release from the subcellular organelles serves as intracellular Zn^{2+} signaling [64] and is involved in object recognition via dentate gyrus LTP expression at non-zincergic synapses [65]. Zn^{2+} levels except for vesicular Zn^{2+} are estimated to be less than 5 % of the total Zn^{2+} amount in the hippocampus and cerebral cortex [66]. ZnT-1 is

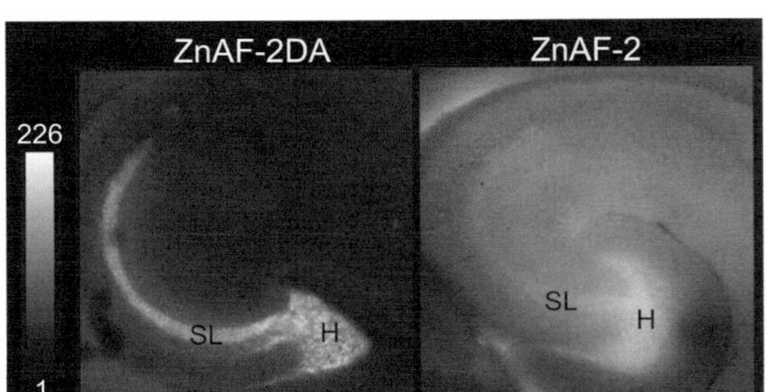

Fig. 15.4 Hippocampal Zn^{2+} imaging under the static condition. Intracellular (cytosolic) and extracellular Zn^{2+} levels are imaged with ZnAF-2DA (*left*) and ZnAF-2 (*right*), respectively. The fluorescence is strongly observed in the hilus (H) and the stratum lucidum (SL)

a major Zn^{2+} transporter on the plasma membrane and involved in cytosolic Zn^{2+} homeostasis in neurons by transporting Zn^{2+} from the somata to the extracellular space [67]. It has been reported that ZnT-1 restricts excessive accumulation of Zn^{2+} in the cytosolic compartment [68], resulting in protecting neurons from Zn^{2+} toxicity in neurological disorders such as transient forebrain ischemia [69]. Tissue plasminogen activator, a secreted serine protease, is excitotoxic and increases lysosomal sequestration of the increased Zn^{2+} in the cytosolic compartment through interaction with ZIP4, which might also lead to protecting neurons from Zn^{2+} toxicity [60]. The spatiotemporal control of intracellular Zn^{2+} signaling via ZIP and ZnT is important for the steady environment in the cytosolic compartment.

Zinc concentration in the brain extracellular fluid, which was measured by the in vivo microdialysis, is increased along with brain development, implying that extracellular Zn^{2+} is increased for developing synaptic functions associated with cognition, as well as the increase in zinc concentration in the synaptic vesicles [37]. On the other hand, the increase in extracellular zinc is suppressed under chronic zinc deficiency [70]; in the hippocampus, extracellular zinc concentration is significantly lower than in the hippocampus of the control rats after 4-week zinc deprivation [71, 72]. Intensity of Timm's stain is also lower in zinc-deficient rats than in the control rats. The decrease in plasma zinc suppresses the increase in zinc concentration in the brain extracellular fluid under dietary zinc deficiency, followed by suppression of the increases in zinc concentration in the synaptic vesicles and synaptic Zn^{2+}.

15.4 Functional Zn^{2+} Signaling to Toxic Zn^{2+} Signaling

Zn^{2+} release from zincergic neuron terminals is dependent on frequency of depolarizing stimulation [53]. The optimal range of extracellular Zn^{2+} concentration after depolarizing stimulation is a matter of debate. When extracellular Zn^{2+} is increased in the hippocampus, it modulates glutamatergic and GABAergic neuron activity; the increase decreases extracellular glutamate concentration, while increasing extracellular GABA concentration [73]. Extracellular Zn^{2+} can suppress glutamatergic excitation. Zn^{2+} potentiates AMPA/kainate receptors, while blocking Ca^{2+} current through calcium channels such as NNDA receptors and voltage-dependent calcium channels (VDCC). Zn^{2+} released from zincergic neuron terminals serves as a negative feedback factor against glutamate release [54, 32] (Fig. 15.5). When cytosolic Zn^{2+} is increased in the hippocampus, it can decrease not only the basal Ca^{2+} level [31] but also Ca^{2+} levels increased by metabotropic glutamate receptors. Crosstalk of synaptic Zn^{2+} signaling to intracellular Ca^{2+} signaling via calcium channels may modify presynaptic and postsynaptic neuronal activity (Fig. 15.5) [74, 75]. Intracellular Zn^{2+} signaling might also modify intracellular Ca^{2+} signaling via Zn^{2+}-dependent system.

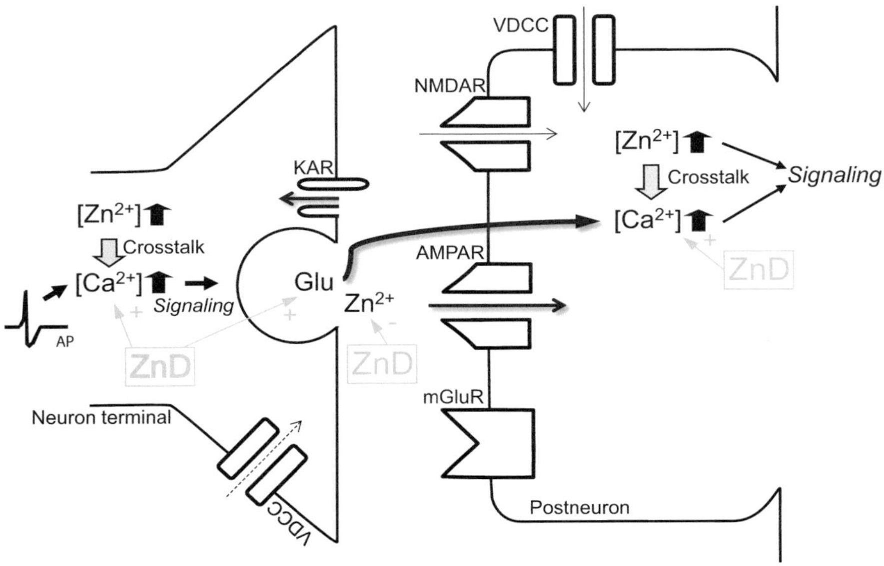

Fig. 15.5 Crosstalk of Zn^{2+} signaling to Ca^{2+} signaling at zincergic synapses and effect of zinc deficiency. Extracellular Zn^{2+} concentration is increased by the release from zincergic neuron terminals, followed by the increase in cytosolic Zn^{2+} concentration through calcium channels (AMPA receptors (calcium-permeable subtype), NMDA receptors, kainite (KA) receptor, and VDCC). The increase in Zn^{2+} concentration in both the extracellular and intracellular (cytosolic) compartments, which serves as a signaling factor, modulates presynaptic and postsynaptic neuronal activity via not only Zn^{2+}-dependent system but also the crosstalk to Ca^{2+} influx through calcium channels and intracellular Ca^{2+} signaling. Zinc deficiency (ZnD) increases cytosolic Ca^{2+} concentration via the action of glucocorticoids, followed by the enhanced presynaptic and postsynaptic activity. Chronic zinc deficiency can decrease synaptic Zn^{2+} signaling. AP, action potential; mGluR, metabotropic glutamate receptor

Glutamate receptor activation by excess extracellular glutamate leads to a number of deleterious consequences, including impairment of calcium buffering, generation of free radicals, activation of the mitochondrial permeability transition, and secondary excitotoxicity [76, 77]. Glutamate excitotoxicity, a final common pathway for neuronal death, is observed in numerous pathological processes such as stroke/ischemia, temporal lobe epilepsy, Alzheimer's disease, and amyotrophic lateral sclerosis. Zn^{2+} plays a neuroprotective role in glutamate-induced excitotoxicity [78, 79]. However, excess extracellular Zn^{2+}, which is induced by excess glutamatergic (zincergic) excitation, is harmful; excess Zn^{2+} influx into postsynaptic neurons, for which calcium-permeable AMPA receptors may play a key role, is involved in neurodegeneration [80–82]. Therefore, synaptic Zn^{2+} signaling is involved in both functional and pathological aspects in synaptic neurotransmission.

15.5 Zn^{2+} Signaling in Cognition

The importance of Zn^{2+} release in LTP induction has been reported in the hippocampal CA3 and CA1, based on the modulatory action of Zn^{2+} against NMDA receptors [83, 84]. Zn^{2+} released is quickly taken up into presynaptic and postsynaptic neurons and astrocytes during synaptic excitation through calcium channels such as calcium-permeable AMPA/kainate receptors [85, 31, 17]. Subsequently, it has been reported that intracellular Zn^{2+} signaling is involved in CA1 LTP [86] and mossy fiber LTP. In mossy fiber LTP, Zn^{2+} transactivates tropomyosin-related kinase B (TrkB) [87]. The involvement of synaptic Zn^{2+} in LTP has been usually examined in hippocampal slices bathed in agents such as Zn^{2+} chelators. Therefore, it is impossible to clarify region-specific Zn^{2+} action in LTP induction in the in vitro slice experiments. It is also impossible to assess the direct involvement of in vitro LTP induction in cognitive performance.

ZnAF-2DA is a useful tool to examine the direct involvement. When ZnAF-2DA is locally injected into the hippocampus, intracellular ZnAF-2 taken up into neurons and glial cell region specifically blocks intracellular Zn^{2+} signaling. The concurrent evaluation of in vivo LTP and learning behavior in separated experiments answers whether the in vivo LTP is involved in learning behavior. The influx of extracellular Zn^{2+} into CA1 neurons, which is linked to Zn^{2+} release from the Schaffer collateral, is required for object recognition via in vivo CA1 LTP [88] (Fig. 15.6). The basal Zn^{2+} levels in the cytosolic compartment are increased during learning and cognitive performance and the increase is required for memory. Intracellular Zn^{2+} signaling in the dentate granule cells, which originates in the internal stores containing Zn^{2+}, is also required for object recognition via LTP at medial perforant pathway-dentate granule cell synapses [65] (Fig. 15.6). The mechanism maintaining LTP via intracellular Zn^{2+} signaling in dentate granule cells, which may be involved in the formation of F-actin, also retains space recognition memory [89]. Interestingly, excess intracellular Zn^{2+} signaling in dentate granule cells, which originates in the influx of extracellular Zn^{2+} via AMPA receptor activation, affects object recognition [90]. Even in the dentate gyrus where it is scarcely innervated by zincergic neurons, the regulation of Zn^{2+} dynamics in the cytosolic compartment, which is closely linked to homeostasis of extracellular Zn^{2+}, is critical for learning and memory retention. In the CA3, block of Zn^{2+} released from mossy fibers has no effect on the consolidation of spatial learning, while Zn^{2+} is required for the consolidation of contextual fear conditioning [91]. Zn^{2+} released from mossy fibers may play a role in rapid storage of emotional memory [92].

Targeted deletion of ZnT-3 reduces Zn^{2+} release into the extracellular space by action potentials [27, 93]. There is the close correlation between vesicular zinc level and ZnT-3 protein expression [66, 26]. Zn^{2+} transport into the synaptic vesicle that is ZnT-3-dependent constitutes a major pool for synaptic Zn^{2+} signaling (Fig. 15.6). [94] reported that ZnT-3KO mice recognize novel or displaced objects. In the later experiments, however, ZnT-3KO mice exhibit significant cognitive deficits at 6-month-old age, but not at 3-month-old age [95]. Furthermore, 3- to 4-month-

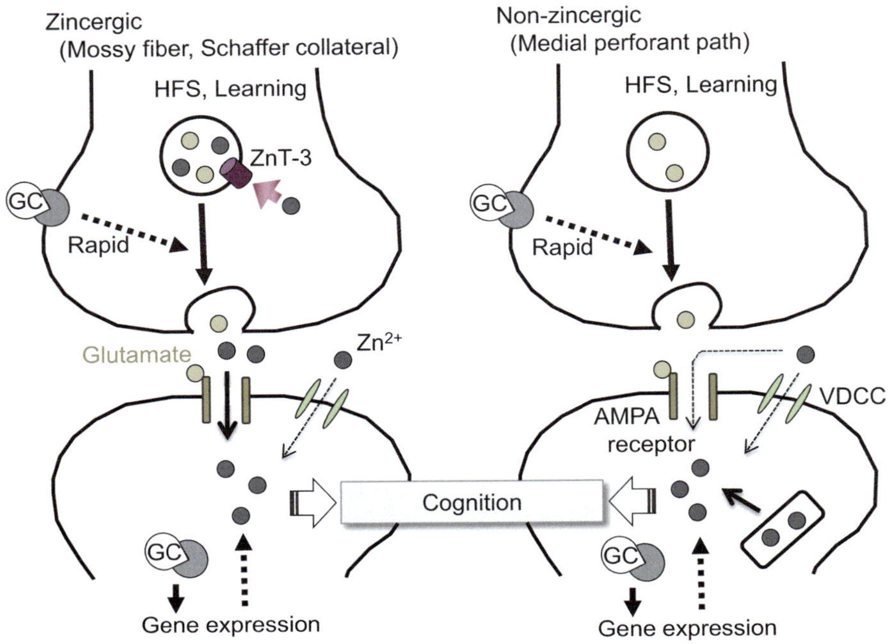

Fig. 15.6 Synaptic Zn^{2+} signaling in cognition. In zincergic synapses, the release of glutamate and Zn^{2+} during cognitive activity is modified via presynaptic glucocorticoid signaling and leads to Zn^{2+} influx into the postsynaptic neurons. This process induces intracellular Zn^{2+} signaling in the postsynaptic neurons. In non-zincergic synapses, glutamate release during cognitive activity is also modified via presynaptic glucocorticoid signaling. Extracellular glutamate signaling induces Zn^{2+} influx from the extracellular compartment and Zn^{2+} release from the internal stores. Zn^{2+} release from the internal stores is critical for intracellular Zn^{2+} signaling. Extracellular glutamate signaling may also be linked to intracellular glucocorticoid signaling in the postsynaptic neurons in both zincergic and non-zincergic neurons, which might induce sustained Zn^{2+} signaling. Calcium-permeable receptors and channels are involved in the rapid influx of Zn^{2+} during synaptic excitation, while Zn^{2+} transport system such as ZIP may be involved in the delayed influx of Zn^{2+} after synaptic excitation. GC, glucocorticoid

old ZnT-3KO mice exhibit significant deficits in contextual discrimination and spatial working memory [96, 97]. ZnT-3 is also involved in associative fear memory and extinction, but not in innate fear, consistent with the role of synaptic Zn^{2+} in amygdala synaptic plasticity [98]. Even in ZnT-3KO mice, synaptic Zn^{2+} signaling seems to be required for cognitive and emotional behavior, although it is unclear where it originates in. It is estimated that the amount of Zn^{2+} functioning as synaptic Zn^{2+} signaling is less available in ZnT-3KO mice than in wild-type mice.

Age-dependent reduction of Zn^{2+} level in the synaptic vesicle of the mossy fibers, which is linked to low expression of ZnT-3, causes the deterioration of learning and memory in senescence-accelerated mouse prone 10 (SAMP10) [99]. There is a significant age-related decline in cortical ZnT-3 level from age

48 to 91 in healthy people [95]. It is possible that the decrease in Zn^{2+} levels in the synaptic vesicles, which is linked to the decrease in ZnT-3 expression along with aging, is involved in age-dependent decline of cognitive performance [97]. However, there is no clear evidence on the relationship between the amount of Zn^{2+} release from zincergic neuron terminals or the increased amount of intracellular Zn^{2+} and cognitive ability.

15.6 Crosstalk of Zn^{2+} Signaling to Glucocorticoid Signaling in Cognition

The hypothalamic-pituitary-adrenal (HPA) axis activity, i.e., glucocorticoid secretion, is closely linked to hippocampal functions and involved in cognitive and emotional behavior [100]. The HPA axis activity can potentiate glutamatergic neuron activity, followed by the modification of synaptic Zn^{2+} signaling by the HPA axis activity [74, 75, 101]. It is well known that the HPA axis activity is enhanced along with aging and that the enhancement also appears in neurological disorders in addition to stress.

The hippocampus is enriched with corticosteroid receptors and is the major target region of corticosteroids [102]. Mineralocorticoid receptors and glucocorticoid receptors are colocalized in CA1 and CA2 pyramidal cells and in dentate gyrus granule cells. In CA3 pyramidal cells, on the other hand, mineralocorticoid receptors are abundantly expressed and glucocorticoid receptors are much less expressed [103]. Mineralocorticoid receptors are extensively occupied with low levels of corticosterone and glucocorticoid receptors are particularly activated after exposure to stress [104, 100]. Dentate granule cells require corticosterone levels at the physiological range; a selective degeneration of dentate gyrus granule cells is observed in the hippocampus after adrenalectomy [106].

An increase in plasma corticosterone level induces a rapid increase in corticosterone level in the hippocampus in parallel with an increase in extracellular glutamate level [107]. Corticosterone-induced increase in extracellular glutamate levels in the hippocampus seems to be exerted through the action of membrane-associated mineralocorticoid receptors and/or glucocorticoid receptors, which increases glutamate release probability in synaptic activation (Fig. 15.6) [108, 109]. The rapid effects of corticosterone on glutamatergic transmission seem to be linked to the diverse effects on synaptic plasticity and memory processes in the hippocampus. Corticosterone can contribute to increase the efficacy of glutamatergic transmission by AMPA receptor insertion at synaptic sites through both the rapid and the delayed (genomic) effects (Fig. 15.6). These effects are of advantage to the process of synaptic plasticity such as LTP and also learning and memory [100]. Corticosterone also increases Zn^{2+} release probability from zincergic neuron terminals through the rapid non-genomic effect in the hippocampus [105], while requiring intracellular Zn^{2+} signal for the genomic effect [101]. It is possible that synaptic Zn^{2+} signaling cooperates with corticosteroid signaling in learning and memory (Fig. 15.6).

In contrast, glutamate accumulates in the extracellular compartment at high levels through corticosterone-mediated blockade of glutamate transporter activity when corticosterone is excessively secreted under severe stress circumstances. The excess corticosterone secretion is also involved in excess glutamate release from neuron terminals. These effects facilitate long-term depression (LTD) and impair LTP and memory processing [110, 111]. Excess synaptic Zn^{2+} signaling induced by corticosterone and/or stress is involved in impairment of LTP [105], possibly followed by the impairment of learning and memory. Furthermore, prolonged exposure to a high level of corticosterone presumably makes dentate granule cells more vulnerable to delayed cell death in experimental animals [106]. Stress and corticosterone strongly inhibit adult hippocampal neurogenesis [112], which is involved in learning and memory, and decreased neurogenesis has been implicated in the pathogenesis of anxiety and depression. Neurogenesis-deficient mice show increased food avoidance in a novel environment after acute stress, increase behavioral despair in the forced swim test, and decrease sucrose preference, a measure of anhedonia; adult hippocampal neurogenesis buffers stress responses and depressive behavior [113]. The hippocampus provides negative control of the HPA axis [114–116]. The HPA axis activity is disrupted in approximately 50 % of human depressives. Interestingly, zinc-deficient diet is a stressor and feeding a zinc-deficient diet elevates the HPA axis activity (Fig. 15.7).

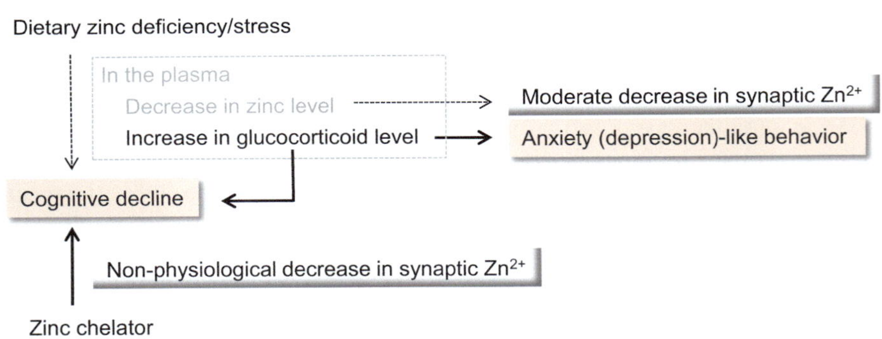

Fig. 15.7 Cognitive decline induced by dietary zinc deficiency. Dietary zinc deficiency induces cognitive decline, which is linked to the increase in plasma glucocorticoid level rather than the moderate decrease in synaptic Zn^{2+}. The increase in plasma glucocorticoid level induces anxiety (depression)-like behavior. In contrast, administration of Zn^{2+} chelators markedly (nonphysiologically) decreases synaptic Zn^{2+} level, followed by cognitive decline

15.7 Neuropsychological Symptoms and Cognitive Dysfunction in Zinc Deficiency

Food intake is reduced in young mice and rats within approximately 3 days after the start of feeding a zinc-deficient diet, and plasma zinc level is also reduced at this time (Fig. 15.7) [117]. Zinc homeostasis in the brain is not easily disrupted by dietary zinc deficiency [118]. However, zinc deficiency decreases extracellular and vesicular zinc in the hippocampus, followed by the decrease in hippocampal zinc concentration [48], suggesting that Zn^{2+} is more susceptible to zinc deficiency than ligand-bound zinc (Fig. 15.5). On the other hand, zinc deficiency increases plasma corticosterone concentration prior to decreasing extracellular and vesicular zinc concentrations in the hippocampus [9, 119]. Abnormal glucocorticoid secretion has been reported in many neuropsychiatric disorders including depression [120–122]. Human depressives are zinc deficient [123, 124] and depression-like behavior is increased in zinc-deficient animals (Fig. 15.7) [34, 35]. Zinc deficiency impairs neuronal precursor cell proliferation in the hippocampus [125, 126]. It is likely that abnormal corticosterone secretion is associated with depressive symptoms in zinc deficiency through the changes in hippocampal function (Fig. 15.7) [74, 127].

In the index of cognitive function, animals and human after weaning are more resistant to zinc deficiency than those before weaning [128–130]. Object recognition and dentate gyrus LTP were not affected when 4-week-old rats were fed a zinc-deficient diet for 4 weeks [55]. The decrease in synaptic Zn^{2+}, which is induced by dietary zinc deficiency, is moderate and does not affect object recognition (Fig. 15.7). Cognitive decline induced by zinc deficiency might be due to excess glucocorticoid secretion rather than decrease in zinc concentration in synaptic vesicle. In contrast, the decrease in Zn^{2+} by administration of Zn^{2+} chelators, which is nonphysiological, induces cognitive decline (Fig. 15.7) [65, 88].

15.8 Zn^{2+} Signaling in Cognitive Decline and Its Relation to Alzheimer's Disease

When excess excitation of glutamatergic synapses is induced by the local injection of high K^+, which induces excess intracellular Zn^{2+} signaling, subsequent object recognition is transiently affected; excess intracellular Zn^{2+} signaling in CA1 pyramidal cells, which is linked with excess Zn^{2+} release from neuron terminals, affects CA1 LTP and object recognition [131, 88]. Homeostasis of intracellular Zn^{2+}, which is closely linked to extracellular Zn^{2+} dynamics, is critical for cognition. Even in non-zincergic synapses, excess intracellular Zn^{2+} signaling in dentate granule cells affects object recognition via attenuation of dentate gyrus LTP [65].

Cognitive function normally declines along with aging and is believed to decline initially due to changes in synaptic function rather than loss of neurons [132]. Some individuals are led to develop Alzheimer's disease (AD) with progressive neurodegeneration. AD is the most common cause of dementia in the elderly and the pathogenesis is thought to need the progress for 20–30 years before clinical onset [133, 134]. β-Amyloid (Aβ) is the major component of neuritic senile plaques, which plays a key role in AD pathogenesis. The amyloid cascade hypothesis has proposed that the key event in AD development is the extracellular accumulation of insoluble, fibrillar Aβ [135, 136]. This hypothesis has been later modified to acknowledge soluble Aβ oligomers as pathogenic agents [137, 138]. One of the pathophysiological mechanisms mediated with Aβ is to alter synaptic functions such as LTP and LTD, prior to visible appearance of neuronal loss. It has been reported that soluble Aβ oligomers inhibit LTP induction and facilitate LTD induction [139–141]. The cognitive deficit and memory loss occurring before any prominent neuronal loss are observed in patients with mild cognitive impairment (MCI) or early-phase AD. Individuals with MCI have 32 % fewer neurons in the entorhinal cortex and show synaptic loss in the dentate gyrus, which correlates with cognitive deficits [142–144]. This cellular disconnection suggests that the dentate gyrus may be one of the earliest sites to display synaptic dysfunction [145]. Post-mortem studies also suggest that the hippocampus and entorhinal cortex are the first brain regions to be affected [146]. On the other hand, Aβ is normally produced in the brain, where the in vivo concentration in the rodent has been estimated to be in the picomolar range [147]. It is reported that Aβ levels in the brain extracellular fluid are linked to cognitive activity [148, 149].

The formation and accumulation of Aβ oligomers are enhanced at synaptic terminals after electrical and chemical stimulation when Aβ1-42 oligomers (500 nM) were added to hippocampal slices from rats and mice. Aβ oligomers are colocalized at the cell surface with NR2B receptor subunits. Clioquinol, an 8-hydroxyquinoline, is a lipophilic chelator for Zn^{2+} and Cu^{2+} and reduces the enhanced colocalization of Aβ oligomers [150]. Clioquinol has been shown to decrease Aβ deposits and to improve cognitive deficits in AD animal models. Clioquinol does not interact directly with Aβ, while it is selectively bound to the Aβ-metal ion complexes and dissociates the metal ion from Aβ [151]. The colocalization is also reduced in brain slices from vesicular zinc transporter ZnT-3 knockout mice [150]. The evidence suggests that synaptic Zn^{2+} and Cu^{2+} play critical roles in the formation of Aβ oligomers (Fig. 15.8), followed by glutamate excitotoxicity via activation of glutamate receptors containing NR2B subunits.

It is likely that crosstalk between metals and Aβ occurs under dynamic synapse environment. However, the involvement of the crosstalk in synaptic plasticity and cognitive activity is poorly understood, while it has been reported that preclinical and clinical data show the potential for metal chelation-based drug therapy for AD; clioquinol reduces zinc accumulation in neuritic plaques and inhibits the amyloidogenic pathway in the AβPP/PS1 transgenic mouse brain [152]. Clioquinol

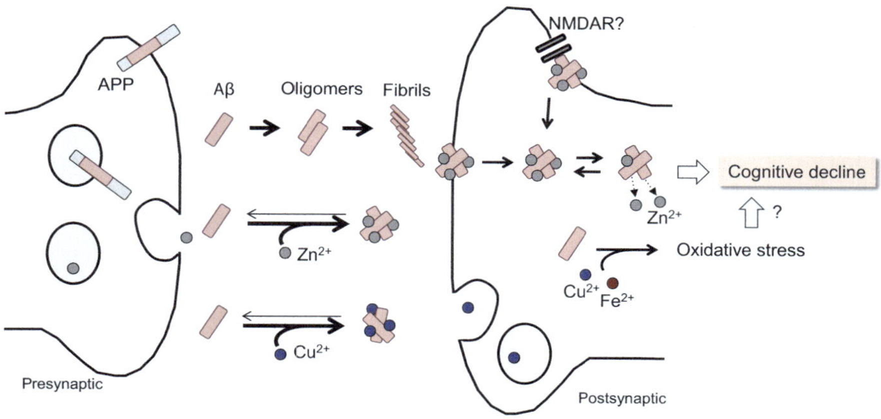

Fig. 15.8 Cognitive decline via interaction of metals and Aβ at the synapses. Aβ can be bound to Zn^{2+} and Cu^{2+} in the extracellular compartment, which often originates in the release from the presynaptic and postsynaptic terminals and from oligomers. Zn-Aβ1-42 oligomers may be taken up into postsynaptic neurons and transiently induce cognitive decline. Increase in intracellular Zn^{2+}, which is induced by the release from Zn-Aβ1-42, may lead to cognitive decline. Cu-Aβ1-42 oligomers may potentiate membrane disruption. Intraneuronal Aβ can interact with Cu^{2+} and Fe^{2+} and generate reactive oxygen species, potentially followed by the involvement in cognitive decline

also promotes the degradation of metal-dependent Aβ oligomers to restore endocytosis and ameliorates Aβ toxicity [153]. Multitarget-directed selenium-containing clioquinol derivatives show higher hydrogen peroxide scavenging and intracellular antioxidant activity than clioquinol, suggesting that the clioquinol derivatives are promising for the treatment of AD [154]. Furthermore, PBT2, a copper/zinc ionophore and second-generation 8-hydroxyquinoline analog, significantly lowers Aβ levels in the cerebrospinal fluid and improves cognitive performance over baseline in several key executive function tests [155, 156].

It is important to assess crosstalk between metals and Aβ in short-term memory loss for the prevention of AD pathogenesis. We examined an idea that short-term cognition is transiently affected by a state of confusion in Zn^{2+} transport system due to local increase in Aβ1-42 concentration in the normal brain. Aβ-mediated Zn^{2+} influx into dentate granule cells transiently induces a short-term cognitive deficit via the attenuation of dentate gyrus LTP [157]. In the attenuated LTP, the cognitive deficit is rescued by co-injection of either extracellular or intracellular Zn^{2+} chelators into the dentate gyrus, suggesting that extracellular Zn^{2+} in the dentate gyrus may play a key role in Aβ-induced cognition deficits (Fig. 15.8) [158] and is a potential target to prevent short-term cognitive deficits in the pre-dementia stage of AD.

15.9 Cu^{2+} Signaling in Cognition and Its Relation to Alzheimer's Disease

Cu^{2+} dynamics at the synapses and its involvement in cognition are poorly understood. Cu^{2+} is released postsynaptically after NMDA receptor activation (Fig. 15.8), which causes the rapid and reversible trafficking of Menkes ATPase to neuronal processes [159]. Menkes ATPase is directly required for the copper efflux. Synaptosomal copper release [160] and copper-mediated modulation of neuronal excitability [161] suggest that Cu^{2+} concentration is increased in the synaptic cleft during synaptic excitation. It is possible that extracellular Cu^{2+} and/or intracellular Cu^{2+} is involved in synaptic function such as changes in plasticity. Thus, extracellular Cu^{2+} levels might also be strictly regulated in the hippocampus for synaptic neurotransmission.

It is unknown whether crosstalk between Cu^{2+} and Aβ is involved in synaptic plasticity (Fig. 15.8). Cu^{2+} accentuates distinct misfolding of Aβ1-40 and Aβ1-42 peptides and potentiates membrane disruption via Cu-Aβ1-42 oligomers [162]. Inhibition of human high-affinity copper importer Ctr1 orthologous in the nervous system of drosophila ameliorates Aβ1-42-induced AD-like symptoms, suggesting that Cu^{2+} may play a causative role in AD pathogenesis, as either Aβ oligomers or aggregates are less toxic in the reduced copper environment or one with less copper binding [163]. A new Cu^{2+}-specific chelating agent, a bis-8-aminoquinoline PA1637, efficiently induces episodic memory recovery in a non-transgenic AD model after a single intracerebroventricular injection of Aβ1-42 [164], suggesting that Cu^{2+} play a key role in Aβ-induced cognition deficits (Fig. 15.8). In both the normal brain and the brain in the pre-dementia stage of AD, crosstalk between Cu^{2+} and Aβ in cognition is an important issue to be clarified.

On the other hand, copper and iron are redox active and generate reactive oxygen species via Fenton reaction and the Haber–Weiss reaction. Aβ become prooxidant when bound to copper or iron, which can generate hydrogen peroxide [165]. APP or Aβ interactions with Cu^{2+} induce reduction to Cu^{+} in vitro, promoting neurotoxic hydrogen peroxide production [166]. Interaction of Cu^{2+} and Aβ in the extracellular compartment reduces bioavailability of intracellular Cu^{2+}, followed by increase in oxidative stress via the reduced expression of antioxidative proteins such as superoxide dismutase [167], which may influence cognitive activity [168]. Although hydrogen peroxide has a bidirectional role in the regulation of hippocampal LTP [169], it is possible that the interaction of intraneuronal Aβ with copper and iron affects synaptic functions such as LTP via generation of reactive oxygen species (Fig. 15.8). Furthermore, extracellular Aβ oligomers with Zn^{2+} and Cu^{2+} may directly interact with astrocytes and microglia [167]. Thus it is also possible that inflammatory cytokines and chemokines, which are produced by their interaction, affect synaptic functions such as LTP.

15.10 Perspective

The crosstalk of synaptic Zn^{2+} signaling to intracellular Ca^{2+} signaling is involved in cognitive activity and seems to be influenced by the HPA axis activity. The HPA axis activity is increased by aging and zinc deficiency, and the increases are superimposed in neurological disorders such as depression and Alzheimer's disease [74, 75]. It is likely that that the modification of synaptic Zn^{2+} signaling through the HPA axis activation is involved in not only cognitive function but also cognitive decline. Molecular mechanism of intracellular Zn^{2+} signaling in cognition seems to be important to understand cognitive decline in normal aging and also in neurological symptoms such as dementia. An important issue that remains to be addressed is how homeostasis of Zn^{2+} is controlled in the brain and what determines the threshold between adaptive and maladaptive Zn^{2+} signaling under diverse circumstances. Significance of synaptic Cu^{2+} signaling in cognition remains to be clarified.

References

1. Deshpande A, Kawai H, Metherate R, Glabe CG, Busciglio J (2009) A role for synaptic zinc in activity-dependent Aβ oligomer formation and accumulation at excitatory synapses. J Neurosci 29:4004–4015
2. Ozawa H (2005) Steroid Hormones, their receptors and neuroendocrine system. J Nippon Med Sch 72:316–325
3. Lannfelt L, Blennow K, Zetterberg H, Batsman S, Ames D, Harrison J, Masters CL, Targum S, Bush AI, Murdoch R, Wilson J, Ritchie CW (2008) Safety, efficacy, and biomarker findings of PBT2 in targeting Abeta as a modifying therapy for Alzheimer's disease: a phase IIa, double-blind, randomised, placebo controlled trial. Lancet Neurol 7:779–786
4. Sindreu C, Palmiter RD, Storm DR (2011) Zinc transporter ZnT-3 regulates presynaptic Erk1/2 signaling and hippocampus-dependent memory. Proc Natl Acad Sci U S A 108:3366–3370
5. Gellein K, Skogholt JH, Aaseth J, Thoresen GB, Lierhagen S, Steinnes E, Syversen T, Flaten TP (2008) Trace elements in cerebrospinal fluid and blood from patients with a rare progressive central and peripheral demyelinating disease. J Neurol Sci 266:70–78
6. Martel G, Hevi C, Kane-Goldsmith N, Shumyatsky GP (2011) Zinc transporter 3 is involved in learned fear and extinction, but not in innate fear. Behav Brain Res 223:233–238
7. Chowanadisai W, Kelleher SL, Lönnerdal B (2005) Zinc deficiency is associated with increased brain zinc import and LIV-1 expression and decreased ZnT-1 expression in neonatal rats. J Nutr 135:1002–1007
8. Faux NG, Ritchie CW, Gunn A, Rembach A, Tsatsanis A, Bedo J, Harrison J, Lannfelt L, Blennow K, Zetterberg H, Ingelsson M, Masters CL, Tanzi RE, Cummings JL, Herd CM, Bush AI (2010) PBT2 rapidly improves cognition in Alzheimer's Disease: additional phase II analyses. J Alzheimers Dis 20:509–516
9. Takeda A, Tamano H (2012) Proposed glucocorticoid-mediated zinc signaling in the hippocampus. Metallomics 4:614–618
10. Bancila V, Nikonenko I, Dunant Y, Bloc A (2004) Zinc inhibits glutamate release via activation of pre-synaptic K_{ATP} channels and reduces ischaemic damage in rat hippocampus. J Neurochem 90:1243–1250

11. Takeda A, Tamano H (2014) Cognitive decline due to excess synaptic Zn^{2+} signaling in the hippocampus. Front Aging Neurosci 6:26
12. Gahtan E, Auerbach JM, Groner Y, Segal M (1998) Reversible impairment of long-term potentiation in transgenic Cu/Zn-SOD mice. Eur J Neurosci 10:538–544
13. Takeda A, Takada S, Nakamura M, Suzuki M, Tamano H, Ando M, Oku N (2011) Transient increase in Zn^{2+} in hippocampal CA1 pyramidal neurons causes reversible memory deficit. PLoS One 6:e28615
14. Barnham KJ, Bush AI (2008) Metals in Alzheimer's and Parkinson's diseases. Curr Opin Chem Biol 12:222–228
15. Gómez-Isla T, Price JL, McKeel DW, Jr Morris JC, Growdon JH, Hyman BT (1996) Profound loss of layer II entorhinal cortex neurons occurs in very mild Alzheimer's disease. J Neurosci 16:4491–4500
16. Cherny RA, Atwood CS, Xilinas ME, Gray DN, Jones WD, McLean CA, Barnham KJ, Volitakis I, Fraser FW, Kim Y, Huang X, Goldstein LE, Moir RD, Lim JT, Beyreuther K, Zheng H, Tanzi RE, Masters CL, Bush AI (2001) Treatment with a copper-zinc chelator markedly and rapidly inhibits beta-amyloid accumulation in Alzheimer's disease transgenic mice. Neuron 30:665–676
17. Snyder JS, Soumier A, Brewer M, Pickel J, Cameron HA (2011) Adult hippocampal neurogenesis buffers stress responses and depressive behaviour. Nature 476:458–461
18. Weiss JH (2011) Ca permeable AMPA channels in diseases of the nervous system. Front Mol Neurosci 4:42
19. Noh KM, Yokota H, Mashiko T, Castillo PE, Zukin RS, Bennett MV (2005) Blockade of calcium-permeable AMPA receptors protects hippocampal neurons against global ischemia-induced death. Proc Natl Acad Sci U S A 102:12230–12235
20. Nolte C, Gore A, Sekler I, Kresse W, Hershfinkel M, Hoffmann A, Kettenmann H, Moran A (2004) ZnT-1 expression in astroglial cells protects against zinc toxicity and slows the accumulation of intracellular zinc. Glia 48:145–155
21. Takeda A, Nakamura M, Fujii H, Uematsu C, Minamino T, Adlard PA, Bush AI, Tamano H (2014) Amyloid β-mediated Zn^{2+} influx into dentate granule cells transiently induces a short-term cognitive deficit. PLoS One 9:e115923
22. Whiteford HA, Peabody CA, Thiemann S, Kraemer HC, Csernansky JG, Berger PA (1987) The effect of age on baseline and postdexamethasone cortisol levels in major depressive disorder. Biol Psychiatry 22:1029–1032
23. Frederickson CJ, Giblin LJ, Krezel A, McAdoo DJ, Muelle RN, Zeng Y, Balaji RV, Masalha R, Thompson RB, Fierke CA, Sarvey JM, Valdenebro M, Prough DS, Zornow MH (2006) Concentrations of extracellular free zinc (pZn)e in the central nervous system during simple anesthetization, ischemia and reperfusion. Exp Neurol 198:285–293
24. Takeda A, Suzuki M, Tamano H, Takada S, Ide K, Oku N (2012) Involvement of glucocorticoid-mediated Zn^{2+} signaling in attenuation of hippocampal CA1 LTP by acute stress. Neurochem Int 60:394–399
25. Takeda A, Nakamura M, Fujii H, Tamano H (2013) Synaptic Zn^{2+} homeostasis and its significance. Metallomics 5:417–423
26. Matheou CJ, Younan ND, Viles JH (2015) Cu2+ accentuates distinct misfolding of Aβ(1–40) and Aβ(1–42) peptides, and potentiates membrane disruption. Biochem 466:233–242
27. Abbot NJ (2005) Dynamics of CNS barriers: evolution, differentiation, and modulation. Cell Mol Neurobiol 25:5–23
28. Cole CR, Lifshitz F (2008) Zinc nutrition and growth retardation. Pediatr Endocrinol Rev 5:889–896
29. Ohinata K, Takemoto M, Kawanago M, Fushimi S, Shirakawa H, Goto T, Asakawa A, Komai M (2009) Orally administered zinc increases food intake via vagal stimulation in rats. J Nutr 139:611–616
30. Frederickson CJ (1989) Neurobiology of zinc and zinc-containing neurons. Int Rev Neurobiol 31:145–238
31. Joëls M (2007) Role of corticosteroid hormones in the dentate gyrus. Prog Brain Res 163:355–370

32. Hyman BT, Van Hoesen GW, Damasio AR, Barnes CL (1984) Alzheimer's disease: cell-specific pathology isolates the hippocampal formation. Science 225:1168–1170
33. Takeda A (2011) Zinc signaling in the hippocampus and its relation to pathogenesis of depression. Mol Neurobiol 44:167–174
34. Black MM (1998) Zinc deficiency and child development. Am J Clin Nutr 68:464S–469S
35. Takeda A, Fuke S, Tsutsumi W, Oku N (2007) Negative modulation of presynaptic activity by zinc released from Schaffer collaterals. J Neurosci Res 85:3666–3672
36. Markesbery WR, Ehmann WD, Alauddin M, Hossain TIM (1984) Brain trace element concentrations in aging. Neurobiol Aging 5:19–28
37. Joëls M (2008) Functional actions of corticosteroids in the hippocampus. Eur J Pharmacol 583:312–321
38. Takeda A, Minami A, Seki Y, Oku N (2003) Inhibitory function of zinc against excitation of hippocampal glutamatergic neurons. Epilepsy Res 57:169–174
39. Takeda A, Tamano H, Ogawa T, Takada S, Nakamura M, Fujii H, Ando M (2014) Intracellular Zn^{2+} signaling in the dentate gyrus is required for object recognition memory. Hippocampus 24:1404–1412
40. Martel G, Hevi C, Friebely O, Baybutt T, Shumyatsky GP (2010) Zinc transporter 3 is involved in learned fear and extinction, but not in innate fear. Learn Mem 17:582–590
41. Black MM (2003) The evidence linking zinc deficiency with children's cognitive and motor functioning. J Nutr 133:1473S–1476S
42. Jia Y, Jeng JM, Sensi S, Weiss JH (2002) Zn^{2+} currents are mediated by calcium-permeable AMPA/kainite channels in cultured murine hippocampal neurons. J Physiol Lond 543:35–48
43. Ueno S, Tsukamoto M, Hirano T, Kikuchi K, Yamada MK (2002) Nishiyama N., Nagano T., Matsuki N. and Ikegaya Y., Mossy fiber Zn^{2+} spillover modulates heterosynaptic *N*-methyl-*D*-aspartate receptor activity in hippocampal CA3 circuits. J Cell Biol 158:215–220
44. Cirrito JR, May PC, O'Dell MA, Taylor JW, Parsadanian M, Cramer JW, Audia JE, Nissen JS, Bales KR, Paul SM, DeMattos RB, Holtzman DM (2003) *In vivo* assessment of brain interstitial fluid with microdialysis reveals plaque-associated changes in amyloid-β metabolism and half-life. J Neurosci 23:8844–8853
45. Cirrito JR, Yamada KA, Finn MB, Sloviter RS, Bales KR, May PC, Schoepp DD, Paul SM, Mennerick S, Holtzman DM (2005) Synaptic activity regulates interstitial fluid amyloid-beta levels in vivo. Neuron 48:913–922
46. Corniola RS, Tassabehji NM, Hare J, Sharma G, Levenson CW (2008) Zinc deficiency impairs neuronal precursor cell proliferation and induces apoptosis via p53-mediated mechanisms. Brain Res 1237:52–61
47. Howland JG, Wang YT (2008) Synaptic plasticity in learning and memory: stress effects in the hippocampus. Prog Brain Res 169:145–158
48. Bobilya DJ, Gauthier NA, Karki S, Olley BJ, Thomas WK (2008) Longitudinal changes in zinc transport kinetics, metallothionein and zinc transporter expression in a blood–brain barrier model in response to a moderately excessive zinc environment. J Nutr Biochem 19:129–137
49. Wong TP, Howland JG, Robillard JM, Ge Y, Yu W, Titterness AK, Brebner K, Liu L, Weinberg J, Christie BR, Phillips AG, Wang YT (2007) Hippocampal long-term depression mediates acute stress-induced spatial memory retrieval impairment. Proc Natl Acad Sci U S A 104:11471–11476
50. Bryan J, Osendarp S, Hughes D, Calvaresi E, Baghurst K, van Klinken JW (2004) Nutrients for cognitive development in school-aged children. Nutr Rev 62:295–306
51. Takeda A, Yamada K, Tamano H, Fuke S, Kawamura M, Oku N (2008) Hippocampal calcium dyshomeostasis and long-term potentiation in 2-week zinc deficiency. Neurochem Int 52:241–246
52. Weiss JH, Sensi SL, Koh JY (2000) Zn(2+): a novel ionic mediator of neural injury in brain disease. Trends Pharmacol Sci 21:395–401

53. Takeda A, Hirate M, Tamano H, Oku N (2003) Release of glutamate and GABA in the hippocampus under zinc deficiency. J Neurosci Res 72:537–542
54. Bush AI (2013) The metal theory of Alzheimer's disease. J Alzheimers Dis 33:S277–S281
55. Smart TG, Xie X, Krishek BJ (1994) Modulation of inhibitory and excitatory amino acid receptor ion channels by zinc. Prog Neurobiol 42:393–441
56. Vogt K, Mellor J, Tong G, Nicoll R (2000) The actions of synaptically released zinc at hippocampal mossy fiber synapses. Neuron 26:187–196
57. Sekler I, Moran A, Hershfinkel M, Dori A, Margulis A, Birenzweig N, Nitzan Y, Silverman WF (2002) Distribution of the zinc transporter ZnT-1 in comparison with chelatable zinc in the mouse brain. J Comp Neurol 447:201–209
58. Takeda A, Tamano H, Ogawa T, Takada S, Ando M, Oku N, Watanabe M (2012) Significance of serum glucocorticoid and chelatable zinc in depression and cognition in zinc deficiency. Behav Brain Res 226:259–264
59. Minami A, Sakurada N, Fuke S, Kikuchi K, Nagano T, Oku N, Takeda A (2006) Inhibition of presynaptic activity by zinc released from mossy fiber terminals during tetanic stimulation. J Neurosci Res 83:167–176
60. Mocchegiani E, Bertoni-Freddari C, Marcellini F, Malavolta M (2005) Brain, aging and neurodegeneration: role of zinc ion availability. Prog Neurobiol 75:367–390
61. Sandi C (2011) Glucocorticoids act on glutamatergic pathways to affect memory processes. Trends Neurosci 34:165–176
62. Querfurth HW, LaFerla FM (2010) Alzheimer's disease. N Engl J Med 362:329–344
63. Saito T, Takahashi K, Nakagawa N, Hosokawa T, Kurasaki M, Yamanoshita O, Yamamoto Y, Sasaki H, Nagashima K, Fujita H (2000) Biochem Biophys Res Commun 279:505–511
64. Dong XX, Wang Y, Qin ZH (2009) Molecular mechanisms of excitotoxicity and their relevance to pathogenesis of neurodegenerative diseases. Acta Pharmacol Sin 30:379–387
65. Brouillette J, Caillierez R, Zommer N, Alves-Pires C, Benilova I, Blum D, De Strooper B, Buée L (2012) Neurotoxicity and memory deficits induced by soluble low-molecular-weight amyloid-β1-42 oligomers are revealed in vivo by using a novel animal model. J Neurosci 32:7852–7861
66. Yokel RA (2006) Blood–brain barrier flux of aluminum, manganese, iron and other metals suspected to contribute to metal-induced neurodegeneration. J Alzheimers Dis 10:223–253
67. Seed JA, Dixon RA, McCluskey SE, Young AH (2000) Basal activity of the hypothalamic-pituitary-adrenal axis and cognitive function in anorexia nervosa. Eur Arch Psychiatry Clin Neurosci 250:11–15
68. Wang Z, Wang Y, Li W, Mao F, Sun Y, Huang L, Li X (2014) Design, synthesis, and evaluation of multitarget-directed selenium-containing clioquinol derivatives for the treatment of Alzheimer's disease. ACS Chem Neurosci 5:952–962
69. Valente T, Auladell C (2002) Developmental expression of ZnT3 in mouse brain: correlation between the vesicular zinc transporter protein and chelatable vesicular zinc (CVZ) cells. Glial and neuronal CVZ cells interact. Mol Cell Neurosci 21:189–204
70. Kamsler A, Segal M (2003) Paradoxical actions of hydrogen peroxide on long-term potentiation in transgenic superoxide dismutase-1 mice. J Neurosci 23:10359–10367
71. Takeda A, Itoh H, Imano S, Oku N (2006) Impairment of GABAergic neurotransmitter system in the amygdala of young rats after 4-week zinc deprivation. Neurochem Int 49:746–750
72. Mathie A, Sutton GL, Clarke CE, Veale EL (2006) Zinc and copper: pharmacological probes and endogenous modulators of neuronal excitability. Pharmacol Ther 111:567–583
73. Takeda A (2001) Zinc homeostasis and functions of zinc in the brain. BioMetals 14:343–352
74. Hopt A, Korte S, Fink H, Panne U, Niessner R, Jahn R, Kretzschmar H, Herms J (2003) Methods for studying synaptosomal copper release. J Neurosci Methods 128:159–172
75. Adlard PA, Parncutt JM, Finkelstein DI, Bush AI (2010) Cognitive loss in zinc transporter-3 knock-out mice: a phenocopy for the synaptic and memory deficits of Alzheimer's disease? J Neurosci 30:1631–1636

76. Ceccom J, Coslédan F, Halley H, Francès B, Lassalle JM, Meunier B (2012) Copper chelator induced efficient episodic memory recovery in a non-transgenic Alzheimer's mouse model. PLoS One 7:e43105
77. Palmiter RD, Cole TB, Quaife CJ, Findley SD (1996) ZnT-3, a putative transporter of zinc into synaptic vesicles. Proc Natl Acad Sci U S A 93:14934–14939
78. Prasad AS (2008) Zinc in human health: effect of zinc on immune cells. Mol Med 14:353–357
79. Wang T, Wang CY, Shan ZY, Teng WP, Wang ZY (2012) Clioquinol reduces zinc accumulation in neuritic plaques and inhibits the amyloidogenic pathway in AβPP/PS1 transgenic mouse brain. J Alzheimers Dis 29:549–559
80. Walsh DM, Klyubin I, Fadeeva JV, Cullen WK, Anwyl R, Wolfe MS, Rowan MJ, Selkoe DJ (2002) Naturally secreted oligomers of amyloid β protein potently inhibit hippocampal long-term potentiation in vivo. Nature 416:535–539
81. Tamano H, Kan F, Kawamura M, Oku N, Takeda A (2009) Behavior in the forced swim test and neurochemical changes in the hippocampus in young rats after 2-week zinc deprivation. Neurochem Int 55:536–541
82. Capasso M, Jeng JM, Malavolta M, Mocchegiani E, Sensi SL (2005) Zinc dyshomeostasis: a key modulator of neuronal injury. J Alzheimers Dis 8:93–108
83. Hershey CO, Hershey LA, Varnes A, Vibhakar SD, Lavin P, Strain WH (1983) Cerebrospinal fluid trace element content in dementia: clinical, radiologic, and pathologic correlations. Neurology 33:1350–1353
84. Sensi SL, Canzoniero LMT, Yu SP, Ying HS, Koh JY, Kerchner GA, Choi DW (1997) Measurement of intracellular free zinc in living cortical neurons: routes of entry. J Neurosci 15:9554–9564
85. Choo XY, Alukaidey L, White AR, Grubman A (2013) Neuroinflammation and copper in Alzheimer's disease. Int J Alzheimers Dis 2013:145345
86. Takeda A, Sawashita J, Okada S (1994) Localization in rat brain of the trace metals, zinc and manganese, after intracerebroventricular injection. Brain Res 658:252–254
87. Wessells KR, Brown KH (2012) Estimating the global prevalence of zinc deficiency: results based on zinc availability in national food supplies and the prevalence of stunting. PLoS One 7:e50568
88. Schlief ML, Craig AM, Gitlin JD (2005) NMDA receptor activation mediates copper homeostasis in hippocampal neurons. J Neurosci 25:239–246
89. Colvin RA, Fontaine CP, Laskowski M, Thomas D (2003) Zn2+ transporters and Zn2+ homeostasis in neurons. Eur J Pharmacol 479:171–185
90. Takeda A, Sawashita J, Okada S (1995) Biological half-lives of zinc and manganese in rat brain. Brain Res 695:53–58
91. Matlack KE, Tardiff DF, Narayan P, Hamamichi S, Caldwell KA, Caldwell GA, Lindquist S (2014) Clioquinol promotes the degradation of metal-dependent amyloid-β (Aβ) oligomers to restore endocytosis and ameliorate Aβ toxicity. Proc Natl Acad Sci U S A 111:4013–4018
92. Lang M, Fan Q, Wang L, Zheng Y, Xiao G, Wang X, Wang W, Zhong Y, Zhou B (2013) Inhibition of human high-affinity copper importer Ctr1 orthologous in the nervous system of Drosophila ameliorates Aβ42-induced Alzheimer's disease-like symptoms. Neurobiol Aging 34:2604–2612
93. Takeda A (2011) Insight into glutamate excitotoxicity from synaptic zinc homeostasis. Int J Alzheimers Dis 2011:491597
94. Starkman MN, Gebarski SS, Berent S, Schteingart DE (1992) Hippocampal formation volume, memory dysfunction, and cortisol levels in patients with Cushing's syndrome. Biol Psychiatry 32:756–765
95. Rosenblum WI (2014) Why Alzheimer trials fail: removing soluble oligomeric beta amyloid is essential, inconsistent, and difficult. Neurobiol Aging 35:969–974

96. Cohen-Kfir E, Lee W, Eskandari S, Nelson N (2005) Zinc inhibition of gamma-aminobutyric acid transporter 4 (GAT4) reveals a link between excitatory and inhibitory neurotransmission. Proc Natl Acad Sci U S A 102:6154–6159
97. Multhaup G, Schlicksupp A, Hesse L, Beher D, Ruppert T, Masters CL, Beyreuther K (1996) The amyloid precursor protein of Alzheimer's disease in the reduction of copper(II) to copper (I). Science 271:1406–1409
98. Aguilar-Alonso P, Martinez-Fong D, Pazos-Salazar NG, Brambila E, Gonzalez-Barrios JA, Mejorada A, Flores G, Millan-Perezpeña L, Rubio H, Leon-Chavez BA (2008) The increase in zinc levels and upregulation of zinc transporters are mediated by nitric oxide in the cerebral cortex after transient ischemia in the rat. Brain Res 1200:89–98
99. Arana GW, Baldessarini RJ, Ornsteen M (1985) The dexamethasone suppression test for diagnosis and prognosis in psychiatry. Arch Gen Psychiatry 42:1193–1204
100. Takeda A, Fujii H, Minamino T, Tamano H (2014) Intracellular Zn^{2+} signaling in cognition. J Neurosci Res 92:819–824
101. Nowak G, Szewczyk B, Pilc A (2005) Zinc and depression. An update. Pharmacol Rep 57:713–718
102. Takeda A, Minami A, Takefuta S, Tochigi M, Oku N (2001) Zinc homeostasis in the brain of adult rats fed zinc-deficient diet. J Neurosci Res 63:447–452
103. Takeda A, Tamano H, Tochigi M, Oku N (2005) Zinc homeostasis in the hippocampus of zinc-deficient young adult rats. Neurochem Int 46:221–225
104. Prasad AS (2014) Impact of the discovery of human zinc deficiency on health. J Trace Elem Med Biol 28:357–363
105. Takeda A, Tamano H (2015) Regulation of extracellular Zn^{2+} homeostasis in the hippocampus as a therapeutic target for Alzheimer's disease. Expert Opin Ther Targets 19:1–8
106. Puzzo D, Privitera L, Fa' M, Staniszewski A, Hashimoto G, Aziz F, Sakurai M, Ribe EM, Troy CM, Mercken M, Jung SS, Palmeri A, Arancio O (2011) Endogenous amyloid-β is necessary for hippocampal synaptic plasticity and memory. Ann Neurol 69:819–830
107. Takeda A, Fuke S, Ando M, Oku N (2009) Positive modulation of long-term potentiation at hippocampal CA1 synapses by low micromolar concentrations of zinc. Neuroscience 158:585–591
108. Suzuki M, Fujise Y, Tsuchiya Y, Tamano H, Takeda A (2015) Excess influx of Zn^{2+} into dentate granule cells affects object recognition memory via attenuated LTP. Neurochem Int 87:60–65
109. Venero C, Borrell J (1999) Rapid glucocorticoid effects on excitatory amino acid levels in the hippocampus: a microdialysis study in freely moving rats. Eur J Neurosci 11:2465–2473
110. Karst H, Berger S, Turiault M, Tronche F, Schütz G, Joëls M (2005) Mineralocorticoid receptors are indispensable for nongenomic modulation of hippocampal glutamate transmission by corticosterone. Proc Natl Acad Sci U S A 102:19204–19207
111. Takeda A, Hirate M, Tamano H, Nishibaba D, Oku N (2003) Susceptibility to kainate-induced seizures under dietary zinc deficiency. J Neurochem 85:1575–1580
112. Izumi Y, Auberson YP, Zorumski CF (2006) Zinc modulates bidirectional hippocampal plasticity by effects on NMDA receptors. J Neurosci 26:7181–7188
113. Joëls M, Karst H, DeRijk R, de Kloet ER (2008) The coming out of the brain mineralocorticoid receptor. Trends Neurosci 31:1–7
114. Pan E, Zhang XA, Huang Z, Krezel A, Zhao M, Tinberg CE, Lippard SJ, McNamara JO (2011) Vesicular zinc promotes presynaptic and inhibits postsynaptic long-term potentiation of mossy fiber-CA3 synapse. Neuron 71:1116–1126
115. Cole TB, Martyanova A, Palmiter RD (2001) Removing zinc from synaptic vesicles dose not impair spatial learning, memory, or sensorimoter functions in the mouse. Brain Res 891:253–265
116. Sensi SL, Ton-That D, Sullivan PG, Jonas EA, Gee KR, Kaczmarek LK, Weiss JH (2003) Modulation of mitochondrial function by endogenous Zn2+ pools. Proc Natl Acad Sci U S A 100:6157–6162

117. Takamura Y, Ono K, Matsumoto J, Yamada M, Nishijo H (2014) Effects of the neurotrophic agent T-817MA on oligomeric amyloid-β-induced deficits in long-term potentiation in the hippocampal CA1 subfield. Neurobiol Aging 35:532–536
118. Johansson PA, Dziegielewska KM, Liddelow SA, Saunders NR (2008) The blood-CSF barrier explained: when development is not immaturity. Bioessays 30:237–248
119. Takeda A, Tamano H (2009) Insight into zinc signaling from dietary zinc deficiency. Brain Res Rev 62:33–34
120. Nestor PJ, Scheltens P, Hodges JR (2004) Advances in the early detection of Alzheimer's disease. Nat Med 10(Suppl):S34–S41
121. Takeda A, Tamano H, Kan F, Itoh H, Oku N (2007) Anxiety-like behavior of young rats after 2-week zinc deprivation. Behav Brain Res 177:1–6
122. Ceccom J, Bouhsira E, Halley H, Daumas S, Lassalle JM (2013) Differential needs of zinc in the CA3 area of dorsal hippocampus for the consolidation of contextual fear and spatial memories. Learn Mem 20:348–351
123. Ceccom J, Halley H, Daumas S, Lassalle JM (2014) A specific role for hippocampal mossy fiber's zinc in rapid storage of emotional memories. Learn Mem 21:287–297
124. Cousins RJ, Liuzzi JP, Lichten LA (2006) Mammalian zinc transport, trafficking, and signals. J Biol Chem 281:24085–24089
125. Crews L, Masliah E (2010) Molecular mechanisms of neurodegeneration in Alzheimer's disease. Hum Mol Genet 19:R12–R20
126. Magneson GR, Puvathingal JM, Ray WJ Jr (1987) The concentrations of free Mg2+ and free Zn2+ in equine blood plasma. J Biol Chem 262:11140–11148
127. Maret W, Sandstead HH (2008) Possible roles of zinc nutriture in the fetal origins of disease. Exp Gerontol 43:378–381
128. Qian J, Noebels JL (2005) Visualization of transmitter release with zinc fluorescence detection at the mouse hippocampal mossy fibre synapse. J Physiol 566:747–758
129. Danbolt NC (2001) Glutamate uptake. Prog Neurobiol 65:1–105
130. Vallee BL, Falchuk KH (1993) The biological basis of zinc physiology. Physiol Rev 73:79–118
131. Dufner-Beattie J, Kuo YM, Gitschier J, Andrews GK (2004) The adaptive response to dietary zinc in mice involves the differential cellular localization and zinc regulation of the zinc transporters ZIP4 and ZIP5. J Biol Chem 279:49082–49090
132. Emmetsberger J, Mirrione MM, Zhou C, Fernandez-Monreal M, Siddiq MM, Ji K, Tsirka SE (2010) Tissue plasminogen activator alters intracellular sequestration of zinc through interaction with the transporter ZIP4. J Neurosci 30:6538–6547
133. Cole TB, Wenzel HJ, Kafer KE, Schwartzkroin PA, Palmiter RD (1999) Elimination of zinc from synaptic vesicles in the intact mouse brain by disruption of the ZnT3 gene. Proc Natl Acad Sci U S A 96:1716–1721
134. Fukada T, Kambe T (2011) Molecular and genetic features of zinc transporters in physiology and pathogenesis. Metallomics 3:662–674
135. Li S, Hong S, Shepardson NE, Walsh DM, Shankar GM, Selkoe D (2009) Soluble oligomers of amyloid β protein facilitate hippocampal long-term depression by disrupting neuronal glutamate uptake. Neuron 62:788–801
136. Liu S, Lau L, Wei J, Zhu D, Zou S, Sun HS, Fu Y, Liu F, Lu Y (2004) Expression of Ca(2+)-permeable AMPA receptor channels primes cell death in transient forebrain ischemia. Neuron 43:43–55
137. Maes M, D'Haese PC, Scharpé S, D'Hondt P, Cosyns P, De Broe ME (1994) Hypozincemia in depression. J Affect Disord 31:135–140
138. Stork CJ, Li YV (2010) Zinc release from thapsigargin/IP3-sensitive stores in cultured cortical neurons. J Mol Signal 5:5
139. Lee JY, Kim JS, Byun HR, Palmiter RD, Koh JY (2011) Dependence of the histofluorescently reactive zinc pool on zinc transporter-3 in the normal brain. Brain Res 1418:12–22
140. Bhatnagar S, Taneja S (2001) Zinc and cognitive development. Br J Nutr 85:S139–S145

141. Colvin RA, Laskowski M, Fontaine CP (2006) Zinquin identifies subcellular compartmentalization of zinc in cortical neurons. Relation to the trafficking of zinc and the mitochondrial compartment. Brain Res 1085:1–10
142. Colvin RA, Bush AI, Volitakis I, Fontaine CP, Thomas D, Kikuchi K, Holmes WR (2008) Insights into Zn2+ homeostasis in neurons from experimental and modeling studies. Am J Physiol Cell Physiol 294:C726–C742
143. Suh SW, Won SJ, Hamby AM, Fan Y, Sheline CT, Tamano H, Takeda A, Liu J (2009) Decreased brain zinc availability reduces hippocampal neurogenesis in mice and rats. J Cereb Blood Flow Metab 29:1579–1588
144. Morrison JH, Hof PR (1997) Life and death of neurons in the aging brain. Science 278:412–419
145. Takeda A, Akiyama T, Sawashita J, Okada S (1994) Brain uptake of trace metals, zinc and manganese, in rats. Brain Res 640:341–344
146. Holsboer F (1983) The dexamethasone suppression test in depressed patients: clinical and biochemical aspects. J Steroid Biochem 19:251–257
147. Salgueiro MJ, Zubillaga MB, Lysionek AE, Caro RA, Weill R, Boccio JR (2002) The role of zinc in the growth and development of children. Nutrition 18:510–519
148. Danscher G, Stoltenberg M (2005) Zinc-specific autometallographic in vivo selenium methods: tracing of zinc-enriched (ZEN) terminals, ZEN pathways, and pools of zinc ions in a multitude of other ZEN cells. J Histochem Cytochem 53:141–153
149. De Felice FG, Vieira MN, Bomfim TR, Decker H, Velasco PT, Lambert MP, Viola KL, Zhao WQ, Ferreira ST, Klein WL (2009) Protection of synapses against Alzheimer's-linked toxins: insulin signaling prevents the pathogenic binding of Abeta oligomers. Proc Natl Acad Sci U S A 106:1971–1976
150. Jou MY, Hall AG, Philipps AF, Kelleher SL, Lönnerdal BJ (2009) Tissue-specific alterations in zinc transporter expression in intestine and liver reflect a threshold for homeostatic compensation during dietary zinc deficiency in weanling rats. Nutr 139:835–841
151. Valente T, Auladell C, Pérez-Clausell J (2002) Postnatal development of zinc-rich terminal fields in the brain of the rat. Exp Neurol 174:215–229
152. Musazzi L, Milanese M, Farisello P, Zappettini S, Tardito D, Barbiero VS, Bonifacino T, Mallei A, Baldelli P, Racagni G, Raiteri M, Benfenati F, Bonanno G, Popoli M (2010) Acute stress increases depolarization-evoked glutamate release in the rat prefrontal/frontal cortex: the dampening action of antidepressants. PLoS One 5:e8566
153. Takeda A, Tamano H (2010) Zinc signaling through glucocorticoid and glutamate signaling in stressful circumstances. J Neurosci Res 88:3002–3010
154. Nakashima AS, Dyck RH (2009) Zinc and cortical plasticity. Brain Res Rev 59:347–373
155. Hardy JA, Higgins GA (1992) Alzheimer's disease: the amyloid cascade hypothesis. Science 256:184–185
156. Frederickson CJ, Danscher G (1990) Zinc-containing neurons in hippocampus and related CNS structures. Prog Brain Res 83:71–84
157. Frederickson CJ, Koh JY, Bush AI (2005) The neurobiology of zinc in health and disease. Nat Rev Neurosci 6:449–462
158. Ballabh P, Braun A, Nedergaard M (2004) The blood–brain barrier: an overview: structure, regulation, and clinical implications. Neurobiol Dis 16:1–13
159. Scheff SW, Price DA, Schmitt FA, Mufson EJ (2006) Hippocampal synaptic loss in early Alzheimer's disease and mild cognitive impairment. Neurobiol Aging 27:1372–1384
160. Mirescu C, Gould E (2006) Stress and adult neurogenesis. Hippocampus 16:233–238
161. Takeda A, Fuke S, Minami A, Oku N (2007) Role of zinc influx via AMPA/kainate receptor activation in metabotropic glutamate receptor-mediated calcium release. J Neurosci Res 85:1310–1317
162. Takeda A, Suzuki M, Tempaku M, Ohashi K, Tamano H (2015) Influx of extracellular Zn^{2+} into the hippocampal CA1 neurons is required for cognitive performance via long-term potentiation. Neuroscience 304:209–216

163. Tamano H, Minamino T, Fujii H, Takada S, Nakamura M, Ando M, Takeda A (2015) Blockade of intracellular Zn^{2+} signaling in the dentate gyrus erases recognition memory via impairment of maintained LTP. Hippocampus 25:952–962
164. Michalke B, Nischwitz V (2010) Review on metal speciation analysis in cerebrospinal fluid-current methods and results: a review. Anal Chim Acta 682:23–36
165. Carroll BJ, Feinberg M, Greden JF, Tarika J, Albala AA, Haskett RF, James NM, Kronfol Z, Lohr N, Steiner M, de Vigne JP, Young E (1981) A specific laboratory test for the diagnosis of melancholia: standardization, validation and clinical utility. Arch Gen Psychiatry 38:15–22
166. Belloni-Olivi L, Marshall C, Laal B, Andrews GK, Bressler J (2009) Localization of zip1 and zip4 mRNA in the adult rat brain. J Neurosci Res 87:3221–3230
167. Takeda A (2000) Movement of zinc and its functional significance in the brain. Brain Res Rev 34:137–148
168. Selkoe DJ (1991) The molecular pathology of Alzheimer's disease. Neuron 6:487–498
169. Greenough MA, Camakaris J, Bush AI (2013) Metal dyshomeostasis and oxidative stress in Alzheimer's disease. Neurochem Int 62:540–555

Chapter 16
Frontier of Development for Metallodrugs on the Basis of Metallomic Pharmacology and Medicinal Inorganic Chemistry

Hiroyuki Yasui and Yutaka Yoshikawa

Abstract Biological trace metals (bio-metals) such as iron (Fe), zinc (Zn), copper (Cu), and manganese (Mn) are essential to life and health of humans, and the success of platinum (Pt) drugs in the cancer chemotherapy has rapidly grown interest in developing inorganic pharmaceutical agents (metallodrugs) in medicinal chemistry, that is, medicinal inorganic chemistry, using either biologically essential metals or other trace metals with pharmacological functions. Transition metal complexes with unique chemical structures may be useful alternatives to the drug available to address some of the incurable diseases. Over thousand years after ancient people practiced many types of inorganic compounds, modern development of metallopharmaceutics has been achieved by P. Ehrlich and his coworkers, who found arsenic-containing arsphenamine (As) and established the concept of chemotherapy in 1910. Since the year, a wide variety of metallodrugs have been developed by many researchers, and some of them are clinically used worldwide involving cisplatin (Pt), auranofin (Au), and polaprezinc (Zn). In this chapter, we emphasize that metal complexes are an expanding of interest in the research field of treatment of lifestyle diseases such as diabetes mellitus and several cancers. Especially, we would like to review the current state for the research of metal-based nonclinical medicines for treatment of cancer and diabetes mellitus.

Keywords Metallodrugs • Medicinal inorganic chemistry • Permitted daily exposure (PDE)-based metallodrug design • Cancer chemotherapy • Pharmacotherapy of diabetes mellitus • Pharmacometallomics

---

H. Yasui (✉)
Department of Analytical and Bioinorganic Chemistry, Division of Analytical and Physical Sciences, Kyoto Pharmaceutical University, Kyoto, Japan
e-mail: yasui@mb.kyoto-phu.ac.jp

Y. Yoshikawa
Department of Health, Sports and Nutrition, Faculty of Health and Welfare, Kobe Women's University, Hyōgo, Japan

Y. Ogra, T. Hirata (eds.), *Metallomics*, DOI 10.1007/978-4-431-56463-8_16

16.1 Introduction

Since prehistoric periods, inorganic compounds have been used for many disease therapies after a wide variety of experiences, attempts, and practice by ancient people, for instance, mercury (Hg) for syphilis, magnesium (Mg) salts for intestinal treatments, and iron (Fe) salts for anemia. Such most crude preparations in those days were obtained from mineral, plant, and animal sources, and thus the people in those days could not believe that inorganic compounds will have the potency to treat diseases.

The considerable advances in organic chemistry since the last half of the eighteenth century made more interest in inorganic compounds bound with organic compounds (ligands). The most important concept has been established in 1909 and named as "chemotherapy." Paul Ehrlich (1854–1915) and Sahachiro Hata (1873–1938) discovered arsphenamine "*Salvarsan*," which is an organometallic compound with inorganic arsenic (As)–carbon bond and spirocheticidal activity. The *Salvarsan* was actually the first artificially and clinically used pharmaceutics [1].

Through the time of Paracelsus (1493–1541), who proposed to apply several inorganic drugs containing Hg, Sb, Pb, Cu, As, S, B, and Ag to treat diseases, interest in inorganic drugs has slowly increased in European countries. Following the establishment of "coordination theory" by Alfred Werner (1866–1919), a new concept "metals and life" has started with finding of urease, which was the first nickel (Ni) enzyme crystallized in 1926, and Ni was found at the active center in 1975. Approximately over thousands metalloproteins and metalloenzymes are now crystallographically analyzed.

On the other hand, the concept of essential trace metals in humans and animals has been proposed, and disorders of health due to deficiency of metals such as Fe, Zn, and Cu in humans have been observed in several countries and local area. In most cases, supplementation of these trace metals improved the disorders of human and animal health. On the basis of such background, a wide variety of metal-containing agents, metallodrugs, have been proposed, and some of them have been clinically used since the twentieth century, as shown in Fig. 16.1.

The most well-known example for using coordination compounds in disease treatment is the application of platinum (Pt) complexes in cancer chemotherapy. In 1969, Barnett Rosenberg and his coworkers observed the unusual phenomenon of filamentous growth of *E. coli* bacteria. This metamorphosis was induced by Pt^{2+} and Pt^{4+} ammine chloride complexes, which were generated in situ during electrolysis at the Pt electrodes, because the medium contained ammonium chloride. The compound responsible for the filamentation was due to the cis-diamminedichloroplatinum^{2+} (cisplatin). This cisplatin was found to be highly effective to treat the tumors in not only animals but also humans with testicular, ovarian, lung, bladder, heart, neck, and cervical cancers. The mechanism of the anticancer activity of cisplatin has been actively studied for many years; however, it is not yet fully understood. Nonetheless, new Pt complexes of second and third

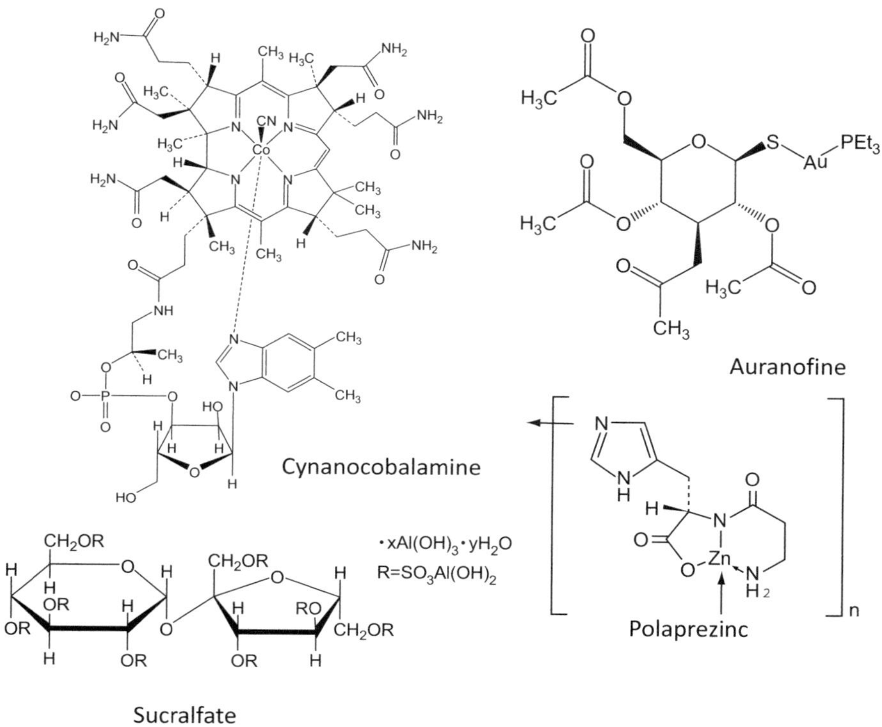

Fig. 16.1 Clinically used metallopharmaceutics

generations involving polynuclear Pt complexes against cisplatin-resistant tumors have been proposed, as shown in Fig. 16.2.

16.2 PDE-Based Concept for the Design and Evaluation of Metallodrugs

When a compound is aimed to develop as a clinically useful drugs, the following five points will be the most important factors; (1) occurrence of good leading compound, (2) high pharmacological activity, (3) low toxicity and low side effects, (4) proper pharmacokinetic (PK) and pharmacodynamic (PD) features in terms of its effective concentration in the blood or targeted organ, and (5) evidence for the pharmacological action mechanism. Among these points, the factors (1), (2), and (5) seem to be established in the frame of the research described in the review. However, the research on (3) and (4) should be important to be intensified in the future. In fact, approximately 30 % of the candidate organic compounds have failed to be developed as clinical drugs owing to the lack of suitable PK and PD features or

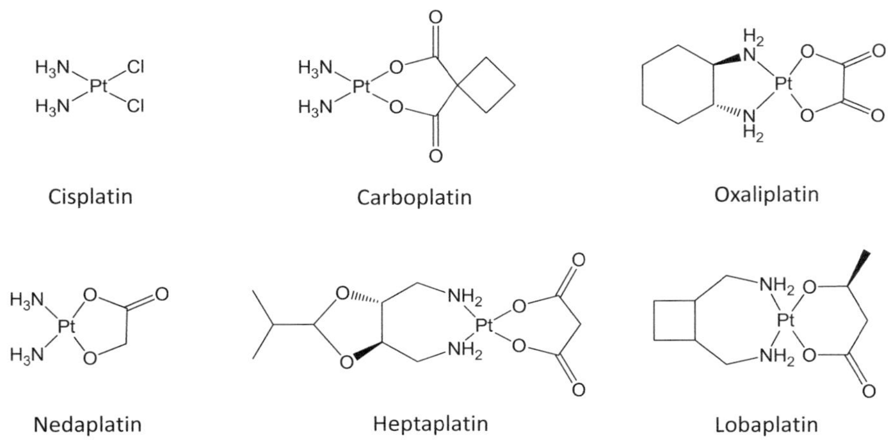

Fig. 16.2 Clinically used anticancer Pt complexes

the in vivo toxicity or side effects. On PK study, the analytical method and instrument have been requested to determine metallodrugs in the blood and organs of animals using atomic absorption spectroscopy or ICP-MS in terms of active metal elements of drugs, which will be important not only to know evidence for the pharmacological activity of metallodrugs but also to predict new metallodrug design.

On the toxicity and adverse effects of metallodrugs, careful and continuous observation of the animals with long-term administrations of drugs at least 1 or 2 years should be necessary. Then, as the index of safety margin for long term using metallodrugs in disease treatment, we would like to propose the permitted daily exposure (PDE)-based strategy for developing metallodrugs. Trace metal elements included as impurity have been classified as five levels (Classes 1, 2A, 2B, 3, S where we named the Class S originally as the lowest toxic classification) according to their PDE values in the guideline produced by the International Conference Harmonization for new drug approval among the USA, EU, and Japan (ICH) [2], as shown in Table 16.1. Briefly, the Class 1 includes the most toxic elements such as cadmium (Cd), lead (Pb), arsenic (As), and mercury (Hg). The Class 2 is divided to higher 2A [cobalt (Co), vanadium (V), and nickel (Ni)] and lower 2B [thallium (Tl), gold (Au), palladium (Pd), iridium (Ir), osmium (Os), rhodium (Rh), ruthenium (Ru), selenium (Se), silver (Ag), platinum (Pt)] toxic groups. The Class 3 has the relatively low toxic elements such as lithium (Li), antimony (Sb), barium (Ba), molybdenum (Mo), copper (Cu), tin (Sn), and chromium (Cr). Finally, the Class S which we named originally includes almost clean ten elements such as aluminum (Al), boron (B), calcium (Ca), iron (Fe), potassium (K), magnesium (Mg), manganese (Mn), sodium (Na), tungsten (W), and zinc (Zn). With respect to these PDE-dependent criteria, we would like to recommend the most little toxic Class S used as the treatment of lifestyle-related diseases such as diabetes mellitus and hypertension needed for longer medication period, while the secondary toxic Class

Table 16.1 Permissible daily exposure (PDE) values based on element contamination in pharmaceutics

Safety classes	Elements	PDE values (micro g/day)		
		Oral formula	Intravenous formula	Inhalation formula
Class 1	Cd	5	2	2
	Pb	5	5	5
	As	15	15	2
	Hg	30	3	1
Class 2A	Co	50	5	3
	V	100	10	1
	Ni	200	20	5
Class 2B	Tl	8	8	8
	Au	100	100	1
	Pd	100	10	1
	Ir	100	10	1
	Os	100	10	1
	Rh	100	10	1
	Ru	100	10	1
	Se	150	80	130
	Ag	150	10	7
	Pt	150	10	1
Class 3	Li	550	250	25
	Sb	1200	90	20
	Ba	1400	700	300
	Mo	3000	1500	10
	Cu	3000	300	30
	Sn	6000	600	60
	Cr	11,000	1100	3
Class S	Al	Not restricted		
	B			
	Ca			
	Fe			
	K			
	Mg			
	Mn			
	Na			
	W			
	Zn			

2 used as antimicrobial agents "original chemotherapy" and anticancer agents "modern chemotherapy."

Actually, for clinical pharmacotherapy as shown in Fig. 16.1, cynanocobalamin as Co-complex is vitamin B_{12}, auranofin as phosphine–Au–thiosugar complex is used for treatment of rheumatology, and sucralfate as Al–sugar complex and polaprezinc as Zn–carnosine complex have been used for gastric mucosal inflammation. Recently, zinc acetate is clinically applied to the treatment of Wilson diseases because of inhibiting copper absorption from gastrointestine. Topically, lanthanum (La) carbonate is newly approved to the treatment of hyperphosphatemia in patients with renal failure on the basis of the potent binding of lanthanum ions with phosphate anions. Additionally, for cancer chemotherapy as shown in Fig. 16.2, several platinum (Pt) complexes are used for treatment of testicular, ovarian, lung, bladder, heart, neck, and cervical cancers.

16.3 Metal-Based Nonclinical Medicines

16.3.1 The Target for Cancers

16.3.1.1 Introduction

The American Cancer Society (ACS) generates a summary containing current ACS cancer screening rates, as well as disease prevalence. Aside from skin cancer, the most common forms of cancer reported in 2015 were cancer of the female breast, prostate cancer in men, colorectal cancer, and lung cancer [3]. Moreover, according to a study by Bray et al. reported in Lancet Oncology, the collective incidence of all cancers will increase from 12.7 million new cases in 2008 to 22.2 million by 2030 [4]. Therefore, cancer is currently not only considered a leading cause of mortality but also a life-threatening disease for future generations [5, 6]. Decades of research have demonstrated that outcomes of cancer treatment vary widely in relation to the location of patient care and the widespread concerns and costs associated with treatment [7, 8].

Under the circumstances, Pt-containing medicines are one of the most commonly used chemotherapy drugs, applied for the treatment of a wide range of cancer types. Cisplatin is commonly used against several solid tumors, and oxaliplatin is an effective cytotoxic drug used in colorectal cancer. Their structures are shown in Fig. 16.2. A major clinical issue affecting 10–40 % of patients treated with cisplatin or oxaliplatin is severe peripheral neuropathy, causing sensory, motor, and autonomic dysfunction, with symptoms including cold sensitivity and neuropathic pain. Furthermore, several complexes, including cis-[PtCl $(NH_3)_2$(naph)]$^+$, [salophen]Pt, and [Pt(en)]$_2$ZL (ZL = 1-hydroxy-3-(1H-imidazol-1-yl)ethane-1,1-diylbisphosphonic acid), have been synthesized in recent years and found to produce anticancer or antiproliferative effects [9–11]. Unfortunately, drug resistance frequently occurs upon cancer recurrence, despite an initial response to the Pt treatment (Fig. 16.3).

For all of the above stated reasons, anticancer metal complexes have been a recent topic of investigation for cancer research. Using the PubMed database, we conducted a literature review on the topic by using the keywords "metal complex" and "anticancer." The results are shown in Fig. 16.4. Several articles indicate the potential clinical use of novel metal complexes as a possible alternative to cisplatin

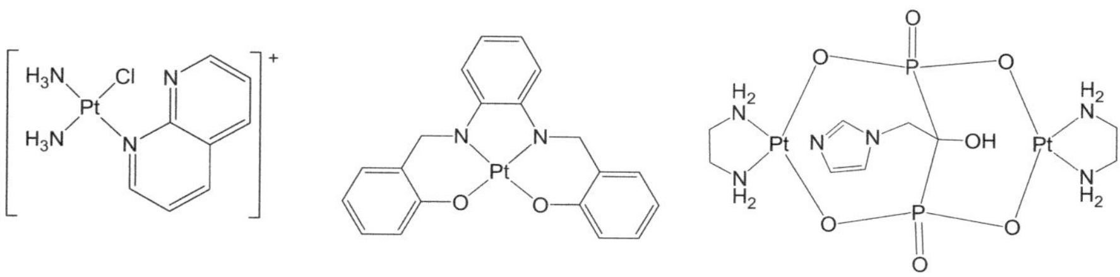

Fig. 16.3 Anticancer Pt complexes of various ligands with new concept

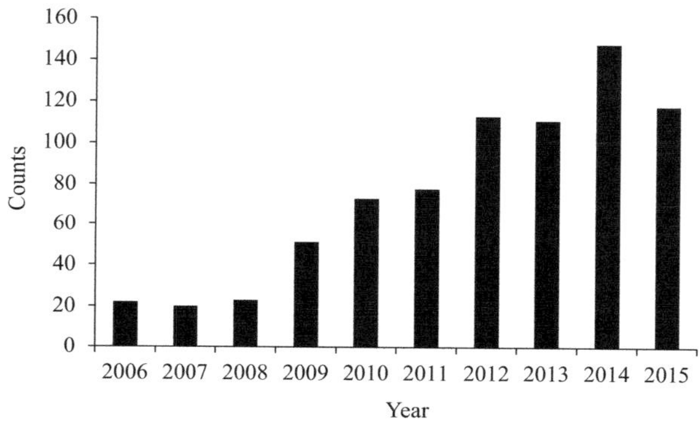

Fig. 16.4 The count of the papers by using the keywords "metal complex" and "anticancer" in the PubMed database

and the associated side effects. We provided the following example of anticancer metal complexes except for Pt.

16.3.1.2 Anticancer Class 1 Metal Complexes

In the Class 1, the metals are Cd, Pb, As, and Hg. It has been found that these elements have high toxicity since early times. Moreover, Cd has been classified by the International Agency for Research on Cancer as a human carcinogen. However, it has been reported that the dinuclear complex of Cd(II) with the condensation product of 2-acetylpyridine and malonic acid dihydrazide, N', N'2-bis[(1E)-1-(2-pyridyl)ethylidene]propanedihydrazide, showed the anticancer effect against the HeLa and murine melanoma B16 cells. This cytotoxic effect is caused by the interaction for minor groove of the double-helical linear and the supercoiled plasmid DNA [12]. Nowadays, Cd(II) complex with 2-formylpyridine selenosemicarbazone was the most effective in inhibiting HeLa cell invasion. This apoptotic mechanism operates in cancer cells via activation of the mitochondrial pathway [13].

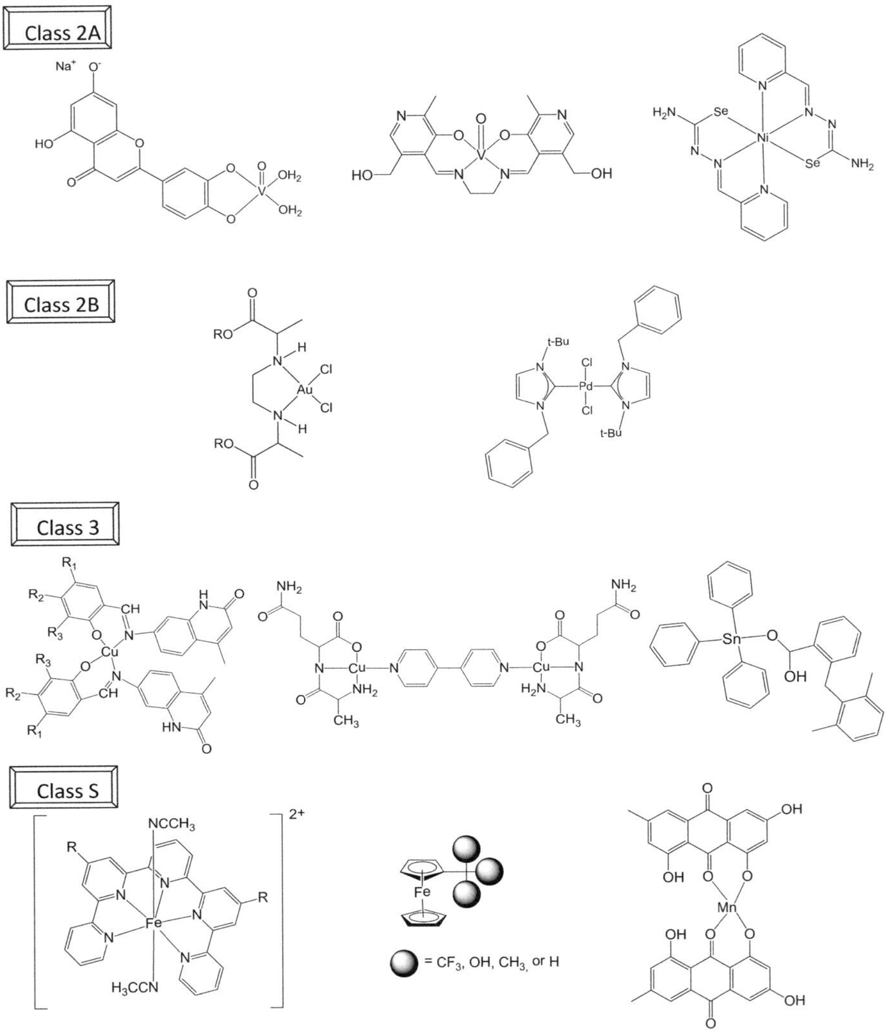

Fig. 16.5 The examples for the metal complexes of the Class 2A, 2B, 3, and S group with the anticancer activity in the nonclinical study

16.3.1.3 Anticancer Class 2A Metal Complexes (Fig. 16.5)

The anticancer activity, together with the low toxicity in humans, suggests vanadium as a candidate antineoplastic agent. Naso et al. reported that the complex of luteolin (lut) and VO(IV) produces anticancer effects. The antitumor effects were evaluated on MDAMB231 (human breast cancer cell line) cells and A549 lung cancer cell lines. When compared with the ligand, this complex exhibited higher inhibition of MDAMB231 cells viability. Oxidative stress processes were active during cancer cell killing. When metals chelated through the carbonyl group and one adjacent OH group of the flavonoid, an effective and functional improvement of the biological properties was observed [14]. Strianese et al. reported that pyridoxal-based VO(IV) complex shows selective cytotoxicity for cancer versus healthy cells. This effect could contribute the pro-apoptotic activity of the VO (IV) complex [15].

Regarding Ni(II) complexes, Zec et al. reported the below results. In in vitro transmembrane assays, an Ni(II) complex with 2-formylpyridine selenosemicarbazone was the most effective in inhibiting invasion of MDAMB361 (human breast cancer cell line) cells. This apoptotic mechanism operates in cancer cells via activation of the mitochondrial pathway [13].

16.3.1.4 Anticancer Class 2B Metal Complexes (Fig. 16.5)

Au(III) complexes have long been sought for anticancer treatments. Many Au(III) complexes have shown interesting anticancer potencies, but their pharmacological applications have always been hampered by their poor stability in solution [16]. Pantelića et al. synthesized five novel Au(III) complexes with the general formula $[AuCl_2\{(S,S)\text{-}R_2eddip\}]PF_6$ (where (S,S)-eddip represents (S,S)-ethylenediamine-N,N'-di-2-propanoate and R represents n-Bu, n-Pe, i-Bu, i-Am, or cPe). In vitro anticancer activity of the Au(III) complexes against human cervix adenocarcinoma HeLa, human myelogenous leukemia K562, and human melanoma Fem-x tumor cell lines, as well as against noncancerous human embryonic lung fibroblast cell line MRC-5, was determined using an MTT assay. Pantelića et al. revealed that the $[AuCl_2\{(S,S)\text{-}R_2eddip\}]PF_6$ (R = i-Am) complex showed the highest activity for their cancer cell lines [17].

Many new mononuclear, dinuclear, and multinuclear Pd complexes with reduced cross-resistance to cisplatin, decreased toxicity, and high specificity have been developed. Similar to platinum agents, DNA is also their major target in the cell. The Pd(II) ions are capable of interacting with DNA, thus enabling cross bindings and inhibiting its synthesis as well as inducing apoptosis [18]. Specifically, bis(*N*-heterocyclic carbene-Pd(II)) complex showed more cytotoxicity than cisplatin against HeLa, breast cancer (MCF-7), and colon carcinoma (HCT 116) cell lines. Further studies showed that this complex inhibited tumor cell proliferation by arresting the cell cycle progression at the G2 phase, preventing the mitotic entry of the cell [19].

16.3.1.5 Anticancer Class 3 Metal Complexes (Fig. 16.5)

The Cu(II) complex, $[Cu_2(C_{10}H_8N_2)(C_{12}H_{11}O_4N)_2(CH_3OH)_2]\cdot 2H_2O$, was developed by Qin et al. Within this structure, $C_{10}H_8N_2$ is structured as 4,4′-bipyridine, and H_2-($C_{12}H_{11}O_4N$) is structured as beta-[(3-formyl-5-methyl-2-hydroxy-benzylidene)-amino]propionic acid. Both of these structures have a Schiff base ligand derived from β-alanine [20]. This complex consists of two Cu(II) cations, one 4,4′-bipyridine ligand molecule, two coordinated methanol molecules, two Schiff base ligand anions, and two crystal water molecules. Moreover, the Cu (II) cation has the potential to form five bonds and is coordinated to form a slightly distorted square pyramidal arrangement. Qin et al. investigated the cytotoxicity of this compound on human umbilical vein endothelial cells (HUVECs), HeLa cells, and human-derived hepatic carcinoma cell (HepG2) by using an MTT assay to determine its potential anticancer properties and to evaluate its biological function. The results indicated that this Cu(II) complex inhibited the growth of HUVECs and HeLa cells. Furthermore, Duff et al. reported another Cu(II) complex. Their study determined the cytotoxic, cyto-selective, and mutagenic properties of the novel quinolinone Schiff base ligand ((7E)-7-(3-ethoxy-2-hydroxybenzylideamino)-4-methylquinolin-2(1H)-one) and its corresponding Cu(II) complexes in HepG2 cells and nonmalignant human-derived hepatic cells [21]. These results indicated that complexation of quinolinone Schiff bases with Cu(II) significantly enhanced cytotoxicity. The Cu(II) complex exhibited the same therapeutic effect as did cisplatin. Cu(II) complexes are expected to be the next generation of Pt medicine, with effective anticancer potential, similar to that of cisplatin.

On the other hand, it has been reported that organotin(IV) compounds have emerged as potentially biologically active compounds among non-platinum chemotherapeutic metallopharmaceuticals. Dakorou et al. reported that triphenyltin (IV) complexes of NSAID derivatives showed the strong anticancer activity (nM order) against A-549 (adenocarcinomic human alveolar basal epithelial cells) and MCF-7 (human breast adenocarcinoma cell line) cancer cell lines [22]. The cytotoxic mechanism of Sn complexes will be elucidated in the near future.

16.3.1.6 Anticancer Class S Metal Complexes (Fig. 16.5)

Fe complexes have long been used for the treatment of various diseases. Fe–bleomycin (FeBLM) [23] is the most notable example, producing oxidative DNA damage in the presence of O_2 and H_2O_2 [24]. Sun et al. have prepared Fe (II) complexes that contain a pentadentate pyridyl ligand and are stable under physiological conditions. These Fe(II) complexes exhibit acute cytotoxicity over a series of select carcinoma cell lines. By means of MTT assays, they were found to be more potent than the clinically used FeBLM [25]. Moreover, Sun et al. found that the mechanism of cell death for these Fe(II) complex-treated HeLa cells was apoptosis. This complex also induced the generation of reactive oxygen species (ROS) in vitro, as measured by the formation of green fluorescent rhodamine 123^+ from the nonfluorescent probe dihydrorhodamine 123 [25]. Moreover, Maschke et al. reported that trifluoromethyl-containing metallocenes with Fe(II) atoms have

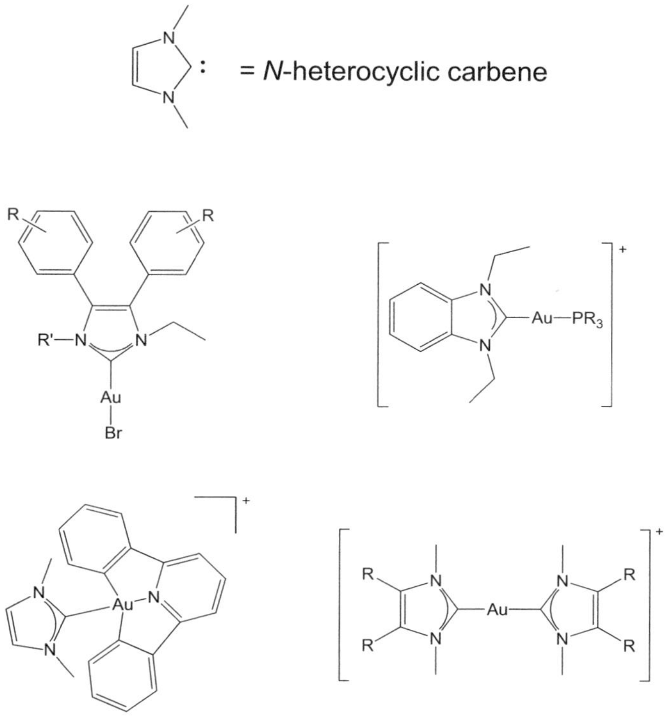

Fig. 16.6 Structures of the *N*-heterocyclic carbene and Au complexes with its derivatives

cytotoxic effects on a range of cancer cell lines [26]. Considering the efficacy of these complexes on a variety of cancer cell lines, it is expected that these Fe (II) complexes will be used for therapeutic purposes in clinical settings in the future.

In 2014, Li et al. reported that the emodin–Mn(II) complex can inhibit the proliferation of cancer cells. Its IC_{50} values is 2–3 times less than cisplatin. Moreover, the complex can make cell cycle arrest at G0/G1 phase in vitro, which leads to inhibit the proliferation of cancer cells [27]. The essential elements like Cu or Mn have a chance of anticancer medicines in the future.

16.3.1.7 Anticancer Au-*N*-Heterocyclic Carbene Complexes (Fig. 16.6)

As potential anticancer medications, metal–NHCs (NHC: *N*-heterocyclic carbene) constitute a recent and very rapidly growing field of research interest [28]. In general, metal–NHCs are efficient as apoptosis-triggering species. Wanzlick et al. reported the first synthesis of stable metal–NHCs in 1968 [29]. Moreover, metal–NHCs have established their selective performance for cancer cells excluding normal healthy cells based on which they are widely utilized for targeting cancer cells specifically which leads to cell death or cell growth inhibition [30]. Subsequently, various metal–carbene complexes were synthesized by many researchers [31, 32].

Especially, Au(I) species may display non-cisplatin-like pharmacokinetic and pharmacodynamic properties and thus have attracted great interest because they serve as potential treatments for cisplatin-resistant cells. Nowadays, several Au (I) complexes, and more recently, Au(III) complexes based on different ligand families, have attracted interest as potential antitumor agents. Their mechanisms of action mainly revolve around the mitochondrion, even though other targets of activity, including DNA, cannot be completely excluded. Many Au(I)–NCH or Au(III)–NHC complexes (cation type, neutral type, etc.) were synthesized by researchers and were examined in in vitro and in in vivo studies [33–37]. According to a report recently released by Rubbiani et al., Au(I)–NHC complexes with phosphane ligands have an inhibitory effect on thioredoxin and trigger cytotoxicity in cancer cells [38]. Some metal–NHC complexes exhibit cytotoxic potencies similar to and frequently greater than that of cisplatin. These metal–NHC complexes are widely anticipated for clinical application.

Many metal complexes have reportedly displayed anticancer activity. Generally, it has been suspected that the other elements are better than essential bio-elements for the chemotherapeutic agents, because, when we used the essential bio-elements, the living body showed the effect of homeostasis. Thus, it is difficult to increase the bio-element concentration in the living body. Interestingly, Au complexes may be promising new inorganic compound. While many of the complexes are currently still being studied through in vitro experiments, if they display cancer cell-specific activity, there is potential for their use in clinical medicine.

16.3.2 The Target for Lifestyle-Related Diseases

16.3.2.1 Introduction

The number of patients with lifestyle-related diseases has been increasing as a result of drastic changes in lifestyle, diet, and car ownership rate. Specifically, the prevalence of diabetes mellitus (DM), hypertension, and osteoporosis has rapidly increased worldwide. DM is associated with increased risks of several serious complications, such as retinopathy, renal disorders, and nerve disorders [39]. Individuals with type 2 DM exhibit impaired insulin sensitivity or insulin resistance. The long-term complications of this disease include the development of vascular disorders such as retinopathy, nephropathy, and angiopathy [40]. Moreover, hypertension is a pathological state independently associated with an increased risk of cardiovascular diseases [41, 42]. Many epidemiological studies have also demonstrated a correlation between insulin resistance and blood pressure (BP), suggesting that insulin resistance is a risk factor in the development of hypertension in humans [43].

On the other hand, there is also concern regarding the increase in the incidence of osteoporosis as the population ages [44]. The World Health Organization defines osteoporosis as a systemic degeneration of the skeleton, characterized by the loss of bone mass (in both organic and mineral bone components) and the structural degeneration of bone tissue, leading to an increase in bone fractures. The risk of fracture is also increased by factors such as lifestyle, drug treatments, family history, and other conditions that cause secondary osteoporosis. About 40 % of women in developed countries will experience an osteoporosis-related fracture in the course of their lifetime, with men facing approximately one-third to one-half the risk of women [45]. In recent years, it has been reported that adequate calcium and vitamin D intake must be coupled with any established treatment. Insufficient calcium intake and poor vitamin D status represent the most common factors contributing to the unresponsiveness of patients to conventional antiosteoporotic therapies [46].

In order to identify possible drugs for the treatment of lifestyle-related diseases, most researchers have investigated organic compounds [47, 48]. However, these compounds sometimes produce severe side effects, and thus, many researchers sought the development of new treatments for these diseases. Zn with atomic number 30, atomic weight 65.39, and oxidation state +2 has been established as an essential element in all living systems and plays a structural role in many proteins and enzymes. Zn(II), an essential metal, is a critical component of Zn (II)-binding proteins such as Zn(II) fingers, Zn(II) enzymes, and metallothioneins. Many proteins have been found to have a Zn(II)-containing motif embedded in their structure. Moreover, Zn(II) is known to be one of the most important essential trace elements of all biological systems and is less toxic than other trace elements [49, 50].

16.3.2.2 The Structure of Zn(II) Complexes Aimed at Improving Lifestyle-Related Diseases

Many researchers revealed the structures of elements within Zn(II) complexes by using an X-ray spectrometer. A colorless single crystal of di(maltolato)Zn(II) ([Zn (ma)$_2$]) was determined to be *t*-[Zn(ma)$_2$(H$_2$O)$_2$][Zn(ma)$_2$(H$_2$O)] · (H$_2$O)$_2$ by X-ray structure analysis [51]. Two different geometries around the Zn(II)s in the complex *t*-[Zn(ma)$_2$(H$_2$O)$_2$][Zn(ma)$_2$(H$_2$O)] · (H$_2$O)$_2$ were revealed to be in a unit cell, where two maltol and two H_2O molecules combine with a Zn(II) (arranged in a *trans* form in an octahedral conformation) and two maltol molecules and an apical H_2O molecule coordinate with Zn(II) (in a square pyramidal conformation) (Fig. 16.7a). The structure of di(thiomaltolato)Zn(II) ([Zn(tma)$_2$]) also shows a distorted tetrahedral coordination sphere (Fig. 16.7b) due to the differing Zn–S and Zn–O bond lengths. The ([Zn(tma)$_2$]) is distinct from the structurally characterized [Zn(ma)$_2$(H$_2$O)$_2$], which is comprised of both six- and five-coordinate complexes [52]. Presumably, the strong donor capacity of thiomaltol precludes the acquisition of additional solvent ligands in the solid state.

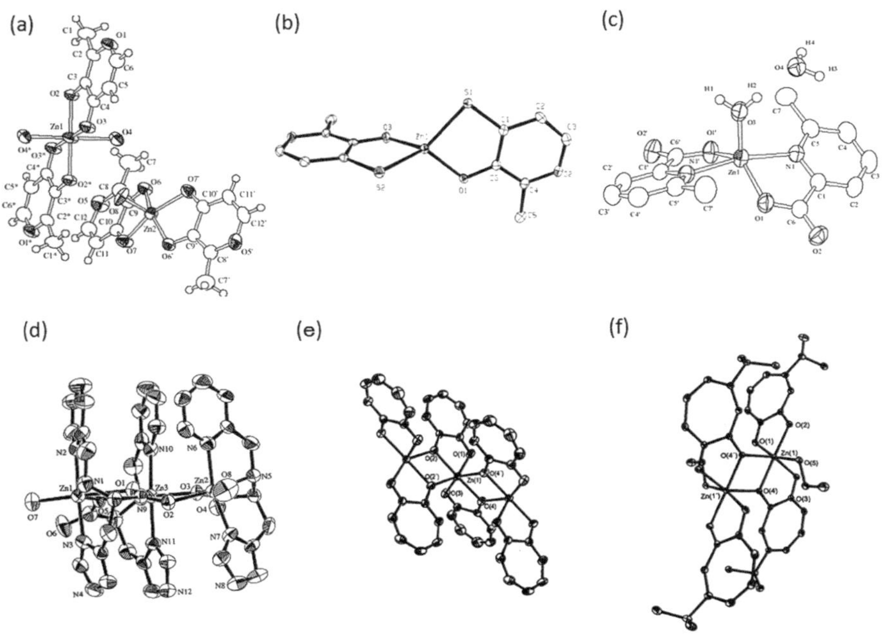

Fig. 16.7 Structural diagrams of Zn(II) complexes with various physiological activity

Another Zn(II) complex was prepared, and the colorless single crystal of di (6-methyl picolinic acidato)Zn(II) ([Zn(6mpa)$_2$]) obtained from a methanol solution was also analyzed [53]. The coordination geometry around the Zn(II) ion in [Zn (6mpa)$_2$(H_2O)] · H_2O was evaluated and found to be a distorted trigonal bipyramidal structure (Fig. 16.7c). The Zn(II) ion lies in an equatorial plane (O1, O1′, and O3) with two apical atoms (N1 and N1′). The Zn(II)(pmH) complex was found to be a trinuclear complex, [Zn_3(pmH)$_3$(H_2O)$_2$] (ClO_4)$_3$/2H_2O. Figure 16.7d shows the oak ridge thermal ellipsoid plot program (ORTEP) view of a trinuclear complex cation of [Zn_3(pmH)$_3$(H_2O)$_2$]. Coordination geometries for the three Zn(II)s of [Zn_3(pmH)$_3$(H_2O)$_2$] exhibit distorted octahedral configurations with N_3O_3 donor sets. The coordination spheres of the three Zn(II)s, except for the coordination of pyridyl and imidazolyl groups of pmH, are formed within one plane [54].

The structure of di(toropolonato)Zn(II) ([Zn(trop)$_2$]) is shown in Fig. 16.7e. The result is a coordination polymer in which Zn_2O_2 rhomboids are linked in an orthogonally alternating manner. The coordination sphere around each Zn(II) is that of a very distorted octahedron. The structure of the di(hinokitiolato)Zn(II) ([Zn (hkt)$_2$]) closely resembles that of [Zn(trop)$_2$] (Fig. 16.7f). The two hinokitiol ligands of [Zn(hkt)$_2$] orthogonally chelate the Zn(II), and the octahedral coordination about Zn(II) in [Zn(hkt)$_2$] is slightly more regular than that in [Zn(trop)$_2$] [55]. Zn(II) complexes with different conformation and coordination modes were

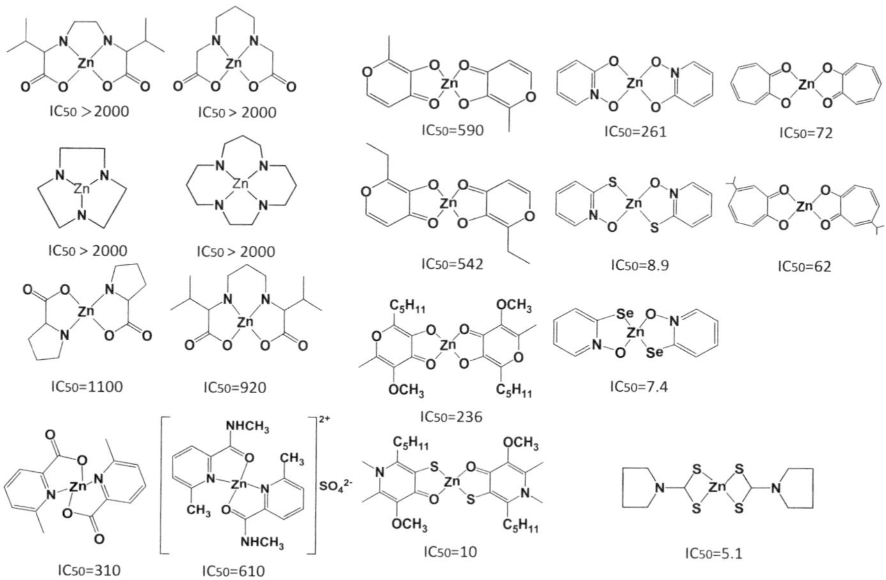

Fig. 16.8 Structures of Zn(II) complexes with insulin-like inhibitory effect on the free fatty acids release (IC_{50}; μM) from isolated rat adipocytes

synthesized, and they have been used as the major compound for the treatment of lifestyle diseases in in vitro or in vivo studies.

16.3.2.3 Antidiabetic Activity and the Antihypertension Effect of Zn (II) Complexes in KK-A^y Mice

Since 2000, we have attempted to identify alternative therapeutic compounds and have discovered that metal complexes, such as Zn(II) complexes with various coordination modes of N_2O_2, O_4, S_2O_2, and N_4 around the metal center, exhibit in vitro insulin-mimetic activity and in vivo antidiabetic effects in experimental animals [1, 56, 57]. In in vitro experiments, we first reported that the compound [Zn (ma)$_2$] with a Zn(O_4) coordination mode exhibits higher in vitro anti-DM activity than that exhibited by a free Zn(II) ion, with respect to the inhibition of free fatty acid (FFA) release in isolated rat adipocytes [58]. Subsequently, many Zn (II) complexes were evaluated for their in vitro anti-DM activity by our group (Fig. 16.8).

One of the most interesting Zn(II) complexes, we focused on [Zn(hkt)$_2$], a complex with a Zn(O_4) coordination mode that has been reported to have higher in vitro anti-DM activity relative to Zn(II) complexes with the Zn(O_4) coordination mode [59]. In in vivo study, [Zn(hnk)$_2$] showed blood glucose-lowering effect and hypoadiponectinemia by intraperitoneal injection. And nowadays, we revealed that

[Zn(hnk)$_2$] has antidiabetic effects in the insulin signaling pathway by inducing phosphorylation of Akt indirectly and, alternatively, inhibiting GSK3β after intracellular Zn uptake. Thus, we concluded that these events may involve the translocation of GLUT4 to the membrane and may be dependent on glucose uptake [60]. Hinokitiol is known to be a tropolone-related compound with an aromatic seven-membered ring. It was reported that the hinokitiol and tropolone obtained from Aomori Hiba have antibacterial, antioxidant, and antitumor activity [61, 62]. In our study, Zn(II) complexes with Zn(O_4) coordination modes that were tropolone derivatives, including hinokitiol or tropolone, were converted into complexes with Zn(S_2O_2) or Zn(S_4) coordination modes, and their insulin-mimetic activity was evaluated by measuring the inhibition of FFA release and enhancement of glucose uptake in isolated rat adipocytes. Zn(II) complexes with coordinated sulfur ligands displayed high insulin-mimetic activity, making them appropriate candidates for a new class of antidiabetic drugs [63].

In recent literature, di(2-selenopyridine-N-oxidato)Zn(II) ([Zn(spo)$_2$]) with a Zn (Se_2O_2) coordination mode exhibited an in vivo blood glucose-lowering effect in KK-A^y mice [64]. When the mice were given these complexes at a dose of 1.25–2.5 mg Zn/kg body weight by oral gavage, the glucose levels were approximately maintained at about 200 mg/dL for 2 weeks. Furthermore, the changes of glycated hemoglobin (HbA1c) in untreated KK-A^y mice and KK-A^y mice treated with these complexes were examined. In untreated KK-A^y mice, the HbA1c level was found to be 8 % following administration. However, the HbA1c level of KK-A^y mice treated with this complex was measured to be 6 % after administration. These results show that the diabetes mellitus and the glucose metabolism in KK-A^y mice treated with these complexes were significantly improved compared with those of untreated KK-A^y mice.

The relationship between trace elements and cardiovascular diseases, including high blood pressure, has been the subject of numerous scientific investigations over the past 40–50 years. In hypertension patients, it has been reported that serum Zn (II) levels were significantly reduced when compared with the control group [65]. Moreover, recent papers reported that hypoadiponectinemia contributed partially to the development of obesity-linked hypertension, such as that in KK-A^y mice [66]. Thus, the evaluation of the relationship between Zn(II) and hypertension was one of critical importance. Our three synthesized complexes, di(1,6-dimethyl-3-hydroxy-5-methoxy-2-pentyl-1,4-dihydropyridine-4-thionato)Zn(II) ([Zn (tanm)$_2$]), di(pyrrolidine-N-dithiocarbamato)Zn(II) ([Zn(pdc)$_2$]), and di (aspirinato)Zn(II) ([Zn(asp)$_2$]), showed an anti-DM effect that may be mediated through a mechanism regulating hypoadiponectinemia [67–69]. Furthermore, these complexes produce an antihypertension effect. Briefly, we concluded that normalization of adipocytokine levels by administration of Zn(II) complexes led to a decrease in blood pressure levels.

In conclusion, the Zn(II) complexes proposed here have been found to produce effective blood glucose-lowering and antihypertension effects in a type 2 DM animal model. Following the careful examination of possible toxicity, bioavailability, and pharmacokinetics, these complexes may be developed as clinically effective therapeutic options for the improvement of lifestyle-related diseases.

16.3.2.4 In Vivo Evaluation of Bone Metabolism in Osteoporosis with Zn Compounds

In recent years, it has been reported that various metal ions are heavily involved in bone metabolism. Tucker et al. reported that Mg(II) intake was significantly associated with bone mineral density in 2001, and Zhu et al. also reported that K (I) intake shows a positive correlation with bone mineral density in 2009 [70, 71]. In addition, many researchers have investigated the influence of Zn(II) on bone metabolism [72, 73]. For example, the bone mineral content of the jawbone and bone mechanical strength of the femur in Zn(II)-deficient Sprague Dawley rats with osteoporosis were reportedly restored by 12-week administration of Zn (II) [74]. Yamaguchi and other investigators have reported that the amounts of DNA, an indicator of bone growth, and alkaline phosphatase (ALP) increase after Zn(II) administration and have proven that Zn(II) effectively treats abnormal bone metabolism caused by diabetes [75, 76]. Hie et al. reported that Zn administration decreased osteoclastogenesis by inhibiting the receptor activator of nuclear factor κB expression through suppression of the production of reactive oxygen species and extracellular signal-regulated kinase activation [77].

In a recent study, when we bred SAMP6 mice (a senescence-accelerated model mouse) treated with zinc sulfate (25 mg Zn/kg body weight) from 7 weeks to 15 weeks of age, the $ZnSO_4$-treated group exhibited improved bone strength, increased ALP activity, and increased bone Zn(II) density when compared to that of the non-treated group [78]. This result is considered to be related to the osteoplasty promotion effect of Zn(II) and the reinforcement of material strength of the osseous tissue by Zn(II) incorporation into the bone. It was reported that the stability of the collagenous triple helix is increased by metals in artificial collagen model peptides [79]. Thus, enhancement of Zn(II) uptake stabilizes bone collagen, and Zn(II) may strengthen bone that has thinned due to osteoporosis. The world awaits the development of Zn(II) compounds with an antiosteoporotic effect.

In animal studies, Zn administration was shown to various anti-lifestyle-related diseases or bone metabolism (Table 16.2). Many Zn complexes were evaluated in in vitro systems, but not many Zn complexes showed the effectiveness in in vivo systems. In the future, researchers should search the Zn complexes with high safety and potent effect.

Table 16.2 In vivo studies in animal models treated with Zn(II) compounds

Zn compounds	Mode of administration	Doses	Model used in study	Major physiological effects of treatment
Zn carbonate [80]	Incorporated in diet	1–1000 ppm for 2 weeks	Congenitally diabetic BioBreed Wistar rats	Lower incidence of diabetes
				Higher pancreatic and serum insulin levels
				Decreased serum glucose and triglycerides
				Improved glucose tolerance
				Decrease in islet inflammation
Zn–Cyclo (His-Pro) [81]	Gastric gavage or drinking water	10 mg Zn/kg body weight or 10 mg Zn/L, for 2 weeks	Goto-Kakizaki rats	Decreased insulin resistance, serum insulin, and glucose levels
				Improved oral glucose tolerance
Zn(3-hydroxy flavone) [82]	Oral administration	5 mg complex/kg, for 30 days	STZ injection-induced diabetes in rats	Normalization of blood glucose
				Decrease in $HbA1_C$ levels
				Improvement in plasma insulin and C-peptide levels
Zn complex with mixed ligand [83]	Drinking water	10 mg/kg body weight	High fructose-induced hypertension Sprague Dawley rats	Antihypertensive activity
Zn acexamate [75]	Oral administration	2.5, 5, and 10 mg Zn/100 g body weight	STZ injection-induced diabetes in rats	Restored effect of bone loss or bone component
Zn(β-alanyl-L-histidine) [84]	Oral administration	10, 30, and 100 mg/kg body weight, for 3 months	Ovariectomized rats	Prevention of bone loss
				Improvement of alkaline phosphatase activity and calcium contents
Zn sulfate [85]	Drinking water	25 mmol/L, for 1 week	Multiple low-dose STZ (MLD–STZ) injection-induced diabetes in c57/BL6 and B6SJL/F_1 mice	Increased levels of metallothionein
				Prevention of MLD–STZ-

(continued)

Table 16.2 (continued)

Zn compounds	Mode of administration	Doses	Model used in study	Major physiological effects of treatment
				induced diabetes and loss of β-cell function
Zn ion [86]	Incorporated in diet	1000 mg Zn/kg diet	Ob/ob mice	Attenuation of exaggerated insulin secretion
Bis(maltolato) Zn [87]	Intraperitoneal injection	68.8 μM Zn $(Mal)_2$/kg body weight	Type 2 diabetic KK-A^y mice	Normalization of blood glucose
				Improvement in serum insulin and triglyceride levels
				Improved oral glucose tolerance
Bis(1-oxy-2-pyridinethiolato) Zn complex [88]	Oral administration	7.5 mg Zn/kg, for 16 days	Type 2 diabetic KK-A^y mice	Normalization of blood glucose
				Decrease in $HbA1_C$ levels
				Improved oral glucose tolerance
				Decrease in serum insulin and triglyceride levels
				Increase in serum adiponectin levels
				Decrease in blood pressure

16.4 Future for Metal-Based Medicines

When highly potent metallodrugs are aimed to develop as clinically useful metal-based medicines, the following three points on the basis of the translational research will be the most important factors: (1) low toxicity and low side effects, (2) evidence and clear target for the in vivo pharmacological action (proof of concept), and (3) good property and relationship between pharmacokinetic (PK) and pharmacodynamic (PD) features in terms of the effective concentrations in the blood and targeted organs.

In fact, approximately 30 % of the candidate organic compounds have failed to be developed as clinical medicines because of the lack of suitable PK and PD features. On PK study, the analytical methods are very important to determine the metal concentrations in all biological samples. On the toxicity and adverse effects of metallodrugs, careful observation of the animals treated with long-term administrations of the candidate metallodrugs should be necessary for least 1 or 2 years.

Thus the comprehensive understanding of the abovementioned three factors in each candidate complex will be essential to produce clinically active oral antidiabetic zinc complexes in the twenty-first century.

Acknowledgments This review was financially supported in part by a grant from the Ministry of Education, Culture, Sports, and Technology of Japan (MEXT) Supported Program for the Strategic Research Foundation at Private Universities, 2012–2016 (S1201008).

References

1. Uchiyama S, Yamaguchi M (2003) Alteration in serum and bone component findings induced in streotozotocin-diabetic rats is restored by zinc acexamate. Int J Mol Med 12:949–954
2. Bakale RP, Naik GN, Mangannavar CV, Muchchandi IS, Shcherbakov IN, Frampton C, Gudasi KB (2014) Mixed ligand complex via zinc(II)-mediated in situ oxidative heterocyclization of hydrochloride salt of 2-chlorobenzaldehyde hydralazine hydrazone as potential of antihypertensive agent. Eur J Med Chem 73:38–45
3. Koya D, King GL (1998) Protein kinase C activation and the development of diabetic complications. Diabetes 47(6):859–866
4. Ohashi K, Kihara S, Ouchi N, Kumada M, Fujita K, Hiuge A, Hibuse T, Ryo M, Nishizawa H, Maeda N, Maeda K, Shibata R, Walsh K, Funahashi T, Shimomura I (2006) Adiponectin replenishment ameliorates obesity-related hypertension. Hypertension 47:1108–1116
5. Yoshikawa Y, Ueda E, Kawabe K, Miyak H, Takino T, Sakurai H, Kojima Y (2002) Development of new insulinomimetic zinc(II) picolinate complexes with a Zn(N_2O_2) coordination mode: structure characterization, in vitro, and in vivo studies. J Biol Inorg Chem 7:68–73
6. Adams C, Grey N, Magrath I, Miller A, Torode J (2010) The world cancer declaration: is the world catching up? Lancet Oncol 11:1018–1020
7. Liu S, Yamauchi H (2006) Hinokitiol, a metal chelator derived from natural plants, suppresses cell growth and disrupts androgen receptor signaling in prostate carcinoma cell lines. Biochem Biophys Res Commun 351:26–32
8. Liu W, Gust R (2013) Metal N-heterocyclic carbene complexes as potential antitumor metallodrugs. Chem Soc Rev 42:755–773
9. Sharma AM, Chetty VT (2005) Obesity, hypertension and insulin resistance. Acta Diabetol 42 (Suppl 1):S3–S8
10. Yoshikawa Y, Yasui H (2012) Zinc complexes developed as metallopharmaceutics for treating diabetes mellitus based on the bio-medicinal inorganic chemistry. Curr Top Med Chem 12 (3):210–218
11. Barnard PJ, Baker MV, Berners-Price SJ, Skelton BW, White AH (2004) Dinuclear gold (I) complexes of bridging bidentate carbene ligands: synthesis, structure and spectroscopic characterisation. Dalton Trans 7(7):1038–1047
12. Pantelića N, Stanojkovićb TP, Zmejkovskic BB, Sabod TJ, Kaluđerović GN (2015) In vitro anticancer activity of gold(III) complexes with some esters of (S, S)-ethylenediamine-N, N'-di-2-propanoic acid. Eur J Med Chem 90:766–774
13. Atik OS, Uslu MM, Eksioglu F, Satana T (2006) Etiology of senile osteoporosis: a hypothesis. Clin Orthop Relat Res 443:25–27

14. Barnard PJ, Wedlock LE, Baker MV, Berners-Price SJ, Joyce DA, Skelton BW, Steer JH (2006) Luminescence studies of the intracellular distribution of a dinuclear gold(I) - N-heterocyclic carbene complex. Angew Chem Int Ed Engl 45:5966–5970
15. Manrique C, Lastra G, Whaley-Connell A, Sowers JR (2005) Hypertension and the cardiometabolic syndrome. J Clin Hypertens 7(8):471–476
16. Hoehn ST, Junker HD, Bunt RC, Turner CJ, Stubbe J (2001) Solution structure of Co(III)-bleomycin-OOH bound to a phosphoglycolate lesion containing oligonucleotide: implications for bleomycin-induced double-strand DNA cleavage. Biochemistry 40:5894–5905
17. Hie M, Tsukamoto I (2011) Administration of zinc inhibits osteoclastogenesis through the suppression of RANK expression in bone. Eur J Pharmacol 668:140–146
18. Tokudome Y, Otsuka M (2012) Possibility of alveolar bone promoting enhancement by using lipophilic and/or hydrophilic zinc related compounds in zinc-deficient osteoporosis rats. Biol Pharm Bull 35:1496–1501
19. Fujimoto S, Yasui H, Yoshikawa Y (2013) Development of a novel antidiabetic zinc complex with an organoselenium ligand at the lowest dosage in KK-A^y mice. J Inorg Biochem 121:10–15
20. Murakami H, Yasui H, Yoshikawa Y (2012) Pharmacological and pharmacokinetic studies of anti-diabetic tropolonato–Zn(II) complexes with $Zn(S_2O_2)$ coordination mode. Chem Pharm Bull 60:1096–1104
21. Yanovski SZ, Yanovski JA (2014) Long-term drug treatment for obesity: a systematic and clinical review. JAMA 311(1):74–86
22. Inzucchi SE, Zinman B, Wanner C, Ferrari R, Fitchett D, Hantel S, Espadero RM, Woerle HJ, Broedl UC, Johansen OE (2015) SGLT-2 inhibitors and cardiovascular risk: proposed pathways and review of ongoing outcome trials. Diab Vasc Dis Res 12(2):90–100
23. Krishnamurthy D, Karver MR, Fiorillo E, Orrú V, Stanford SM, Bottini N, Barrios AM (2008) Gold(I)-mediated inhibition of protein tyrosine phosphatases: a detailed in vitro and cellular study. J Med Chem 51:4790–4795
24. Proetto MT, Liu W, Molchanov A, Sheldrick WS, Hagenbach A, Abram U, Gust R (2014) Synthesis, characterization, and in vitro antiproliferative activity of [salophene]platinum (II) complexes. Chem Med Chem 9:1176–1187
25. Shaw CF III (1999) Gold-based therapeutic agents. Chem Rev 99:2589–2600
26. Ohly P, Dohle C, Abel J, Seissler J, Gleichmann H (2000) Zinc sulphate induces metallothionein in pancreatic islets of mice and protects against diabetes induced by multiple low doses of streptozotocin. Diabetologia 43:1020–1030
27. Naito Y, Yoshikawa Y, Yasui H (2011) Cellular mechanism of zinc–hinokitiol complexes in diabetes mellitus. Bull Chem Soc Jpn 84:298–305
28. Yamaguchi M, Kishi S (1993) Prolonged administration of β-Alanyl-L-Histidinato Zinc prevents bone loss in Ovariectomized rats. Jpn J Pharmacol 63:203–207
29. Yoshikawa Y, Adachi Y, Yasui H, Hattori M, Sakurai H (2011) Oral administration of Bis (aspirinato)zinc(II) complex ameliorates hyperglycemia and metabolic syndrome-like disorders in spontaneously diabetic KK-A^y mice: structure-activity relationship on zinc-salicylate complexes. Chem Pharm Bull 59:972–977
30. Murphy SL, Xu J, Kochanek KD (2013) Deaths: final data for 2010. Natl Vital Stat Rep 61 (4):1–117
31. Margiotta N, Savino S, Gandin V, Marzano C, Natile G (2014) Monofunctional platinum (II) complexes with potent tumor cell growth inhibitory activity: the effect of a hydrogen-bond donor/acceptor N-heterocyclic ligand. Chem Med Chem 9:1161–1168
32. Rubbiani R, Salassa L, de Almeida A, Casini A, Ott I (2014) Cytotoxic gold(I) N-heterocyclic carbene complexes with phosphane ligands as potent enzyme inhibitors. Chem Med Chem 9:1205–1210
33. Sun RWY, Ma DL, Wong ELM, Che CM (2007) Some uses of transition metal complexes as anti-cancer and anti-HIV agents. Dalton Trans 21:4884–4892

34. Karieb S, Fox SW (2012) Zinc modifies the effect of phyto-oestrogens on osteoblast and osteoclast differentiation in vitro. Br J Nutr 108:1736–1745
35. Nakai M, Watanabe H, Fujisawa C, Kakegawa H, Satoh T, Takada J, Matsushita R, Sakurai H (1995) Mechanism on insulin-like action of vanadyl sulfate: studies on interaction between rat adipocytes and vanadium compounds. Biol Pharm Bull 18:719–725
36. Barret MC, Mahon MF, Molloy KC, Steed JW, Wright P (2001) Synthesis and structural characterization of tin(II) and zinc(II) derivatives of cyclic alpha-hydroxyketones, including the structures of Sn(maltol)$_2$, Sn(tropolone)$_2$, Zn(tropolone)$_2$, and Zn(hinokitiol)$_2$. Inorg Chem 40(17):4384–4388
37. Kanis JA, Johnell O, Oden A, Sembo I, Redlund-Johnell I, Dawson A, De Laet C, Jonsson B (2000) Long-term risk of osteoporotic fracture in Malmö. Osteoporos Int 11:669–674
38. Hu C, Li X, Wang W, Zhang R, Deng L (2014) Metal-N-heterocyclic carbene complexes as anti-tumor agents. Curr Med Chem 21:1220–1230
39. Tobia MH, Zdanowicz MM, Wingertzahn MA, McHeffey-Atkinson B, Slonim AE, Wapnir RA (1998) The role of dietary zinc in modifying the onset and severity of spontaneous diabetes in the BB Wistar Rat. Mol Genet Metab 63:205–213
40. Ball D, Blanchard J, Jacobson-Kram D, McClellan R, McGovern T, Norwood DL et al (2007) Development of safety qualification thresholds and their use in orally inhaled and nasal drug product evaluation. Toxicol Sci 97:226–236
41. Strianese M, Basile A, Mazzone A, Morello S, Turco MC, Pellecchia C (2013) Therapeutic potential of a pyridoxal-based vanadium(IV) complex showing selective cytotoxicity for cancer versus healthy cells. J Cell Physiol 228:2202–2209
42. Solanki V, Solanki M, Khubchandani A, Parmar V, Parmar U, Shah P (2015) Significance of serum levels of copper and zinc in hypertensive patients. Int J Res Med 4:137–139
43. Clough JD, Patel K, Riley GF, Rajkumar R, Conway PH, Bach PB (2015) Wide variation in payments for medicare beneficiary oncology services suggests room for practice-level improvement. Health Aff (Millwood) 34(4):601–608
44. Tucker KL, Hannan MT, Kiel DP (2001) The acid-base hypothesis: diet and bone in the Framingham Osteoporosis study. Eur J Nutr 40:231–237
45. Yoshikawa Y, Ueda E, Miyake H, Sakurai H, Kojima Y (2001) Insulinomimetic bis(maltolato) zinc(II) complex: blood glucose normalizing effect in KK-A^y mice with type 2 diabetes mellitus. Biochem Biophys Res Commun 281:1190–1193
46. Duff B, Thangella VR, Creaven BS, Walsh M, Egan DA (2012) Anti-cancer activity and mutagenic potential of novel copper(II) quinolinone Schiff base complexes in hepatocarcinoma cells. Eur J Pharmacol 689:45–55
47. Begin-Heick N, Dalpe-Scott M, Rowe J, Heick HM (1985) Zinc supplementation attenuates insulin secretory activity in pancreatic islets of the ob/ob mouse. Diabetes 34:179–184
48. Gavin JR, Alberti KGMM, Davidson MB, DeFronzo RA, Drash A, Gabbe SG, Genuth S, Harris MI, Kahn R, Keen H, Knowler WC, Lebovitz H, Maclaren NK, Palmer JP, Raskin P, Rizza RA, Stem MP (1997) Report of the expert committee on the diagnosis and classification of diabetes mellitus. Diabetes Care 20:1183–1197
49. Birkmeyer NJ, Goodney PP, Stukel TA, Hillner BE, Birkmeyer JD (2005) Do cancer centers designated by the National Cancer Institute have better surgical outcomes? Cancer 103 (3):435–441
50. Gao EJ, Liu C, Zhu MC, Lin HK, Wu Q, Liu L (2009) Current development of Pd (II) complexes as potential antitumor agents. Anticancer Agents Med Chem 9:356–368
51. Maschke M, Alborzinia H, Lieb M, Wölfl S, Metzler-Nolte N (2014) Structure-activity relationship of trifluoromethyl-containing metallocenes: electrochemistry, lipophilicity, cytotoxicity, and ROS production. Chem Med Chem 9:1188–1194
52. Mir LM, Tounekti O, Orlowski S (1996) Bleomycin: revival of an old drug. Gen Pharmacol 27:745–748

53. Qiu L, Lv G, Cao Y, Chen L, Yang H, Luo S, Zou M, Lin J (2015) Synthesis and biological evaluation of novel platinum complexes of imidazolyl-containing bisphosphonates as potential anticancer agents. J Biol Inorg Chem 20:1263–1275
54. Adachi Y, Yoshida J, Kodera Y, Kiss T, Jakusch T, Enyedy EA, Yoshikawa Y, Sakurai H (2006) Oral administration of a zinc complex improves type 2 diabetes and metabolic syndromes. Biochem Biophys Res Commun 351:165–170
55. Naso LG, Lezama L, Valcarcel M, Salado C, Villacé P, Kortazar D, Ferrer EG, Williams PA (2016) Bovine serum albumin binding, antioxidant and anticancer properties of an oxidovanadium(IV) complex with luteolin. J Inorg Biochem 157:80–93
56. Dokorou V, Primikiri A, Kovala-Demertzi D (2011) The triphenyltin(VI) complexes of NSAIDs and derivatives. Synthesis, crystal structure and antiproliferative activity. Potent anticancer agents. J Inorg Biochem 105:195–201
57. Ahera SB, Muskawara PN, Thenmozhib K, Bhagata PR (2014) Recent developments of metal N-heterocyclic carbenes as anticancer agents. Eur J Med Chem 81:408–419
58. Trust TJ, Coombs RW (1973) Antibacterial activity of beta-thujaplicin. Can J Microbiol 19:1341–1346
59. Yoshikawa Y, Murakami H, Fujimoto S, Michigami K, Yasui H (2013) Zinc and lifestyle – related disease – with focus on diabetes mellitus and osteoporosis. Vitam Miner S6:002
60. de Frémont P, Stevens ED, Eelman MD, Fogg DE, Nolan SP (2006) Gold(I) NHC complexes bearing biologically compatible moieties. Organometallics 25:5824–5828
61. Kadowaki S, Munekane M, Kitamura Y, Hiromura M, Kamino S, Yoshikawa Y, Saji H, Enomoto S (2013) Development of new zinc dithiosemicarbazone complex for use as oral antidiabetic agent. Biol Trace Elem Res 154:111–119
62. Lewis JA, Tran BL, Puerta DT, Rumberger EM, Hendrickson DN, Cohen SM (2005) Synthesis, structure and spectroscopy of new thiopyrone and hydroxypyridinethione transition-metal complexes. Dalton Trans 7(15):2588–2596
63. Yoshikawa Y, Kawabe K, Tadokoro M, Suzuki Y, Yanagihara N, Nakayama A, Sakurai H, Kojima Y (2002) New zinc(II) complexes with tetradentate amino acid derivatives: structure characterization, solution chemistry, and in vitro insulinomimetic activity. Bull Chem Soc Jpn 75:2423–2432
64. Yamaguchi M, Uchiyama S, Ishiyama K, Hashimoto K (2006) Oral administration in combination with zinc enhances beta-cryptoxanthin-induced anabolic effects on bone components in the femoral tissues of rats in vivo. Biol Pharm Bull 29:371–374
65. Teyssot ML, Jarrousse AS, Manin M, Chevry A, Roche S, Norre F, Beaudoin C, Morel L, Boyer D, Mahiou R, Gautier A (2009) Metal-NHC complexes: a survey of anti-cancer properties. Dalton Trans 21:6894–6902
66. Vujčić M, Lazić M, Milenković M, Sladić D, Radulović S, Filipović N, Anđelković K (2011) A comparative study of DNA binding and cell cycle phase perturbation by the dinuclear complex of Cd(II) with the condensation product of 2-acetylpyridine and malonic acid dihydrazide N′, N′ 2-bis[(1E)-1-(2-pyridyl)ethylidene]propanedihydrazide. J Biochem Mol Toxicol 25:175–182
67. Yamane M, Adachi Y, Yoshikawa Y, Sakurai H (2005) A new anti-diabetic Zn(II)-hinokitiol (β-Thujaplicin) complex with $Zn(O_4)$ coordination mode. Chem Lett 34:1694–1695
68. Ray S, Mohan R, Singh JK, Samantaray MK, Shaikh MM, Panda D, Ghosh P (2007) Anticancer and antimicrobial metallopharmaceutical agents based on palladium, gold, and silver N-heterocyclic carbene complexes. J Am Chem Soc 129:15042–15053
69. Yoshikawa Y, Murayama A, Adachi Y, Sakurai H, Yasui H (2011) Challenge of studies on the development of new Zn complexes ($Zn(opt)_2$) to treat diabetes mellitus. Metallomics 3:686–692
70. Body JJ, Bergmann P, Boonen S, Devogelaer JP, Gielen E, Goemaere S, Kaufman JM, Rozenberg S, Reginster JY (2012) Extraskeletal benefits and risks of calcium, vitamin D and anti-osteoporosis medications. Osteoporos Int 23(Suppl 1):S1–S23

71. Koide T, Yuguchi M, Kawakita M, Konno H (2002) Metal-assisted stabilization and probing of collagenous triple helices. J Am Chem Soc 124:9388–9389
72. Yoshikawa Y, Ueda E, Kawabe K, Miyake H, Sakurai H, Kojima Y (2000) New insulin-mimetic zinc(II) complexes; Bis-maltolato zinc(II) and bis-(2-hydroxypyridine-N-oxido) zinc (II) with $Zn(O_4)$ coordination mode. Chem Lett 29:874–875
73. Yamaguchi M, Ozaki K, Suketa Y (1989) Alteration in bone metabolism with increasing age: effects of zinc and vitamin D_3 in aged rats. J Pharmacobiodyn 12:67–73
74. Yan L, Tan J, Wan BC, Zhu LC (2014) Synthesis, characterization, and anti-cancer activity of emodin-Mn(II) metal complex. Chin J Nat Med 12:937–942
75. Song MK, Hwang IK, Rosenthal MJ, Harris DM, Yamaguchi DT, Yip I, Go VL (2003) Anti-hyperglycemic activity of zinc plus cyclo (his-pro) in genetically diabetic Goto-Kakizaki and aged rats. Exp Biol Med 228:1338–1345
76. Kaplan NM, Opie LH (2006) Controversies in hypertension. Lancet 367(9505):168–176
77. Qin XY, Yang LC, Le FL, Yu QQ, Sun DD, Liu YN, Liu J (2013) Structures and anti-cancer properties of two binuclear copper complexes. Dalton Trans 42:14681–14684
78. Underwood EJ (1977) In trace elements in human and animal nutrition, 4th edn. Academic, New York, p 545
79. Vijayaraghavan K, Iyyam Pillai S, Subramanian SP (2012) Design, synthesis and characterization of zinc-3 hydroxy flavone, a novel zinc metallo complex for the treatment of experimental diabetes in rats. Eur J Pharmacol 680:122–129
80. Sakurai H, Yoshikawa Y, Yasui H (2008) Current state for the development of metallopharmaceutics and anti-diabetic metal complexes. Chem Soc Rev 37:2383–2392
81. Yoshikawa Y, Adachi Y, Sakurai H (2007) A new type of orally active anti-diabetic Zn(II)-dithiocarbamate complex. Life Sci 80:759–766
82. Zhu K, Devine A, Prince RL (2009) The effects of high potassium consumption on bone mineral density in a prospective cohort study of elderly postmenopausal women. Osteoporos Int 20:335–340
83. Sakurai H, Kojima Y, Yoshikawa Y, Kawabe K, Yasui H (2002) Antidiabetic vanadium (IV) and zinc(II) complexes. Coord Chem Rev 226:187–198
84. Rubbiani R, Kitanovic I, Alborzinia H, Can S, Kitanovic A, Onambele LA, Stefanopoulou M, Geldmacher Y, Sheldrick WS, Wolber G, Prokop A, Wölfl S, Ott I (2010) Benzimidazol-2-ylidene gold(I) complexes are thioredoxin reductase inhibitors with multiple antitumor properties. J Med Chem 53:8608–8618
85. Zec M, Srdic-Rajic T, Konic-Ristic A, Todorovic T, Andjelkovic K, Filipovic-Ljeskovic I, Radulovic S (2012) Anti-metastatic and anti-angiogenic properties of potential new anti-cancer drugs based on metal complexes of selenosemicarbazones. Anticancer Agents Med Chem 12:1071–1080
86. Smith RA, Manassaram-Baptiste D, Brooks D, Doroshenk M, Fedewa S, Saslow D, Brawley OW, Wender R (2015) Cancer screening in the United States, 2015: a review of current American cancer society guidelines and current issues in cancer screening. CA Cancer J Clin 65:30–54
87. Bray F, Jemal A, Grey N, Ferlay J, Forman D (2012) Global cancer transitions according to the Human Development Index (2008–2030): a population-based study. Lancet Oncol 13:790–801
88. Wanzlick HW, Schoenherr HJ (1968) Chemistry of nucleophilic carbenes. XIV. Direct synthesis of a mercury salt-carbene complex. Angew Chem Int Ed Engl 7:141–142

MIX
Papier aus verantwortungsvollen Quellen
Paper from responsible sources
FSC® C105338

If you have any concerns about our products,
you can contact us on
ProductSafety@springernature.com

In case Publisher is established outside the EU,
the EU authorized representative is:
Springer Nature Customer Service Center GmbH
Europaplatz 3, 69115 Heidelberg, Germany

Printed by Libri Plureos GmbH
in Hamburg, Germany